C TABLE

...eses are mass
...otopes.

			III A	IV A	V A	VI A	VII A	O
								2 **He** 4.003
			5 **B** 10.81	6 **C** 12.011	7 **N** 14.007	8 **O** 15.999	9 **F** 18.998	10 **Ne** 20.179
	I B	II B	13 **Al** 26.982	14 **Si** 28.086	15 **P** 30.974	16 **S** 32.06	17 **Cl** 35.453	18 **Ar** 39.948
28 **Ni** 58.71	29 **Cu** 63.546	30 **Zn** 65.38	31 **Ga** 69.72	32 **Ge** 72.59	33 **As** 74.922	34 **Se** 78.96	35 **Br** 79.904	36 **Kr** 83.80
46 **Pd** 106.4	47 **Ag** 107.868	48 **Cd** 112.40	49 **In** 114.82	50 **Sn** 118.69	51 **Sb** 121.75	52 **Te** 127.60	53 **I** 126.905	54 **Xe** 131.30
78 **Pt** 195.09	79 **Au** 196.967	80 **Hg** 200.59	81 **Tl** 204.37	82 **Pb** 207.2	83 **Bi** 208.981	84 **Po** (210)	85 **At** (210)	86 **Rn** (222)

65 **Tb** 158.925	66 **Dy** 162.50	67 **Ho** 164.930	68 **Er** 167.26	69 **Tm** 167.26	70 **Yb** 173.04	71 **Lu** 174.97

97 **Bk** (247)	98 **Cf** (252)	99 **Es** (254)	100 **Fm** (257)	101 **Md** (257)	102 **No** (255)	103 **Lr** (256)

INDUSTRIAL APPLICATION OF RADIOISOTOPES

EDITED BY

G. FÖLDIÁK

INSTITUTE OF ISOTOPES OF THE
HUNGARIAN ACADEMY OF SCIENCES,
BUDAPEST, HUNGARY

AKADÉMIAI KIADÓ, BUDAPEST 1986

This monograph is published by Elsevier Science Publishers as vol. 39 in the series of "Studies in Physical and Theoretical Chemistry."

The revised and enlarged English version is based on the Hungarian "Az izotópok ipari alkalmazása", published by Műszaki Könyvkiadó, Budapest

Translated by L. Bartha, I. György, L. Markó, Z. Paál, F. Réz, P. Roboz and G. Uhrin

CHEM

ISBN 963 05 3500 9

Joint edition published by Akadémiai Kiadó, The Publishing House of the Hungarian Academy of Sciences, Budapest, Hungary and Elsevier Science Publishers, Amsterdam, The Netherlands

LIST OF CONTRIBUTORS

Bálint, T., Hungarian Oil and Gas Research Institute, H-8201 Veszprém, Wartha Vince utca 1, Hungary

Balla, B., Institute of Isotopes of the Hungarian Academy of Sciences, H-1525 Budapest, P.O. Box 77, Hungary

Bányai, É., Institute of General and Analytical Chemistry, Technical University of Budapest, H-1521 Budapest, Hungary

Bartha, L., Research Institute for Technical Physics of the Hungarian Academy of Sciences, H-1325 Budapest, P.O. Box 76, Hungary

Biró, T., Institute of Isotopes of the Hungarian Academy of Sciences, H-1525 Budapest, P.O. box 77, Hungary

Czeglédi, T., Hungarian National Oil and Gas Trust, H-1502, Budapest, P.O. Box 22, Hungary

Elekes, I., Institute for Drug Research, H-1325 Budapest, P.O. Box 82, Hungary

Földiák, G., Institute of Isotopes of the Hungarian Academy of Sciences, H-1525 Budapest, P.O. Box 77, Hungary

Hirling, J., Institute of Isotopes of the Hungarian Academy of Sciences, H-1525 Budapest, P.O. Box 77, Hungary

Horváth, L., Institute of Isotopes of the Hungarian Academy of Sciences, H-1525 Budapest, P.O. Box 77, Hungary

Keömley, G., Training Nuclear Reactor, Technical University of Budapest, H-1521 Budapest, Hungary

Lengyel, T., Institute of Isotopes of the Hungarian Academy of Sciences, H-1525 Budapest, P.O. Box 77, Hungary

Nagy, L. Gy., Institute of Applied Chemistry, Technical University of Budapest, H-1521 Budapest, Hungary

Paál, Z., Institute of Isotopes of the Hungarian Academy of Sciences, H-1525 budapest, P.O. Box 77, Hungary

Pozsár, B., Institute of Isotopes of the Hungarian Academy of Sciences, H-1525 Budapest, P.O. Box 77, Hungary

Rózsa, S., Institute of Isotopes of the Hungarian Academy of Sciences, H-1525 Budapest, P.O. Box 77, Hungary

Szabó, E., Central Research Institute for Physics of the Hungarian Academy of Sciences, H-1525 Budapest, P.O. Box 49, Hungary

CONTENTS

1. BASIC DATA AND DEFINITIONS

In order to make easier a survey of methods used in the industrial applications of isotopes, this chapter will give a brief review of the fundamental concepts, terms and symbols, a summary of the main characteristics of the most important radioactive isotopes (nuclides), and a reference to the general characteristics of radioactive preparations and the problems arising at their use.

1.1. SYMBOLS

The rapid but rather isolated development of the different fields of application of radioactive isotopes, e.g., nuclear physics and chemistry, gave rise to a non-uniform or even contradictory and incorrect terminology and notation. The situation has become, at least temporarily, even more confused by the introduction of SI (Système International d'Unités) units.

The conversion factors between SI and "traditional" units are tabulated in the Appendix. Physical quantities in the text will be given in SI and many times also in conventional units (in brackets).

Nearly one hundred symbols will be used for different terms relating to the industrial applications of isotopes and it should be mentioned that, unfortunately, there is a great diversity in the interpretation of these abbreviations in different fields.

Unusual, i.e. gothic letters were avoided owing to the compromise accepted by the editor and the authors, namely, in this section will be summarized the most widely used notations, those having the same interpretation throughout the whole book, but at the same time reference will be made on their units at the particular place of occurrence if required for a clear understanding of the text. To call the Reader's attention to these symbols they will be marked with **bold italics**.

Dimensions or units of quantities not listed below will be given at the appropriate place in the text: these symbols will be denoted with *simple italics*; in order to avoid clumsy notation, indices will occasionally be applied.

This means that the same letters printed in bold or thin italics may have different meanings which should always be taken into account.

A activity, e.g., in Bq (Ci)

$\mathscr{A}$ mass number

a specific activity, e.g., in Bq/kg (Ci/g)

c concentration, e.g., in mole/m^3, mg/cm^3

D dose (as to the units, reference is made in the index or in the text), e.g., in Gy (rd)

E energy of radiation (the subscript e.g., E_γ, refers to the type of radiation), e.g., in MeV

h height, e.g., in m (cm)

I dose rate, dose intensity, e.g., in Gy/s, A/kg (rd/s, R/s)

I_{rel} relative dose rate, relative dose intensity, relative activity, relative counting rate, e.g., in 1/s (cps = counts per second)

K dose rate constant (a subscript, if used, refers to the type of radiation), e.g., in aA m^2/(kg Bq), [R m^2/(h Ci)]

l length, distance, e.g., in m (cm)

m mass, e.g., in kg (g)

n counts

R geometrical radius, e.g., in m (cm)

t time, e.g., in s (h)

$t_{1/2}$ half-life, e.g., in s (h)

U potential, e.g., in V

V volume, e.g., in m^3 (ml)

v velocity, e.g., in m/s

$\mathscr{Z}$ atomic number

λ decay constant, 1/s

μ linear absorption coefficient, e.g., in 1/m (1/cm)

μ_m mass absorption coefficient, e.g., in m^2/kg (cm^2/g)

ρ density, e.g., in kg/m^3 (g/cm^3)

σ standard deviation

Φ flux density (subscript refers to the type of radiation), e.g., in 1/(m^2s) [1/(cm^2s)]

1.2. BASIC DEFINITIONS USED IN ISOTOPE TECHNIQUES

Many of the definitions and expressions to be encountered in later chapters will not be defined separately because either they are taken as known or they are excluded to make the text more lucid. The most frequently used terms and their definitions are listed below in alphabetical order.

Absorbed dose
Radioactive radiations impart energy to the medium when attenuated in it. The mean energy absorbed by unit mass or volume of the irradiated medium is called absorbed dose and expressed in units of *gray*: 1 gray = 1 Gy = 1 J/kg. The unit that was used earlier is *rad*; in order to differentiate between this unit and that of the planar angle the commonly used abbreviation of rad is rd. 1 rd = 0.01 Gy.

Absorbed dose rate
The absorbed dose related to unit time is the absorbed dose rate that is to be given correctly in Gy/s.

Accumulation factor
See Build-up factor.

Activity
Radioactive decay is a spontaneous process that cannot be influenced from the exterior and that leads to the formation of new nuclei from the originally existing unstable ones. The rate of decay is characterized by the activity.

The unit of activity is the *becquerel* (Bq). A source has an activity of one Bq if one disintegration takes place in it in one second; that is, 1 Bq = 1 1/s.

The traditional unit of activity, *curie*, was defined as a quantity of a radioactive nuclide in which the number of transformations or disintegrations per second — transformations/s (tps), disintegrations/s (dps) — is 3.70×10^{10}. This unit has a historical background, as it corresponds to the activity of 1 g radium in secular equilibrium with its radioactive daughters.

This unit, abbreviated Ci, as well as its fractions and multiples, is still in use. The relationship with the Bq and its multiples is as follows:

$$1 \text{ MCi} = 10^6 \text{ Ci} = 3.7 \times 10^{16} \text{ Bq} = 37 \text{ PBq}$$
$$1 \text{ kCi} = 10^3 \text{ Ci} = 3.7 \times 10^{13} \text{ Bq} = 37 \text{ TBq}$$
$$1 \text{ mCi} = 10^{-3} \text{ Ci} = 3.7 \times 10^7 \text{ Bq} = 37 \text{ MBq}$$
$$1 \text{ } \mu\text{Ci} = 10^{-6} \text{ Ci} = 3.7 \times 10^4 \text{ Bq} = 37 \text{ kBq}$$
$$1 \text{ nCi} = 10^{-9} \text{ Ci} = 3.7 \times 10^1 \text{ Bq} = 37 \text{ Bq}$$
$$1 \text{ pBq} = 10^{-12} \text{ Ci} = 3.7 \times 10^{-2} \text{ Bq} = 37 \text{ mBq}$$

Activity is often characterized by the number of disintegrations per minute — transformations/min (tpm) — that is: $1 \text{ Ci} = 2.22 \times 10^{12}$ disintegrations/min. Taking the degree of efficiency of the measurement to be η, we can formulate the following relationship between the counts per minute (cpm) and the activity in Ci: $1 \text{ Ci} = 2.22 \times 10^{12} \times 1/\eta$ counts/min.

α-Radiation

α-Radiation is that type of decay in which the nucleus emits an α-particle that has two units of positive charge and four units of atomic mass and thus corresponds to the nucleus of a helium atom. This corpuscular radiation possesses discrete energy. The initial energies of the particles emitted by the nucleus range from 3 to 8 MeV.

Annihilation radiation

The positively charged particles emitted by the nucleus in the positron decay interact with the negatively charged electrons surrounding the nucleus. The interaction of the opposite charges results in the formation of two photons each having an energy of 0.511 MeV. The electromagnetic radiation thus produced is called annihilation radiation.

Atomic mass

The atomic mass of an element is the sum of the numbers of protons and neutrons in the nucleus. Since the atoms of most of the elements are built up of several nuclides the atomic mass of the element is the weighed average of those of the isotopic nuclides, according to their natural abundances.

Atomic number

The atomic number of a given element equals the number of protons in the nucleus and so indicates its place in the periodic table.

β-Radiation

The transformation of the nucleus associated with the emission of high energy electrons (β-radiation) or positrons (β^+-radiation) from the nucleus is called β-radiation. The emergence of electrons is accompanied by the emission of an uncharged antineutrino while that of the positron by the emission of a neutral neutrino. The latter process is attended by the emission of *annihilation radiation* as a secondary process.

In the case of both types of corpuscular radiations a continuous energy spectrum is observed as a result of the statistical distribution of excess energy causing instability between the β-particle and the neutrino or the antineutrino, respectively.

Bremsstrahlung (X-ray)

When β^-- and β^+-particles emitted by radioactive nuclides are slowing down in the force field of the atoms of an absorbent, electromagnetic radiation is brought about that has a continuous energy spectrum and is called bremsstrahlung. The intensity and maximum energy of this radiation is determined by the primary kinetic energy responsible for the process, as well as the atomic number and density of the absorbent.

Beside β-particles, bremsstrahlung can be brought about by the shift of charge occurring subsequent to electron capture.

Build-up factor

When measuring the absorption of γ-radiation in thick absorbents a smaller amount of decrease in intensity is observed than that calculated from the exponential law of attenuation. This is the result of the scattering effects taking place in the target and can be taken into account in the calculations with the use of the build-up factor.

Carrier-free isotopes

The radioactive product is called carrier-free (c.f.) if it does not contain any active or inactive isotopes of the element, except the radioactive isotope of the given element. As the term "carrier-free" can only approximately be true, the preparations from which the carriers were removed with special care are, in general, considered to be carrier-free.

Compton effect

The scattering of the γ-photons present in the electromagnetic field causes a decrease in the frequency of the quanta; this phenomenon is called Compton effect.

Conversion electron radiation

A γ-photon of the electromagnetic radiation emitted by nuclear isomers can eject an electron from the electron shell surrounding the nucleus and make it move out of the force field of the atom with a certain amount of energy; the electron deficiency caused by the emission brings about the emission of characteristic X-rays by the product nuclide.

Corpuscular radiation

The type of nuclear radiations during which the radioactive nucleus becomes stabilized through the emission of particles (i.e., not electromagnetic radiations) is called corpuscular radiation.

Counting rate

In practical activity measurements usually only relative values are needed. The relative activity is mostly characterized by giving the so-called counting rate expressed in counts per unit time (imp/s, imp/min, cps, spm, ctc.). In order to calculate absolute activity different correction factors (referring to absorption, backscattering, geometric arrangement, resolution of the detector, etc.) must be taken into account.

Cross-section

The probability of nuclear processes is characterized, in general, by the cross-section, that is, essentially, a measure of the interaction between the corpuscular or electromagnetic radiation and the medium. As the interaction can be characterized, quantitatively, by the surface perpendicular to the direction of radiation, the cross-section has a dimension of area and its usual unit is 10^{-28} m^2 = 10^{-24} cm^2 = 1 *barn*, a value comparable with the real average cross-section of the atomic nuclei. It is recommended that the cross-section hereafter be expressed in m^2.

In accord with the generalized concept, a cross-section can be defined, in a similar way, for *scattering, absorption, activation* and *fission* processes, as well.

Dead time

The number of pulses recorded by a detector and counter system is always smaller than that entering the detector. The reason for this is that closely spaced events cannot be resolved either by the detector or the counting instrument, due to the dead time. This may cause discrepancies in the correct determination of activity especially in the case of high counting rates.

Decay constant

The fraction of radioactive nuclei transformed in unit time can be characterized by the decay constant. The following relationship holds between the decay constant and the *physical half-life:*

$$\lambda = \frac{\ln 2}{t_{1/2}} = \frac{0.693}{t_{1/2}}$$

Dose

See Exposure, Equivalent dose, Absorbed dose.

Dose constant

The dose rate (dose intensity) brought about at unit distance from a source of unit activity is called dose constant. Depending on the type of radiation, α-, β-, γ- and neutron-dose constants can be distinguished; these are denoted by K_α, K_β, K_γ and K_n, respectively. From a practical point of view, the role of K_γ is the most important; using a more correct terminology it can be named as *specific exposure rate* and given in SI units as aA m^2/(kg Bq), which can be converted to a more practicable unit, viz. μGy$_{air}$ m^2/(GBqh). Conversion: 1 aA m^2/(kg Bq) = 121.27 μGy$_{air}$ m^2/(GBq h). The earlier used unit was R m^2/(h Ci). The conversion of units can be made using the expression

$$1 \text{ R m}^2/(\text{h Ci}) = 1.937 \text{ aA m}^2/(\text{kg Bq})$$

The main advantage in using the dose constant K_γ lies in the fact that the *exposure rate* (*I*), in A/kg, can be calculated from it by the expression

$$I_\gamma = \frac{10^{-18} K_\gamma A}{I^2} \text{ A/kg}$$

where A is the activity of the radiation point source, in Bq, and I is the distance from it, in m.

Effective half-life

The rate of depletion of a radioactive material incorporated into human body follows an exponential law. The time during which one half of the incorporated radioactive material leaves the organism is called *biological half-life*. Parallel with the above, a diminution of the activity takes place that is described by the law of radioactive decay (physical half-life);

the half-life observed corresponds to the joint effect and is known as the effective half-life. The following relationship holds for the concepts mentioned above:

$$t_{\text{eff}\,1/2} = \frac{t_{1/2}\,t_{\text{biol}\,1/2}}{t_{1/2} + t_{\text{biol}\,1/2}}$$

Electromagnetic radiation

The radiation that is made up of quanta (has discrete energies), has no charge and is of rather short wavelength (of the order of nm) is called electromagnetic radiation. For practical reasons X-rays and γ-radiations are jointly termed electromagnetic radiations. X-rays have their origin from the electron shell(s), γ-radiation, however, is emitted from the nucleus.

Electron

The atomic nucleus is surrounded by negatively charged particles, electrons, that have a mass of 9.108×10^{-28} g. β^--Radiation is also composed of high energy electrons, but these are formed in the decay of neutrons of the nucleus along with the by-products: proton and antineutrino.

Electron capture

A particular type of radioactive decay involves the capture of an electron by the nucleus from the nearest to the nucleus (K-shell). This process results in the transformation of a proton of the nucleus into a neutron. The energy released in the exothermic process can contribute, in part, to the excitation of the nucleus and, in part, to the emission of a neutrino. Electron capture is accompanied by the emission of characteristic X-rays from the nucleus left behind with an excess amount of energy.

Electron volt

The energy of particles and photons produced in nuclear decay can be related to the amount of energy gained by an electron when passing through an electric field of 1 V potential. Outside the field of nuclear physics it is straightforward to express energy in units of J. 1 electron volt = 1 eV = 1.602×10^{-19} J = 0.1602 aJ and 1 J = 6.242 EeV.

In practical use, mainly the multiples of the electron-volt are applied, for example: 10^3 eV = 1 keV, 10^6 eV = 1 MeV.

Equivalent dose (Dose equivalent)

The absorbed dose is, mainly in the practice of health physics, substituted by the concept of equivalent dose, considering the type of radiation, the relative biological efficiency (RBE) and other factors (e.g., the so-called quality factor, Q). This is characterized by the dose that has the same biological effect on the human organism as X-rays.

The new unit of equivalent dose is sievert (Sv). 1 Sv = 1 J/kg.

The unit used earlier was the "rad equivalent man", or rem; 1 Sv = 100 rem, or 1 rem = 0.01 Sv, respectively.

Equivalent thickness

See Surface mass

Exposed dose

The radioactive radiation that cannot ionize directly can bring about charged particles as a result of the dose given to the irradiated medium. The unit of the amount of dose (also known as kerma) is J/kg, similarly to that of the absorbed dose.

Exposure

The electromagnetic radiation causes ionization in air. The measure of the interaction is characterized by the exposure (or exposure dose, used earlier) in units of coulomb per *kilogram*. The ionizing radiation of constant intensity is of unit exposure if it produces one coulomb electric charges of either sign in one kilogram of air.

The former name of exposure, *roentgen* (R), is still in use, however the gradual elimination of this unit is recommended. The amount of radiation that forms 1 e.s.u. of ions of either sign in 1.293 mg dry air (1 cm^3 of dry air at standard state) is 1 R.

The conversion between the two units is performed using the expression:

$$1 \text{ R} = 2.5798 \times 10^{-4} \text{ C/kg}$$

Exposure rate

The exposure rate, i.e., the exposure related to unit time, is denoted by I and should be given in A/kg. Earlier, the widely used units were R/s and/or R/h.

$$1 \text{ R/s} = 258 \text{ }\mu\text{A/kg} \quad \text{and} \quad 1 \text{ R/h} = 71.66 \text{ nA/kg}$$

γ-Radiation

See Electromagnetic radiation. The main feature of the term γ-radiation refers to the radiation emitted from the nucleus.

G-value

The yield of chemical reactions brought about by nuclear radiations is given by the so-called G-value. This is equal to the number of species (molecules, atoms, ions, radicals, etc.) decomposed or formed as a result of the absorption of 100 eV radiation energy, i.e., the radiation chemical yield.

Considering the new system of units this quantity should be given in 1/J:

$$1/100 \text{ eV} = 6.242 \times 10^{16} \text{ J}^{-1}$$

However, the earlier definition of the G-value is still valid.

Half-life

The time needed for the decrement of the number of nuclides to one-half of the original value in a homogeneous radioactive material, as a result of radioactive decay, is called the *physical half-life* of the given isotope, or shortly its half-life. The half-life can be given in any kind of time-units; it is denoted by $t_{1/2}$. (See also Decay constant.)

Half-thickness of absorbent

The thickness of an absorbent, usually given in units of g/m^2, that decreases the intensity of radiation to one-half of its original value is called half-thickness of the given absorbent. In addition to the half-thickness, the *tenth-thickness* is also used, mainly in the case of γ-radiations.

In practice, the distance corresponding to eight times the half-thickness is looked upon as the *maximum range* of a given radiation.

Isotope effect
The physical and sometimes the chemical properties of isotopes or compounds containing isotopes of the same element differ, especially if relatively large differences are seen in the mass numbers. This phenomenon is called isotope effect.

LET value
See Specific ionization.

Mass-absorption coefficient
The absorption of radioactive radiation in matter can be calculated, in general, by the expression

$$\frac{I}{I_0} = \exp(-\mu_m d)$$

which states that the intensity of radiation decreases from the original value of I_0 to a value of I after proceeding through a layer of the absorbent of thickness d; and the decrease in the given medium is determined by the constant of proportionality in the exponent μ_m. If d is taken as equivalent thickness (see also Surface mass) and given in kg/m^2 (or g/cm^2), the unit of the mass absorption coefficient μ_m should be, correspondingly, expressed as m^2/kg (or cm^2/g). With the value of d given in units of length, e.g., in m or cm, the linear absorption coefficient μ having a dimension of $1/m$, or $1/cm$ is to be used instead of the mass absorption coefficient.

The two constants can be converted easily to each other knowing the density of the absorbent, ρ (kg/m^3), by the expression

$$\mu_m \equiv \frac{\mu}{\rho}$$

In practice, the use of μ_m is of advantage because the dependence of the absorption on energy can be described in such a manner in a much simpler way.

Mean energy of nuclear radiation
The energy spectrum of β^--radiation emitted by radionuclei is continuous up to the limit $E_{\beta_{max}}$. The mean energy of the particles can roughly be estimated from the relationship $E_\beta \approx 1/3\ E_{\beta_{max}}$. The coefficient of proportionality is a function of the atomic number of the radionuclide: it is smaller or greater than 1/3 in the case of greater or smaller atomic numbers, respectively, and approximately equal to 1/2 when the above expression is used for very high energy β^+-radiations.

The mean energy of γ-radiation can be estimated from the percentage distribution of the constituent γ-photons forming the weighed average, however, serious errors are expected due to the complications involved in the interaction of the radiation with matter.

Mössbauer effect
Mössbauer showed that in the interaction of low energy γ-photons with suitably selected absorbing crystals a *resonance scattering* without recoil can be achieved and the scattering cross-section is close to the maximum value. The nuclear excitation levels can be prevented from thermal broadening by an appropriate selection of experimental conditions (first of all, by cooling the system).

If we compare the spectrum of the electromagnetic beam of radiation that passed through an absorbing crystal surrounding the source with the spectrum of a beam that proceeded through a substance containing the same atoms as did the former absorber but in different chemical bonding, the changes observed due to the different electron structures allow conclusions to be drawn regarding the isomer shift. The absorption lines can be resolved if we put the nucleus studied in an inhomogeneous electric field: the *quadrupole splitting* thus observed can give valuable information on the structure and symmetry.

Neutron

The neutron is a neutral elementary particle that participates in the construction of every atomic nucleus, except ^1H. The rest mass of the neutron is 1.6747×10^{-24} g; this is equivalent to an energy of 939.5 MeV $= 0.15011$ μJ. Neutrons are regarded as thermal if their energy equals about 0.025 eV.

Neutron flux density

The product of the mean velocity of neutrons derived from the Maxwell distribution and the density of the neutrons proceeding through an *l* area perpendicular to the direction of the beam in unit time is called neutron flux density, denoted by Φ or Φ_n and measured in units of $1/(m^2s)$. In an analogous way, the concept can be applied to the beam of particles or photons of electromagnetic radiations.

Nuclear fission

Nuclei that have a high mass number and an excess amount of neutrons decompose spontaneously or upon bombardment with particles into two primary nuclei, with a ratio of mass numbers approximately equal to $3:2$, while elementary particles (neutrons) are emitted. Nuclear fission forms stable products very rarely in a direct way, the nuclides produced stabilize through a series of decompositions (mostly by consecutive β-decays).

Nuclear isomer

The absorption of nuclear radiation or the radioactive decay can result in the production of an excited nuclide the half-life of which is relatively long (of the order of ns); this species is called the nuclear isomer of the corresponding ground state nuclide. The nuclear isomer is denoted by ^{m} in the right upper index, e.g., $^{99}Tc^m$.

Nuclide

The atomic species having a given atomic and mass number and a given state of energy is called nuclide. Among the nuclides stable and radioactive ones should be distinguished. In addition, another basis of grouping may be according to the same atomic number (the same number of protons) but different mass numbers (*isotopic nuclides*), the same number of neutrons (*isotonic nuclides*), the same mass number (*isobaric nuclides*), etc.

In agreement with widespread usage, the term "*isotope*" is used throughout this book in place of the correct word "nuclide".

Pair production

A special result of the interaction of γ-radiation with matter is pair production, i.e., the γ-photon is converted in the electric field of the nucleus into a positron and electron. The *threshold energy* for this interaction is the sum of the energy equivalents of the corresponding rest masses, i.e., 1.02 MeV.

Particle flux density
The number of particles proceeding through a surface of unit area in unit time is called particle flux density. Its unit is $1/(m^2s)$.

Photoelectric effect
Low-energy γ-photons interact with matter most probably in such a way that the whole energy is imparted to an electron while the excited nucleus is usually stabilized by the emission of characteristic X-rays.

Positron radiation
Some radioactive nuclei stabilize through the decomposition of a proton of the nucleus into a neutron and a particle called a positron, that has a mass equal to that of the electron but a positive charge. The nuclear process is accompanied by the formation of a neutrino. A secondary result of positron radiation is *annihilation radiation*.

Proton
The proton, the constituent particle of every atomic nucleus, has a unit positive charge and forms the nucleus of the lightest isotope of hydrogen, ^1H. The number of protons is equal to the atomic number in each element, thus, the chemical properties can be deduced from the number of protons. The rest mass of the proton equals 1.6724×10^{-24} g, corresponding to an energy of 938.2 MeV $= 0.15011$ µJ.

Radioactive concentration
The activity of the isotope solute per unit volume of the solution is called radioactive concentration. It is recommended to give this quantity in units of GBq/dm^3, instead of mCi/ml used earlier. 1 mCi/ml $= 3.7 \times 10^{10}$ $Bq/dm^3 = 37$ GBq/dm^3.
    It should be noted that this term was treated incorrectly as specific activity, especially in earlier works.

Radiochemical purity
The fraction of the total activity present in the unsealed radioactive preparation in the chemical form covered by the name of the product is called radiochemical purity and is expressed, in general, in percent.

Radionuclidic purity
The term radionuclidic (or *radioisotopic*) purity covers the percentage contribution of the activity of the given nuclide to the total activity.

Radium equivalent
The activity of sealed sources is in an old-fashioned way still sometimes today given in radium equivalents. Using this term the activity is related indirectly to the dose constant (exposed dose rate) of ^{226}Ra. A source has an activity of 1 g-radium-equivalent, if the dose rate measured at a distance from it is the same as in the case of 1 g of ^{226}Ra being in equilibrium with its decay products.
    In practice, the activity given in g-radium-equivalents is converted into units of Bq by multiplication of the activity of unit mass of ^{226}Ra with the ratio of the dose constants K_γ of ^{226}Ra and the given isotope. Thus, for example, the activity of an ^{192}Ir source of activity

of 1 g-radium-equivalent is converted into units of Bq as follows:

$$\frac{K_{\gamma 226_{Ra}}}{K_{\gamma 192_{Ir}}} = \frac{1.59 \text{ aA m}^2/(\text{kg Bq})}{0.97 \text{ aA m}^2/(\text{kg Bq})} \times 3.7 \times 10^{10} \text{ Bq} = 60.6 \text{ GBq}$$

Saturation activity
The maximum activity that can be achieved, in principle, in the given activating nuclear reaction, at given values of the flux and the cross-sections is called saturation activity. The saturation activity can be reached using an "infinitely long" time of activation.

Self-absorption
Unless the source of activity is infinitely thin, a part of the radiation will be absorbed by the source itself thus resulting in the decrease of the output of particles and/or photons. Self-absorption is of special importance in the case of soft β-emitters.

Specific activity
In the case of a radiating material, the activity of the radioactive isotope characterizing the preparation related to unit mass of the element labelled with the radioactive isotope is called the specific activity. The activity used to be related rather, to the relative molecular mass of the compound, especially in the case of labelled organic compounds, when it is, however, more correct to talk about *molar activity* of the preparation.

   It is expedient to give the activity in units of Bq/kg or analogously in PBq/g (GBq/g), etc.

   The activity related to unit volume denotes *radioactive concentration* rather than specific activity.

Specific ionization
Ionizing particles bring about ion-pairs along their track when penetrating into the medium. The number of ion-pairs related to unit path-length is called linear ion density or specific ionization. The energy delivered to the medium in unit path-length is given in eV/nm (keV/μm) and called *linear energy transfer (LET)*.

   In studies outside the field of nuclear physics the linear energy transfer should be given in J/m.

$$1 \text{ keV/μm} = 0.1602 \text{ nJ/kg}$$

Standard deviation
If the experimental data follow a Gaussian distribution, the error is given most frequently by the standard deviation corresponding to the statistical confidence of 0.683; this means that 68.3% of the experiments will yield a smaller error than that declared by the standard deviation. The standard deviation is denoted by σ.

   The value of the standard deviation can be calculated for a great number of experiments, m, using the expression:

$$\sigma = \pm \left[\frac{\sum_{i=1}^{m} (\xi_i - \zeta)^2}{m-1} \right]^{1/2}$$

where ξ_i stands for the individual data and ζ for their arithmetic mean $\sigma \approx \sqrt{\xi}$.

Surface mass (Equivalent thickness)

Mainly in nuclear techniques, thickness is given usually in equivalent thickness (e.g., in g/m^2) instead of distance units. The concept of equivalent thickness (named earlier also as *surface density*) finds its application in the calculation of the absorption of radiation, when considering the mass-absorption coefficient.

Szilard–Chalmers effect

The new nuclei formed in nuclear decay are produced in excited states, and the kinetic energy of the *recoil* can alter the form and nature of the original chemical bonds. This phenomenon is called *Szilard–Chalmers effect*; it enables us, in certain cases, to use radiochemical methods for the *separation of products* formed in nuclear reactions revealing no changes in the atomic numbers.

X-ray fluorescence

When electromagnetic radiations of wavelengths in the region of X-rays interact with matter, X-rays are formed which are characteristic of the composition of the substance and this phenomenon is called X-ray fluorescence. Characteristic radiations of lower intensity can be brought about using other types of ionizing radiations, as well.

1.3. THE MAJOR PROPERTIES OF THE 100 MOST IMPORTANT RADIOISOTOPES

The 100 radioisotopes described below represent a — more or less — arbitrary selection. When collecting the data on radiations (β^-- and γ-energies) the list was restricted to the most important values; this becomes clear from the incomplete sequence of indices in the appropriate column of Table 1.1. (e.g., β^-_4, γ_{13}).

The percentage distribution of radiation energies as well as the number of γ-quanta emitted in 100 disintegrations are given in brackets, if known.

Table 1.1. The most important properties of radioisotopes

Isotope	$t_{1/2}$	Energy (MeV) and abundance % of characteristic radiations	Dose constant, K_r	
			$\dfrac{aA\ m^2}{(kg\ Bq)}$	R m²/(Ci h)
³H	12.26 y	β^-_1 0.019 (100)		
¹⁴C	5730 y	β^-_1 0.156 (100)		
¹⁸F	110 min	β^+ 0.635 (97)	1.10	0.57
		K		
²²Na	2.62 y	β^+_1 0.54 (89)	2.32	1.20
		γ_1 1.275 (99.9)		
		K		
²⁴Na	15.05 h	β^-_1 1.389 (100)	3.66	1.89
		γ_1 1.369 (100); γ_2 2.754 (100)		

Table 1.1 (cont.)

Isotope	$t_{1/2}$	Energy (MeV) and abundance % of characteristic radiations	Dose constant, K_r	
			$\dfrac{\text{aA m}^2}{\text{(kg Bq)}}$	R m²/(Ci h)
^{28}Mg	21.2 h	β_1^- 0.46 (100) γ_1 0.03 (97); γ_2 0.40 (29); γ_3 0.45 (29); γ_4 1.35 (71)	1.41	0.73
^{31}Si	2.62 h	β_1^- 1.48 (100) γ_1 1.26 (0.07)		
^{32}P	14.28 d	β_1^- 1.71 (100)		
^{33}P	24.4 d	β_1^- 0.25		
^{35}S	87.9 d	β_1^- 0.167 (100)		
^{36}Cl	3.1×10^5 y	β_1^- 0.714 (98) K		
^{39}Ar	269 y	β_1^- 0.565 (100)		
^{42}K	12.36 h	β_1^- 1.99 (18); β_2^- 3.52 (82) γ_1 0.31 (0.2); γ_2 1.52	0.15	0.08
^{45}Ca	165 d	β_1^- 0.252 (100)		
^{47}Ca	4.54 d	β_1^- 0.66 (83); β_2^- 1.98 (17) γ_1 0.49 (6); γ_2 0.82 (6); γ_3 1.31(76)	0.99	0.51
^{46}Sc	83.9 d	β_1^- 0.357 (100); β_1^- 1.48 (0.004) γ_1 0.889 (100); γ_2 1.120 (100)	2.15	1.11
^{51}Cr	27.8 d	γ_1 0.320 (9) K	0.04	0.02
^{54}Mn	303 d	γ_1 0.835 (100) K	0.93	0.48
^{56}Mn	2.58 h	β_1^- 0.72 (20); β_2^- 1.04 (30); β_3^- 2.85 (50) γ_1 0.85 (99); γ_2 1.81 (26); γ_3 2.11 (14)	1.61	0.83
^{55}Fe	2.60 y	K 0.006		
^{59}Fe	45.6 d	β_1^- 0.27 (46); β_2^- 0.46 (53.7); β_3^- 1.57 (0.3) γ_1 0.19 (2.5); γ_2 1.10 (56); γ_3 1.29(44)	1.32	0.68
^{57}Co	267 d	γ_1 0.014 (9); γ_2 0.122 (88); γ_3 0.136 (12); γ_4 0.692 (0.1) K	0.12	0.06
^{58}Co	71.3 d	β_1^+ 0.47 (15) γ_1 0.81 (100); γ_2 0.87 (1.4) K	1.07	0.55

Table 1.1 (cont.)

Isotope	$t_{1/2}$	Energy (MeV) and abundance % of characteristic radiations	Dose constant, K_r	
			$\dfrac{\text{aA m}^2}{\text{(kg Bq)}}$	R m²/(Ci h)
^{60}Co	5.26 y	β_1^- 0.312 (99) γ_1 1.173 (99); γ_2 1.332 (100)	2.52	1.30
^{63}Ni	92 y	β_1^- 0.067 (100)		
^{65}Ni	2.56 h	β_1^- 0.65 (29); β_2^- 1.02 (1.4); β_3^- 2.14 (57) γ_1 0.37 (11); γ_2 1.12 (19); γ_3 1.48 (11)	0.43	0.22
^{64}Cu	12.8 h	β_1^- 0.57 (38); β_1^+ 0.66 (19) γ_1 1.33 (0.05) K	0.21	0.11
^{67}Cu	61 h	β_1^- 0.40 (45); β_2^- 0.48 (35); β_3^- 0.57 (20) γ_1 0.09 (23); γ_3 0.18 (23)	0.12	0.06
^{65}Zn	245 d	β_1^+ 0.33 (1.7) γ_1 1.12 (49) K	0.52	0.27
^{67}Ga	77.9 h	γ_1 0.09 (42); γ_3 0.18 (26); γ_5 0.30 (19); γ_6 0.39 (5) K	0.17	0.09
^{72}Ga	14.1 h	β_1^- 0.64 (42); β_2^- 0.96 (31); β_3^- 1.51 (10); β_4^- 2.53 (9); β_5^- 3.15 (8) γ_2 0.63 (21); γ_7 0.84 (97); γ_{12} 1.04 (9); γ_{14} 1.27 (5); γ_{19} 1.46 (3); γ_{29} 2.20 (39); γ_{31} 2.50 (30)	3.04	1.57
^{71}Ge	11.4 d	K		
^{76}As	24.6 h	β_1^- 0.36 (9); β_2^- 1.76 (4); β_3^- 2.41 (31); β_4^- 2.97 (56) γ_1 0.56 (38); γ_2 0.66 (6.3); γ_3 1.22 (5.8); γ_6 2.10 (0.9)	0.46	0.24
^{75}Se	120.4 d	γ_3 0.10 (3); γ_4 0.12 (15); γ_5 0.14 (54); γ_7 0.27 (56); γ_8 0.28 (23); γ_{10} 0.40 (13) K	0.29	0.15
^{82}Br	35.3 h	β_1^- 0.44 (100) γ_1 0.55 (67); γ_2 0.62 (42); γ_3 0.70 (24); γ_4 0.78 (80); γ_5 0.83 (24); γ_6 1.04 (30); γ_7 1.32 (31); γ_8 1.48 (19)	2.81	1.45
^{85}Kr	10.76 y	β_1^- 0.15 (0.7); β_2^- 0.67 (99.3) γ_1 0.514 (0.4)	0.004	0.002

Table 1.1 (cont.)

Isotope	$t_{1/2}$	Energy (MeV) and abundance % of characteristic radiations	Dose constant, K_r	
			$\dfrac{\text{aA m}^2}{\text{(kg Bq)}}$	$\text{R m}^2/(\text{Ci h})$
^{86}Rb	18.7 d	β_2^- 0.71 (9); β_3^- 1.78 (91) γ_1 1.08 (9)	0.12	0.06
^{85}Sr	64 d	γ_1 0.514 (99) K	0.56	0.29
^{89}Sr	52.7 d	β_1^- 1.462 (100)		
^{90}Sr	28.1 y	β_1^- 0.546 (100)		
^{88}Y	108.1 d	β_1^+ 0.76 γ_1 0.91 (90); γ_2 1.86 (99)	2.73	1.41
^{90}Y	64 h	β_1^- 2.27 (100) β_2^- 0.36 (43); β_3^- 0.40 (45)		
^{95}Zr	65.5 d	β_4^- 0.89 (2) γ_1 0.724 (55); γ_2 0.756 (42)	0.81	0.42
^{95}Nb	35 d	β_1^- 0.16 (99); β_2^- 0.92 (1) γ_1 0.77 (100)	0.85	0.44
^{99}Mo	66.7 h	β_1^- 0.41 (14); β_3^- 1.18 (85) γ_2 0.18 (4); γ_3 0.37 (1.3); γ_4 0.74 (12)	0.23	0.12
^{99}Tc^m	6.05 h	γ_1 0.140 (89); γ_3 0.142 (0.03) K 0.018	0.12	0.06
^{106}Ru	1.0 y	β_1^- 0.039 (100)		
^{110}Ag^m	255 d	β_1^- 0.086 (65); β_2^- 0.536 (33); β_4^- 2.87 (2) γ_2 0.556 (8); γ_5 0.677 (95); γ_7 0.706 (17); γ_9 0.762 (21); γ_{10} 0.818 (8); γ_{11} 0.885 (69); γ_{12} 0.937 (29); γ_{13} 1.384 (26); γ_{15} 1.506 (14)	2.77	1.43
^{111}Ag	7.5 d	β_1^- 0.69 (6); β_2^- 0.79 (1); β_3^- 1.05 (93) γ_1 0.247 (1); γ_2 0.342 (5.6)	0.02	0.01
^{115}Cd^m	43 d	β_1^- 0.70 (3.3); β_2^- 1.62 (96.7) γ_1 0.49 (0.3); γ_2 0.94 (2.3); γ_3 1.29 (1.0)	0.04	0.02
^{115}Cd	53 h	β_1^- 0.59 (24); β_2^- 0.63 (13); β_4^- 1.11 (62) γ_1 0.23 (10); γ_3 0.33 (52); γ_4 0.53 (25)	0.45	0.23
^{113}In^m	1.66 h	γ_1 0.393 (64)	0.31	0.16
^{114}In^m	50.0 d	γ_1 0.19 (18.2); γ_2 0.56 (3.5); γ_3 0.72 (3.5) K	0.08	0.04

BASIC DATA AND DEFINITIONS

Table 1.1 (cont.)

Isotope	$t_{1/2}$	Energy (MeV) and abundance % of characteristic radiations	Dose constant, K_r	
			$\dfrac{\text{aA m}^2}{(\text{kg Bq})}$	R m²/(Ci h)
^{113}Sn	115 d	γ_1 0.257 (0.25); γ_3 0.650 (0.18)		
^{122}Sb	2.8 d	β_1^- 0.74 (4); β_2^- 1.40 (63); β_3^- 1.97 (30); β^+ 0.56 γ_1 0.57 (68.5); γ_2 0.69 (3.5); γ_3 1.14 (1.0); γ_4 1.26 (0.7) K		
^{124}Sb	60.9 d	β_1^- 0.25 (9); β_2^- 0.62 (53); β_3^- 0.94 (9); β_4^- 1.60 (7); β_5^- 2.31 (22) γ_1 0.609 (98.6); γ_2 0.646 (7.5); γ_4 0.723 (10.0); γ_5 0.969 (2.5); γ_9 1.370 (3.6); γ_{11} 1.69 (50.0); γ_{12} 2.088 (6.5)		
^{132}Te	77.7 h	β_1^- 0.22 (100) γ_1 0.053 (17); γ_2 0.230 (90)	0.25	0.13
^{125}I	60.2 d	γ_1 0.035 (7) K 0.028	0.14	0.07
^{131}I	8.05 d	β_1^- 0.25 (2.8); β_2^- 0.34 (9.3); β_3^- 0.61 (87.2); β_4^- 0.81 (0.7) γ_2 0.284 (5.0); γ_3 0.364 (78.4); γ_4 0.637 (9.0); γ_5 0.722 (3.0)	0.45	0.23
^{132}I	2.26 h	β_1^- 0.73 (15); β_2^- 0.90 (20); β_3^- 1.16 (23); β_4^- 1.53 (24); β_5^- 2.12 (18) γ_1 0.375 (4.8); γ_3 0.521 (21.0); γ_4 0.645 (30.0); γ_5 0.670 (100); γ_6 0.773 (86); γ_7 0.950 (23); γ_8 1.500 (5.8)	2.19	1.13
^{133}Xe^m	2.3 d	γ_1 0.233 (14)	0.02	0.01
^{134}Xe	5.27 d	β_1^- 0.35 (100) γ_1 0.08 (35.6); γ_2 0.16 (1.6)		
^{131}Cs	9.7 d	K 0.03		
^{134}Cs	2.05 y	β_1^- 0.08 (32); β_2^- 0.31 (5); β_3^- 0.66 (50); β_4^- 0.68 (13) γ_2 0.56 (10); γ_3 0.57 (14); γ_5 0.61 (95); γ_6 0.80 (90)	1.72	0.89
^{137}Cs	30 y	β_1^- 0.514 (92.4); β_2^- 1.18 (7.6)		
^{133}Ba	10.7 y	γ_2 0.08 (38); γ_3 0.16 (3); γ_6 0.30 (12); γ_7 0.36 (64); γ_8 0.39 (4); γ_9 0.63 (6) K	0.46	0.24
^{137}Ba^m	2.6 min	γ_1 0.662 (85)	0.66	0.34

Table 1.1 (cont.)

Isotope	$t_{1/2}$	Energy (MeV) and abundance % of characteristic radiations	Dose constant, K_r	
			$\dfrac{aA\ m^2}{(kg\ Bq)}$	$R\ m^2/(Ci\ h)$
^{140}Ba	12.8 d	β_1^- 0.48 (25); β_2^- 0.60 (10); β_3^- 0.90 (5); β_4^- 1.02 (60) γ_1 0.03 (16); γ_3 0.16 (5); γ_4 0.30 (5); γ_5 0.44 (5); γ_6 0.54 (25)	0.48	0.25
^{140}La	40.2 h	β_1^- 0.42 (16); β_2^- 0.86 (12); β_3^- 1.15 (20); β_4^- 1.36 (30); β_5^- 1.69 (14); β_6^- 2.18 (8) γ_1 0.323 (20); γ_3 0.491 (40); γ_5 0.815 (19); γ_6 0.868 (5); γ_7 0.923 (9); γ_8 1.596 (95); γ_9 2.53 (3)	2.32	1.20
^{141}Ce	33.1 d	β_1^- 0.44 (75); β_2^- 0.58 (25) γ_1 0.145 (48)	0.10	0.05
^{143}Ce	33.4 h	β_1^- 0.22 (6); β_2^- 0.50 (12); β_3^- 0.74 (5); β_4^- 1.13 (40); β_5^- 1.40 (38) γ_1 0.057 (6); γ_3 0.294 (23); γ_4 0.351 (23); γ_6 0.565 (6); γ_9 0.861 (6); γ_{10} 1.10 (6)	0.37	0.19
^{143}Pr	13.6 d	β_1^- 0.93 (100)		
^{147}Pm	2.62 y	β_1^- 0.224 (100)		
^{153}Sm	46.8 h	β_2^- 0.64 (38); β_3^- 0.71 (40); β_4^- 0.81 (22) γ_1 0.070 (8); γ_2 0.103 (34); γ_4 0.545 (0.2)	0.04	0.02
$^{152}Eu^m$	9.3 h	β_1^- 0.56 (2); β_2^- 1.55 (2); β_3^- 1.87 (74); β_1^+ 0.82 (0.01) γ_1 0.12 (15); γ_3 0.85 (14); γ_4 0.97 (7); γ_5 1.33 (2)	0.27	0.14
^{152}Eu	12.2 y	β_1^- 0.22 (2); β_2^- 0.36 (3); β_3^- 0.71 (12); β_4^- 1.05 (2); β_5^- 1.47 (7); β_1^+ 0.47 (0.01); β_2^+ 0.70 (0.01) γ_1 0.12 (59); γ_2 0.24 (9); γ_4 0.87 (6); γ_5 0.96 (14); γ_6 1.09	1.12	0.58
^{154}Eu	16 y	β_1^- 0.12 (2); β_2^- 0.25 (27); β_3^- 0.59 (35); β_4^- 0.83 (23); β_5^- 1.60 (3); β_6^- 1.84 (10); γ_1 0.12 (14); γ_2 0.25 (6); γ_3 0.59 (4); γ_4 0.73 (21); γ_5 0.87 (13); γ_6 1.00 (31); γ_7 1.28 (42)	1.20	0.62

BASIC DATA AND DEFINITIONS

Table 1.1 (cont.)

Isotope	$t_{1/2}$	Energy (MeV) and abundance % of characteristic radiations	Dose constant, K_r	
			$\dfrac{\text{aA m}^2}{\text{(kg Bq)}}$	R m²/(Ci h)
^{155}Eu	1.81 y	β_1^- 0.15 (70); β_2^- 0.19 (10); β_3^- 0.25 (20) γ_1 0.061 (25); γ_2 0.087 (73); γ_3 0.100 (30); γ_4 0.106 (28); γ_5 0.125 (17); γ_6 0.132 (6)	0.06	0.03
^{160}Tb	72.1 d	β_1^- 0.30 (12); β_2^- 0.46; β_3^- 0.58 (37.6); β_4^- 0.87 (31); β_5^- 1.77 (0.4) γ_1 0.087 (8); γ_4 0.298 (23); γ_5 0.878 (26); γ_6 0.962 (35); γ_7 1.179 (15); γ_8 1.273 (8)	1.07	0.55
^{170}Tm	134 d	β_1^- 0.884 (24); β_2^- 0.968 (76) γ_1 0.084 (2.5) K	0.006	0.003
^{175}Hf	70 d	γ_1 0.09 (3.4); γ_5 0.34 (75.5); γ_6 0.43 (1.4) K	0.41	0.21
^{181}Hf	42.5 d	β_1^- 0.34 (1); β_2^- 0.41 (97); β_3^- 0.55 (2) γ_1 0.133 (48); γ_3 0.346 (14); γ_5 0.482 (86)	0.60	0.31
^{182}Ta	115.1 d	β_1^- 0.18 (38); β_2^- 0.25 (5); β_3^- 0.33 (2); β_4^- 0.36 (20); β_5^- 0.44 (23); β_6^- 0.48 (4); β_7^- 0.51 (8); β_8^- 0.59 (0.9) γ_1 0.16; γ_5 0.22; γ_7 0.26; γ_8 1.12; γ_9 1.16; γ_{10} 1.19; γ_{11} 1.22	1.18	0.61
^{185}W	75 d	β_1^- 0.430 (100)		
^{187}W	24 h	β_1^- 0.34 (10); β_2^- 0.63 (7); β_3^- 1.33 (20) γ_1 0.07 (10); γ_2 0.13 (9); γ_{11} 0.48 (23); γ_{13} 0.55 (7); γ_{14} 0.62 (8); γ_{16} 0.69 (31); γ_{18} 0.78 (5)	0.60	0.31
^{192}Ir	74.2 d	β_2^- 0.24 (15); β_3^- 0.54 (38); β_4^- 0.67 (41) γ_4 0.30 (59); γ_6 0.32 (81); γ_9 0.47 (49); γ_{12} 0.61 (15) K	0.97	0.50
^{198}Au	2.7 d	β_1^- 0.29 (1); β_2^- 0.96 (99) γ_1 0.41 (95); γ_2 0.68 (1); γ_3 1.09 (0.2)	0.45	0.23

Table 1.1 (cont.)

Isotope	$t_{1/2}$	Energy (MeV) and abundance % of characteristic radiations	Dose constant, K_r	
			$\dfrac{\text{aA m}^2}{\text{(kg Bq)}}$	R m²/(Ci h)
^{199}Au	3.15 d	β_1^- 0.25 (24); β_2^- 0.30 (69); β_3^- 0.46 (7) γ_2 0.16 (47); γ_3 0.21 (11)	0.17	0.09
^{197}Hg^m	24 h	γ_2 0.13 (36); γ_3 0.16 (6) K	0.06	0.03
^{197}Hg	65 h	γ_1 0.077 (28.6); γ_2 0.191 (0.06); γ_3 0.268 (0.15) K	0.08	0.04
^{203}Hg	46.9 d	β_1^- 0.214 (100) γ_1 0.279 (77)	0.29	0.15
^{204}Tl	3.80 y	β_1^- 0.766 (98) K		
^{210}Bi	5.01 d	α_1 4.65 (2×10^{-4}) β_1^- 1.16 (100)		
^{210}Po	138.4 d	α_1 5.30 (100) γ_1 0.80 (10^{-3})		
^{226}Ra	1602 y	α_1 4.50 (5.7); α_2 4.78 (94.3) γ_2 0.242 (10.5); γ_4 0.295 (18.9); γ_5 0.352 (37.7); γ_{13} 0.609 (47.1); γ_{26} 1.120 (16.6); γ_{29} 1.238 (6.0); γ_{39} 1.764 (16.3); γ_{46} 2.204 (5.2)	1.82 Through Pt filter of 0.50 mm thickness: 1.59	0.94 0.82
^{238}U	4.5×10^9 y	α_1 4.15 (25); α_2 4.20 (75) γ_1 0.048 (18.7); γ_2 0.112 (0.02)		
^{239}Pu	2.44×10^4 y	α_1 5.11 (11); α_2 5.16 (88) γ_1 0.039 (0.007); γ_2 0.052 (0.02); γ_3 0.129 (0.005); γ_4 0.375 (0.001)		
^{241}Am	458 y	α_2 5.38 (1.7); α_3 5.44 (12.6); α_4 5.48 (85); α_5 5.50 (0.23); α_6 5.53 (0.35) γ_2 0.067 (0.53); γ_3 0.070 (0.53); γ_9 0.304	0.12	0.06
^{252}Cf	2.65 y	α_1 6.08 (15); α_2 6.12 (82) γ_1 0.043; γ_2 0.100; γ_3 0.16 n (spontaneous fission)		

Special attention was paid to make sure that selection was from data determined most recently and considered most reliable. The fact that several data in Table 1.1 differ somewhat from those published in earlier compilations reflects this effort.

1.4. MAIN PROPERTIES OF RADIOISOTOPES AND ISOTOPIC PREPARATIONS

The knowledge of the main features of the isotopic products is of primary importance when selecting the isotopic method for a given task. The compilation given below claims to compare the needs arising in the application of radioisotopes with the possibilities of the production, by listing the more important preparations and by submitting useful information generally not available from commercial catalogues.

In order to facilitate a survey of the field, the radioisotopes emitting β^--radiation only are tabulated in Table 1.2; a classification of the isotopes according to the half-lives and γ-photon energies is in Table 1.3.

Virtually no discussion of the properties of radioisotopes and radioactive preparations can cover every feature partly because of the vast data and partly because the actual values of certain parameters are strongly dependent on the commercial source.

The main data (e.g., half-life, energy of radiation) mentioned here are usually rounded values; the exact figures are compiled in Table 1.1 and Table 1.2.

1.4.1. GENERAL PROPERTIES OF RADIOISOTOPES

^{3}H(T)

Tritium, like ^{14}C, is used mainly for tracing organic compounds thus its main field of application is in the *pharmaceutical* and in the *organo–chemical industry*; it is also used for research in physical chemistry.

Table 1.2. Main properties of frequently used isotopes emitting only β-particles, in the order of increasing $E_{\beta_{max}}$

Isotope	$E_{\beta_{max}}$ MeV	$t_{1/2}$	Note
^{3}H	0.019	12.26 y	Carrier-free; theoretical value of specific activity ~ 360 PBq/kg (9.7 kCi/g)
^{106}Ru	0.039	1.0 y	The hard γ-emitting ^{106}Rh^m and ^{106}Rh are formed in the decay with half-lives 2.2 h and 30 s, and $K_\gamma = 2.60$ and 0.29, respectively
^{63}Ni	0.067	92 y	The commercial product is usually contaminated with minor amounts of ^{59}Ni (half-life 7.5×10^4 y) that decays with K-capture
^{14}C	0.156	5730 y	Carrier-free; theoretical value of specific activity ~ 170 TBq/kg (~ 4.6 Ci/g)
^{35}S	0.167	87.9 d	Carrier-free; theoretical value of specific activity ~ 1.6 EBq/kg (~ 43 kCi/kg)

Table 1.2 (cont.)

Isotope	$E_{\beta_{max}}$ MeV	$t_{1/2}$	Note
^{147}Pm	0.224	2.62 y	Carrier-free; theoretical value of specific activity ~ 35 PBq/kg (~ 946 Ci/g)
^{45}Ca	0.252	165 d	It can also be prepared in carrier-free form with a theoretically attainable specific activity of ~ 670 PBq/kg (18 kCi/g)
^{185}W	0.430	75 d	Available with limited specific activity
^{137}Cs	0.514	30 y	The commercially available product is nearly carrier-free (theoretical value of the specific activity ~ 3.6 PBq/kg = ~ 97 Ci/g, but is generally contaminated with minor amounts of the γ-emitting ^{134}Cs, having a half-life of 2.2 y. The ^{137}Ba^m nuclide (half-life 2.6 min) is formed in the decay of the ^{137}Cs, and it emits 0.662 MeV γ-photons; $K_\gamma = 0.66$
^{90}Sr	0.546	28.1 y	Carrier-free; theoretical value of the specific activity ~ 526 PBq/kg (14.2 kCi/g). The daughter nuclide of ^{90}Sr is ^{90}Y (as to its properties see there)
^{39}Ar	0.565	269 y	The commercially available product is generally contaminated with ^{37}Ar (half-life 34.3 y) which decays with K-capture
^{85}Kr	0.670	10.76 y	The β^--emission is accompanied by minor amounts of 0.514 MeV energy γ-radiation
^{36}Cl	0.714	3.1×10^5 y	The nuclide can be prepared only with low specific activity
^{204}Tl	0.766	3.80 y	The β^--radiation is accompanied by minor amounts of K-capture. One of the radioisotopes formed in the decay is ^{204}Pb^m (half-life 70 min), which emits hard γ-rays ($K_\gamma = 1.32$)
^{143}Pr	0.930	13.6 d	The product is nearly carrier-free the theoretically attainable specific activity is ~ 2.48 EBq/kg = 67 kCi/g, but the commercially available preparation is of variable radionuclidic purity
^{89}Sr	1.462	52.7 d	The commercial product is generally contaminated with ^{85}Sr, which has a half-life of 65 y, and decays with γ-emission and K-capture
^{32}P	1.707	14.28 d	Carrier-free; the theoretically attainable specific activity is ~ 10.8 EBq/kg (~ 290 kCi/g)
^{90}Y	2.273	64.0 h	It can be prepared in carrier-free form, too; the theoretical value of the specific activity is, then, ~ 20 EBq/kg (~ 540 kCi/g)

Table 1.3. The most important radioisotopes emitting γ-photons in addition to other types of radioactive radiation classified according to their half-lives and characteristic γ-energies

E_γ MeV $\diagdown$ $t_{1/2}$	<0.1	0.1–0.5	0.5–1.0	1.0–1.5	>1.5
<1 d		^{28}Mg ^{99}Tc^m ^{113}In^m ^{132}I ^{197}Hg^m	^{18}F ^{56}Mn ^{64}Cu ^{72}Ga ^{132}I ^{137}Ba^m	^{24}Na ^{28}Mg ^{31}Si	^{24}Na ^{42}K ^{56}Mn ^{72}Ga ^{132}I
1 d–10 d	^{67}Cu ^{67}Ga ^{133}Xe ^{153}Sm ^{197}Hg	^{47}Ca ^{65}Ni ^{67}Cu ^{67}Ga ^{99}Mo ^{111}Ag ^{115}Cd ^{132}Te ^{131}I ^{133}Xe^m ^{131}Cs ^{140}La ^{143}Ce ^{153}Sm ^{187}W ^{198}Au ^{199}Au	^{47}Ca ^{76}As ^{82}Br ^{99}Mo ^{115}Cd ^{122}Sb ^{131}I ^{140}La ^{187}W	^{47}Ca ^{65}Ni ^{76}As ^{82}Br	^{140}La
10 d–50 d	^{140}Ba	^{51}Cr ^{71}Ge ^{114}In^m ^{141}Ce ^{181}Hf ^{102}Hg	^{84}Rb ^{95}Nb ^{115}Cd^m ^{115}In^m ^{140}Ba	^{59}Fe ^{86}Rb ^{115}Cd^m	
50 d–365 d	^{125}I ^{160}Tb ^{170}Tm	^{57}Co ^{75}Se ^{113}Sn ^{160}Tb ^{175}Hf ^{182}Ta ^{192}Ir	^{46}Sc ^{54}Mn ^{58}Co ^{65}Zn ^{85}Sr ^{95}Zr ^{110}Ag^m ^{113}Sn ^{124}Sb ^{160}Tb ^{192}Ir ^{210}Po	^{46}Sc ^{65}Zn ^{110}Ag ^{160}Tb ^{182}Ta	^{124}Sb
>1 y	^{55}Fe ^{133}Ba ^{155}Eu ^{232}Th ^{238}U ^{239}Pu ^{241}Am	^{133}Ba ^{152}Eu ^{154}Eu ^{155}Eu ^{226}Ra	^{22}Na ^{85}Kr ^{134}Cs ^{133}Ba ^{152}Eu ^{154}Eu ^{226}Ra	^{60}Co ^{134}Cs ^{152}Eu ^{154}Eu ^{226}Ra	^{226}Ra

Tritium is produced in the most economical way by the $^6Li(n, \alpha)^3H$ nuclear reaction; the product is commercially available in the form of tritium gas and ^3H_2O. During preparation and storage special care should be taken to keep humidity of the air from the product to prevent it from becoming diluted with inactive hydrogen.

The theoretically achievable *maximum specific activity* is 360 PBq/kg = 9.74 kCi/g referring to carrier-free tritium, this corresponds to 100 PBq/kg in the case of T_2O, that is, water that contains hydrogen exclusively in the form of ^3H.

The concept of concentration expressed in units of TU (*tritium unit*) is widely used mainly in environmental research. There is a tritium concentration of 1 TU in a sample, if out of 10^{18} hydrogen atoms one is tritium. 1 TU = 120 Bq/m^3.

Tracing with tritium is most often performed using the so-called *Wilzbach method*. This involves the establishment of contact between the substance to be tritiated and a high pressure tritium gas in a closed system; the isotope exchange results in the formation of a *non-specifically* labelled compound. *Specific* labelling can be brought about by catalytic exchange processes, by suitably modified versions of commonly used syntheses or sometimes by *biosynthesis*.

In tritium tracer studies attention should be paid to the fact that this isotope gives rise to the greatest isotope effect and this possibility should always be borne in mind when interpreting the results.

The detection of the very soft β^--radiation is a rather difficult task but sufficiently accurate routine work can be done using *liquid scintillation techniques*. The activity can be measured using appropriately prepared samples and "in situ" detection can be resolved only quite exceptionally.

^{14}C

^{14}C is the most widely used radioisotope in research. Since it is applicable primarily for tracing organic compounds it finds its application mainly in the *pharmaceutical* and, to a lesser extent, in the *organic chemical industry*, where it enables the mechanism of reactions and biological processes to be discovered.

The *preparation* of ^{14}C products is based on the nuclear reaction $^{14}N(n, p)^{14}C$. During the preparation and especially the transformation of the raw-material, special attention should be given to keep off the air and carbon dioxide present in it, because even traces of CO_2 can effectively decrease the specific activity of the product by isotope exchange.

The theoretically available *maximum specific activity* equals 170 TBq/kg = 4.59 Ci/g; the molar activity is 2.37 TBq/mole of carbon (64 mCi/mmole), if a 100% efficiency is attributed to the monoatomic labelling. However, these values are not reached in practice, since the specific activity of $Ba^{14}CO_3$ used as the raw-material in organic syntheses is 10 TBq/kg $BaCO_3$ at best, instead of the theoretical value of 11.8 TBq/kg $BaCO_3$.

The *measurement* of the soft β^--radiation of ^{14}C was carried out earlier with rather low efficiency; up-to-date flow counters as well as the liquid scintillation techniques used recently provide reasonably accurate data.

^{22}Na

There are certain cases, especially in the *silicate industry* and, in general, in long-term studies when the otherwise rather widely used ^{24}Na cannot meet the requirements, due to its relatively short half-life of 15 h. ^{22}Na, with a half-life of 2.6 years, finds its application primarily in these kinds of works and has, thus, become one of the few *cyclotron products* having widespread usage.

In addition to positron radiation and a decay associated with electron capture, ^{22}Na emits medium energy γ-photons; its dose constant is nearly the same as that of ^{60}Co: $K_y = 2.32$ aA m^2/(kg Bq).

For its *production* the ^{24}Mg(d, α)^{22}Na nuclear reaction is used; owing to the longer half-life, longer periods of times of irradiation are necessary making production costly due to the expensive operation of the cyclotron.

The radionuclidic purity of the product is generally high because the half-lives of the by-products are shorter than that of ^{22}Na. However, in the processing and purification of the product, it will inevitably be contaminated with inactive impurities, and thus the appearance of certain amounts of salts is to be expected.

^{24}Na

^{24}Na is a commonly used tracer in *industry*. The reason for its widespread application lies, in part, in its favourable half-life and detectability and, in part, in the fact that it is inexpensive. As a matter of fact, ^{24}Na is not only a specific tracer for the labelling of sodium atoms and ions, but it is helpful in studies of general interest, such as *diffusion, mixing, mass-transport*, etc.

It can be *prepared* by the ^{23}Na(n, γ)^{24}Na reaction, which allows a high specific activity to be produced, owing to the favourable value of the cross section.

The short half-life and the high energy γ-radiation can be of advantage and disadvantage, as well. Because of the rapid decay, generally high activities are needed for the performance of an experiment, and therefore special care should be exercised in the design of radiation shielding.

^{32}P

Although, radioactive phosphorus is used primarily in the investigation of *biological processes*, the high energy β^--radiation makes it a good applicant as a tracer in studies on industrial operations and numerous other subjects in research.

In some industrial applications ^{32}P is used in the form of elementary red phosphorus produced in the (n, γ) reaction, however, most of the users need *carrier-free* phosphate preparations. These latter ones are prepared in the ^{32}S(n, p)^{32}P nuclear reaction followed by separation using extraction or distillation. The primary product is, in general, hydrochloric acid solution of H$_3$^{32}PO$_4$, while the solutions of different pH-values contain different ligands depending on the value of the deprotonation constant of the conjugated acid. Thus, the ligands H$_2$PO$_4^-$, HPO$_4^{2-}$ and PO$_4^{3-}$ exist in the ranges, pH = 4–5, 9–10 and > 13, respectively.

Because of the competitive nuclear reaction, the formation of ^{33}P cannot be avoided, however the presence of this nuclide, that emits low-energy β^--particles and has a somewhat longer half-life, usually does not cause any inconveniences in the measurements, as its contribution to the total activity is very little.

^{35}S

The sulphur isotope emits low-energy β^--particles; it is used in the *pharmaceutical industry* and in *agriculture*.

Its *production* by the ^{34}S(n, γ)^{35}S reaction proceeds with rather low efficiency, therefore, the carrier-free product formed in the ^{35}Cl(n, p)^{35}S reaction is commercially available. The high specific activity is favourable in the detection of the 0.17 MeV energy β^--particles, that would otherwise cause serious difficulties due to *self-absorption*. The isotope producing centres offer ^{35}S in the form of different compounds corresponding to the demands ($Na_2{}^{35}S$, $Na_2{}^{35}SO_4$, $H_2{}^{35}SO_4$, etc.).

^{42}K

The only isotope of potassium commonly used is *produced* in the (n, γ) reaction; essentially the same statements are true for this isotope as those mentioned earlier regarding ^{24}Na: it finds its application in tracer studies of *general* interest. ^{42}K emits γ-radiation of much lower intensity than does ^{24}Na, but it has a β^--radiation of remarkably high energy: the major component of the radiation has an energy equal to $E_{\beta-max} = 3.52$ MeV.

^{42}K, likewise ^{24}Na, is sold mostly in an aqueous solution of its chloride salt.

^{45}Ca

This isotope of calcium has a half-life of 165 days; it is used mainly in *human diagnostics* and less often in the *silicate industry*. It is *produced* in the (n, γ) reaction with a medium specific activity and in the ^{45}Sc(n, p) ^{45}Ca reaction in *carrier-free* form. The latter process is certainly much more expensive, but in certain cases the highly efficient detection of the 0.25 MeV energy β^--particles may be necessary in order to decrease the effect of self-absorption.

The calcium isotope plays a secondary role among the commonly used radioisotopes of alkaline earth metals (^{28}Mg, ^{45}Ca, ^{85}Sr, ^{89}Sr, ^{90}Sr, ^{133}Ba and ^{140}Ba), as in most experiments it can be substituted with an isotope that is easier to detect; the γ-emitting ^{47}Ca, having a half-life of 4.54 days, can eventually also be considered as a tracer.

^{51}Cr

Radioactive chromium is used in *metallurgical* and *corrosion* studies, but its application in *wearing tests* is also of growing importance. These works generally do not require high specific activities, as opposed to diagnostic applications. Therefore, commercially available solutions of ^{51}CrCl$_3$ and $Na_2{}^{51}CrO_4$ are *prepared* by irradiating targets enriched in ^{50}Cr; their *specific activity* can reach or even exceed a value of 1 PBq/kg.

The yield of the 0.32 MeV γ-photons is low, but the radiation is still, in general, well detectable.

For explicitly industrial uses or in sealed sources metallic chromium is applied, the specific activity of which is certainly considerably lower.

^{55}Fe

From the two isotopes of iron, the one *produced* in the (n, γ) reaction, ^{55}Fe, and that has the longer half-life, has obtained the less widespread application as a tracer, because of the low energy of its radiation; it is used rather as a filling of special sealed sources.

In sealed sources, the iron is fixed generally using an electroplating method. In order to decrease the extent of self-absorption, a layer of the smallest possible thickness is prepared for the electrolysis, and the cover is designed with the same requirements kept in mind.

^{59}Fe

In industrial, primarily *metallurgical*, investigations, ^{59}Fe is used for the labelling of iron, with a half-life of 45.6 days. This isotope can be *prepared* with high specific activity in the (n, γ) reaction using a target enriched in ^{58}Fe, or in a *carrier-free* form in the ^{59}Co(n, p) ^{59}Fe reaction. The disadvantage of the (n, p) reaction lies in the fact that the competitive (n, γ) reaction gives rise to the formation of ^{60}Co with several orders of magnitude greater yield. The separation of ^{59}Fe, as well as the accurate determination of the amount of ^{60}Co contaminant in the product is rather difficult, due to the high degree of similarity between the γ-spectra of the two isotopes. The end-product is commercially available in the form of ^{59}FeCl$_3$.

The products of both methods are rather expensive. Therefore, in some studies mixtures of ^{55}Fe and ^{59}Fe are used, which are certainly much cheaper.

^{58}Co

^{58}Co is an appropriate substitute of the much cheaper ^{60}Co in certain *tracer* studies, where the short half-life is an important requirement in labelling cobalt.

The carrier-free isotope is prepared in the ^{58}Ni(n, p) ^{58}Co nuclear reaction. Both the 0.81 MeV energy γ-photons emitted by the nuclide and the annihilation radiation following positron decay can be detected with sufficient accuracy and, moreover, the half-life of 71.3 days enables long-term studies to be performed. The preparation is sold generally in the form of solution of the nitrate or chloride salt.

^{60}Co

Of the artificial radioisotopes, those radiation sources containing ^{60}Co have perhaps the longest history and obtained the greatest attention. Owing to the low costs and the almost unlimited range of activity, these kinds of sources have thrust radium sources into the background. Numerous *large radiation sources* (so-called *irradiation facilities*) are equipped with ^{60}Co filling of several PBq activity.

One more advantage of ^{60}Co over radium lies in the fact that the dose rate can be evaluated more easily and accurately, the reason being that only two discrete γ-energies (1.17 and 1.33 MeV) of the radioactive cobalt have to be taken into account. The value of the dose-constant is 2.52 aA m^2/kg that is, about 150% of that of the radium.

The ^{60}Co is produced by the (n, γ) reaction. The target used for the production of smaller sources is an alloy composed of 50% Co and 50% Ni, while higher activity sources are prepared from pure metallic cobalt. In order to keep active dimensions of the source at the smallest possible value, the specific activity should be of the order of PBq/kg.

^{64}Cu

The radioisotopes of copper are of use, primarily, in the *electroplating industry*. Of the two useful copper isotopes, the one having the shorter half-life, ^{64}Cu, is *prepared* in *carrier-free* form in the ^{64}Zn(n, p) ^{64}Cu reaction, or with high specific activity in the (n, γ) reaction. A nearly carrier-free product is obtained using the *Szilard–Chalmers* effect, however this method is diminishing in importance due to the small yields achieved with it.

The nuclide emits β^-- and β^+-particles in addition to γ- and annihilation radiation; it depends on the nature of investigations as to which of the radiations is the most straightforward to detect.

The preparations (the hydrochloro-acidic solution of ^{64}CuCl$_2$ or pure metallic ^{64}Cu) are of high purity, and contain only minor amounts of ^{67}Cu contaminant.

^{67}Cu

^{67}Cu is the sole radioisotope of copper having a half-life in excess of several hours ($t_{1/2} = 61$ h). The *nearly carrier-free* product is *prepared* in the (n, p) reaction. As one can get rid off the ^{64}Cu produced in the competing reaction only by cooling the products, the activity achievable is limited and the high purity ^{67}Cu is rather expensive.

^{67}Cu offers some advantages over ^{64}Cu chiefly because of its relatively longer half-life.

^{82}Br

This halogen isotope has a half-life of one day and a half, and emits γ-radiation of remarkably high intensity $[(K_y = 2.81 \text{ aA m}^2/(\text{kg Bq})]$; therefore, it is of use mainly in investigations in which a considerable extent of absorption of the radiation might be expected.

The commonly available *preparation* is Na^{82}Br solution produced in the (n, γ) reaction with a high specific activity, but certain catalogues offer *organic compounds labelled* with ^{82}Br, as well.

^{85}Kr

Radioactive krypton is the most widely used *inert, gaseous tracer*, apart from ^{133}Xe. It is, however, not the best choice for studies in fluid mechanics as its half-life is relatively long. It can be *prepared* by direct (n, γ) process.

Sealed radiation sources containing ^{85}Kr have been growing in importance recently due to the fact that the 0.67 MeV energy β^--radiation is accompanied by hardly any photon emission. These are used especially as substitutes for ^{204}Tl radiation sources because the effects of self-absorption are negligible.

The "clathrate"-type molecular compounds have obtained widespread usage in the last two decades; noble gases can be bound as inclusion in certain organic compounds containing an aromatic ring. The so-called *krypton technique* is a related subject; the release of radioactive krypton enclosed in various compounds and alloys is a function of the experimental conditions — primarily the temperature — and the principle can be useful in numerous investigations (e.g., temperature measurement, corrosion testing, radioanalysis).

^{86}Rb

^{86}Rb is a widely used tracer in studies of general interests, as it emits well detectable γ-rays and has a favourable half-life ($t_{1/2} = 18.7$ days). In many cases it is applied instead of ^{24}Na and ^{42}K, the reason being that its half-life is more advantageous, and its chemical properties are very similar to those of the latter isotopes, and at the same time, it can be *produced* in the (n, γ) reaction comparatively cheaply and with relatively high specific activity.

^{85}Sr

In certain cases, when chemical or other points of view claim for strontium as the most suitable tracer, the γ-emitting ^{85}Sr, having a medium half-life, is applied instead of ^{90}Sr, because lower activities are sufficient when scintillation counting is applied and — at the same time — the radiation hazard is diminished by the shorter half-life.

^{85}Sr is *produced* generally in the (n, γ) reaction using a target enriched in ^{84}Sr. The use of the enriched target is necessary to decrease the activity due to the ^{89}Sr formed in the competing reaction, but at the same time the rather expensive raw-material enhances the costs of the product, as well.

The nearly monoenergetic γ-radiation is *easy to detect* and, consequently, the field of application of this isotope is expected to grow, in spite of its costly production.

^{90}Sr

This isotope has a remarkably long half-life, and this property justifies its application in *sealed sources*.

^{90}Sr is a *fission product* and as its separation is a relatively easy task, the price of the carrier-free isotope is relatively low, as well. Its decay product is ^{90}Y with a half-life of 64 h; thus, in the determination of the activity of ^{90}Sr, the high energy β^--particles of the decay product are always measured.

When designing sealed radiation sources, special care should be exercised on the proper binding of ^{90}Sr, as this isotope falls in the category of "extremely dangerous" ^{90}Sr is generally fixed in glass or enamel of high point of fusion; when the melt is heated above the fusion temperature, the strontium ions insert between the divalent cations of the glass. (See also ^{90}Y.)

^{90}Y

The use of this isotope in *sealed sources* is justified by the high yield of ionization brought about with it. For these purposes, however, it is used together with its parent, ^{90}Sr, because if this is done its apparent half-life equals that of the parent. The energy of the β^--particles renders it possible, for example, to *determine the thickness* of layers of medium surface mass.

^{90}Y emits only β^--particles, the energy of which is the highest among the commonly used β^--emitting isotopes: $E_{\beta^- \text{ max}} = 2.27$ MeV.

When separated from ^{90}Sr, the isotope ^{90}Y can be prepared *carrier-free*, but the specific activity of the product of the (n, γ) reaction can be well in excess of 100 TBq/kg, as well.

^{99}Tc^m

This isotope has a half-life of 6.05 hours, and emits γ-photons of 0.14 MeV; it has obtained a widespread usage in *human diagnostics*. Attention should be given to ^{99}Tc^m in the fields of *industrial tracer studies* and research, when a soft γ-emitting isotope of short half-life is needed and corpuscular radiation is undesirable.

Because of the short half-life, ^{99}Tc^m is obtained from *isotope generators* at the place of the application; the method involves selective elution from a chromatographic column containing ^{99}Mo, with dilute acid. The *procedure* is based on the (n, γ) $\xrightarrow{\beta^-}$ reaction, with Mo as a starting material, or on the

$$^{235}\text{U(n, f)} \, ^{99}\text{Mo} \xrightarrow{\beta^-} {}^{99}\text{Tc}^m$$

nuclear reaction. In the latter reaction the cumulative fission yield exceeds 6%.

When separated from ^{99}Mo, that has a half-life of 67 h, ^{99}Tc^m is obtained in the form of a solution containing *carrier-free* pertechnetate ions. The advantage of the isotope generator lies in the fact that the ^{99}Tc^m can be eluted repeatedly, each time the transient state of equilibrium between parent and daughter has been attained.

^{124}Sb

The application of ^{124}Sb tends to be used when an isotope is needed that has a medium half-life and emits *high energy* γ-photons. These kinds of requirements are fulfilled, for example, when *tracing* larger masses of materials. The isotope can be *prepared* by the (n, γ) reaction choosing proper irradiation time.

The ^{124}Sb can be built into organic compounds, as well (e.g., triphenylstibene); these chemicals are used primarily in tracer studies in the *petroleum industry*. ^{124}Sb is frequently used in *radiation sources* if the detection is to be performed behind thick layers of absorbents. This isotope is one of those relatively cheap γ-emitters whose γ-energy goes beyond the threshold value of 1.6 MeV, and it can thus be utilized as a filling in γ-*neutron sources*, according to the reaction ^{9}Be$(\gamma, n)^8$Be.

^{125}I

^{125}I, which has a half-life of 60.2 days and is known to emit very low energy γ-rays, is growing in importance compared with the "classical" radioisotope of iodine, ^{131}I.

^{125}I is sold in dilute alkaline or thiosulphate solution; in addition to *tracer* studies, it has been applied recently as a *portable X-ray source*.

It is *prepared* from xenon target in the $(n, \gamma) \xrightarrow{K}$ nuclear reaction. The methods of separation applied result in the formation of carrier-free preparation with a radionuclidic purity exceeding 99% and with minor amounts of ^{126}I contaminant.   The ^{125}I solution is generally of higher radiochemical purity than the corresponding ^{131}I products. Because of the longer half-life and the smaller extent of radiolysis, the solutions can be stored and used for longer times.

The soft γ-radiation of the isotope is of advantage in certain experiments but in most cases it is unfavourable because it cannot be measured accurately even with sensitive scintillation detectors, because of the absorption losses.

^{131}I

This is one of the most widely used radioisotopes: both in the form of Na^{131}I and any kinds of labelled organic compounds. It finds widespread applications chiefly in *physiological* studies (diagnostics of thyroid functions) and also in *hydrology* and *geology*.

Methyl iodide labelled with ^{131}I is very easy to evaporate and, therefore, it attracts interest in studies of *gas flow*.

Although ^{131}I is formed in fission reactions with high yields, its *preparation* is based on a tellurium dioxide target, exposed to the $(n, \gamma) \xrightarrow{\beta^-}$ nuclear reaction, which is followed by separation from the target material using distillation.

Because *carrier-free* ^{131}I is extraordinarily volatile, Na^{131}I is sold in a slightly alkaline buffer solution in order to avoid contamination.

In aqueous solution, ^{131}I is found in anionic form and it readily takes part in disproportionation reactions associated with radiation chemical processes. The radiochemical purity of the solution cannot, therefore, be guaranteed for a longer time.

^{133}Xe–^{133}Xe^m

From the noble gas isotopes, the mixture of xenon isotopes is used most often. *The testing for leakage* of tubings as well as investigations on *gas flow* should be mentioned with special emphasis from the wide range of industrial applications.

The mixture of the isotopes can be *produced* both by the (n, γ) and the fission reactions. The two components of the mixture, however, cannot be obtained separately because the half-lives are very much alike.

^{137}Cs

The most widely used radioisotope of caesium is commonly looked upon as a γ-emitter of medium energy, although in its application use is made of the 662 keV energy γ-photons produced by $^{137}Ba^m$ formed in β^--decay. As a result of the relation between the corresponding half-lives, the ^{137}Cs–$^{137}Ba^m$ system is in secular equilibrium, i.e., the activity of the parent is equal to that of the daughter, and the total activity decreases according to the half-life of the parent isotope.

The ^{137}Cs is applied primarily in *sealed sources* and sometimes in high activity irradiation facilities.

The ^{137}Cs is *available* as a *fission product* generally *nearly carrier-free* and/or eventually with barium as a carrier. Its most common marketed form is, however, the hydrochloric and nitric-acid solution.

^{147}Pm

Since promethium is not found in nature, it is not used in chemical tracer studies; however, it offers some advantages in certain fields (*luminous paints, static eliminators, thickness gauges*, etc.) that require isotopes emitting only weak β^--radiation and have longer half-lives.

^{147}Pm can be obtained from *fission products* or produced from neodymium by the (n, γ)$\xrightarrow{\beta^-}$ nuclear reaction and subsequent separation. Both methods give *carrier-free* products that reach the user generally in the form of hydrochloro–acidic or nitric acid solution.

^{192}Ir

^{192}Ir γ-radiation sources have great importance in the field of *non-destructive testing* of materials. Besides the general requirement of being properly sealed, the major demand with respect to the sources is that the specific activity be as high as possible, that is, the sources should have a total activity of the order of TBq, and that the dimensions of a point-like source be approached.

It is a hard task to fulfil these requirements, mainly because the absorption cross-section of the metallic iridium, which is the target material of the (n, γ) reaction during *production*, is comparable with the cross-section of activation. Besides, the γ-radiation intensity of the source is about 20% lower than that calculated from the decay—the reason being that the high density ($\rho = 22420$ kg/m^3) gives rise to a greater extent of self-absorption. Recently, however, the application of reactors with high neutron fluxes as well as the appropriate selection of the dimensions of the targets, have led to the elimination of most of the difficulties.

^{198}Au

It is not only in *therapy* that radioactive gold has obtained widespread usage but also in *tracer* studies where its chemical resistivity offers some advantages, besides the possibilities given by the medium energy γ-radiation and the half-life of 2.7 days. This is illustrated by the investigations carried out on *stream deposits* with fine-grained sand labelled with ^{198}Au.

^{198}Au is *produced* in the (n, γ) reaction with relatively high specific activity due to the great cross-section of activation. During preparation, the successive (n, γ) reactions cannot be avoided therefore the product is always contaminated by 5 to 15% ^{199}Au, which has a similar half-life; this condition, however, does not generally cause problems.

The isotope can be obtained either in the solid state, in the form of metal foil, or as a *metallic colloid* with a grain size of 10 to 100 nm, or dissolved in aqua regia in the form of hydrogen tetrachloro-aurate(III).

^{197}Hg^m–^{197}Hg

As mercury is one of the very few *liquid metals*, its isotope has some special fields of application. Unfortunately, from the currently used mercury isotopes the one having a half-life of 65 h and a relatively high specific activity, ^{197}Hg contains always certain amounts of the isomer, ^{197}Hg^m, with a half-life of 24 h, causing difficulties in the interpretation of data. One has to take into account the presence of a small percentage of ^{203}Hg as well, the relative contribution of which to the over-all activity increases with time.

This isotope is commercially available in the form of the metal, in addition to the acidic (aqueous) solution of its different compounds.

^{203}Hg

^{203}Hg preparations differ from those containing ^{197}Hg^m and ^{197}Hg *formed* in the analogous (n, γ) processes mainly in that ^{203}Hg has a longer half-life (46.9 days), γ-radiation of higher energy, a higher radionuclidic purity and, at the same time, a much lower specific activity. The field of application of ^{203}Hg is determined by the properties mentioned above.

^{204}Tl

^{204}Tl is used mainly as a medium energy *β-radiation source*, although, as a consequence of the relatively low specific activity attainable by the (n, γ) reaction and the significant extent of self-absorption, it is gradually losing its importance compared with the ^{85}Kr sources emitting β^--radiation of similar energy, but only a negligible amount of γ-radiation.

Thallium naphthenate labelled with ^{204}Tl is produced from the solution of ^{204}Tl(NO$_3$)$_3$, making use of the isotope exchange reaction, and is applied in the *petroleum industry* as a tracer.

^{210}Po

Besides the transuranium elements, ^{210}Po is a nearly unique α-emitting isotope in that it is produced artificially and has obtained a wide range of applications.

The isotope, that has a half-life of 138 days, is used, in part, as an α-radiation source and, in part, as a filling in portable *α-neutron sources*, corresponding to the ^{9}Be(α, n)^{12}C nuclear reaction.

It is *produced* in the ^{209}Bi(n, γ)^{210}Bi $\xrightarrow{\beta^-}$ ^{210}Po nuclear transformation and, thus, the reaction results in the formation of carrier-free preparation, a fact that is of great importance when talking about α-emitting isotopes.

^{252}Cf

Of the transuranium elements, ^{252}Cf has become significant in the common practice of isotope techniques, the reason being that it decays via spontaneous fission accompanied by *emission of neutrons*, α-radiation and minor amounts of γ-radiation.

This isotope, with a half-life of 2.65 years, is *produced* by irradiation of ^{239}Pu or ^{244}Cm in a reactor of extremely high neutron flux. Its field of application is primarily related to the development of laboratory scale neutron sources (1 mg of it emits 2.3×10^9 neutrons per second).

1.4.2. MAIN PROPERTIES OF SEALED SOURCES

A special requirement for sealed sources is that the radiating material be properly encapsulated so that direct contact with the environment be avoided and, at the same time, the particles or photons emitted by the isotope enclosed in the source do exert their intended effect more or less unlimitedly, owing to the penetration ability of the radiation. The main features of particular types of radiation sources are determined by the properties of the radiating material (see Sections 1.3. and 1.4.1.).

The sources are composed of the *radiating isotope* contained in the *filling*, the single or double isotope *holder* that partially or totally surrounds the filling, the outer *cover* that contains the parts mentioned above and the capsule closed air-free by welding or some another method. The capsule is to be tested for leakage occasionally (see Section 8.4.1.).

The main features of radiation sources are:

— the characteristics of the active filling (radioisotope);

— the activity or the resulting ionization current or particle flux;

— with α, β and γ-emitting sources, the intensity of radiation; with neutron sources, the neutron emission;

— the outer dimensions of the active filling and the capsule;

— further requirements raised by the particular mode of application (corrosion resistance, resistance to heat, stability, absence of leakage, etc.).

The commonly used *classification* of sources according to the type of radiation and the points of view in the application is:

— α-radiation sources;

— β-radiation sources;

— γ-radiation sources;

— neutron sources;

— other sources for special uses.

1.4.2.1. α-RADIATION SOURCES

The most important requirement is the strong local ionization and the limited penetration range of radiation. An essential point is that there be no or only minor amounts of radiations except α-particles. There is only a very limited number of sources commonly used, the reason being that the energies of the α-particles are very much alike and the half-lives of the isotopes to be taken into consideration are unfavourable.

The homogeneity of the active filling is very important because of the relatively short range of radiation; in order to fulfil this requirement the α-emitting isotope is prepared by electroplating or powder metallurgy.

In order to achieve a proper encapsulation, the active filling is usually covered with a window made of noble metal foil in a thickness of a few μm-s. Autoradiography (see Section 7.6.2.) is used to check for homogeneity.

The following α-emitting isotopes have become commonly used in the sources:

^{210}Po (half-life 138.4 days; $E_\alpha = 5.30$ MeV); the penetration range of the particles in air and aluminium is 38 mm and 59 g/m^2, respectively.

^{241}Am (half-life 458 years; $E_\alpha = 5.53$ MeV); the penetration range in air and aluminium is 41 mm and 63 g/m^2, respectively.

The maximum value of the activity at the surface of the sources mentioned above can be as high as 100 GBq/m^2.

1.4.2.2. β-RADIATION SOURCES

The most important requirements are as follows:

— emission of any kind of radiation except β-particles is undesirable;

— in order for the applicability to be maintained for longer times the half-life of the radioisotope is expected to be neither too short, nor too long, so that a satisfactory particle flux is achievable;

— the fastener, the holder and the window of the source is expected to be totally transparent to the radiation, but not for other types of radiations brought about in the interaction of the β-radiation with matter;

— the radioactive filling is required to be fixed properly to prevent the radiating material from moving on to the surface thus changing the sealed source to an unsealed one.

From the purely β^--radiating isotopes listed in Table 1.2 the aforementioned are used in practice. The design of radioactive fillings as well as the technique of fixing them is largely dependent on the given kind of isotope.

The ^{14}C radiation sources are made of polymethyl-methacrylate (lucite) foils to allow for the preparation of sources with large active surface. As in the processing the monomer is labelled with ^{14}C, after polymerization a homogeneous distribution in the activity is obtained. Similar sources are made with tritium for special needs.

^{63}Ni can be fixed by electroplating on thin foils; alternatively, thin and nearly homogeneous active layers can be produced by thermal treatment and isotope exchange.

In the case of ^{85}Kr, the frozen gas is transferred into a thin-walled (10 to 50 μm) copper or nickel cylinder, which is kept air-tight, then sealed, and placed in a stainless steel box. Using another method, radioactive krypton is adsorbed on activated charcoal or bound with hydroquinone as a clathrate, in the solid state. The common disadvantage of the latter methods lies in the fact that the radiolysis places limits on the time the capsule can exist without leakage.

The construction of ^{90}Sr–^{90}Y sources requires special design because of the higher energy of β^--particles and the extreme radiation hazard. Radioactive strontium is generally built in a glass having a higher melting point and special chemical composition; the process combined with sintering and subsequent heating results in a homogeneous active source.

^{106}Ru is usually electroplated on to silver or platinum plates. As a result of the interaction with platinum minor amounts of bremsstrahlung generally cannot be avoided. It should be noted that the two isotopes formed in the decay of ^{106}Ru, viz., ^{106}Rh^m and ^{106}Rh, emit high energy γ-radiation.

^{147}Pm can be fixed, for example, on ion exchange resins, or in certain other versions, using powder metallurgy.

Radiation sources containing ^{204}Tl are also prepared by electroplating or powder metallurgy. It is very difficult to achieve a satisfactory degree of homogeneity, as the value of the specific activity is relatively low and the metallic thallium is very sensitive to oxidation. For this reason a protective cover is often used.

Other kinds of radiation sources used for special purposes can be built with other methods corresponding to the given requirements.

1.4.2.3. γ-RADIATION SOURCES

The γ-radiation sources are prepared in a broad range of design and activity. Radiation sources having an activity in excess of the order of 100 TBq are used in teletherapeutic devices and in irradiation facilities; these sources have individual design and are therefore not treated in this section (see Section 6.5.1. for details), but a schematic is given here on the methods applied in the construction of some active fillings.

The sources filled with ^{22}Na can be looked upon as γ-radiation sources, however in their field of application one can make use of positron radiations as well, if they are equipped with a window made of mica. The radiating nuclide can be deposited on the plate of the desired dimensions using electroplating or evaporation in vacuum. When it is expected that the source would meet higher temperatures during application, the isotope can be incorporated in enamel.

^{55}Fe sources are used only in special cases when the weak X-rays of 5.9 keV energy are applicable. The active layer made of ^{55}Fe is generally electroplated on to a copper disc. The usual value of the activity is of the order of 100 MBq. A source of 100 MBq(2.7 mCi) activity emits 8×10^6 photons per second into a 2π solid angle.

^{60}Co sources are the most widely used ones from all the known sources. The active part is usually a cylinder made of an alloy composed of 50% cobalt and 50% nickel, or metallic cobalt. Higher activity sources can be made with a specific activity of 4 PBq/kg (≈ 100 Ci/g), in order to decrease the dimensions of the active part.

The relatively inexpensive sources offer a wide range of applications, owing to the considerable dose constant, the favourable half-life and energy of the γ-photons (1.17 and 1.33 MeV).

^{60}Co sources are prepared not only for irradiation plants but also for radiography.

The activity of a source of 1 g-radium-equivalent ^{60}Co equals 24 GBq (0.65 Ci).

The application of ^{124}Sb γ-sources is justified primarily in those cases when a high energy γ-radiation having a long penetration range is necessary to solve a given task. High activity sources filled with metallic antimony can be prepared with relatively low costs, but the application of the source is rather limited in time due to the short half-life of the isotope: 60.9 days.

Radiation sources containing ^{125}I have found widespread application only in the past two decades. The ^{125}I isotope emitting 35 keV energy γ-photons and 28 keV energy X-rays is placed behind a window of 100 μm thickness in the form of an insoluble compound, in a holder having the smallest possible dimensions. ^{125}I sources having an activity of the order of 10 GBq are produced to substitute X-ray machines. In the case of an activity of 10 GBq, the yield of photons is equal to 5.4×10^9 per second in a solid angle of 2π.

The sources containing ^{137}Cs (^{137}Cs–^{137}Ba^m) can be rated between ^{170}Tm and ^{192}Ir. From the point of view of the K_γ dose constant, however, the higher energy of the γ-

photons allows them to pass through a wall of nearly the same thickness as the one that can be handled by the photons of a ^{60}Co source.

In sources built on ^{137}Cs isotope, the carrier-free isotope is incorporated into glass or ceramics. The prices are nearly the same as those of the ^{170}Tm or ^{192}Ir sources of the same activity, but the application of the former sources is more favourable from the economic point of view as the half-life of ^{137}Cs is much longer.

The activity of 1 g-radium-equivalent ^{137}Cs equals 92.5 GBq (2.5 Ci).

^{152}Eu–^{154}Eu γ-sources have a remarkably long half-life. Irradiating sintered europium oxide, high activities can be achieved in relatively small dimensions. The application of these sources has not become particularly widespread, the reason being the too complex γ-radiation emitted.

^{170}Tm sources can be applied for the radiographic study of relatively thin layers because the γ-photons have low energy and the value of the dose constant is also rather small. The advantage in using ^{170}Tm instead of ^{192}Ir lies in the fact that the former nuclide has a longer half-life. However, the smaller specific activity of the sintered ^{170}Tm$_2$O$_3$ filling, as well as the higher price in the world market is a serious drawback.

^{192}Ir sources are applied exclusively in radiography. Correspondingly, the high activity and the small dimensions of the active part (i.e., a nearly point-like source) form the major requirements in manufacturing ^{192}Ir sources. It is not easy to fulfil these requirements, even if the irradiations were carried out in reactors having high neutron fluxes. Given optimum conditions for irradiation, an activity of about 2 TBq (ca. 50 Ci) can be produced using an active filling of 2×2 mm dimensions and about 6 TBq (ca. 160 Ci) with a filling of 3×3 mm, the usual values being, however, much poorer.

The activity estimated from the measured dose and the activity produced in reality can be different, the reason being the significant extent of self-absorption (10–30%).

The activity of 1 g-radium-equivalent of ^{192}Ir is 60.6 GBq (1.6 Ci).

^{241}Am emits α- and γ-radiation when decaying; the ^{241}Am sources made with a shielding for the α-radiation enable the measurement of the surface mass of thin layers, making use of the 60 keV energy γ-photons. The sources have been rather expensive so far, but in spite of this their application is becoming more and more widespread. Sources carrying activities of the order of 10 GBq (0.3 Ci) are prepared nowadays; in the case of 10 GBq (0.27 Ci) activity, the yield of photons is 4×10^9 per second in a solid angle of 2π.

1.4.2.4. NEUTRON SOURCES

The principle data of usual laboratory scale neutron sources are summarized in Table 1.4. The following properties of the sources are given: the nuclear reaction producing neutrons, the half-life of the nuclide that undergoes the reaction, the approximate yield of neutrons, and the mean energy of neutrons emitted by the source.

Neutron generators will not be treated here because this kind of equipment varies greatly depending on the manufacturer (see Section 4.2.2.).

Table 1.4. The most important neutron sources

Source	Nuclear reaction	Half-life of the radioisotope, $t_{1/2}$	Yield of neutrons		Mean energy of neutrons, MeV
			For (α, n) sources: 1/(TBq s); for (γ, n) sources: 1/(TBq s g) target	For (α, n) sources: 1(Ci s); for (γ, n) sources: 1(Ci s g) target	
^{210}Po–Be	^{9}Be(α, n)^{12}C	138.4 d	6.8×10^7	2.5×10^6	4.5
^{210}Po–B	^{11}B(α, n)^{14}N	138.4 d	2.4×10^7	9×10^5	2.5
^{210}Po–F	^{19}F(α,n)^{22}Na	138.4 d	1.1×10^7	4×10^5	1.4
^{210}Po–Li	^{7}Li(α, n)^{10}B	138.4 d	2.4×10^6	9×10^4	0.4
^{226}Ra–Be	^{9}Be(α, n)^{12}C	1602 y	4.6×10^8	1.7×10^7	5
^{226}Ra–B	^{11}B(α, n)^{14}N	1602 y	1.8×10^8	6.8×10^6	3
^{239}Pu–Be	^{9}Be(α, n)^{12}C	2.44×10^4 y	4.8×10^7	1.8×10^6	4
^{241}Am–Be	^{9}Be(α, n)^{12}C	458 y	5.7×10^7	2.1×10^6	4
^{242}Cm–Be	^{9}Be(α, n)^{12}C	163 d	6.8×10^7	2.5×10^6	4.5
^{88}Y–Be	^{9}Be(γ, n)^{8}Be	105 d	2.7×10^6	1×10^5	0.16
^{124}Sb–Be	^{9}Be(γ, n)^{8}Be	60.9 d	5.4×10^6	2×10^5	0.024
^{226}Ra–Be	^{9}Be(γ, n)^{8}Be	1602 y	8.1×10^5	3×10^4	0.5
^{226}Ra–D$_2$O	^{2}H(γ, n)^{1}H	1602 y	2.7×10^5	1×10^4	0.1
^{252}Cf	Spontaneous fission + neutron emission	2.65 y	1.2×10^{11}	4.3×10^9	1

1.4.2.5. SPECIAL RADIATION SOURCES

Only those special sources will be discussed here which are mass produced but which, for some reason, cannot be included in the above grouping.

Table 1.5. Mössbauer radiation sources

Pair of isotopes	$t_{1/2}$	Utilizable photon energy, keV	Absorbent	Temperature appropriate for the experiment
^{57}Co–^{57}Fe	267 d	14.4	Fe, with natural isotopic abundance $K_4Fe(CN)_6 \cdot 3\,H_2O$ $Na_2Fe(CN)_5NO \cdot 2\,H_2O$	Room temperature
^{119}Sn^m–^{119}Sn	245 d	23.9	Sn $BaSnO_3$	Liquid nitrogen temperature
^{125}I–^{125}Te	60.2 d	35	TeO_2	Liquid nitrogen temperature
^{151}Sm–^{151}Eu	93 y	21.6	Eu_2O_3	Room temperature
^{153}Gd–^{153}Eu	236 d	97 103	Eu_2O_3	Liquid helium temperature
^{169}Er–^{169}Tm	9.5 d	8.4	Tm_2O_3	Room temperature
^{241}Am–^{237}Np	458 y	59.6	Np_2O_3	Liquid nitrogen temperature

The operation of *sources emitting bremsstrahlung* is based on the interaction of β-particles with matter: the slowing-down of the particles gives rise to the emission of secondary electromagnetic radiation with *continuous spectrum of energy*. In principle, this kind of radiation can be observed in any β-emitting source; however, using the appropriate β-emitter and absorbent, sources can be manufactured for special uses and with required maximum energy. The sources most commonly used are composed of pure β-emitters ^{3}H, ^{85}Kr, ^{90}Sr–^{90}Y and ^{147}Pm, and absorbents such as aluminium, titanium, zirconium and platinum. The maximum energy of the electromagnetic radiation is in the range 5 to 100 keV, depending on the composition and construction of the source. The activity of the sources can approach the order of PBq; a spectrum of the γ-rays is generally attached to the certificate of the sources.

Mössbauer sources are composed of a weak γ-emitting isotope distributed homogeneously in an appropriate matrix and an absorbent suitable for the investigations to be carried out. The commonly used isotopes in Mössbauer sources, the corresponding photon energies, the requirements raised to the absorbents and the temperature required in the application are summarized in Table 1.5.

Table 1.6. Simulator ("Mock") radiation sources

Isotope to be studied			Major properties of the source		
Symbol	$t_{1/2}$	Major γ-energies, MeV	Isotope or mixture of isotopes	$t_{1/2}$	Major γ-energies, MeV
^{59}Fe	45.6 d	1.10 1.29	^{60}Co	5.26 y	1.17 1.33
^{64}Cu	12.8 h	0.511 1.33	^{22}Na	2.62 y	0.511 1.275
		0.140	^{57}Co	267 d	0.122 0.136
^{99}Tc^m	6.05 h	0.142	^{141}Ce	33.1 d	0.145
			^{144}Ce + ^{144}Pr	285 d	0.134 0.696
^{125}I	60.2 d	0.028 0.035	^{239}Pu	2.44×10^4 y	0.039
^{131}I	8.05 d	0.284 0.364 0.637 0.722	^{133}Ba + ^{137}Cs–^{137}Ba^m	10.7 y and 30 y, resp.	$\begin{cases} 0.30 \\ 0.36 \\ 0.39 \\ 0.66 \end{cases}$
Fission products			^{89}Sr + ^{90}Sr + ^{95}Zr + ^{103}Ru + ^{106}Rh + ^{137}Cs + ^{141}Ce + ^{144}Ce + ^{147}Pm	Variable	Mixed γ-spectrum

Simulator ("Mock") radiation sources are used to facilitate the measurement of the relative activity of short-lived isotopes, as well as for the calibration of instruments. These sources are filled with an isotope having a longer half-life; they can simulate every characteristic of the nuclide to be studied, except its half-life, primarily its γ-spectrum (eventually by proper discrimination carried out using a pulse-height analyser). Some commonly used simulator sources are summarized in Table 1.6 together with data necessary for comparison.

2. NUCLEAR INSTRUMENTS AND MEASUREMENTS

Industrial applications of isotopes utilize the radiation of radioactive materials (radiation sources). Depending on the mode of industrial application, information is in most cases obtained through the *effects of materials on radiation*, about, e.g., the *qualitative and quantitative material properties*. The fundamental requirement to obtain and process the desired information about technological processes is the detection and numerical evaluation of radiation.

Radioactive radiation cannot be sensed in a direct way; however, the interaction between radiation and suitably chosen media placed in the radiation path and within its effective range induces effects which can be used to determine the intensity and, in some cases, the energy of the radiation by electrical or electronic techniques, using appropriate instrumentation. The component of the total measuring system sensing and converting radiation into electrical signals is the *radiation detector*, the unit processing and recording signals from the detector is the *measuring equipment* or *instrument*.

The principal purpose of nuclear measurement technology is to determine either the integral radiation parameters or the partial parameters of particles or quantas constituting certain portions of radiation.

2.1. NUCLEAR RADIATION DETECTORS

The stringent requirements set against radiation detectors for nuclear instruments are:
— sensing radiation with high efficiency;
— performance independent of or proportional to the energy of radiation;
— providing high-level electrical signals to make the electronic equipment as simple as possible;
— requiring the lowest possible supply voltage;
— operation not sensitive to supply voltage fluctuations;
— operation not sensitive to changes of ambient temperature and of other climatic parameters;
— high resistance against shock and vibration;
— long service life with stable operation;
— relatively low costs.

At present, not all the requirements listed above are satisfied simultaneously by the detectors in general use. The type of detector should always be chosen in accordance with the nature of a given problem to meet the most suitable conditions for that particular problem.

The most extensively utilized effects for radioactive radiation detection are:
— *ionization:* The atoms of gases or solids (e.g., semi-conductors) penetrated by radiation are ionized at a degree proportional to the intensity of radiation;
— *luminescence:* The absorption of radiation induces light flashes or scintillations in certain substances.

Table 2.1. Features

Properties	Ionization chamber	Proportional counter
Field of application	Measurements of high activities	Primarily for particle radiation measurements
Measurement efficiency	Energy-dependent efficiency	For particle radiation near 100%; for electromagnetic radiation 1 or 2%
Dead time		$< 10 \ \mu s$
Energy selectivity	Non-selective	Selective
Costs	Low	Low
Other aspects	Without precise calibration, only for high-energy (100 keV) radiation measurements	Well-stabilized power supply required

In addition to the above effects, other physical phenomena can also be used for detecting radiation. The radiation-induced blackening of X-ray *films* (Section 6.2) is applied in nondestructive testing (radiography, see Section 7.1.3) and in radiation protection (Section 8.3.1.1). The track detection methods (Wilson cloud chamber, bubble chamber, nuclear emulsion detector, spark counters, spark chambers, etc., Section 7.6) and *Cherenkov* radiation are not utilized for nuclear measuring instruments.

The nuclear radiation detectors used for industrial isotope measurements can be classified according to their operation principles, i.e., to the nature of the interaction. The main detector types and their important characteristics discussed below are summarized in Table 2.1.

No single radiation detector of the different types in general use can be ranked above any other, as every one of them possesses certain advantages and disadvantages from a technical point of view (e.g., energy selectivity, resolving power, efficiency) that require or exclude their application in the particular case at hand. Besides, for industrial purposes, it is always more expedient to use the most economic detector that suits the given task and can reliably fulfil the technical requirements.

2.1.1. GAS IONIZATION DETECTORS

Two major detector types can be distinguished according to the media where the ionization takes place, viz., gas ionization and solid state (semiconductor) detectors.

The common structural design of the various gas ionization detectors is that they consist of *two metallic electrodes* connected to a power supply *with a gas between them which can be ionized by radioactive radiation.* The neutral gas molecules are ionized by the incoming radiation causing the production of pairs of positively charged ions and negatively charged electrons. In the electric field between the electrodes the positive ions and the electrons migrate to the negatively and positively biased electrodes, respectively.

Their operation is demonstrated by the experimental arrangement shown in (Fig. 2.1). If a positive potential is applied through a resistor R to the central wire electrode (anode) and the negative terminal of the U_0 voltage route is connected to the outer metal case (cathode) then the degree of ionization can be obtained from the voltage U of electrical pulses

of main detector types

GM counter	Scintillation detector	Semiconductor detector
Primarily for particle radiation measurements	Measurements of any radioactive radiation types	Measurements of any radioactive radiation
For particle radiation near 100%; for electromagnetic radiation 1 or 2 %	Generally good	Generally good strongly temperature dependent at some types
<1 ms	<1 μs	<0.1 μs
Non-selective	Selective	Very selective
Low	High, due to accessories	High
Limited but usually long life	High counting rates	For drifted semiconductors, cooling required both for measurement and storage

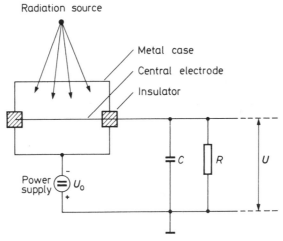

Fig. 2.1. Gas ionization detector arrangement for experimental investigations

appearing across the resistor shunted by a capacitor C. The time constant RC is higher than the time needed to collect the overwhelming majority of charged particles produced in the gas space.

The performance of gas ionization detectors depends strongly on the *electric field between the electrodes*. Figure 2.2 shows the operational diagram of a detector presenting the supply voltage (U_0) dependence of the average number of ion pairs $\bar{N}$ produced directly or indirectly by impact ionization by a single ionizing particle. Curves *1* and *2* show the numbers of ion pairs generated by the weakly ionizing β-particles and the strongly ionizing α-particles, respectively, as long as they induce distinct effects in the detector. From a threshold voltage U_4 the operation is independent of the ionization capability of the incident primary particles (see Curve *3*). From an analysis of Fig. 2.2, six distinct operational regions can be observed, their respective names are indicated. The recombination, the restricted proportional and the self-discharge regions are not used for

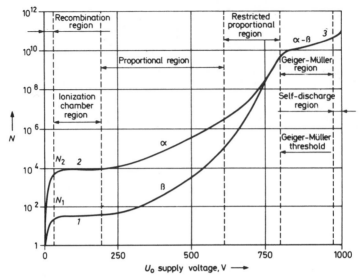

Fig. 2.2. Supply voltage dependence of the number of ion pairs related to a single ionizing particle in a gas ionization detector

measurement purposes. Detectors operating in the three remaining ones will be described later under names similar to the respective regions.

Because of their simple design, gas ionization detectors are currently the most extensively used radiation sensing devices. Though similar in operation principles, actual detector constructions can be widely different from each other in order to achieve optimal behaviour within their voltage ranges.

2.1.1.1. IONIZATION CHAMBERS

Ionization chambers are in most cases cylindrical detectors of different sizes, filled with air or rare gases. They are provided with two electrodes isolated from each other, one is usually the metal wall of the chamber; the other, the so-called collector, is a coaxial metal rod in the central axis of the cylinder. The ionization chamber can be regarded as a condenser charged by radiation-produced ion pairs.

Figure 2.3 shows the structure and circuit scheme of an ionization chamber with plane parallel electron configuration.

As a result of the ionization of gas molecules in the gas space of the chamber by the incoming radiation a current of the order of about 1 μA is induced in the circuit due to the potential difference between the electrodes. For suitably chosen voltages (region between U_1 and U_2 in Fig. 2.2) the current depends only on the intensity of the radioactive radiation causing ionization. The saturation current corresponding to this saturation region belongs to a voltage range where the primarily generated ion pairs from radiation do not recombine and they do not yet undergo an avalanche ion multiplication process due to impact ionization. The metal housing of the chamber is grounded to reduce leakage currents from the signal electrode.

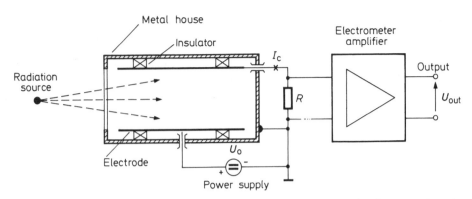

Fig. 2.3. Structure and circuit scheme of an ionization detector. $U_{out} = A \cdot T_C \cdot R$; I_c-chamber current intensity A-amplification of preamplifier

Radiation from radioactive *samples placed either inside or outside* the chamber can be measured by it. For the latter case, however, both the chamber wall thickness dependence of absorption and the secondary effects in the wall and measurement geometry should also be taken into consideration.

The ionization chamber is usually used in the current mode which means that the instrument connected to the detector measures the ionization current induced continuously by radiation.

In the circuit arrangement of ionization chambers shown in Fig. 2.3 the current intensity (I_c) is determined by the equation

$$I_c = k_c \times I_r \tag{2.1}$$

where k_c is the so-called chamber constant representing chamber sizes, gas pressure etc., I_r is the radiation intensity reaching the chamber.

The chamber constant k_c and thus the chamber sensitivity, too, can be increased by high-pressure rare gas filling. Though there are chambers filled with gas up to 150 kbar pressures the optimal pressure range, taking into account leaks and other pressure losses, is 20 to 30 kbar. In Fig. 2.4, a set of characteristics for ionization chambers with the dose rate as parameter is shown. The dotted line represents the minimum working voltage. The figure indicates clearly a long straight (saturation) region of curves implicating high internal resistance which allows the use of high resistive load impedances ($R \approx 10^{12}$ Ω). The ionization chamber sensitivity can be increased by scaling up its volume.

Electrometer type amplifiers are used generally for processing signals from ionization chambers. They can be operated either directly or indirectly. The input of the direct system is a field-effect transistor impedance matching stage. The indirect system converts the d.c. voltage to be measured to an a.c. voltage then the amplified signals are converted back to d.c. by a phase sensitive rectifier circuit.

The extended use of ionization chambers in nuclear industrial measurements is due to the fact that they satisfy most of the requirements of detectors. Its prominent characteristics are the infinite lifetime, on the one hand, and the capacity for measuring unlimited dose rates, on the other, provided, of course, that the construction is adequate for the given purposes. Its design is resistant, it works on a rather low voltage supply (a few

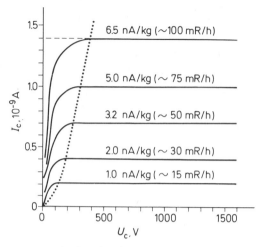

Fig. 2.4. Ionization chamber characteristics: $I_c = f(U_c)$

100 V)—a particularly important feature in industrial applications. Its disadvantage lies in the great internal capacity, on the one hand, and the low sensitivity to γ-radiation, on the other: thus its applicability to count single particles or quanta is very limited.

2.1.1.2. THE PROPORTIONAL COUNTER

Proportional counters belong to the class of ionization detectors, too. Their operational principle differs from that of ionization chambers in that the potential difference between the electrodes of the proportional counter tube (region $U_2 - U_3$ in Fig. 2.2) reaches a value to accelerate electrons produced by radiation to such an extent that they initiate further ionization by collisions with gas molecules. Thus, the primary process, is repeated several times in an enhanced way and it will have been completed when the generated ions and/or electrons reach the electrodes.

The ratio of the number of electrons generated secondarily in the gas space relative to that produced by the primary radiation is called the *multiplication factor*. It depends nearly exponentially on the voltage and the filling gas pressure. Its typical value is between 10^3 and 10^6. The electrical signal height is proportional to the intensity and energy of the ionizing radiation within a proportional counter tube filled with rare gases and some hydrocarbons at a pressure usually lower than atmospheric. It means that instruments connected to the detectors in pulse mode can measure indirectly both the number and energy of the incident particles.

The design and circuit of the proportional detector are shown in Fig. 2.5. The supply voltage U_0 of the counter is chosen to lie in the proportional region of the characteristics and is connected through a load resistor R across which a voltage pulse is induced by the amount of charge of a single pulse. The preamplifier, d.c. isolated by a capacitor C_1, increases the low level (amplitude) of the detector signal to the necessary extent. If the time constant RC_2 of the load resistance R and the combined stray capacity C_2 of the circuit of the capacitor are significantly lower than the positive ion collection time of 0.1 to 1 ms but

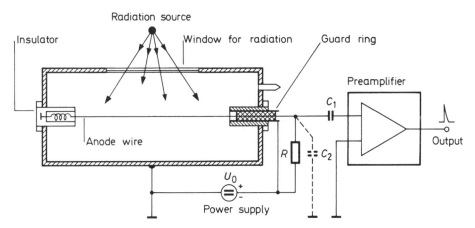

Fig. 2.5. Construction and circuit layout of a cylindrical proportional detector

higher than that of electrons then very short detector pulses can be obtained. The peak amplitude of pulses is given by the equation

$$M_p = \frac{k \times e \times E_p \times A_g}{C_2 \times W} \qquad (2.2)$$

where k is a proportionality factor, e the electron charge, E_p the particle (quantum) energy, A_g the gas amplification, C_2 the stray capacity and W the power needed to produce an ion pair in the gas space.

Proportional counters are extensively used in nuclear instruments designed for *chemical analysis* problems, to detect mainly low-energy (soft) radiation by an energy-selective technique (see Section 2.2.3).

Proportional detectors with BF_3 *gas filling* are used in nuclear moisture gauges (see Section 2.4.6) for detecting *thermal neutrons* produced by α-particle conversion. Figure 2.6 shows the supply voltage dependence of gas amplification factor (A_g) for three different proportional counter types with BF_3 gas filling. According to the manufacturers (e.g., 20th Century Electronics Ltd, UK), the operating point should be chosen in the shaded region for stable operation and long life.

Of the advantageous properties of the proportional counter, the one that should be mentioned with special emphasis is that in spite of the multiplication effect brought about in the gas the pulse produced will have an amplitude proportional to the energy of the incident particle and, thus, it can be used also in energy-selective techniques provided that appropriate electronic signal processing is available. Its time resolution is good, its lifetime is long, and it is particularly applicable to the detection of β- and weak γ-radiations. Its disadvantages lie in the fact that its performance is dependent on the temperature and the voltage supply and that it is working on a power supply providing voltages higher than 1000 V.

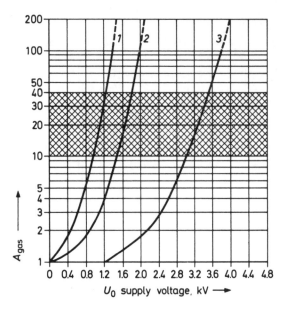

Fig. 2.6. Bias voltage dependence of gas amplification. Curve *1* — ⌀ 25 × 120 mm dimensions, 260 mbar pressure; Curve *2* — ⌀ 25 × 120 mm dimensions, 530 mbar pressure; Curve *3* — ⌀ 50 × 150 mm dimensions, 900 mbar pressure

2.1.1.3. GEIGER–MÜLLER (GM)-COUNTERS

Geiger–Müller (GM)-counters differ from ionization chambers and proportional counters in their *high internal electric field*—this having been chosen to induce high multiplication factors (10^8 to 10^9 in magnitude) within the detector. The quantitative change in the multiplication factor invokes also a qualitative one: the primary ions are accelerated by the voltage difference up to a value where an *avalanche ionization process develops* in the gas space which will propagate along the anode wire to the total length of the tube within several hundred nanosecs.

In addition to ionization, a secondary effect of the *partial recombination* of ions leads to the appearance of *excited gas atoms;* they release their excess energy by *photon emission.* As a result of the combined effect, a gas discharge appears between the electrodes and the pulse height will be independent of the number of primary ions. However, the resistivity between the electrodes will be reduced and if no provisions for maintaining the potential difference were to be made the ionization would continue even without the arrival of new particles or quanta. It means that the tube would be inadequate for further radiation detection because no further ionization process could be initiated in the "*dead time*".

The quenching of the discharge process and thus the *reduction of dead time* can be realized by self-quenching counter tubes, i.e., detectors filled by special gas. They contain, in addition to the high ionization-energy rare gas (usually argon), alcohol or other volatile *organic compounds,* too, in order to ensure the collision of organic molecules with photons emitted after the excitation of rare gas atoms prior to their arrival to the cathode. In this way they release their excess energy without additional electron emission, converting it into chemical energy by dissociation (see Chapter 6). The dead time of this type of tube is

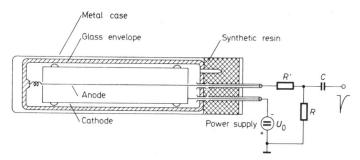

Fig. 2.7. Structure and electronic circuit of a cylindrical Geiger–Müller (GM) counter tube

several hundred microseconds but the life time is limited, after the arrival of about 10^8–10^9 pulses the dissociable molecules vanish from the gas filling.

Tubes containing halide quenching gases are more beneficial with regard to life time because their dissociation is reversible, i.e., the halogen atoms generated are converted back into the original gas molecules.

Figure 2.7 shows the structure and electronic circuit of a cylindrical GM-counter tube. The useful signals are obtained from a load resistor R with a resistance of 0.1 to 3 $M\Omega$. The limiting resistor R' separates the stray capacitances. The GM-tube operates between voltages U_5 and U_6 of Fig. 2.2. The operation of the counter tube is represented by the characteristics shown in Fig. 2.8, similar to the plots for the respective ranges in Fig. 2.2. In the counter tube used in the circuit according to Fig. 2.7, pulses are generated by the radioactive radiation of a given dose rate I_1. The voltage U_t represents the *counting threshold* where the gas amplification starts. By increasing the voltage, the number of detected pulses changes slightly in the region U_1–U_2. This region is called the *plateau* of the counter, its centre should be chosen as the *working point*. The longer the plateau and the less its rise (the pulse number difference $\bar{n}$–$\bar{n}_1$), the better the counter.

The slope of plateau M is expressed numerically by the equation

$$M = \frac{\dfrac{d\bar{n}}{\bar{n}}}{dU} = \frac{\dfrac{\bar{n}_2 - \bar{n}_1}{\bar{n}_2 + \bar{n}_1}}{U_2 - U_1} \tag{2.3}$$

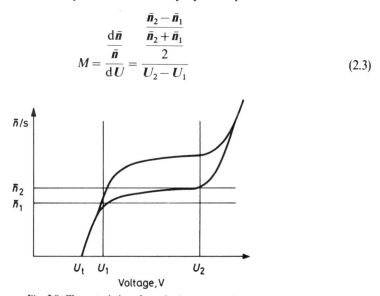

Fig. 2.8. Characteristics of a pulsed counter tube

If the slope is given in percent and is related to 100 volts then we have

$$M = 2 \times 10^4 \times \frac{\bar{n}_2 - \bar{n}_1}{(\bar{n}_2 + \bar{n}_1)(U_2 - U_1)} \%/100 \text{ V} \tag{2.4}$$

EXAMPLE 2.1

The beginning of the plateau of a counter tube is at $U_1 = 400$ V, its end at $U_2 = 700$ V. According to measurements, the pulse numbers for voltages U_1 and U_2 are $\bar{n}_1 = 100$ and $\bar{n}_2 = 103$, respectively, at a dose rate of 7×10^{-11} A/kg (≈ 1 mR/h).

The plateau slope of the GM-tube, according to Eq. 2.4, is

$$M = 2 \times 10^4 \times \frac{103 - 100}{(103 + 100)(700 - 400)} = \frac{6 \times 10^4}{203 \times 300} = 1\%/100 \text{ V}$$

which is acceptable for industrial applications because it means a measurement error of $\pm 0.1\%$ using a supply voltage with an easily achievable stability of $\pm 2\%$.

Figure 2.8 shows the characteristics of a GM-counter for two types of radiation with different dose rates. With increasing intensity the plateau shifts to higher values accompanied by a simultaneous shift in U_2 to the right. The operational characteristics should be defined in accordance with the parameters of a given application.

The dead time (within which the GM-counters cannot detect new particles) depends on the geometrical dimensions, the composition and pressure of the filling gas, the supply voltage at the working point, the circuit components and the rate of detected pulses. Its value under normal operational conditions for a typical counter tube design is in the region from 50 to 200 μs.

The dead time effect is disadvantageous if the number of particles to be detected in unit time is high. If one detects $\bar{m}/s$ particles by a counter then there was no detection for a time of about $\bar{m} \cdot t_d$. If there were no dead time effect then a count number of $\bar{n}/s > \bar{m}/s$ would be detected. Within the period $\bar{m} \cdot t_d$, the number of particles undetected is $\bar{n} \cdot \bar{m} \cdot t_d$. The difference in the two pulse numbers is equal to the number of undetected particles:

$$\bar{n} - \bar{m} = \bar{n} \times \bar{m} \times t_d \tag{2.5}$$

By reducing this equation one can obtain the true particle number $\bar{n}/s$ in terms of the measured count number and dead time:

$$\bar{n} = \frac{\bar{m}}{1 - \bar{m} \times t_d} \tag{2.6}$$

The dose rate dependence of the average pulse number for constant voltage in a typical GM-counter is shown in Fig. 2.9. Obviously, the relationship between the average count number ($\bar{n}/s$) and the dose can be regarded as linear up to about 2000 counts per second. At higher dose rates, however, due to the dead time effect, the slope of the curve decreases strongly before reaching a saturation.

A general drawback of *cylindrical* GM-counters is a reduction of efficiency due to absorption effects of the tube wall. For activity measurement of soft β-ray emitting isotopes the use of special tubes with *end windows* of a few mg/cm^2 thickness, made usually of mica, is recommended (Fig. 2.10) or, alternatively, so-called *flow-through* type tubes (Fig. 2.11) are used with the active sample introduced into the tube itself.

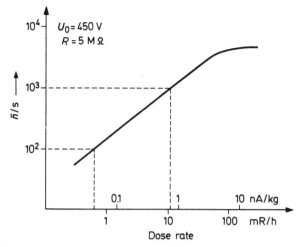

Fig. 2.9. Dose rate dependence of the average pulse number in a GM-counter tube

The detection efficiency of GM-counter tubes for α- and β-particles is close to 100 percent but for electromagnetic radiation (γ- or X-rays) it rarely exceeds 1 or 2 percent.

The life-time of GM-tubes was a major problem at one time. The life time of the present halogen-filled GM-tubes in terms of count number is in the order of 10^{11} pulses, against the former value of 10^8–10^9. This means that, for continuous operation, a life time of about 10 years is feasible which is quite acceptable for practical purposes.

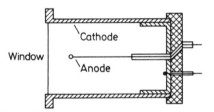

Fig. 2.10. Scheme of end-window type GM-tube

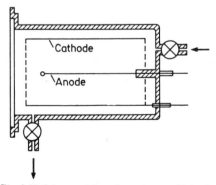

Fig. 2.11. Scheme of flow-through type GM-tube

A feature of GM-counter tubes not yet noted is the *"zero effect"* giving the number of electrical pulses for a specified tube type, induced by natural background radiation.

GM-counters are the most suitable for constructing the simplest detector systems. These counters work on a few 100 V power supply whose instability has an almost negligible effect on the performance of the detector, provided that the instruments are appropriately set to the conditions. The performance of the detector is, to a minor extent, dependent on temperature; it provides reasonably large electric signals, thus the requirements for signal processing are rather moderate. Its disadvantages lie in the low sensitivity for γ-radiation, the limited (and unique for a given type of instrument) range of dose rate, the inapplicability for the discrimination of energy and its finite life-time. In spite of these disadvantages, the GM-counter is the most widely used detector, mainly because of the low costs.

The parameters of an advanced GM-counter (Valvo 18552, Germany, F. R.) are presented as an illustrative example:

Starting voltage	400 V
Plateau	from 450 to 800 V
Plateau slope	2 percent for each 100 V
Dead time	70 μs
Zero effect	30 pulses/min
Tube diameter	18 mm
Tube length	143 mm

2.1.2. SEMICONDUCTOR DETECTORS

The basic principle of radiation sensing by semiconductor detectors is the ionization interaction process identical to that in gas ionization detectors. The semiconductor detector or the *particle-counting diode* as it is called can also be considered as a planar-electrode ionization chamber with a semiconducting crystal substituting the filling gas. The stopping power of *solid-state detectors* prepared from semiconductor materials is several times higher than that of the gas filling in ionization detectors, due to the difference in specific weights of these materials.

An elementary particle incident on a *germanium* or *silicon single crystal* semiconductor detector and causing ionization there is stopped rapidly, within 10^{-11} s. During slowing-down, its energy will be transferred to the electrons of crystal atoms (ionization) which in turn can give rise to secondary ionization if the absorbed energy is high. In the semiconducting material charge carrier pairs (electrons and holes) are produced directly by electrically charged particles (α- or β-particles) or indirectly by X-ray, γ- or neutron radiation, in the latter cases through the charged particles from some radiation interaction process. The energy needed to generate an electron-hole pair is 3.6 eV in silicon and germanium, respectively.

The electron-hole pairs induced by the ionization effect of radiation can be separated and collected in the semiconducting material by the application of an *electric field*. Due to the relatively high conductivity of pure semiconducting substances, the production of an electric field required for carrier collection is difficult, therefore a diode structure with a p-n junction is favoured for this purpose. The resistivity of the barrier is high enough to achieve a suitably high electric field. The free carriers released by the ionization process are separated in the junction layer. According to the arrangement and circuit presented in Fig. 2.12, a current pulse corresponding to the integrated value of charges generated by the absorption of a charged particle in the semiconductor appears. This current pulse causes a voltage drop across the load resistor R; that is, it results in an electrical pulse. The amplitudes of the electrical pulses are low, therefore a special (usually charge sensitive) *low-noise preamplifier* is needed to process the semiconductor detector signals. The linearly (undistorted) amplified detector signals can be counted, integrated, recorded or amplitude-discriminated by suitable electronic equipment.

During carrier production by charged particle or photon absorption the number of resulting carrier pairs will be W/w, where W is the particle energy and w the energy required to produce an electron-hole pair. The charge produced by the collected electrons and holes is Q which is expressed with elementary charge ($e = 1.6 \times 10^{-19}$) as follows:

$$Q = e \times \frac{W}{w} \tag{2.7}$$

This amount of charge can be collected when the collection process is fast (there is no recombination) and *every carrier generated will be collected*. The carrier collection time taking into account the detector geometry and carrier mobility is inversely proportional to the field strength. To increase the field strength, a relatively high detector voltage of several hundred volt is applied. *To reduce noise effects* accompanying the operation of detectors they are, in general, *cooled below −190 °C* by liquid air or liquid nitrogen. Under these conditions the noise equivalent power of a good detector is lower than 200 eV.

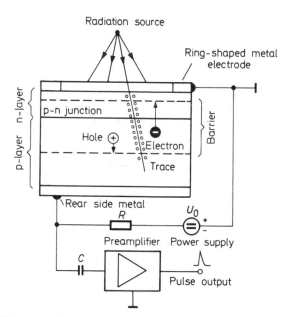

Fig. 2.12. Design and basic circuit of a surface barrier semiconductor detector

Silicon semiconductor detectors can be used at *room temperature* too, but naturally with a higher background noise. Their detection efficiency for α- and β-rays is 100 percent and a high value also for γ-radiation.

In the measurement of monoenergetic radiation by semiconductor detectors the standard deviation in pulse amplitudes is very low because the energy required to produce a single carrier pair is low. Hence, by discriminating the semiconductor detector pulses, good distinction of different radiation energies can be achieved.

The most extensively used semiconductor detector types are summarized in Fig. 2.13.

Figure 2.13/1 shows the structure of a "*homogeneous*" type detector made of a uniform semiconductor material *without a p-n junction*. The carriers generated within a crystalline semiconducting material by ionization are separated and collected by a bias applied to metal electrodes on the two opposite sides of the crystal. Their practical application, however, is limited by several facts, mainly by the inadequately low resistivity of presently available semiconducting materials. Homogeneous detectors can be prepared from gallium arsenide, germanium, silicon and other semiconductors. Detectors from cadmium

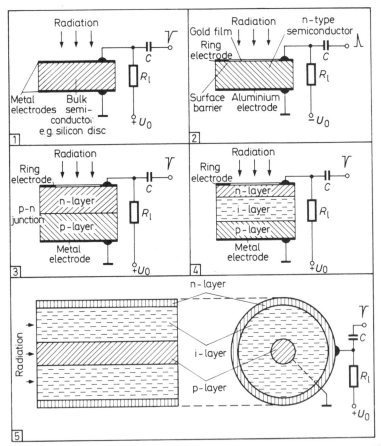

Fig. 2.13. Basic types of semiconductor detectors. 1 — homogeneous; 2 — surface barrier; 3 — diffusion; 4 — planar p-i-n; 5 — coaxial p-i-n semiconductor detector structures

sulphide (CdS) crystals have already found practical application, namely in dose rate measurements of high intensity γ-rays.

The p-n junction type diode semiconductor detectors are produced from silicon or germanium. The properties of these materials are different, the density of silicon of atomic number 14 ($_{14}Si$) is $p = 2330$ kg/m^3, while that of $_{32}Ge$ is $p = 5330$ kg/m^3. They differ also in various other factors, like the ranges of absorbed particles, their penetration depth maxima (e.g., the ranges of 1 MeV electrons or β-particles in silicon and germanium are 1.6 mm and 0.8 mm, respectively), the mobilities of free carriers in the crystal, the temperature dependences, the backward currents, the noises, etc. Hence, the starting material is either silicon or germanium according to the purpose of particular detector use and the preferred properties during their specific applications.

The structure of the *surface barrier* type detector is shown in Fig. 2.13/2. The detector material is a high resistivity, carefully cleaned n-type silicon disc of 1 or 2 mm thickness. The detector surface is covered with a 10 μm thick protective coating of evaporated gold film connected to a conductive metal ring electrode. The surface barrier is formed by the oxidation of the previously polished and etched surface yielding a smooth mirror-like appearance. The effectiveness of the surface barrier forming is enhanced by the negative charge on the n-type semiconductor crystal surface due to internal space charges which alone would be enough to produce a surface barrier. The detector operates under reverse bias.

In the diffusion type semiconductor detectors (Fig. 2.13/3) the p-n junction is formed by a diffusion process. An n-type impurity is diffused into the p-type semiconductor on the side of radiation incidence. The condition of good performance is that a relatively wide, continuous rather than abrupt p-n transition should be located close to the surface of radiation incidence in order to minimize absorption in the overlying dead layer. The active volume of diffusion detectors is the barrier layer widened and depleted by the electric field of the reverse bias. The thickness of the active zone is in the order of 1.5 mm, thus this type of counter is used mainly for charged particle (α- and β-rays) detection.

The introduction of *lithium ion drifting* has made it possible to detect, by total absorption, radiation with ranges exceeding 1.5 mm. The relatively large radiation-sensitive volume having a thickness of at least 10 mm can readily be formed using a *p-i-n detector structure* (Fig. 2.13/4). In this configuration, an *intrinsic layer* (i) has been formed by drifting lithium ions between the p and n layers which constitute part of the active detector volume. The thickness of the i layer in a silicon (*Si/Li*) *detector* is 5–6 mm, in a germanium (Ge/Li) detector it is 10–12 mm. Due to the different absorption properties of silicon and germanium and because the penetration depth of particles or photons in germanium is less than that in silicon at equal energies, the Si(Li) detectors can be used for the measurement of X-rays and are applicable to soft γ-rays; the *Ge(Li) detectors* are applicable also for hard γ-radiation measurement.

The coaxial-structure or coaxially drifted p-i-n detector was developed for those situations where the planar Ge(Li) p-i-n detector does not ensure total absorption (Fig. 2.13/5). The active volume of these p-i-n counters detects radiation arriving from any direction and, though the maximum radial extension of the i layer is 12 mm, the absorption path of radiation entering the front face can be as long as 60 or 70 mm. With these very high sensitive-volume detectors extremely hard γ-radiation can also be measured.

Semiconductor detectors have been developed rapidly in the past years utilizing recent advances in semiconductor device technology. It means that their growing use can be anticipated both in scientific research and engineering applications. Their principal advantages are:

— excellent energy discrimination;
— linear energy-to-pulse height conversion (for total absorption);
— short pulse generation;
— non-sensitivity to magnetic fields;
— low supply voltage requirement;
— vibration and impact resistance;
— long service life.

On the other hand, there are some drawbacks of semiconductor detectors, such as
— relatively high, supply voltage dependent self capacity;
— disturbances in operation by lattice defects in the crystal;
— the signal-to-noise ratio at room temperature is low compared with other detector types;
— surface effects of atmospheric contaminations can increase the inherent noises;
— long term irradiation can cause damage in the crystal structure leading to the degradation of detector characteristics;
— the geometrical dimensions are limited.

In spite of these disadvantages the field of semiconductor detector applications is growing, mainly due to its feature of good energy discrimination for particles or photons constituting the radiation. In other words, the "energy resolution" of semiconductor detectors is the best of all counter types.

Energy resolution as an important detector property can be interpreted according to Fig. 2.14. Even if a monoenergetic radiation of E_0 energy is detected, the pulse heights are unequal but they are located in the vicinity of the amplitude value corresponding to E_0. The pulse numbers stored in the measuring channels differing by small steps during a measurement period t_m are represented on the ordinate. The energy scale on the abscissa expresses also the pulse height relations. The amplitude distribution of pulses gives the shape of a bell diagram shown by the Figure. If n denotes the pulse number at the peak

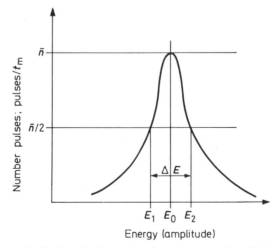

Fig. 2.14. Energy distribution of pulse numbers for monoenergetic radiation measurement

corresponding to E_0, and the energies E_1 and E_2 corresponding to pulse number $n/2$ are recorded then the differential energy resolution ΔE of the detector, the full width at half maximum *(FWHM)* is equal to the difference of the two energies $(\Delta E = E_2 - E_1)$. The detector resolution belonging to *FWHM* can also be given as a percentage of the nominal energy:

$$FWHM = \frac{\Delta E}{E_0} \times 100\% = \frac{E_2 - E_1}{E_0} \times 100\% \qquad (2.8)$$

which is strongly energy (E_0) dependent.

The excellent energy resolution on semiconductor detectors makes them suitable for solving several analytical and material composition determination problems by nuclear methods—as will be discussed in Chapter 4, mainly in Section 4.2. If, for example, there are two different chemical elements with closely spaced nominal characteristic radiation energies, the high resolution of the detector allows the separation and distinction of these two radiations.

Semiconductor detectors can be connected practically to any measuring equipment. Their beneficial properties can, however, be economically utilized mainly for energy selective measurement or sorting of radiation (amplitude analysis).

2.1.3. SCINTILLATION DETECTORS

The operation principle of scintillation detectors is based upon the specific property of *some organic and inorganic substances*, viz., *light-flash generation* or *luminescence* induced by radioactive radiation. The weak flashes or scintillations are detected and amplified by a photomultiplier used as a light-to-electric signal transducer. The scintillation is proportional to incident particle energy; thus, from the voltage output of the photomultiplier, due to its linear characteristics, information on particle energies can be obtained.

Substances possessing this luminescence property are called scintillators in radioisotopic measurement techniques. The scintillation mechanism is a complex process. Initially, the scintillator absorbs radiation energy resulting excited or ionized states of some of the atoms. They will fall back into their ground state in a very short time accompanied by light, or by photon emission. The higher the energy absorption from the traversing particle in the material, the more scintillating atoms will be excited. The energy transfer from individual particles of photons will depend on the interaction process. Figure 2.15 shows the principal interaction processes between scintillator material and α- or β-particles and γ-photons, respectively.

The penetration depths of α-*particles* is small though their energy can be several MeV, therefore the energy transfer takes place in a very thin layer. They ionize atoms along their paths strongly, producing secondary electrons which excite the atoms.

The penetration depths of β-*particles* can be several mm depending on their energies. During penetration they ionize more or less strongly, causing secondary electron production which in turn excite the atoms.

During the well-known interaction processes of γ-*quanta* such as photoeffect, Compton scattering and pair production, electrons are created which ionize and excite the atoms similarly to the direct effect of β-rays. In Compton scattering, also a secondary γ-radiation appears which transfers its total energy through repeated interaction processes

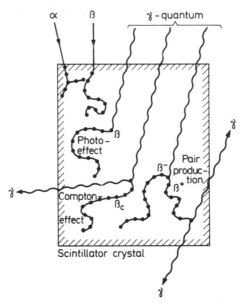

Fig. 2.15. Interaction processes of elementary particles in a scintillator material

to the atoms provided the scintillator is large enough in size. The β^+-particle (positron) generated during pair production ionizes as well but when it combines with an electron a γ-radiation arises *(annihilation radiation)* which can be absorbed through additional interaction processes by the scintillator. Thus, provided the crystal sizes are adequate, the total energy of γ-radiation will be expended to excitation through several intermediate processes.

Neutron radiation can be detected by scintillators containing boron or lithium where the neutrons produce α-radiation which gives rise to scintillation in the above described way.

The total amount of light energy resulting in scintillator materials during the various scintillation processes, perhaps in different ways, is called *light yield.* The reference scintillation light yield is that of an *anthracene* scintillator considered to be unity or 100%. The light output intensity from scintillators rises abruptly up to a maximum followed by an exponential decay. It has a tail so the luminescence is maintained for a while *(afterglow).* The light output *(afterglow)* time constant denoted by τ_0 is the time needed for intensity reduction down to 37% of its maximum (initial) value; normally, its range is from 3 ns to 10 μs.

There are three groups of *scintillator materials,* inorganic single crystals, liquid scintillators and plastic scintillators.

The most frequently used *inorganic scintillator single crystal* is thallium-activated sodium iodide. Due to its relatively high density, it is suitable mainly for the detection of strongly penetrating electromagnetic γ- or χ-radiation (Table 2.2 a).

The most important *liquid scintillator* compounds are 2,5-diphenyloxazol (PPO), 2-phenyl-5-(4-diphenyl)-1,3,4-oxadiazol (PBD), 2-[4'-tert-butylphenyl-5-(4''-diphenyl)]-1,3,4-oxadiazol (butyl-PBD), 1,-4-di (5-phenyloxazol-2-yl) benzene (POPOP), and 1,4-bis 2(4-methyl-5-phenyloxazolite) benzene (dimethyl-POPOP). They are used in solutions of about 1 g/l concentration in cyclic hydrocarbons (mixed with toluene or xylene or cyclohexane, for example). The specimen tested is dissolved in the liquid allowing high efficiency soft β-ray counting, in particular (Table 2.2 (b)).

Plastic scintillators (scintillator materials embedded in polymers) have found wide use due to simpler techniques for obtaining sizes and shapes most suitable for detector operation by controlling the polymerization process than those for inorganic single crystals.

The characteristic data of the significant solid scintillator materials are summarized in Table 2.2 (a, b). The Table shows which radiation types can be detected by the scintillator selected. The light emission spectra of scintillation materials have maxima at well defined wavelengths depending on the host material and activators used; the relevant numerical values are also given in Table 2.2.

The arrangement known as a *scintillation counter* includes a *scintillator* and a *photomultiplier* optically coupled to it (Fig. 2.16). The detection of a particle takes place in several steps:

— the scintillator material absorbs an impinging particle or quantum inducing scintillation, i.e., photon emission;

— the photons produced this way are transferred by light guiding to the cathode of the photomultiplier tube;

— the bombarding photons cause photoelectron emission from the photomultiplier cathode;

— the photoelectrons leaving the cathode are directed by a focusing electrode to the first anode called the dynode.

The dynode anode is prepared from materials possessing the feature of providing several secondary electrons for each single photoelectron reaching its surface. The electrons produced by the voltage dependent secondary emission process are accelerated in the field of the next dynode and this electron multiplication process will be repeated again. On increasing the number of dynodes a significant gain may result, owing to the progressive nature of the multiplication. The voltage steps needed for the dynodes are produced by a voltage dividing resistor chain. The secondary emission factor δ is defined as the number of secondaries produced by a single primary electron. The strongly supply voltage dependent gain of an n-stage multiplier tube with n dynodes is given by the equation

$$G = \delta n \qquad (2.9)$$

In case of the typical values $n = 10$ and $\delta = 4$, the numerical value of G is about 10^6.

As the output of the scintillation detector, a *current pulse* is generated at the last electrode, the anode. Using a load resistor connected to the anode, the current pulses will be converted into voltage pulses with amplitudes from a few millivolts to several volts, depending on the impinging particle energy, taking into account the usual load resistor ratings.

The scintillation spectrometry is based on this effect. The load resistor R_1 in Fig. 2.16 is shunted by various stray capacities, their combined values denoted by C_s. Under practical conditions, the value of R_1 is chosen to satisfy the expression

$$R_1 \times C_s = \tau_0 \qquad (2.10)$$

where τ_0 is the decay time constant of the scintillator (see Table 2.2).

Table 2.2. Physical constants of scintillators

(a) Crystals

Scintillator type	Structure of material	Density, kg/m³	Refraction index	Melting point, °C	Light yield 100% anthracene	t_0 afterglow time constant, ns	Wave length of maximum emission, nm	Detectable radiation
Anthracene	Organic crystal	1250	1.62	217	100	30	447	α–β–γ fast neutron
Stilbene	Organic crystal	1160	1.63	125	50	4.5	410	α–β–γ fast neutron
NaI(Tl)	Inorganic crystal	3670	1.775	650	230	230	413	X-ray γ
NaI (pure)	Inorganic crystal	3670	1.775	651	440*	60*	303*	X-ray γ
CsI/Tl	Inorganic crystal	4510	1.788	620	95	1100	580	γ
CsI/Na	Inorganic crystal	4510	1.787	621	150–190	650	420	γ
CsI/pure	Inorganic crystal	4510	1.788	621	500*	600*	400*	γ
KI(Tl)	Inorganic crystal	3130	–	723	50	500	420	γ
LiI/Eu	Inorganic crystal	4060	1.955	445	75	1200	475	n, γ
CaF₂(Eu)	Inorganic crystal	3170	1.443	1418	110	1000	435	X-ray β–γ
CaWO₄	Inorganic crystal	6100	1.92	1535	36	6000	430	γ
CdWO₄	Inorganic crystal	7900	2.356	1850	300	200	450	α–β–γ neutron

ZnS(Ag)	Polycrystal	4090	2.356	1850	300	200	450	α
⁵Li–ZnS(Ag)	Polycrystal	2360	–	110	300	200	450	slow neutron
NE 905 NE 906	Glass	2500	1.55	1200	25	18–60	395	slow neutron
Plastic	Terphenyl dissolved in polystyrene, etc.	1030	1.58	40–80	50–70	3	420	α–β–γ fast neutron

*at −195 °C

(b) Liquid scintillators

Table 2.2. (cont.)

Compound Abbreviation	Composition	Absorption nm	Emission nm	ϕ Relative fluorescence (Anthracene = 100)	Solubility in toluene, 25 °C, g/l	Usual concentration, g/l
ϕ3	p-Terphenyl		340	90	8	5
PPO	2,5 Diphenyloxazole	301	370	100	270	4
PBD	2-Phenyl-5-(4-diphenyl)--1,3,4-oxadiazole	302	370	83	20	10
Butyl PBD	2-/4'-t-Butylphenyl-5--(4''-diphenyl)-1,3,4--oxadiazole	305	365 385	83	130	7–10
POPOP	1,4-Di(5-phenyloxazol-2-yl) benzene		415	93	0.5	0.1
Dimethyl-POPOP	1,4-bis 2(4-Methyl-5--phenyloxazyl) benzene		430	93	2	0.1–0.5
NPO	2-Naphthyl-5-phenyloxazole Naphthyl		415	93		0.1–1

ϕ Relative fluorescence = the integrated values of the given fluorescence spectra in relation to one another

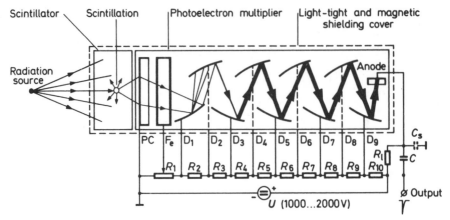

Fig. 2.16. Construction principle and circuit arrangement of a scintillation counter. PC — Photocathode; Fe —
Focusing electrode; $D_1 \ldots D_9$ — Dynodes

Both the number of photons produced in the scintillator and that of the photoelectrons are proportional to the energy of the absorbed particles. At the end of the process an electrical pulse with an amplitude U_0 proportional to the particle energy *E* is induced by the multiple number of photoelectrons (gain). The pulse amplitude will, however, fluctuate statistically about the voltage U_0 due to the fact that the photon production, the photoelectron generation and multiplication are *a priori* statistical phenomena. Categorizing pulses by their amplitudes into channels of ΔU widths, the diagram shown by Fig. 2.17 (a) is obtained. The spectrum of monochromatic γ-rays obtained by pulse-height discrimination is characterized by a so-called *photopeak* defined by the *FWHM* value corresponding to the photoeffect.

The flat region $(U_2 - U_1)$ in the supply voltage dependent characteristics, similar to the plateau of GM-counters, is obtained for scintillation detectors only when the radiation is monochromatic. If that is the case, the working point should be set at voltage U_m to achieve some stabilization of operation.

The efficiency, counting probability of scintillation detectors is high. For α- and β-rays it can be as high as 100% but it can also reach 20 to 30% for γ-radiation. However, the scintillation counter is composed of elements which themselves depend on many factors, too, and it can be operated conveniently only if its components are carefully matched to each other and to the particular measurement problem. The scintillator crystal should be selected according to the actual radiation type (α, β, γ or neutron radiation) and, additionally, a high radiation-to-light conversion efficiency should be ensured.

The crystal size, encapsulation and its incorporation in the sensor are determined by the radiation energy. To achieve suitable performance of the scintillator crystal, the fact should be taken into consideration that the light yield is temperature dependent and that the radiation-to-light conversion process has a maximum at a given wavelength. Additionally, a lossless transfer of light to the photomultiplier *(light coupling)* should be provided for. Only those photomultiplier tubes with cathode-material dependent spectral responses coincident with light yield maxima of the scintillator should be used.

The supply voltage, its stability limits and the voltage dividing resistor chain components should be chosen very carefully. The output of the multiplier tube has to be

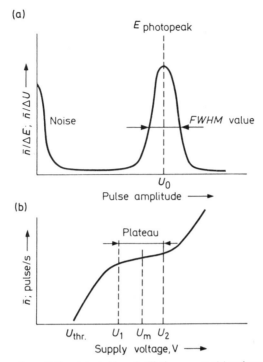

Fig. 2.17. Energy spectrum of γ-radiation detected by scintillation counter (a) and counting characteristics of the detector (b)

matched electrically to the measuring unit characteristics. The electron multiplication process and, consequently, the multiplier performance is sensitive to temperature and magnetic field variations.

The fact that the performance of the scintillation detector tends to be dependent on the working conditions is highly disadvantageous if these disturbing factors are not properly eliminated. Provided that a good protection against these effects has been worked out, the use of the scintillation counter is advantageous, mainly in industry, because of its features listed as follows:

— a quite good time-resolution;
— the proportionality of the signal to the energy;
— the great amplitude of the electric signal;
— the high efficiency detection (also for γ-radiation);
— the (practically) infinite lifetime.

In the foregoing discussions, several important dependences were dealt with, to be borne in mind when using scintillation detectors. The long-term stability of scintillation counters without protective measures (e.g., temperature stabilization) is insufficient both for laboratory spectrometry and industrial measurement purposes. The instabilities result usually in the change of detected average counts in unit time even if the measuring arrangement is unaltered. According to experience, neither the pulse number nor the peak amplitude is stable. To avoid these insufficiencies, various *automatic calibration and/or stabilization* schemes have been developed; the most convenient one is the automatic electronic compensation.

(a)

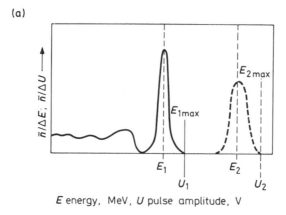

E energy, MeV, U pulse amplitude, V

(b)

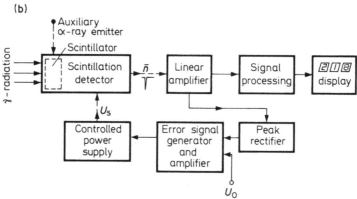

Fig. 2.18. Operation stabilization of scintillation counters by peak rectification. (a) signal spectrum; (b) block diagram

The block scheme of a well-known approach to electronic stabilization is shown in Fig. 2.18, in two versions, with and without an auxiliary α radiation source. The scintillation counter output signals after distortionless amplification are, partly, subjected to further processing or, partly, coupled to a peak rectifier stage. The average output of the latter, corresponding either to the energy $E_{1,\,max}$ of the detected γ-rays or to the energy $E_{2,\,max}$ of the optional α-ray emitter, controls the high-voltage power supply to maintain the pulse height maximum and, simultaneously, the average count number at constant levels.

The integration circuit time constant of the *peak rectifier* is chosen to provide a relatively high value for the high-pulse-rate *monochromatic* γ-rays and, on the other hand, a low value for the infrequent noise pulses. By a well designed circuit array faults resulting from both temperature effects and aging degradation can be *compensated*. The stability maintainable by the system described is in the order of 1 per mil which means that high precision industrial measurement can be performed within an error of ± 1‰, too.

Using a well-adjusted and stabilized scintillator detector with an electronic equipment capable for amplitude discrimination, energy selective measurements can be performed. In Chapter 4 examples for element identification by scintillator spectrometry are given. Figure 2.19 shows a charachteristic K-radiation spectrum of ^{55}Fe recorded by a

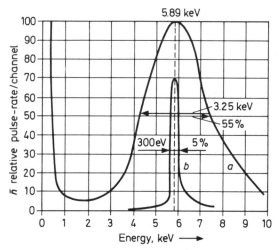

Fig. 2.19. Comparison of energy resolutions of (*a*) scintillation and (*b*) semiconductor detectors at 5.89 keV
^{55}Fe K-radiation

scintillator counter. From the analyses of the spectra, a high *FWHM* value (55%, 3.25 keV) band centered around 5.89 keV is observed which can be explained by the superposition of multiple statistical processes. For comparison, also the spectrum obtained for this particular radiation by a semiconductor detector is plotted in Fig. 2.19 showing an *FWHM* value as low as 300 eV (5%). From an evaluation of these curves the energy resolution of the semiconductor detector an order of magnitude higher in this energy region than that of the scintillation counter is concluded. For γ-radiation with higher energies the relative resolution of scintillator detector improves significantly, accompanied by a simultaneous monotonic growth of the *FWHM* value.

2.2. BASIC TYPES OF RADIATION MEASURING AND ANALYSING INSTRUMENTS

The information carrying output from nuclear detectors as discussed in Section 2.1, appears in the form of d.c. signals or of pulses. Accordingly, the electrical signals can be processed by electrometers or by pulse mode instruments. A condition of reliable and reproducible measurements is the use of *stabilized power supplies* to the detectors. The detector supply voltages lying typically in the range of 0.5 to 5 kV providing adjusted and stable nominal values are generated by so-called high-voltage power supplies. Their normal stability requirement is a voltage variation of less than 10^{-3} to 10^{-5} times the preset value, under any operational conditions.

Electrometers are used to amplify the d.c. output from a detector (ionization chamber) producing a time averaged output proportional to the radiation dose rate.

Measuring instruments connected to pulse-type detector amplify and analyse pulses. They can analyse either *pulse numbers* proportional to dose rates or pulse heights associated with *radiation energy*. The former are called *rate meters* and *scalers* the latter *pulse-height analysers.* In recent years the pulse-type instruments have been used more extensively in practice.

The pulse-height *discriminator* is a usual component of pulse-type equipment: it is used to reduce disturbing low-energy radiation effects sensed by the detector and other noise effects from the amplifier, photomultiplier and other electronic components, commonly known as background effects; additionally, it performs classification of radiation by energy. This latter function of pulse amplitude analysers is unavoidable because their very purpose is to determine the energy distribution of radiation.

Background reduction can be accomplished also by *coincidence circuits;* the function of these is to reject signals not simultaneously arriving from two or more detectors. *Anticoincidence circuits* function in the opposite way: they deliver pulses from a detector to the measuring unit only if no pulse is obtained simultaneously from another one.

2.2.1. COUNTING RATE METERS

Rate meters measure the average number of pulses produced by a detector in a unit time period. The block scheme of a typical measuring configuration is presented in Fig. 2.20. Detector pulses are amplified and occasionally discriminated, followed by a shaping process. The output pulses identical in height and width from the forming stage will be summed by an integrating stage. A d.c. voltage proportional to the pulse number will appear at the integrating stage output which can be continuously measured by an electronic voltage meter and monitored by an instrument or a recorder. An important parameter of the equipment is the time constant defining the time of averaging.

The instruments can indicate the dose rate in A/kg or mR/s units if the combined measurement efficiency is known; commercial portable dose rate meters usually have this feature (see also Section 8.3.2.1).

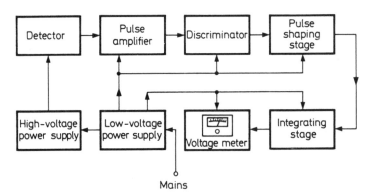

Fig. 2.20. Block scheme of a pulse-rate meter with auxiliary units

2.2.2. SCALERS

For counting detector pulses, scalers operating in conjunction with electronic dividing circuits are used (Fig. 2.21). The output signal from such an equipment following suitable preparation (amplification, discrimination, pulse shaping) is fed into a counting unit

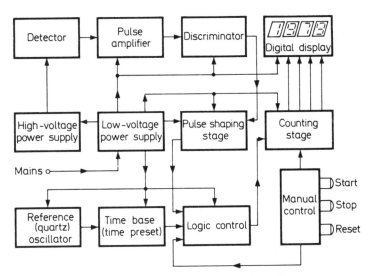

Fig. 2.21. Block scheme of a laboratory pulse scaler

through a logic control device. To this unit, also a time base signal controlled by a reference (quartz) oscillator is delivered.

The zero setting of the scaler is carried out by a manually operated reset control. When the *start control* is switched on pulse counting and time measurement begin. Once the time preset value chosen in time base units is finished, the *logic control stops the operation* of the scaler. The number of pulses that arrived during the measuring period is *displayed digitally.* In some equipment types, *punch tape devices and printers* can be connected to the counting unit for recording.

2.2.3. AMPLITUDE ANALYSERS

For pulse-height sorting of signals from a properly chosen detector, pulse amplitude analysers are used. They collect undistorted amplified pulses characterized by their voltages in one or more channels, the channel voltage range or channel width can be varied as desired.

In the block scheme of a single-channel pulse-height analyser presented in Fig. 2.22, employ a pair of discriminators with different threshold levels. The passing-through level is determined by voltage level to be varied continuously by the potentiometer P_1. The difference between the shifted threshold levels or the so-called *channel width* (or window) can be adjusted by potentiometer P_2 independently of the setting of P_1. The function of the *anticoincidence* stage is to transmit pulses through the lower discriminator but not through the upper one into the shaping stage because their amplitudes lie in the voltage region from U_0 to $U_0 + \Delta U$. The signals from the shaping stage can be evaluated by analogue or digital techniques.

For a better understanding of the operation, the signal shapes at points denoted by letters in the diagram of a *single-channel analyser* are shown in Fig. 2.23. From a comparison of figures it is obvious that signals with voltages higher than U_0 are passed

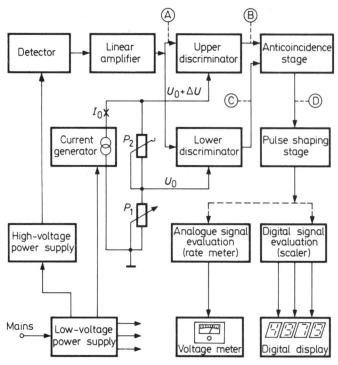

Fig. 2.22. Block scheme of a laboratory single-channel pulse-height analyser. P_1 — Threshold level control; P_2 — Channel width setting

only by the lower discriminator, while those higher than $U_0 + \Delta U$ are passed by the upper one, too. The anticoincidence circuit suppresses all signals transmitted by both discriminators. A single-channel analyser allows spectra to be recorded by subsequent measurements through quantized alterations of the voltage U_0.

In contrast, the *multichannel analyser* classifies detector pulses one by one and collects signals belonging to the same channels in corresponding storage blocks following pulse-height discrimination. In accordance with this, each detector pulse is evaluated individually excluding those which arrive within the time period required for amplitude determination (dead time).

Figure 2.24 illustrates the amplitude determination principle by voltage-to-frequency conversion, as most frequently used in multichannel analysers. The detector signals delivered to input A are processed by a linear amplifier with a *linear gate circuit* connected to its output through a delay line L. The linear gate circuit allows pulses to pass, distortionless, only if it is not prohibited by the discriminator or by the dead-time generator. The discriminator provides protection against long-term paralysis of the system by high amplitude signals beyond the analysed region. Pulses transmitted by the linear gate circuit will charge a condenser C up to the amplitude of the signal from the detector. A constant-current discharging circuit will start to discharge the condenser after the charging has been fully completed (rise time completion). At discharge starting the gate circuit G opens and the frequency signal from a reference oscillator starts flowing to output B as long as the completion of discharge is indicated by a zero-level comparator. In

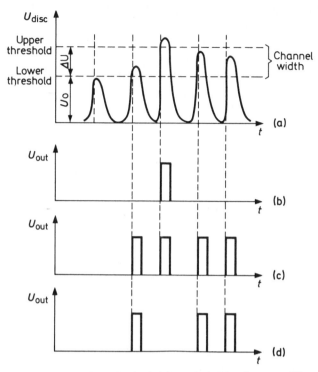

Fig. 2.23. Signal shapes in the single-channel pulse-height analyser (a) at linear-amplifier output; (b) at upper-discriminator output; (c) at lower-discriminator output; (d) at anticoincidence-stage output

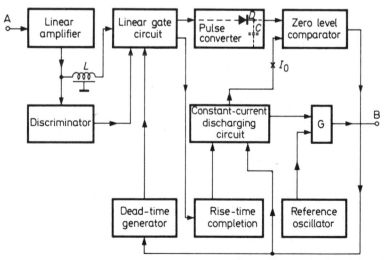

Fig. 2.24. Block scheme of an amplitude-to-frequency converter for a multichannel pulse-height analyser. A — Pulse input; B — Channel number output; L — Delay line; G — Gate circuit; D — Detector; C — Condenser

a suitably designed system, the pulse number transmitted by gate circuit G from the reference oscillator is proportional to the detector pulse amplitude. By counting the transmitted pulses, the number or the address of that relevant storage channel is obtained whose content should be increased by one.

Pulse-height analysers are suitable mainly for measuring radiation energy spectra and for identification or purity monitoring of radionuclides. Therefore, they are especially important tools in radiochemical analysis (see Chapter 4).

2.3. BASIC PRINCIPLES FOR INDUSTRIAL USES OF NUCLEAR MEASUREMENTS

The interaction processes between radioactive radiation and materials tested are the bases for information gathering in measurements with isotopes. The total measuring system is composed of an *isotopic radiation source with a protective shield, a radiation sensing device (detector)* in suitable geometrical arrangement, *electronic unit(s)* processing detector signals and an *instrument* or a *recorder* displaying or storing the measurement results. Isotopic instruments can also be connected to *computers* for data storage or process control as data acquisition elements in the system. The components of the measuring line in the system should be chosen so that any measurement functions induce changes as high as possible in the measured (sensed) radiation signals for unit variations of the parameter to be measured.

Nuclear or radioisotopic industrial measurements are used as advanced monitoring, controlling and automation tools for technological processes. In a measuring system using isotopes the *directed radiation* from a radioisotope source equipped with protective shields penetrates the sample where it will be partly *absorbed* and partly *scattered (reflected)* by the sample. During penetration or scattering, the number, energy and/or direction of elementary particles or quanta of which the radiation is composed, will change due to

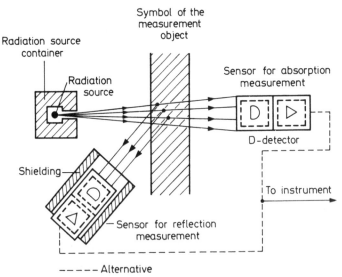

Fig. 2.25. Common basic concept for nuclear measurements

radiation absorption and reflection. From the changes in properties of particles or quanta observed by a radiation sensing detector, the *measuring equipment* determines the values of the parameters to be measured (Fig. 2.25).

Industrial nuclear measurements are characterized by *nondestructive sensing*. The advantages of this are:

— *contactless* detection and measurement—meaning that neither the measuring "signal" emitter, nor its receiver will be in physical contact with the material tested;

— absence of the influence by *environmental factors* (pressure, temperature, vibration, humidity and others) on the origin of radioactive radiation, with no maintenance requirements of the isotopic radiation source used as a signal source in the system;

— data obtained are usually *averaged over large material quantities;*

— measurement results are produced by *electronic equipment* with outputs in digital or analogue forms, according to the particular problems, appropriate for *display, recording* or connecting to *control systems;*

— the resulting values are obtained in *real time*, without any delay during measurement.

In a suitable measurement configuration and with favourably chosen source characteristics in terms of energy, radiation type and activity, together with an appropriate detector system, the *physical* parameters (level, density, thickness, quantity)

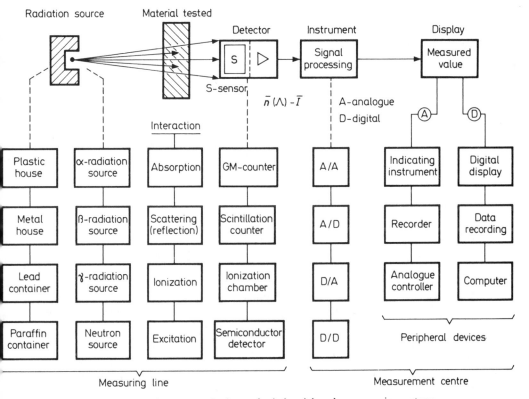

Fig. 2.26. Alternative structural schemes for industrial nuclear measuring systems

and the *chemical* features (moisture, metal content, sulphur content, etc.) of dynamically varying or statistically stored materials in technological processes can be obtained.

The alternative structural schemes for nuclear measuring systems with suitably selected components for various measurement problems in industry are shown in Fig. 2.26. The principal elements of the system are:

— *radiation sources* emitting α, β, γ, or neutron radiation;

— *interaction processes*, such as absorption, scattering (reflection), ionization or secondary radiation generation;

— *detectors*, such as various types of GM-counters, scintillation counters, ionization chambers or semiconductor detectors.

The presented elements of the measuring line, depending on the state of material and its properties to be measure, can be arranged in any combination, the only restriction being that the γ- and neutron radiation do not directily ionize. Bearing in mind that the different radiation types (as far as their energies are concerned), there can be numerous different ways of obtaining information in practice, therefore the requirement arises to select the most appropriate one for a given measurement problem.

2.3.1. RADIATION SOURCES AND CONTAINERS

In nuclear industrial measurements generally *sealed radiation sources* can be utilized. The various radiation sources, their configurations and their other radiation characteristics are discussed in Sections 1.3 and 1.4; the overall radiation protection problems are treated in Section 8.4. No more than a *few dozen* of the thousands of natural and artificial radiation isotopes known in physics and chemistry providing individually different types and energies of radiation are used in industrial measurement technology.

This reduction in choice is caused either by the short half lives of the majority of radiation sources or by the problems associated with the difficulties of their production in satisfactory forms. The most frequently used isotope sources for nuclear measurements are:

— *α-ray emitters:* ^{210}Po, ^{239}Pu, ^{241}Am;

— *β-ray emitters:* ^{85}Kr, ^{90}Sr, ^{147}Pm, ^{204}Tl;

— *γ-ray emitters:* ^{60}Co, ^{137}Cs, ^{170}Tm, ^{241}Am;

— *neutron emitters:* ^{206}Ra–Be, ^{210}Po–Be, ^{241}Am–Be.

Detailed data on the listed sources are given in Section 1.3.

One of the conditions for reliable information is the *constant intensity of radiation entering the measuring line.* It means that, in principle, the half-life dependent and exponentially decreasing intensity of radiation from radioisotopes may cause problems in the measurements. The magnitude of the intensity reduction with respect to half-life in terms of $t/t_{1/2}$ time units is given in relative values in the inside cover.

As shown above for industrial measurements, only isotopes with relatively long half-lives (from 0.5 to 30 years) should be used. However, even when using ^{137}Cs with an approximate half life of 30 years, consideration must be given to the physical effect that an intensity drop, e.g., of 2% ($I/I_0 = 0.98$), appears at $t/t_{1/2} = 0.02$; if the half-life $t_{1/2} = 30$ years, this reduction occurs in $t = 0.02 \times 30 \times 12 = 7.2$ months, but if the half-life is shorter, say 3 years or 0.3 year, the same error occurs after 21 or 2.1 days, respectively. An intensity drop of 2% may result in considerable errors in measurements characterized otherwise by a high accuracy of e.g., 0.1% in order of magnitude.

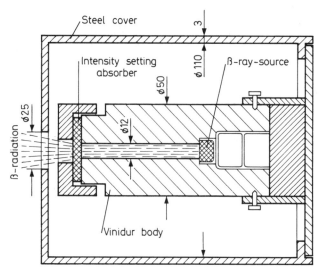

Fig. 2.27. Construction of a β-ray source container

The sealed mechanical structure containing the radiation source is called the *isotope container*. It has the dual function of providing the prescribed radiation protection at the location of radiation source application and, if necessary (e.g., during maintenance), it has the function of preventing the emission of radiation (shut-off) by manually or remotely operated insertion of strongly absorbing shields. The mechanical design of containers depends on the actual radiation type. Radiation protection for α- or β-ray sources can be readily provided for while that for the penetrating γ-ray and neutron sources is accomplished by containers produced from very high-density metal (lead, tungsten, uranium, etc.) content materials as γ-radiation absorbers and from neutron stopping materials to provide environmental protection against neutron radiation.

Figure 2.27 shows the schematic structure of a container for disc-type *β-sources* (see Section 1.4.2.2). The intensity of radiation emitted through a hole *(window)* in the front face of the steel cover serving as a mechanical shield can be varied by changing the absorber discs of different thicknesses. The source is held in position by a plastic (vinidur) structure to minimize bremsstrahlung.

Figure 2.28 shows a commercial container designed for industrial *γ-source* applications (see Section 1.4.2.3). The source is fixed on one end of a spring-loaded, axially movable cylindrical rod protruding from the right-hand side of the lead container. In opened position achieved by pulling and twisting the cylindrical rod and fastening it there, the radiation source is moved in front of the bore and collimated radiation will be emitted. The container is protected mechanically by a welded steel shell. The mechanical driving mechanism defining the radiation source position is covered by a protective cup and secured against unauthorized handling by a lead seal and a lock. To shut off, the source is introduced by pushing into the interior of the container thereby ensuring that no radiation is able to emerge through the aperture. The source container is designed so as to enable the *insertion of standard* absorber plates into the radiation path even in mounted position, in order to provide a controlled reduction of radiation for calibration purposes.

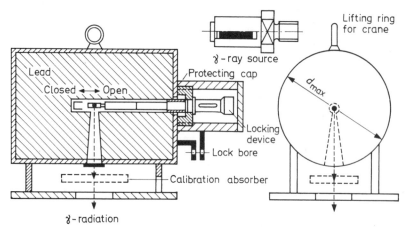

Fig. 2.28. Construction of a γ-ray source container

The geometrical dimensions of the container and the thickness of the built-in lead shields are determined by the activity of the source applied and the energy of radiation emitted (see Section 8.1.4.2). In the particular arrangement shown schematically by the figure, if the lead shield is, for example, 9 cm thick, equivalent to a mass of about 120 kg ($d_{max} \sim 200$ mm), then a ^{60}Co source with an activity up to 400 MBq (~ 10 mCi) or a ^{137}Cs source with an activity up to 200 GBq (~ 5 Ci) can be placed into the container.

The neutron sources are shipped and stored *in paraffin-loaded containers.*

2.3.2. RADIATION DETECTION IN INDUSTRIAL ENVIRONMENTS

Nuclear radiation detectors discussed in detail in Section 2.1 are employed also in the radiation sensing gauges of industrial nuclear instruments (Fig. 2.26). For measurements in industrial environments, however, the actual detector should be coupled to electronic equipment (amplifier, impedance inverter, voltage converter, etc.) providing *remote operation* capabilities and, in addition, it *should be protected against environmental influences by suitable covers.* The protective shield or encapsulation of the industrial radiation detector generally provides protection against dust and water drops though it can also be made explosion-proof by appropriate design. The detector metal housing is utilized to *mount and fasten it in a measuring position* within the total arrangement. For the detection of low-energy or corpuscular radiation (α- or β-rays) a window in the housing is unavoidable. The characteristics (material, thickness, etc.) of the window is determined according to the type and energy of radiation measured.

The industrial radiation detector is often called a *radiation gauge or probe.* The example shown in Fig. 2.29 is a hard γ-ray detector with GM-counter tubes. It is protected against atmospheric dust and dropping water by a double rubber o-ring-sealed cylindrical steel tube envelope. Inside the steel tube housing two GM-counter tubes—for higher sensitivity—with high-voltage power supply and preamplifier are placed. The cable containing wires for supply voltages and for detector signals is introduced across a sealed lead-through.

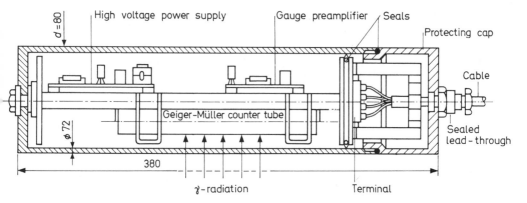

Fig. 2.29. Structural scheme of a γ-radiation detector for industrial applications

For universal industrial measuring systems various gauge types are needed even if different versions of a single radiation detector type (say a GM-tube) is to be used, depending on the actual radiation type (α, β, γ or neutron radiation) to be sensed.

2.3.3. MEASURING SYSTEMS
FOR INDUSTRIAL PROBLEMS

The conceptual construction of most industrial nuclear instruments can be derived from some common measuring arrangements.

The simplest configuration for the measurement of material properties is the *direct measuring system* shown schematically in Fig. 2.30. Radioactive radiation from source following interaction with the material tested, enters detector. Detector signals are processed by electronic unit and converted to results displayed by instrument. The electronic equipment contains amplifiers, discriminator, pulse shaping, integrating or counting circuits; display can be performed by line recorder, compensograph, pointer instrument, etc. The *measured value can be compensated electrically*, in this case the instrument indicates deviations from a nominal value.

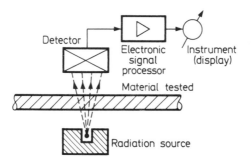

Fig. 2.30. Arrangement of direct measuring system

Differential-type measuring systems can be constructed by using dual measuring positions. They provide results of higher accuracy compared with direct systems and, in addition, the resulting values are free from environmental influences (atmospheric pressure, humidity, etc.). The arrangement in Fig. 2.31 is characterized by the dual detection technique. The radiation from a common source is emitted in two directions: one beam interacting with the substance tested, the other with a reference specimen of standard properties (thickness, density). The difference between radiation levels assessed by detectors is processed by an electronic unit and indicated by an instrument. When suitably designed, the detected differences in radiation levels are proportional to the deviation of the tested sample properties from those of the standard specimen. In practice, another version using two different radiation sources has found application due to its simpler handling. *Automatic servo compensation* versions of the difference technique which are suitable for displaying absolute values will be discussed in Section 2.4.4, in relation to thickness monitors.

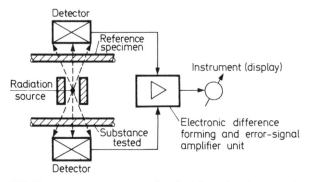

Fig. 2.31. Measurement arrangement for obtaining values by comparison

The nuclear measuring systems for industrial uses have been classified and produced according to their *measurement functions.* Consequently, the following instruments were regarded as different equipment types:

devices for the measurement of
— level (see Section 2.4.1);
— thickness (see Sections 2.4.4 and 2.4.5);
— density (see Section 2.4.2);
— moisture (see Section 2.4.6);
— soil testing devices (see Section 2.5);
— instruments for analytical measurements (see Section 4.1).

For wider application possibilities, a *certain degree of standardization* has been accepted within some instrument types. As an example, four or five types of measurement stations (measuring lines) combining radiation source and detector have been developed for a thickness monitor, each for different problems or measurement ranges.

The advance and miniaturization of electronic components (semiconductor devices, integrated circuits) has provided the possibility, while the need for more economic equipment production has required the solving of different problems by a single universal measuring equipment with the opportunity to connect various information acquisition

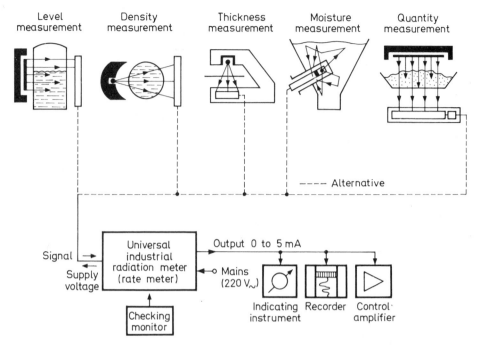

Fig. 2.32. Various measurement configurations for universal industrial nuclear measuring equipment (analogue version)

measuring stations to it as shown in Fig. 2.32. The measuring station including radiation source, and industrial detectors, should be arranged according to the particular measurement function.

The central unit of the universal nuclear measuring system for industrial use can be operated in different ways depending on the preferred mode of signal processing and the detector system chosen (see Fig. 2.26). *For ionization chamber detector* units with analogue-input A/A and A/D systems are applied. The results are displayed by indicating instruments or recorders in the former and in digital form in the latter case. *For pulsed detectors* (GM-counters, scintillation and semiconductor detectors) universal industrial radiation measuring units with digital-input–analogue-output (D/A) or digital-input– digital-output (D/D) can be applied. The central component of the measuring system on Fig. 2.32 is a D/A mode radiation gauge (rate meter) because its output signal set is compatible with numerous industrial needs (measurement, recording, control).

The advance in digital techniques and computerized monitoring and control supports the expansion of digital mode instrumentation. The nuclear digital measurement system with its block scheme (Fig. 2.33) can adopt both pulsed detectors and analogue radiation sensor by inserting an A/D converter. The introduction of *operational units* or modules into the signal processing system operating by the counting principle is useful for the following reasons:

— the characteristics of the measuring system can be linearized by computer techniques;

— the measurement results can be converted into direct physical parameters;

— correction characteristics measured by non-nuclear methods can also be

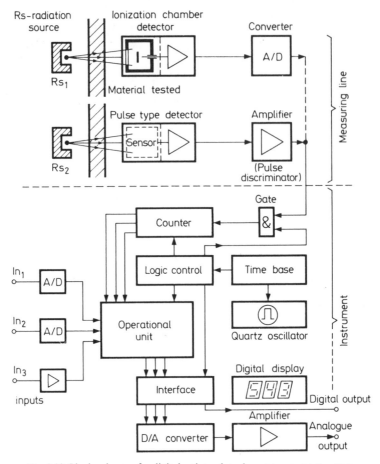

Fig. 2.33. Block scheme of a digital universal nuclear measurement system

considered (temperature, pressure, humidity, flow rate, motion velocity and any other signals to In_1, In_2 and In_3).

The measurement results displayed digitally in physical units (m, kg, etc.) can also be produced as analogue signal forms (current-voltage) through D/A conversion if such need arises.

In the digital nuclear measurement systems presented here the information gathering rate can be limited, for example, by the activity of radiation sources satisfying safety requirements, by the precision needed for the particular measurement and by the operating speed of the detector. The time needed for the specified measurement accuracy is in the order of 1 to 100 s, which is satisfactory from process dynamics aspects but exceeds significantly the signal processing time. In view of this, a single signal processing equipment can serve several measuring stations or points in *multi-channel mode of operation* if the input and output circuits allow it. The multiple utilization ability can, however, present itself as a requirement from the problem side in many cases. In certain technological systems, several identical measurement functions need to be solved simultaneously (for instance, level monitoring or control).

Form the comparison of various views it is concluded that the functions expected from nuclear instruments tend to those requiring *microprocessors*; this implies that in the construction of the equipment for universal nuclear measuring systems the introduction of microprocessors by the vendors is expected. The development of the general industrial measurement and control technology shows also the trend of minicomputerization and microcomputerization (Direct Digital Control, DDC). The need for *nuclear measuring lines connected directly or indirectly through data preparation to minicomputers arises* (Fig. 2.34). In such systems the detector unit contains the power supply and the necessary signal amplification and signal matching stages. For uniform signal processing and multichannel operation the calibration of the measuring lines can be reasonably performed by geometrical adjustment and radiation influencing techniques.

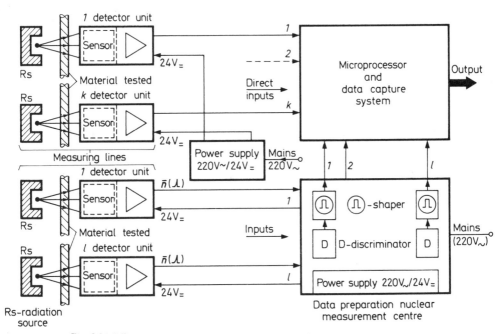

Fig. 2.34. Microprocessor versions for multipoint nuclear measurement systems

Advanced nuclear measurement systems also contain some *auxiliary monitoring elements and devices*. They can include radiation sources for testing, standard absorbing or reflecting elements, electronic devices for checking and dose rate meters. *The electronic checking monitor* is an auxiliary instrument with power supplied either from a battery or from the apparatus tested; it can be a replica of the measurement system except for a pulsed oscillator substituting the measuring line. In the construction shown in Fig. 2.35, the monitor can be used to measure the power supply voltages of the main equipment and the voltage pulse rate from the detector so as to check the operational power of the measuring line. If a failure occurs in the line, the signal processing system can be tested with the aid of the pulsed oscillator in the monitor. The nuclear detectors attached directly to the control equipment can be calibrated as well resulting in uniform output for identical measurement values, by the monitor presented schematically in Fig. 2.35.

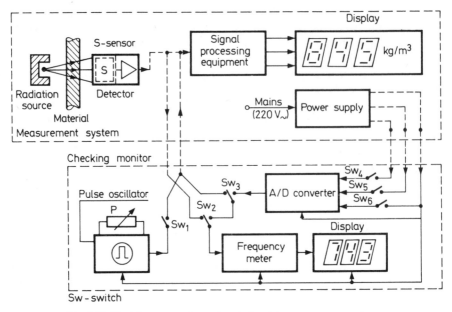

Fig. 2.35. Checking system for nuclear industrial measurement equipment

A separate group within industrial nuclear instruments is the category of *portable* instruments, primarily *soil testers*. Their major requirements are lightweight design, simple operation and battery power supply. With this type of instrument all measurements, to be presented in Section 2.5, can be performed by a single portable control equipment with many different probes. The central unit usually operates in counter mode but it also operates in combination with a rate meter.

2.3.4. STATISTICAL MEASUREMENT ERROR

The concept of statistical error (standard deviation) is treated in Section 1.2. Its degrading effect in industrial measuring systems can be reduced by an appropriate selection of the elements in the measuring line.

The product RC defined by the components of the electronic network is called the effective integration time, t_i, of the *simple averaging* integrating stages. If we introduce the notation $\bar{n}$ for the particle number detected during unit time, an average pulse number $\bar{n}_{inst} = 2t_i\bar{n}$ will be displayed by the instrument. The relative single, double and triple statistical errors (fluctuations) of the mean values are, respectively,

$$\sigma_{rel} = \frac{100}{\sqrt{2t_i n}}, \quad \frac{200}{\sqrt{2t_i n}}, \quad \frac{300}{\sqrt{2t_i n}}\% \qquad (2.11)$$

For simple instrument designs, the double standard deviation on 95.4% of the measurement results, for higher accuracy applications the triple standard deviation on 99.7% of the results should be taken into consideration.

In addition to statistical errors, other types of errors expressed in σ_x relative values may arise, depending on equipment data and on changes in measurement conditions. Their effects can be combined with those of the standard deviation by

$$\sigma_{tot} = \sqrt{\sigma_{rel}^2 + \sigma_x^2} \tag{2.12}$$

The statistical error can arbitrarily be reduced by *increasing the radiation source activity and the detector sensitivity*, by an optimal *measurement station configuration* and by choosing an adequately high *time constant*. In advanced equipment, one third of the total measurement error is the statistical error. Therefore, for fast acting nuclear measuring devices (i.e., assuming a low time constant), relatively high activity radiation sources, in the order of 100 GBq (several Ci) should be employed.

It should be noted here that precise and detailed determinations for the concepts and definitions used in Sections 2.1, 2.2 and 2.3 can be found in the relevant publications of the *International Electrotechnical Commission*. (For example: Publication No 259; Expression of the Functional Performances of Electronic Measuring Equipment; Draft 45–73: General recommendations concerning electrical measuring instruments utilizing radioactive sources.)

2.4. DETERMINATION OF PHYSICAL MATERIAL CHARACTERISTICS BY NUCLEAR MEASUREMENTS

In industry a number of physical parameters are used to characterize material flow and product properties in technological processes, almost any of which can be followed by suitably chosen nuclear measurement configurations. However, usually but a few physical material parameters (level height, bulk density, thickness, moisture, quantity) are determined by isotope techniques because in many other cases conventional measurement methods are preferred (as for instance, in temperature measurements, length determination).

2.4.1. LEVEL HEIGHT DETERMINATION

Level height is an important feature in several pieces of technological apparatus, in material storing or transportation systems, both from process control, either automatic or manual, and from quantity monitoring considerations. For monitoring the material level in a given system, level height indicators and level measuring devices are used. *The level indicator* is suitable for sensing and remotely indicating a specified extreme value (minimum, maximum, etc.). The *level measuring device* can be applied through continuous level height monitoring for measurement and recording as well as for level height telemetering. Both instrument types can be used as *control elements*, typically as process sensors.

Even though reliable level monitoring is extremely important under industrial conditions, *universal techniques for any types of applications* could not have been developed because, due to the diversity in the properties of stored or transported materials (state, physical conditions, temperature, etc.) on one hand, and those of storage vessels, tanks and transportation means on the other, the requirements can only be satisfied by *significantly different equipment constructions.* A wide variety of measurement principles used so far in industrial electronics have already been utilized for the construction of level indicators and level gauges. If the various methods applied in technology are compared with each other and with the isotope technique only the latter can be considered as the *most universal* one because, assuming the choice of devices needed for realization, only nuclear measurements can satisfy all of the following requirements:

— *discrete value indication or continuous level monitoring* by the same measurement principle;

— applicability in *closed and open* systems;

— level monitoring of material in *different states* (solid, liquid, particulate, etc.);

— resistance to *aggressive properties of the measured material;*

— negligibility of *environmental influences* (temperature, pressure, humidity variations and supply voltage fluctuations);

— transmission of *proportional electrical output* for control purposes.

"Conventional" level measurements can possibly be realized in a simpler, more accurate, easier and, perhaps, in a more economical way than those by nuclear techniques if but some of the requirements listed are to be met. However, isotopic techniques should be preferred for problems where this more expensive method is safer from operational aspects and provides more reliable information than its conventional counterparts. Taking this into account, nuclear techniques have found wide applications for level monitoring problems under *hard operational conditions* like in mines, raw material transportation, ore processing, chemical technology, coal supply for power stations and others.

Isotope level sensors contain a shielded *radioactive source* as a signal source (transmitter) and a *detector* converting radioactive radiation to electrical pulses (receiver). The basic requirement of radioisotope level monitoring is the implementation of measurement configuration capable either of provoking *abrupt changes or gradual variations* (level indication and level measurement, respectively) in the radioactive radiation interacting with the measured material when the level has been altered.

2.4.1.1. LEVEL HEIGHT (EXTREME VALUE) INDICATION

The basic version of the *isotope level switch (γ-relay)* measurement arrangement is presented in Fig. 2.36. These measurement systems, called absorption types after the particular interaction process involved, offer an indication of appearance or disappearance of material level in the measuring line defined by the geometrical positions of radiation source and detector. The advantage of nuclear sensing in this case is that the indication depends only on *relative absorption variations* caused by material level changes. The influence of tank walls in the radiation path presenting a steady absorption can be compensated by increasing the radiation source activity.

The sensing system is coupled with the material in the monitored vessel via radiation thus the sensing is physically *contactless*. Figure 2.36 shows both a *total transradiation* (using external source) measuring station arrangement and an *immersed source* version used mainly for very large tanks.

The type and activity of *radiation sources* is determined by the measurement function. For isotope level gauges, usually γ-rays from sealed ^{60}Co or ^{137}Cs radiation sources are employed. Their activity values in industrial applications range from about 40 MBq to 4 GBq (1 to 100 mCi). For measurement of small bulk density substances (like foams) or

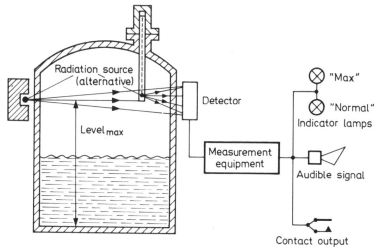

Fig. 2.36. Schematic principle of nuclear level indication

with a small quantity of material available for measurement (as in the case of dental paste quantity indication in the tube), the use of *β-rays* is recommended. As a *β-radiation* source, typically ^{90}Sr isotope with an activity range from 40 to 400 MBq (1 to 10 mCi) is used. Level indicators for high hydrogen-content substances such as petroleum products are activated with *neutron* radiation sources, as a consequence of the fact that in some cases the specific neutron moderation characteristics of hydrogen make the measurement more selective than that using *γ*-radiation. The source used is ^{241}Am–Be, with activities in the range of about 1 to 10 GBq (30 to 300 mCi).

The detector type used more frequently is the Geiger–Müller-counter but also scintillator counters with higher sensitivities than the former and ionization chambers with virtually unlimited life are applied as well. For level indication, the use of semiconductor detectors has been rapidly growing, too. The power levels of electronic signals are low, therefore they are inappropriate for driving instruments or trigger switching relays directly. In engineering practice, *analogue or digital electronic equipment* types are used *to amplify and convert* electronic signals from the detectors. Their functions are to determine the variation of level from stepwise or continuous change of radiation and to produce an output in any desired form like a pointer instrument deviation, a light or sound indication, contact configuration change and so on.

The block scheme of a level indicator is shown in Fig. 2.37. According to the most extensively used solution in industrial practice, the pulses from a GM-counter trigger a signal shaping circuit through an amplifier or an impedance inverter, producing pulses of identical amplitudes and defined widths. The shaping stage is followed by an integrating stage providing an output signal with a voltage proportional to the number of pulses, to drive a flip-flop circuit. The contacts of an electromechanical relay in the flip-flop can be coupled either to light or sound indicators or to control elements. The power supply provides d.c. power and stabilization for the equipment to avoid disturbances by the voltage fluctuation of the mains.

The operational diagram is given in the coordinate system of Fig. 2.38. The pulse number corresponding to the dose rate at the detector site is converted to a proportional

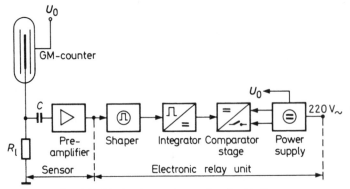

Fig. 2.37. Block scheme of an nuclear level indicator

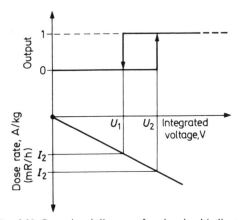

Fig. 2.38. Operational diagram of nuclear level indicators

d.c. voltage by the integrating circuit. The level indicator output, up to the integrated d.c. voltage U_2 corresponds to the dose rate I_2, 0. At this critical point the relay will be switched over and its output turned to the state designated by logical 1. Switching back takes place at a lower U_1 voltage and I_1 dose rate, due to the *hysteresis* effect of the electromechanical system. The hysteresis is an inherent property of every piece of equipment and can be decreased or increased electronically to an arbitrary extent. Though, for accuracy of level indication by relay switches a lower hysteresis would be preferable, it is a necessary effect, too, because of statistical fluctuations in radioactive radiation. If there were no hysteresis, the device would be switched perpetually from one state to the other, as a consequence of statistical fluctuation. In properly designed equipment, U_2 or a dU fluctuation belonging to U_1 or U_2 would never reach a value of U_2-U_1, even in extreme cases, so the device could not be triggered by statistical fluctuation.

The transition state, that is, if the detector is exposed to an intermediate radiation level between I_1 and I_2 which can last a fairly long time due to technological reasons (tank refill is slow), this state is the most critical situation from the point of view of *erroneous readings*. Under such circumstances a "*stammering*" of the relay occurs frequently, i.e., switching on and back for positive and negative fluctuations, respectively. This error is less pronounced in equipment with higher hysteresis and using radiation sources with collimated beams.

A *complete measurement arrangement* is characterized by the following parameters:

— *sensitivity*: minimum dose rate measured at the detector site needed for the reliable switching of a built-in electromechanical relay (e.g., 0.7 to 3.5×10^{11} A/kg = 0.1 to 0.5 mR/h);

— *hysteresis*: dose rate ratio required for the switching over and back of the level indicator device (e.g., 3 : 1, 4 : 1);

— *time constant*: the time constant of the RC integrating circuit in the level indicator; its value is related to the sensitivity: due to the statistical character of radiation, a high time constant, e.g., between 2 and 30 s, is needed to achieve a high sensitivity (to avoid erroneous switching).

Other important parameters are data on climatic resistance, on power fluctuation toleration and on output signals which are characteristic of parameters of other kinds of industrial electronic equipment, too. In certain industrial environments, including mines, equipment with explosion-proof housing should be used. The wide application range and rapid installation of equipment can be achieved by a *complete set of accessories*. Major accessories are the various source containers, detectors, indicator units and dedicated monitoring instruments.

The starting point in the *systems engineering assessment of measuring stations* is that both the multipositional level monitoring and the simple level indication functions occur usually in the forms of multiple, similar requirements, even within the same plant. This implies a technological and economic justification of *multipoint, multichannel level monitoring equipment*. From the instrumentation point of view, the same conclusion can be drawn because a centralized measurement system has several merits compared with a system with a number of individual devices, viz.

— elements of the measuring system can be *combined*;

— combined units can be constructed to have *better performance*;

— *stand-by elements* can be included in case of failures;

— equipment production and installation is *easier*;

— *monitoring systems* can be developed to reduce failure elimination time and for systematic maintenance;

— *equipment maintenance* is simpler.

There are two basic versions for *multipoint measuring systems* in electronic measurement techniques. Also for a combined solution of *n* different measurement problems, one needs *n* radiation sources and *n* detectors. In a *time division multiplex* multichannel system, a synchronized switching device connects sequentially, in a specified order, the sensors to the input of a single central unit and, simultaneously, the relay units indicating the results from each channel and storing them until the next cycle, are coupled to the output. The measurement cycle time per channel should be chosen so as to ensure unambiguous determination of the level state or its shift in the measuring line. The drawbacks of the system with a high number of channels are, from measurement technological and instrumental aspects, the time delay for rapid changes and the insufficient reliability of a simple synchronous switching device or, for higher reliability, it is expensive. In addition, interfacing single channels of the system to different measurement functions is difficult or impossible.

Parallel multichannel systems can be characterized by parallel operations of stages for independent measurement functions in each channel. In such a system, the auxiliary or peripheral units and the mechanical framework can be combined. The system design allows individual adjustments and applications of each channel, rendering possible their most flexible matching to industrial requirements.

Multichannel equipment types are especially suited for *centralized (coordinated) control of technological processes*. A typical application is the multipositional fullness indication of large vessels designed for storing several hundred cubic meters of loose material. The use of a single measuring line for detecting changes of material quantity in high storage-capacity vessels is not reliable, due to different ramp inclinations, different

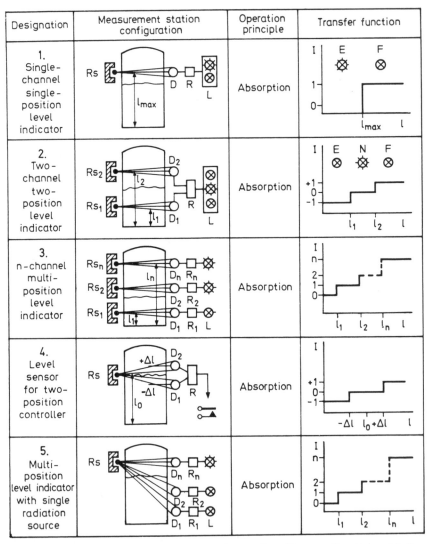

Fig. 2.39. **Basic versions of nuclear absorption level indicators. D — Detector; R — Relay unit; L — Lamp indication; Rs — Radiation source; N — Normal; E — Empty; I — Output information; F — Full; 1 — Level**

downflow, replenishment on many inlets, material sticking, etc., making the information inadequate for actual level state determination.

Relay operated instruments are mostly used for level indication problems. These can be single-, double- or multipositional (quasicontinuous) systems depending on flow, storage or processing conditions of the material. Control signals for two-position switches are generated by a single indicator or several ones. Measurement station configurations, operational principles and transfer functions of problems occurring most frequently in technology are tabulated in Fig. 2.39. *The γ-ray absorption measurement arrangements* presented are used for extreme-level indication of material undergoing processes in closed or open storage systems or tanks, for technological or safety purposes (e.g., filling of propane-butane gas tanks, level indication of chlorine storage tanks).

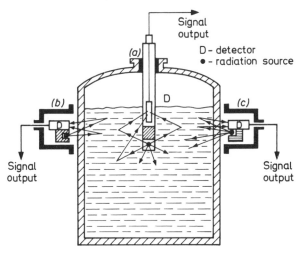

Fig. 2.40. Level sensing with special measuring line arrangements. *(a)* γ-ray reflection intruding probe version; *(b)* γ-ray reflection surface probe version; *(c)* neutron moderation version for hydrogen containing media

Measurement station versions for level indication with special measuring line arrangements are presented in Fig. 2.40. Versions *(a)* and *(b)* respectively, use intruding and surface probes fastened on the tank surface operating by *γ-ray reflection*. If no material appears in the sensing zone of the probes, the reflection is low, due to interaction restricted to the protective cover of the probe or to the tank wall. With rising material level, the reflection increases significantly, an electronic relay connected to the probes indicates the level-height extreme. The level indication accuracy of γ-ray reflection gauges is lower compared with absorption gauges but in many instances installation of the former type is easier. Version *(c)* is the *neutron moderation* level indicator mentioned earlier used for substances with high hydrogen content (petroleum products).

In addition to storage vessel level indication, relay operated nuclear measurement systems can be used for several other performance monitoring purposes, extreme-value indication or simple control purposes. Loaded and unloaded *conveyor belts* can be distinguished by a γ-relay and a suitable absorption-type measuring line. The apparatus makes the unloaded belt stop for power saving. With an absorption measuring line, material jamming at reload points can be sensed and indicated.

Charge level height in melting furnaces should be kept at a steady value for uniform operation. As the level moves downward corresponding to the charge reload increment, refeed is needed. Instead of a visual inspection for judging the amount, a γ-ray level gauge can be used. Transmitted radiation below the charge-level maximum corresponds to the feeder vessel volume. When the sensing zone becomes unblocked with diminishing charge level the γ-relay instantaneously triggers the control of feeder equipment to discharge the content of the feeder into the furnace. The charging is stopped when the radiation path is blocked again. Obviously, here the γ-relay controls only the feeder equipment; forwarding, discharging, returning and reloading of the vessel is controlled by its own control unit.

A three-positional level indicator of *high-capacity storage bunkers for lumpy solid material* is shown in Fig. 2.41. The full load—when charging has to be stopped—is sensed by the measuring line D_1–Rs_1; the line D_2–Rs_2 induces a signal at an intermediate state when charging can be restarted due to free storage capacity in the bunker. The line D_3–Rs_3 built into the walls indicates level height minimum when discharging should be stopped or reloading should be started immediately.

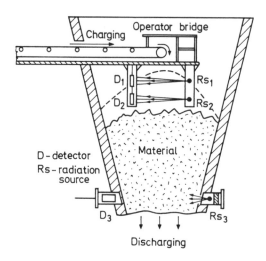

Fig. 2.41. Three-positional level indication for high-capacity material storage bunkers

Charging of *shuttle truck transportation systems* from bunkers can be automated by γ-relays (Fig. 2.42). The isotopically or by other means sensed position of an empty truck moved into the charging station induces a signal which starts the loading conveyor belt if the line D_1–Rs_1 indicates the presence of an adequate amount of material in the bunker. The γ-relay situated at a level lower by Δh than the allowable charge level height maximum (D_2–Rs_2 line) stops the loading belt when this height is achieved. The Δh value should be chosen so that the additional charge loaded during the period between the instances of switching and actual belt stop could find room in the car.

The most extensive field of application of β-ray relay systems is *piece counting*. The nuclear radiation characteristics of β-rays allow the counting even of extremely small cases, foils, etc., due to total absorption occurring when the objects block the radiation from the source. Large objects like coal-loaded mine shuttle trucks are counted by γ-rays to distinguish between empty and full-loaded trucks while the β-ray equipment is used for shuttle truck traffic monitoring.

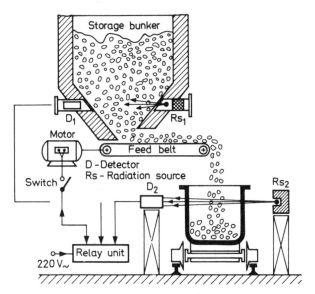

Fig. 2.42. Automatic control of shuttle truck charging by γ-relays

EXAMPLE 2.2

Calculate the γ-ray source activity needed for level height indication using the following data (based on Fig. 2.43):

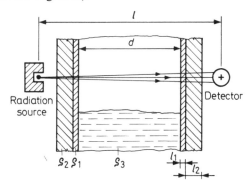

Fig. 2.43. Geometrical and material characteristics of a level indication station necessary for calculating the required radiation source activity

dose rate required at the detector for an empty tank:

$$I_1 = 3 \times 10^{-11} \, \text{A/kg} \, (\approx 0.4 \, \text{mR/h});$$

instrument hysteresis is 4:1, hence $I_2 = 0.75 \times 10^{-11}$ A/kg (0.1 mR/h);
wall thickness of a steel plate tank: $l_1 = 1$ cm $= 10^{-2}$ m;
wall density: $\rho_1 = 7800$ kg/m³;
tank cover thickness: $l_2 = 10$ cm $= 0.1$ m;
cover density: $\rho_2 = 2300$ kg/m³;
inner diameter of tank: $d = 1$ m;

7*

density of aqueous solution stored: $\rho_3 = 1000 \text{ kg/m}^3$;
estimated distance of radiation source and
detector: $l = 1.5$ m.

As a first step in the calculation, the activity value of the source yielding the specified dose rate at the detector site in the empty state should be determined. Next, the expected change in the absorption when the tank is full, should be checked from the instrument hysteresis.

For the solution of the problem, a ^{60}Co γ-radiation source will be chosen taking into account the relatively large source-to-detector separation and the long absorption path.

Data needed for the measurement design are: dose constant of ^{60}Co:

$$K_\gamma = 22.509 \times 10^{-18} \text{ Cm}^2/(\text{kg Bq})$$

$$(1.29 \text{ R} \times \text{m}^2/\text{h} \cdot \text{Ci});$$

mass absorption coefficient of iron:

$$\mu_{m,1} = 519 \times 10^{-5} \text{ m}^2/\text{kg}$$

$$(0.0519 \text{ cm}^2/\text{g});$$

mass absorption coefficient of concrete:

$$\mu_{m,2} = 541 \times 10^{-5} \text{ m}^2/\text{kg}$$

$$(0.0541 \text{ cm}^2/\text{g});$$

mass absorption coefficient of water:

$$\mu_{m,3} = 620 \times 10^{-5} \text{ m}^2/\text{kg}$$

$$(0.0620 \text{ cm}^2/\text{g}).$$

The dose rate at the detector site with free radiation transmission (without tank walls in between) can be obtained from the equation

$$I_0 = K_\gamma \frac{A}{l^2} \text{ A/kg} \tag{2.13}$$

where A is the source activity in Becquerels (Bq).

The relation between the specified dose rate I_1 and I_0 is, due to absorption by substances present, exponential:

$$I_1 = I_0 \exp\left[-(\mu_{m,1} \times \rho_1 \times 2l_1 + \mu_{m,2} \times 2l_2)\right] \quad \text{A/kg} \tag{2.14}$$

With equations (2.13) and (2.14),

$$I_1 = K_\gamma \frac{A}{l^2} \exp\left[-(\mu_{m,1} \times \rho_1 \times 2l_1 + \mu_{m,2} \times \rho_2 \times 2l_2)\right] \quad \text{A/kg} \tag{2.15}$$

is obtained, yielding an activity of

$$A = \frac{I_1 l^2}{K_\gamma} \exp\left(\mu_{m,1} \times \rho_1 \times 2l_1 + \mu_{m,2} \times \rho_2 \times 2l_2\right) \quad \text{Bq} \tag{2.16}$$

Substituting the data given, one obtains

$$A = 728 \text{ MBq}$$

Obviously, the same result can be obtained from the calculation with conventional units:

$$A = 19.7 \text{ mCi}.$$

If the specified dose rate of 3×10^{-11} A/kg (0.42 mR/h) does not belong to the relay activation in the most sensitive range of the level indicator then, in practice, a radiation source with an activity of 1000 MBq (~ 25 mCi) should be chosen instead. In the opposite case, when the sensitivity maximum is equal to the specified dose rate, a radiation source of approximately double activity (1.5 GBq, ≈ 40 mCi) is required to cope with the source decay, in order to have a level indicator with a reliable operational life of about 5 years, taking into account the half life of ^{60}Co sources. The higher dose rates occurring initially could be reduced, for example, by the insertion of a lead plate (see Fig. 2.28).

For comparison, if a ^{137}Cs source preferred because of its longer half life is applied, an activity of about 20 GBq (500–600 mCi) is required for the same problem.

The hysteresis can also be verified by extending the above equations to include the terms ρ_3 and d. A simpler way of doing this is, however, the computation using the half layer thickness.

The γ-radiation dose rate from a ^{60}Co source is reduced to half of its initial value by an about 20 cm thick layer of water-density solution. With an absorption path of 1 metre, the radiation will be halved five times yielding an intensity reduction ratio of $2^5 = 32$. Hence, the ratio of dose rates is $I_1 : I_2 = 32$, a value significantly higher than the specified ratio of $4:1$. It means that the level indicator will work correctly because, with full tank, the dose rate will be reduced to 3% of its initial value which is obviously lower than the 25% limit.

2.4.1.2. CONTINUOUS LEVEL HEIGHT MEASUREMENT

Several nuclear measurement techniques have been developed for continuous level height determination. They are, however, different versions of three basic types:

— absorption measurement methods;
— servo-controlled hunting level measurements;
— γ-ray location profile measurements.

The absorption type level gauges with suitable measurement configurations are used to convert level height changes into proportional absorption variations so that the level of stored material could be obtained from the intensity reading of the instrument attached to the radiation gauge.

The multiple (n position) level indicators can be regarded as an approach to continuous *level monitoring* (see Fig. 2.39, version 3). Normally, of the n gauges applied only those are exposed, in practice, to radiation which have been fixed at positions above the actual material level, therefore the approximate level height can be deduced from their number.

In principle, the most accurate level measurement can be implemented by an array consisting of an *infinite number of level switches*. However, in a real system, a single linear source made of, for example, a ^{60}Co wire with a specific activity between 4 and 40 MBq (0.1 and 1 mCi/(cm) substituting $n = \infty$ radiation sources and a linear array of gauges for the equal number of detectors is applied (see Fig. 2.39, version 3). The material to be measured (i.e., the tank) is between the *vertically assembled source and the sensor array*. In the absence of material in the tank and with full charge in it, a radiation dose rate of I_1 and I_2 is detected, respectively. If the level of the monitored material is at a height of h_x of the total height h in the system and assuming a proportional screening of the radiation source, we have

$$h_x = h\frac{I_1 - I_x}{I_1 - I_2} \tag{2.17}$$

where I_x is the dose rate measured.

The dose rate I_2 can be made arbitrarily small either by nuclear or by electronic methods relative to I_1, hence

$$h_x = h\frac{I_1 - I_x}{I_1} = h\left(1 - \frac{I_x}{I_1}\right) \tag{2.18}$$

This system has proved to be suitable for industrial purposes up to 50–100 cm ranges. The linear source can be substituted by several point sources because of radiation scattering effects (Fig. 2.44). For high absorption path lengths even a single point source (the uppermost one) could be adequate, in this case the instrument scale will be nonlinear.

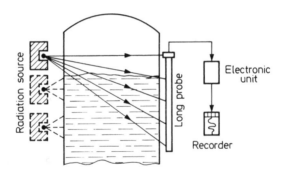

Fig. 2.44. Continuous level monitoring with point-type radiation sources and linear detector

According to the dual version of the measurement scheme (Fig. 2.45), a linear radiation source and a point-type detector is used. In this technically simpler solution a linear source made of a wire with a length equal to the measurement range is placed on one side of the tank, or possibly inside a tube. The detector, a counter tube for lower accuracy or a scintillation detector for higher accuracy requirements, is mounted at the top end of the range.

The variation of the radiation intensity reaching the detector is plotted versus the level height as shown at the right-hand side of Fig. 2.45 (heights are equal to those of levels in the tank, for illustrative reasons, the coordinate system is rotated). According to curve *b*, the relationship deviates from linear towards the lower end of the range. It can be corrected by using an additional point source (curve *a*). The continuous level gauges produced by the *Instrument Factory Laboratory Berthold* (Germany, F. R.) have ^{60}Co wire γ-sources with

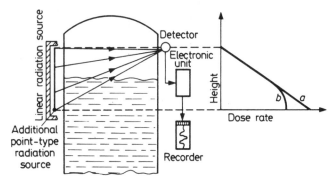

Fig. 2.45. Continuous level monitoring with linear source and point-type detector

downward denser thread pitches coiled around a core rod to achieve complete linearity up to about 3 m ranges. For still higher ranges, e.g., up to 6 m, a source with linear dimension equal to the range and two detectors are applied with one of them placed in the centre of the range. The detector pulses are processed by a common electronic unit. In such equipment a zero shift and a deviation extension (electronic scale correction) is required to provide matches between the ranges and scale calibration of instruments or recorders.

EXAMPLE 2.3.

Let us take $h = 1.5$ m in an absorption measurement system. If the tank is empty, a dose rate meter coupled to a linear detector indicates a voltage of 32 V proportional to the intensity; for a full tank, it reads 2 V. Let the measurement system be compensated by a voltage of 2 V with reverse polarity yielding 0 V for the empty tank. Determine the level position and the measurement accuracy when the output voltage fluctuation is ± 0.5 V for a mean value of $kI_x = 21$ V.

Substituting the known data into Eq. (2.18), we have

$$h_x = h\left(1 - \frac{I_x}{I_1}\right) = 1.5\left(1 - \frac{21}{30}\right) = 1.5\,(1 - 0.7) = 0.45 \text{ m}$$

To calculate the accuracy the heights corresponding to the extreme values (lower, upper limit) around the mean measured value should be determined as well:

$$h_{x1} = 1.5\left(1 - \frac{21.5}{30}\right) = 0.475 \text{ m}$$

$$h_{xu} = 1.5\left(1 - \frac{20.5}{30}\right) = 0.425 \text{ m}$$

The height measurement error is half of the difference of the two extremes:

$$h_x = \frac{h_{x1} - h_{xu}}{2} = \pm\,\frac{0.475 - 0.425}{2} = \pm 0.025 \text{ m} = \pm 2.5 \text{ cm}$$

Hence, the average level position (45 cm) is obtained with an accuracy of ± 2.5 cm or $\pm 5.5\%$.

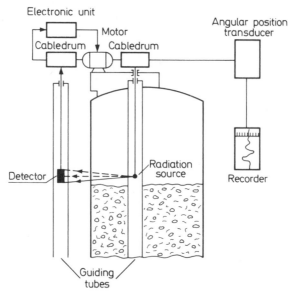

Fig. 2.46. Principle of hunting level gauges

The measurement uncertainty of absorption level gauges is always higher at one end of the range than at the other, because the pulse number referring to the integrating time which determines the accuracy is not constant. The so-called hunting level gauges are free from this drawback and also from the restricted extent of the range.

Hunting level gauges operating with *isotope gauges and servo systems* track the level of material and if found they stop there. The tracking and servo systems are controlled by the nuclear sensing and signal processing; the actual level height can be read locally or remotely from the servoposition. This type of level monitors is characterized by a constant absolute measurement error (± 0.5–2 cm), equal for each point within the range. The level limit depends only on the servomechanism, there are types with ranges up to 40–50 m. Their design principle is shown in Fig. 2.46. The detector and the source are moved synchronously in vertical guiding tubes. If their common position is above the level, the servomotor is driven downwards by the control unit to a point where the detector will be exposed to a specified radiation dose, e.g., half of the maximum. On the other hand, if the sensor submerges in the material, because no radiation is observed by the detector, then the servo motor is driven upwards by electronic control to the point where the specified radiation level is reached. The level height is transferred from the cabledrum to the recorder by an angular position transducer. It follows from the *self-compensating* nature of the measurement that there exists a constant instrumental error depending on the vertical dimensions of the detector as well as on the radiation sensing and signal processing parameters. The level tracking speed is about 5 to 20 cm/min. The system can be operated by the γ-ray reflection principle, too.

If the measurement range is smaller, it is sufficient to *move the detector alone* (Fig. 2.47). If this case exists the detector is exposed to a varying though always high radiation field in the space above the substance level and to practically no radiation below it. The servo-motor drives the detector into a partially immersed position in the monitored material to a point until it detects a preadjusted dose rate. The advantage of this technique is to be able

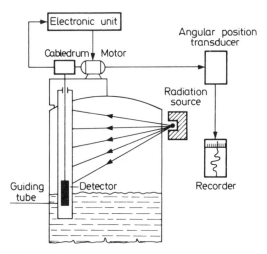

Fig. 2.47. Hunting level monitoring with fixed radiation source

to monitor a number of level heights with one or two fixed radiation sources even for horizontally large tank dimensions.

The USSR-made URMS-2 charge-level hunting monitor with fixed radiation sources is used to measure *charge levels in blast furnaces* as a special dedicated instrument. In addition to charge level measurement it can also be used as a recorder or for feed control. Its range extends from 0 to 5 m with an accuracy of ± 2.5 cm at any point in the range, the level tracking speed is 3 m/min.

The operation principle of the equipment is as follows. Two ^{60}Co radiation sources, of about 20–25 GBq (600 mCi) activity each, are placed at the upper part of the furnace shaft (at the top of the measuring height) at two opposite end points of a selected diameter (Fig. 2.48). The directional radiation from the isotope sources enters the interior of the blast furnace. To guide the detectors, four tubes surrounded by cooling jackets are built into the furnace walls, two of them facing each source. The detectors are introduced into the guiding tubes by a cabledrum under electromechanical control of the tracking mechanism. The signals from the GM-counter are fed into an electronic comparator

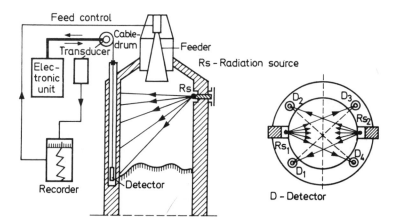

Fig. 2.48. Scheme for incorporating the (USSR) hunting level monitor type URMS–2 into the blast furnace

following integration providing three types of output depending on the relative positions of the detector and the charge level: commands to raise or to lower the detector or standstill. According to the signal received, the relay amplifier controls the hunting mechanism either to raise or to lower the detector until the zero output of the comparator is reached (standstill position).

The self-contained automatic four-channel measuring system with the configuration outlined can track the charge level in the blast furnace. Four measuring channels are needed to cope with the spatially nonuniform distribution of the charge level as a consequence of the large (4 to 6 m) inner diameter. The transition state between total absorption and free radiation is determined solely by the geometrical dimensions of the GM-counter which are negligible compared with the monitored level heights. The "standstill" output of the comparator unit can be set somewhere between the two states inducing a stop when the detector is immersed to about half of its size into the charge.

The position of the cable drum driving the detectors is in all cases proportional to the charge level in the blast furnace. The cabledrums are equipped with induction transducers to transmit the instantaneous positions of the cabledrum, i.e., of the detector, to a remote level recorder in the control room of the plant. The remote level recording system can both indicate an instantaneous situation and monitor the charge level variation in the blast furnace. Under normal furnace and instrumentation operation conditions a saw-tooth shaped diagram appears on the recorder with a period of 5–6 minutes. For automatic feed control a switching system can be included in the equipment which can be assigned to any point within the recorder's range.

The stringent requirements of the URMS-2 equipment are the consequences of experience with continuous blast furnace operation for several years under hard conditions. The detectors are exposed to 600 °C ambient temperature so they need water cooling. Should a failure occur in the water supply, the detectors have to be removed from the furnace walls immediately, otherwise the whole electrical installation together with the cable insulation will burn within minutes. The tracking mechanism at the top of the furnace should be water cooled as well. An additional difficulty is that the tracking system will sometimes be completely covered by flying dust or falling charge material. Also the amplifiers and recorders in the control room work under severe conditions, in dust.

The level measuring techniques described above are not suitable for accurate level determinations in large volume or large cross-section storage systems of solid materials, the level information supplied by them are averages and, normally, restricted to relatively small portions of the material surface area.

The γ-location technique is used for spatial determination of stored lumpy solid material surfaces by γ-ray scattering measurements. If this method is used both the mean value and the profile of the surface along a diameter in simple cases or throughout the whole surface area with more complex equipment can be obtained. The principle for such measurements is presented in Fig. 2.49. A radiation source providing a collimated beam and a direction sensitive detector is placed at the top part of the storage vessel. The directional sensitivity is achieved also by strong collimation. The radiation impinging on the surface is scattered from it in every direction. By changing the angular position of the detector the most intensively reflecting point of the surface can be observed which may be regarded as the point of radiation incidence. Using appropriate servosystems, the radiation source can be moved automatically and the incident radiation can be tracked at the surface by a sensor system applying self-adjustment or optimum finding. A suitable mechanical system plots the surface profile of the solid material at given periods from the

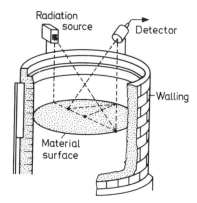

Fig. 2.49. Principle of γ-location (profile measurement) technique

angular coordinates of the source and the detector. This technique has been successfully utilized for level measurements and to achieve stable operation of huge blast furnaces in the volume range of 1000 to 1500 m³.

2.4.2. DENSITY MEASUREMENT

The density of substances is a significant factor in numerous chemical engineering processes, in raw material processing technology, in the building industry, in hydraulic material transport systems, etc. The condition of automation is to carry out continuous, satisfactorily accurate and possibly contactless density measurements. In a number of problems the most convenient techniques are nuclear measurements. Nuclear density measurements are based on the absorption of radiation in matter.

During absorption measurement, the radiation passing through the material is weakened by an extent depending on the elementary composition, thickness and density of the substance in the radiation path, according to the equation

$$I = I_0 \exp(-\mu \times l) \tag{2.19}$$

where I_0 is the dose rate of the incoming radiation in A/kg (mR/h), μ is the linear absorption coefficient in 1/m, and l is the material thickness in m (cm).

Introducing the concept of mass absorption coefficient $\mu_m = \dfrac{\mu}{\rho}$, we have

$$I = I_0 \exp(-\mu_m \times \rho \times l) \tag{2.20}$$

where ρ is the density in kg/m³.

According to Eq. (2.20), the degree of radioactive radiation absorption is determined for a constant material composition by the product of thickness and density. Using a *suitably arranged measuring station,* one can always ensure a constant absorption path length of radiation in the substance so that the measured radiation intensity *depends exponentially only on the material density.*

The problem is usually to *measure continuously* and perhaps to record the density of a substance flowing in a pipe. An example for solving the problem is to apply a radiation source in a lead container at one side of the pipe and a detector at the other (Fig. 2.50). The electronic unit amplifies, integrates and converts the signals from the detector into a form

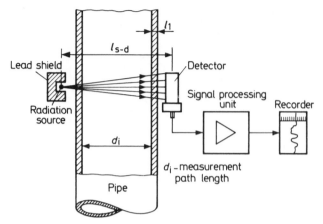

Fig. 2.50. Configuration for absorption density measurement

appropriate for recording. The measured values which are proportional to the density can be recorded, e.g., by a *line plotter*.

The measurement sensitivity depends on the extent of dose rate changes produced by density variations. To determine how strict this relationship is, the density derivative by the measurement path length of Eq. (2.20) is given as

$$\frac{dI}{d\rho} = -\mu_m \times l \times I_0 \times \exp\left(-\mu_m \times \rho \times l\right) \tag{2.21}$$

or, substituting the expression for I from (2.20),

$$\frac{dI}{d\rho} = -\mu_m \times l \times I \tag{2.22}.$$

It follows from Eq. (2.22) that, for a given variation in density, high changes in intensity can be obtained if a *well absorbing radiation* has been chosen, i.e., by using a high μ_m value, a long measurement path length and a high radiation intensity. Obviously, the minus sign indicates a decrease in radiation intensity with increasing density. The value of the mass absorption coefficient μ_m is specified by the isotope used or, rather, by the type and energy of radiation from the isotope source. For density measurements, mostly γ-ray absorption methods are used though β-ray absorption techniques exist, too.

As γ-ray sources, taking into account half-lives, generally the ^{137}Cs isotope sources can be of best use in nuclear density gauges; however, for large diameter pipes, also ^{60}Co sources can be applied while for small (a few cm in diameter) pipes ^{241}Am sources can be applied.

The consequence of increasing the *measurement path length* is an improvement in the accuracy; in practice, path lengths from 300 to 600 mm are normally employed. If the pipe diameter is less than that required by the minimum I value for a reasonable accuracy then care should be taken to provide a suitably high path length for interaction between radiation and matter, either by direction breaking or by an increase in the measurement station dimensions.

The increase of dose rate has its practical constraints. For high resolution detectors, the limiting factors are associated with radiation protection and price. For a GM-counter, the operation is limited by the device life and by the radiation maximum with counting losses taken into consideration (practically 600 to 1000 counts/s if the dose rate at the detector is $3.5-7 \times 10^{-10}$ A/kg or $\approx 5-10$ mR/h).

Compensation techniques are also used for density measurements. The absolute value of density can be observed by finding the zero position of a compensating device. The principle of the *double path* density measurement compensated by radioactive radiation is presented in Fig. 2.51. The source is made to emit radiation in two directions, one to be absorbed by the material tested, the other by a compensating wedge made of a high density substance. The difference in the current intensities corresponding to the two components of radiation detected by ionization chambers induces a voltage drop across the load resistor R. By *wedge position setting*, the current difference and thus the voltage drop can be reduced to zero; hence, for identical measuring vessels or pipes, the wedge adjustment mechanism can be calibrated to absolute density values. The voltage difference across the load resistor is amplified prior to measurement to ensure a more precise zero setting.

With fixed wedge setting in an otherwise similar arrangement, the value read on the instrument scale is proportional to the difference between the actual density and the preset nominal value. The equipment can be used for absolute density readings as well, with a *motor* connected to the amplifier output to drive the compensating wedge in a correct phase for eliminating the difference.

Any of the three principal detector types (ionization chamber, GM-counter, scintillation detector) can be used for isotope density gauges. Bearing in mind the detector resolution and response time, GM-counters are preferred for moderate accuracy density monitoring; ionization chambers and scintillation counters in proper measurement station configurations are suitable for high accuracy density measurements. Moderate accuracy equipment can detect density changes of 1–2%, those detected by high accuracy equipment can be as low as 0.05–0.1%.

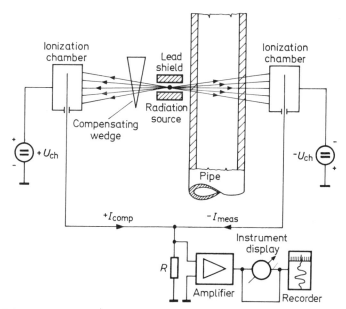

Fig. 2.51. Schematic principle for a density measurement equipment with radiation compensation

A given density measurement problem can be accomplished optimally by a *suitably arranged measuring scheme*. The measuring vessel or pipe section in the measurement station should be fully charged all the time to avoid erroneous results caused by, for instance, air bubbles. Therefore, if possible, an overflow bin should be used with the measuring pipe or vessel or, alternatively, they should be positioned vertically. The bubble formation resulting from possible pump failures should be eliminated to avoid measurement errors and damage to the pump. Material deposits at the measurement station and wear of the measuring pipe can cause disturbances leading to shifts in the set (electrically or radiation compensated) zero position and to *sensitivity changes* due to absorption path length changes. Small deviations can be overcome by recalibration but an extensive deposition (solid precipitation) prevents the measurement completely. Isotope slurry bulk density measurement *during flotation* can effectively be performed using the measuring vessel shown in Fig. 2.52.

High accuracy isotope density measurements have found wide application in the *petroleum industry*. In the pipe lines between the oil refinery or, considering sea transport, between the port and the distribution plant, petroleum product batches of more or less different densities are forwarded successively or alternately which, obviously, should be stored at the receiving station separately. To eliminate interference by air pockets and air bubbles moving in the pipe, it should be irradiated horizontally at the measurement station. For oil density monitoring in large pipes of 800 to 1000 mm in diameter, equipment with ^{60}Co radiation source and scintillation detector has been found most suitable because of the much lower efficiency for γ-ray detection of the other two detector types, in addition to the inadequate resolution of the GM-counter for that purpose, with the consequence of requiring a radiation source with unreasonably high activity.

The alteration of petroleum product batch transported in pipe lines can easily be monitored by density measuring equipment with recorders if a density difference exists between two successive batches (Fig. 2.53). Knowing the speed of the recording paper type and based on the recorded densities, the mass of material transported in unit time, together with the material mixed during transportation, can be calculated.

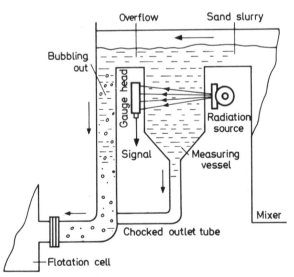

Fig. 2.52. Measuring vessel and measurement configuration for technological density measurement applications

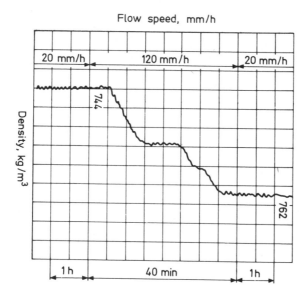

Fig. 2.53. Record from a density gauge used for a petroleum pipe line

Moderate accuracy density gauges are used mainly for density and solid content monitoring of *slurries* handled by slime pumps; such apparatus has proved to be advantageous also for automatic control of suction dredger slime pumps. The accuracy required for slurry bulk density monitors is an order of magnitude lower than that for petroleum product monitoring.

γ-ray absorption density gauges can be applied effectively for process measuring purposes in *chemical engineering*, the building materials industry, etc. Figure 2.54. shows an absorption density measuring scheme equipped with a rod driving mechanism to move the source (^{137}Cs) between its shut and fixed open positions, used for blending vessels in chemical engineering processes. Shut-off is needed for repair and maintenance of the vessel. At the outer wall plane of the tank the dose rate measured by detector is a function of density, because the absorption path length inside the vessel is constant. The instantaneous densities at any time can be determined by density gauge, for recording and control purposes.

For density measuring of material transported by *conveyor belts*, reflection-type density gauges with ^{137}Cs radiation sources and scintillator detectors are recommended. For reflection measurements on conveyor belts either a radiation source for which the layer thickness is infinite or a layer thickness kept constant is needed.

The density gauge can be used as a sensing element in *automatic control loops* (Fig. 2.55). In binary systems, blends with required densities or compositions (e.g., for dilution), can be produced by controlling the feed of one of the components. In the outlined arrangement the output from the density gauge is fed by base value compensation into a control amplifier. Its output controls a regulator valve used as an actuator. The recorder coupled to the equipment monitors the results and effectiveness of the control.

The γ-ray density gauges discussed so far are suitable for measurements and recording in engineering facilities. *Portable instruments* applied in field measurements for *soil density* and *soil compactness* testing will be discussed below, in Section 2.5.

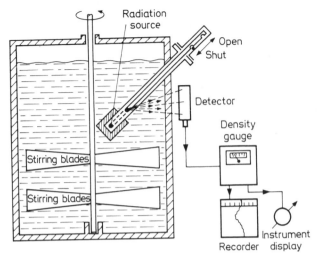

Fig. 2.54. γ-ray absorption density measurement in large blending vessels

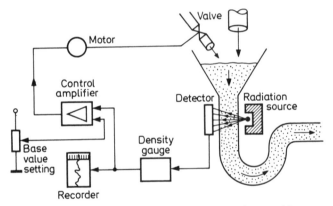

Fig. 2.55. Isotope density gauge applied in automatic control loops

The only extensive application field of *β-ray absorption* density measurement is the continuous monitoring of tobacco trunk density in *cigarette manufacturing machines*. The charge packing can be automatically controlled through the measurement.

EXAMPLE 2.4.

Density gauges for technological applications satisfy their specified accuracies only if two conditions are met. First, the dose rate measured at the detector site should be equal to or slightly higher than a predetermined value (I_{min}) and, second, the numerical value of radiation attenuation by the material tested (ΔI_1) should be lower than a specified value (ΔI_{max}), that is, $S \to \Delta I > S_1 \to \Delta I_1$, $\Delta I_{max} > \Delta I_1$. If S_1 is extremely low then, according to Eq. (2.21), the sensitivity becomes lower than required.

Let the specified dose rate and radiation values be I_{min} and $\Delta I_1 < 1/4$, respectively, in the scheme shown by Fig. 2.49, as a basis for source activity calculations.

The dose rate at the detector location supplied by a source of an activity A and a dose constant K_γ, with a source-to-detector distance l_{s-d}, is

$$I_0 = K_\gamma \frac{A}{l^2}. \tag{2.13}$$

The inner diameter of the measuring pipe defining the measurement path length is d_i its outer diameter is d_0. In our case, the measurement path length for a substance-filled pipe is equal to d_i. The mean bulk density of the substance investigated is ρ_1 and its mass absorption coefficient is $\mu_{m,2}$. The bulk density and the mass absorption coefficient of the pipe wall are ρ_2 and $\mu_{m,2}$, respectively. The numerical values of $\mu_{m,1}$ and $\mu_{m,2}$ can be obtained from nuclear data manuals (or Fig. Appendix A/5) as functions of γ-ray energies of the particular source (^{137}Cs, for the present case). The effects of the material in the measuring line and the tube walls on the radiation are considered by the exponential equations

$$I_{min} = I_0 \exp\left[-\mu_{m,1} \times \rho_1 \times d_i - \mu_{m,2} \times \rho_2 \times (d_0 - d_i)\right] \tag{2.23}$$

$$I_{min} = K_\gamma \frac{A}{l_{s-d}^2} \exp\left[-\mu_{m,1} \times \rho_1 \times d_i - \mu_{m,2} \times \rho_2 \times (d_0 - d_i)\right] \tag{2.24}$$

$$I_{min} = K_\gamma \frac{A}{l_{s-d}^2} \times \Delta I_1 \times \Delta I_2 \tag{2.25}$$

with $\Delta I_1 = \exp\left(-\mu_{m,1} \times \rho_1 \times d_i\right)$ \hfill (2.26)

and $\Delta I_2 = \exp\left[-\mu_{m,2} \times \rho_2 \times (d_0 - d_i)\right]$ \hfill (2.27)

During rating, factors ΔI_1 and ΔI_2 or the substance flowing in the available pipe and for the walls are determined, respectively, then the activity of the required radiation source, according to the condition for I_{min}, will be calculated. If the $\Delta I_{max} > \Delta I_1$ condition cannot be satisfied, then the measurement path length should be increased, using a special measurement station configuration, because the required ΔI value can be provided by increasing the absorption path length.

In the present case, the density of a salt solution circulating in a steel pipe with inner and outer diameters of $d_i = 0.2$ m and $d_0 = 0.22$ m, respectively, should be determined continuously. The accuracy of the measurement system is satisfactory if the radiation dose rate at the detector site is $I_{min} \geq 3.6 \times 10^{-10}$ A/kg (~ 5 mR/h) and the radiation attenuation in the measured substance (ΔI_1) is numerically less than $\Delta I_{max} = 0.25$.

Due to the longer isotope half-life and to the higher absorption, the system is operated with a ^{137}Cs radiation source ($K_\gamma = 6.59 \times 10^{-19}$ A m^2/(kg Bq) or 0.34 mR $\times$ m^2/(h mCi). The mass absorption coefficient of the salt solution with a density of $\rho_1 = 1200$ kg/m^3 (1.2 g/cm^3) is $\mu_{m,1} = 9 \times 10^{-3}$ m^2/kg (0.009 cm^2/g), for ^{137}Cs γ-radiation.

With the data given, the value of the coefficient ΔI_1 will be

$$\Delta I_1 = \exp\left(-9 \times 10^{-3} \times 1200 \times 0.2\right) = 0.115 \tag{2.26}$$

which is lower than the specified value of 0.25, implying that the measurement is feasible using a ^{137}Cs source. It is noted, however, that the real attenuation value of radiation differs from the calculated theoretical result, due to the strong absorption dependence on the measurement configuration.

Considering the theoretical values alone, the pipe diameter minimum can also be determined by reversing Eq. (2.26):

$$\Delta I_{max} = 0.25 = \exp(-9 \times 10^{-3} \times 1200 \times d_{i,\,min})$$

$$\exp 10.8 \times d_{i,\,min} = 4$$

$$d_{i,\,min} = \frac{\ln 4}{10.8} = 0.128 \text{ m}$$

The source activity can be obtained from Eq. (2.24):

$$A = \frac{I_{min} \times I_{s-d}^2}{K_\gamma} \exp[\mu_{m,\,1} \times \rho_1 \times d_i + \mu_{m,\,2} \times \rho_2 (d_0 - d_i)]$$

The distance between detector and source is chosen to be $I_{s-d} = 40$ cm $= 0.4$ m taking into account the gaps for container and mounting. The ρ_2 density and $\mu_{m,\,2}$ mass absorption coefficient of the measuring pipe walls are 7800 kg/m^3 and 7.5×10^{-3} m^2/kg (0.075 cm^2/g), respectively. Substituting the known numerical values,

$$A = \frac{3.58 \times 10^{-10} \times 0.4^2}{4.6 \times 10^{-19}} \exp[9 \times 10^{-3} \times 1200 \times 0.2 +$$

$$+ 7.5 \times 10^{-3} \times 7800 (0.22 - 0.2)] = 3.50 \text{ GBq} (\sim 96 \text{ mCi})$$

for which, in practice, a ^{137}Cs source activity of 4 GBq (≈ 100 mCi) is applied. In this case, the dose rate at the detector is

$$I_{det} = I_{min} \frac{3.7}{2.43} = 3.6 \times 10^{-10} \times 1.06 \text{ A/kg} = \sim 3.81 \times 10^{-10} \text{ A/kg} (\sim 5.3 \text{ mR/h})$$

a more preferable value than the starting condition. The same results are obtained if the conventional units given in parentheses are used during the calculations.

2.4.3. QUANTITY MEASUREMENT

À frequently occurring problem is that of measuring material flow in transportation systems. The problem can be solved by nuclear means for solid material transportation by belt conveyors, hydraulic or pneumatic (pipe line) technology. The measurement is based upon the radiation absorption dependence on the $\rho \times I$ product according to Eq. (2.20), the product being equal to the mass of material per unit surface area in the radiation path.

The nuclear conveyor belt weigher used for quantity measurement at *belt transporters* consists of a linear radiation source with its length equal to the belt width and a detector array of the same length. In the fork-shaped measuring line arranged perpendicularly to the belt advance direction, the conveyor belt loaded with material moves between the source and the detector. The system does not contain any mechanical parts and there is no contact with the transported material.

The ^{60}Co or ^{137}Cs γ-*source* of the isotopic belt weigher is placed underneath the belt and the detector is positioned above it. The radiation emitted by the source (assumed to be constant in intensity from the measurement aspect) is partly absorbed by the substance transported on the belt. An unloaded belt absorbs approximately 5 to 10% of the radiation and the variation in radiation attenuation by the transported material depends on the instantaneous thickness (or, sometimes, density) of the layer at the measurement point. Therefore, the detected radiation intensity is proportional to the transported material

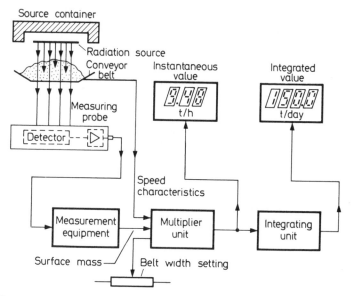

Fig. 2.56. Measurement station configuration and block scheme of an isotopic belt weigher system

thickness at the measurement point or, indirectly, to the instantaneous quantity passing through the measurement lines.

In the isotopic belt weigher shown schematically in Fig. 2.56, the instantaneous quantity value is determined by the "instrument" unit. Its output is fed into the "multiplier unit" where it is multiplied by the motion speed and, simultaneously, the belt width correction is set. The output from the multiplier/unit yields the transported quantity in transportation power (t/h) units which can be integrated over arbitrary time periods to obtain material quantities handled by each production cycle or shift.

The isotopic belt weigher has found extended use *in coal mines and coal-fired power plants* to measure the quantity of produced and consumed fuel, respectively.

The solid material quantity handled by *hydraulic pipe transportation systems* can be obtained by γ-ray absorption density gauges in conjunction with electronic equipment. The output from the nuclear density gauge can be made zero by electrical compensation (zero shift) if there is no solid content in the carrier medium (e.g., water). In order to have a correspondence between the output signal maxima (20 mA or 10 V, for example) and the transportable solid quantity maxima, the instrument scale should be extended or reduced. The output signal of the density monitor is proportional to the instantaneous value of the forwarded solid content if the measuring system contains a linearizing circuit.

Figure 2.57 shows the measurement station configuration and the block scheme of the total system. The carrier media motion speed is measured by an induction-type flow velocity meter to obtain transportation rate and transported material quantity from instantaneous solid content values, through multiplying them by velocity data and, in the latter case, by integrating them over given periods of time. The meter M_1 coupled to density gauge and meter M_2 connected to the electronic unit of the induction flow velocity gauge indicate the instantaneous *solid content* values and speed, respectively. The multiplier unit produces an electrical signal proportional to the transportation rate by multiplying the readings of density gauge and electronic velocity measurement unit; their product will be indicated by the instrument M_3 calibrated in t/h units. The multiplier output is delivered also to the integrators, the one denoted by I for transportation output indication by shifts, the other denoted by II for that by months or decades.

For the determination of the produced solid quantities by *floating suction dredgers* and of the disposed quantities by hydraulic fly ash removal at *power plants*, the apparatus illustrated in Fig. 2.57 is used extensively.

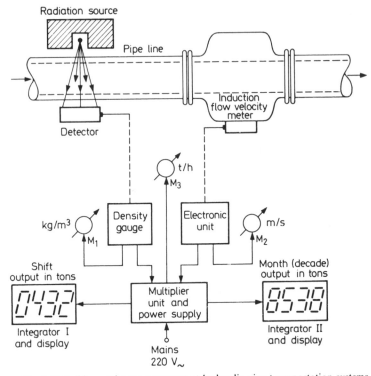

Fig. 2.57. Solid quantity measurement at hydraulic pipe transportation systems

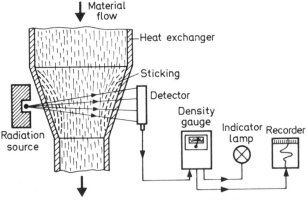

Fig. 2.58. Continuous monitoring of sticking and average quantity of flowing material in a counter-flow-type heat exchanger

In pneumatic pipe-line transportation systems, the solid quantity flowing in the gas (air) carrier can also be obtained by radiation absorption techniques. The instantaneously flowing masses are determined by the instruments based on β-ray and γ-ray absorption for low-mass and high-mass material transportation, respectively. The construction of the equipment is similar to the measuring system presented in Fig. 2.57.

A well-known application field for nuclear quantity measuring devices is the continuous monitoring of gravitationally flowing crude powder quantity in *cement manufacturing heat exchangers* (Fig. 2.58). Due to gravitational material delivery, no velocity measurement is needed, the output of the nuclear density gauge is proportional to the falling solid quantity in the heat exchanger. The continuous material transfer monitoring offers the capability of predicting the tendency of the material to stick on the walls and to estimate the extent of sticking which could have taken place.

2.4.4. THICKNESS MEASUREMENT

The surface mass or, for constant density, the thickness of various materials should be instrumentally monitored during continuous technological processes, in order to comply with quality requirements (tolerance limits). This is motivated also by the need to save *raw materials*. The accuracy of sampling techniques can be improved almost arbitrarily— though due to its batch character, it is not suitable for continuous production control or for automatic control purposes.

Judging by experience, of all known continuous thickness measurement techniques the one using isotopes meets the most general requirements and, consequently, is the most suitable solution in an industrial environment. Concerning the principle of measurement, two basic types of isotope thickness gauge exist: absorption gauges and reflection gauges. In *thickness measurements by absorption*, the source and the detector are positioned at opposite sides of the tested material [Fig. 2.59(a)], whereas in thickness measurements by reflection they are placed at the same side of the specimen [Fig. 2.61(b)].

The fundamental principle of nuclear thickness measurement is expressed by Eqs (2.19) and (2.20) in Section 2.42. Accordingly, provided *the density and the elementary composition of a substance are constant*, a reasonable requirement for most *rolled products* and for those produced by other processes like *casting* or *material deposition* by spreading, then the degree of absorption depends on the *thickness of material* alone. For some substances, e.g., paper, instead of its thickness in mm units, the *surface mass* expressed in g/m^2 units, or the surface mass (mass per unit area) is the significant term because the products are specified commercially by this value ($\Delta I = \rho \times d$).

The reflection arrangement [Fig. 2.59(b)] is favoured in frequent practical situations when access to both sides of the tested substance is impossible. Reflection-type gauges can also be used under the special conditions when the dead time resulting from the great distance between the measurement point and the actuator device needs to be reduced for proper control; reflection sensors can be placed near the point of action, e.g., on the spreading roller mantle. The electronic equipment of the reflection principle is essentially similar to that of the corresponding absorption gauges.

A crucial point in the solution of thickness measurement problems is the correct choice of the particular *radiation source*. The options available, restricted in the first place by short life times, are further reduced by the specified measurement accuracy, due to radiation energy requirements. Satisfactory systematic error can be obtained only by

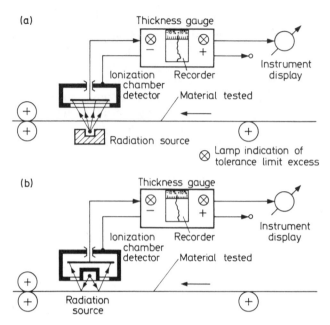

Fig. 2.59. Basic versions of nuclear thickness measurement techniques. (a) absorption arrangement, (b) reflection arrangement

using an optimally absorbing kind of radiation and, consequently, for gradually increasing thicknesses, radiation with increasingly higher penetration power (higher quantum energy) should be applied. The absorption curves (I/I_0) of radiation sources for thickness monitors and the characteristic relations of relative measurement errors $\dfrac{dI/I_0}{dS/S_0}$ versus the surface mass S are summarized in Fig. 2.60, in the order of growing penetration power: the maxima of part (b) of the Figure correspond to the optimum surface weight region measurable by a given radiation source.

α-radiation can be used for the thickness measurement of very thin samples or, in other words, substances in the thickness range of 5 to 50 g/m^2 can be measured only by α-ray emitting isotopes (for instance, ^{210}Po, ^{239}Pu).

Thickness gauges with *β-ray emitting isotopes* (^{85}Kr, ^{90}Sr, ^{147}Pm, ^{204}Tl) can be prepared for the thickness range of 50 to 10 000 g/m^2, corresponding to steel plates with a thickness up to 1.2 mm.

The thickness range of 10^4 to 10^5 g/m^2 can be covered by the absorption *of γ-rays* (from ^{60}Co, ^{137}Cs, ^{241}Am sources) *or of β-ray-induced bremsstrahlung.*

Ionization chambers, scintillation counters or proportional counters can be used as *detectors* for thickness gauges. The most extensively used detector for nuclear thickness measurement is, however, the ionization chamber, due to its infinite life and high resolution.

It is obvious from the basic expression (2.20) of nuclear thickness measurement that the dose rate varies exponentially with the surface mass. To achieve high accuracy and quasilinear operation, the electronic equipment of most thickness monitors uses compensation—enabling electrical and isotope compensation to be distinguished. The

(a)

(b)

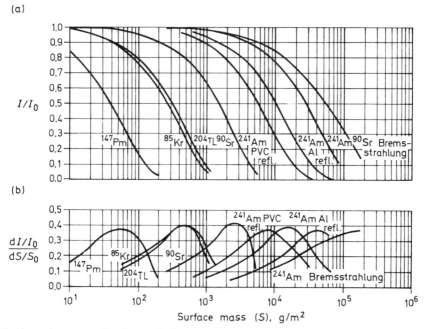

Surface mass (S), g/m²

Fig. 2.60. Absorption curves of various radiation sources (a) and relative measurement sensitivities (b) versus surface mass

operational principles of nuclear thickness gauges are tabulated in Fig. 2.61. It follows from the table that eight versions exist for continuous thickness gauges, two for periodic types. Of the eight versions, four are absorption-type, four are reflection-type gauges. Within each group, both relative and absolute value indication can be accomplished either

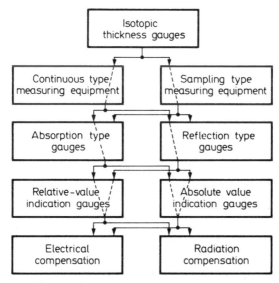

Fig. 2.61. Classification of nuclear thickness gauges by principle of operation

by electrical or isotopic compensation. Because of the similar design of absorption and reflection measurement systems, the operation principles of an absorption-type system will be described.

Thickness gauges utilizing the *compensation* principle and providing results in relative values, *indicate the deviation from a manually adjusted nominal value of the measured parameter*. Both the radiation and the electrical compensation types of equipment with absolute readings control an automatic compensation system by the deviation from the value indicated. The effect of the motor-driven regulation is to eliminate that deviation. The instrument in this case indicates the *position of the compensating system* which is, when in equilibrium, proportional to the absolute value, because the condition for the system standstill is that the compensating dose rate or voltage be equal to the dose rate or voltage proportional to the measured parameter.

The common construction principle of nuclear thickness monitors operating with ionization chamber detectors and providing *relative readings* is shown in Fig. 2.62, both for electrical and radiation compensation. The current from the ionization chamber IC_1 depending on the thickness or surface mass of the specimen flows into the current-summing amplifier A_1. A compensating direct current with opposite sign flowing into the same amplifier and corresponding to the nominal thickness or surface mass is produced either by a helical potentiometer H (*electrical compensation*) or by the irradiation of the ionization chamber IC_2 at appropriate level (*radiation compensation*). If a difference exists between the absolute values of currents flowing into A_1, the amplifier output will be in

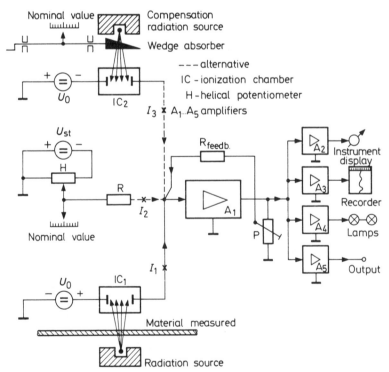

Fig. 2.62. Construction principle of relative-value indication thickness monitors with electrical or radiation compensation using ionization chamber detector

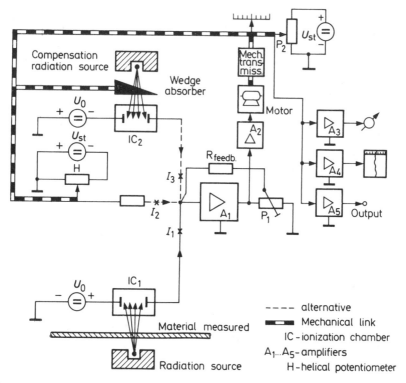

Fig. 2.63. Construction principle of absolute-value indication thickness monitors with electrical or radiation compensation using ionization chamber detector

phase (with positive or negative deviation) with and proportional to the difference from the nominal value. The separating and range-setting amplifiers $A_2 \ldots A_5$ are used for providing a more effective display of the results. In the described measurement system the absolute thickness or surface mass values can be read from the compensating device position calibrated in suitable units, if the deviation indicator meter M reads zero.

The same system in a modified design shown by Fig. 2.63 is suitable for continuous *absolute value* indication. A phase-sensitive motor (M) is connected through amplifier A_2 to the output of current-summing amplifier A_1 with the motor mechanically linked to the compensating system. If a deviation from the instantly indicated value occurs, motor M sets the compensation system back to the neutral position through phase sensing (plus or minus deviation). Here again, the compensation can be electrical (helical potentiometer, H) or isotope (IC_2, Rs_{comp}). The absolute specimen thickness or surface mass is read in either case from the mechanical or geometrical position of the compensating system. Potentiometer, P_2, in conjunction with the mechanical system, can be used for absolute value telemetering. The voltage proportional to the absolute value and generated across potentiometer P_2 is applied through amplifiers $A_3 \ldots A_5$ for remote indication, recording and/or for control purposes.

The major advantage of isotopically compensated thickness monitors is that they are less sensitive to environmental factors (atmospheric pressure, temperature, humidity, etc.). Additionally, the correspondence between specimen thickness or surface mass and the

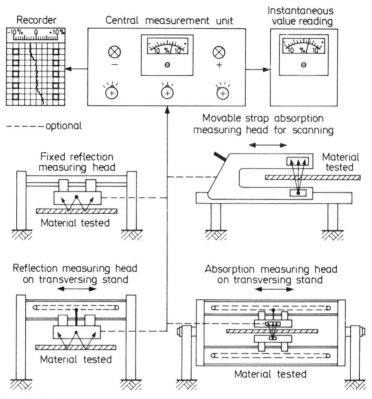

Fig. 2.64. Measurement station versions of nuclear thickness measuring equipment

preset nominal value is not influenced by isotope decay effects (contrary to electrically compensated systems). The isotope decay results, however, in a sensitivity reduction, i.e., errors occurring less indication than those with initial activity of the source.

Thickness measurements for *industrial* purposes can be performed with any types of the above discussed thickness measuring equipment if they are supported by the peripheral units (gauge heads, radiation sources and mechanical framework) required in an industrial environment. Thickness gauges for industrial uses are presented by a system engineering approach in Fig. 2.64. *Universal measuring systems* are equipped with various types of detector and racks. The information gathering line (detector + source) can be arranged to suit both the absorption and the reflection version. The mounting of the measuring line allows either *fixed or transverse regimes*, in the latter it can be moved back and forth, transversely to the direction of advance. The strap versions of absorption-type lines are produced with different intrusion depths.

The second indicator instrument provides measurement information at remote locations. Complex measurement stations (assuming transversing detector systems) can perform profile measurements at preselected time intervals (say, each 15 minutes) with individual interim results displayed independently by a built-in second recorder or stored until the next run. During profile measurement, *the sensor scans* as quickly as possible the specimen thickness across its widths followed by its return to a selected stationary position where the measurement is continued longitudinally.

The most difficult problem of a thickness gauge installation is the proper *measurement station* set-up. It is especially difficult to find a suitable position for incorporating the device in technological apparatus of older construction types, taking into account both radiation protection and instrumentation requirements. Most thickness gauges use β-ray sources, their sealings cannot be oversized mechanically because it may result in no "dose rate ejection". With intrusion into the radiation path, significant doses can be obtained due to the absorption of all elementary particles. The measurement requirements include strictly keeping to the source-to-detector distance, to guide the material to be tested all the time as centrally as possible and to protect the equipment from impurity, dust and dirt deposition. An additional important requirement is—especially with monitors for low surface mass materials—that of maintaining the ambient temperature and humidity in the measurement gap at constant values to avoid measurement errors due to their variation.

Nuclear thickness monitors are used mainly for *paper manufacture*, for *metal sheet and foil rolling* and for *machines manufacturing rubber foil and plastic foil*—though they have found successful applications in contactless continuous surface mass monitoring in the case of *plate glass, textile, fibre slab*, etc., production technologies.

A β-ray isotope thickness monitor and control system utilized at an *artificial leatherboard manufacturing line* is presented in Fig. 2.65. If there is a deviation from the present nominal surface mass value, the control knife position of the spreading roller will be automatically controlled so as to eliminate the error (deviation). The control system manages to keep the nominal value within the measurement accuracy.

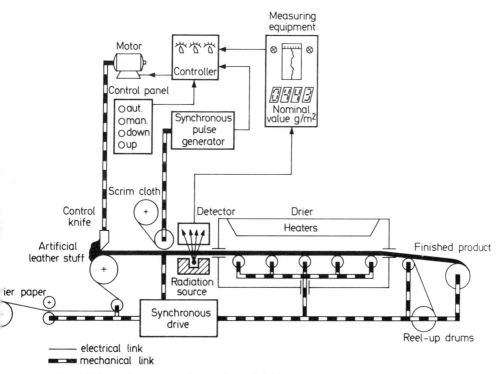

Fig. 2.65. Isotopic control equipment for artificial leatherboard manufacturing line

β-Ray surface mass monitors are utilized for paper sheet manufacturing machines. They are sometimes combined with dielectric moisture gauges using either fixed-strap or transversing type probes, depending on the particular purpose. Similar gauges can be used for rubber or plastic calenders.

To measure metal sheet and foil thicknesses in *cold rolling mills*, mostly β-ray monitors are utilized up to the 5000 g/m^2 square weight range. The hard operating conditions at rolling mills should be considered for the aforementioned appropriate probe selection. Above the mentioned thickness range, mainly bremsstrahlung sources are applied. Recently, the use of ^{241}Am source with a half-life of 458 years has grown rapidly, due to its γ-ray emission in the energy range most suitable for this particular application field. Radiation sources of 10 to 40 GBq (≈ 0.3 to 1 Ci) activities are applied for these measurements.

Figure 2.66 shows the operational principle and actual design of a programmable nuclear thickness measuring and control system which can be mounted on a reversible roll stand. It contains seven individual potentiometers for nominal-value setting (Nv$_1$...Nv$_7$) with a selector switch S$_0$ to preadjust each step according to a rolling schedule. Prior to rolling, the roll throat and the nominal value should be set followed by a draw-in of the particular measurement line corresponding to the rolling direction from the stand-by units (A, B) at the side. When the normal operational speed (revolution) is reached, the control equipment can be turned on by the switch S$_4$ which provides for roll throat correction if necessary. When the deviation indicator of the central measuring unit indicates an error relative to the preset nominal (basic) value (error voltage generation) then the throat will be opened or closed to eliminate the error.

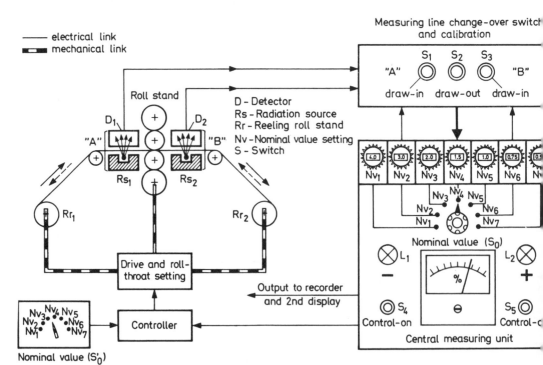

Fig. 2.66. Thickness monitoring and control on reversible roll stand

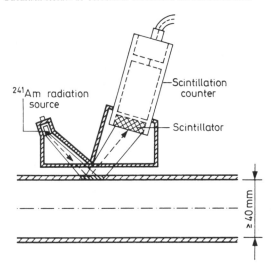

Fig. 2.67. Tube-wall thickness measurement by reflection principle

The extremely hard working conditions at *steel plate hot rolling* plants make the requirements for γ-ray thickness monitors very strict. High-temperature protection of measuring components is accomplished by utilizing the source and the detector at suitable distances. Radiation sources of activities as high as 300 to 600 GBq (ca. 10 to 20 Ci) are located usually beneath the sheets tested while the detector above them in an independent mechanical structure to avoid vibration and shock effects by rolling mills.

The thickness of *plate glass* can be measured by the γ-rays from ^{241}Am. The hard manufacturing conditions mean that the thickness determination of plate glass presents difficulties similar to those of hot-rolled iron plate measurement: the detector works in a high-temperature environment and protection against mechanical damage by glass fragments should be provided for in case of fracturing. The ambient temperature when drawing vertical glass plates is around 150 °C at which the ionization chamber does not need cooling; however, for horizontal glass plate drawing, measurements at temperatures as high as 600 °C should be carried out only where water-cooled detectors can be employed.

The wall thickness of steel pressure bottles can be measured by γ-ray absorption or reflection techniques. Even though, in practice, direct radiation transmission methods have also been applied, their limited accuracies makes them ineffective because during measurement the radiation passes through the bottle walls twice resulting in cumulative errors (deviations of identical magnitudes but opposite signs apparently compensate each other. Wall thicknesses of tubes with diameters greater than 40 mm can be measured, according to the scheme in Fig. 2.67, also by the reflection technique. Provisions for sensing reflections from the tube wall next to the detector are only made by an appropriate measurement configuration. The measurement accuracy for a ^{241}Am source of 10 GBq (200 mCi) activity is approximately 1%, at a time constant of 1 s.

Portable reflection monitors for measuring wall thicknesses of steel plate tanks or vessels will be discussed below, in Section 2.5.

2.4.5. COATING THICKNESS MEASUREMENT

There are several *surface protection* and *surface refining* technologies in industrial manufacturing processes. Due to the growing use of protective coating techniques, an increasing interest for objective layer thickness measurement arises. Such measurements are important both from economical reasons (raw material requirement) and from quality control point of view.

Coating thicknesses can be determined by destructive or nondestructive techniques. The accuracy of measurement for the former is higher than for the latter, though destructive thickness measurements do not yield results immediately because of the necessary preparation of the samples. Film thicknesses can be measured with isotopes in various ways (XRF, activation, scattering, etc.). In this chapter only the most extensively used β-ray reflection techniques for coating thickness determination will be discussed.

During β-ray scattering or reflection, the intensity of scattered particles for homogeneous material composition is *proportional to the thickness,* up to a saturation material thickness; it is the basic law of reflection thickness measurement. The saturation thickness of solids is a few mm or, in other terms, lies in the surface mass region from 500 to 1500 g/m^2, depending on the primary radiation energy and on the material properties.

The β-radiation reflectivity of substances thicker than the saturation value depends on the atomic number $\mathscr{Z}$ of the element (Fig. 2.68). This relationship is expressed mathematically by

$$I_{scat} = k \times I_{direct} \times \mathscr{Z}^{2/3} \tag{2.28}$$

The factor k includes all the "geometrical dependences" of the measurement station and the energy relations of the primary radiation; its numerical value should be determined experimentally in each situation.

The simultaneous *atomic number* and coating thickness dependence of β-ray scattering offers the possibility of thickness measurement of materials composed of different layers.

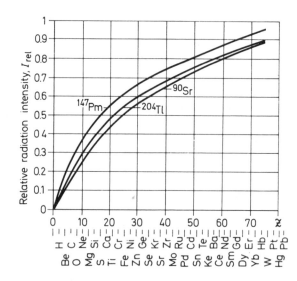

Fig. 2.68. Atomic number dependences of scattered β-ray intensities for various primary energies

The measurement system arrangement at industrial plant laboratory level is shown in Fig. 2.69. The collimated beam from an encapsulated β-ray source incident on a coated substrate will be partially absorbed and partially reflected. The intensity of the reflected fraction for zero coating thickness increases monotonically up to the saturation layer thickness I_∞ of the substrate, above this value the intensity does not vary (Fig. 2.70). Under unaltered measurement conditions, if the thickness of coating, made of a substance with an atomic number differing by at least 2–4 from that of the substrate thicker than L_∞, is increased continuously from its initial zero value, the intensity variation of the scattered radiation will depend on the atomic numbers of the substrate and coating substances. If

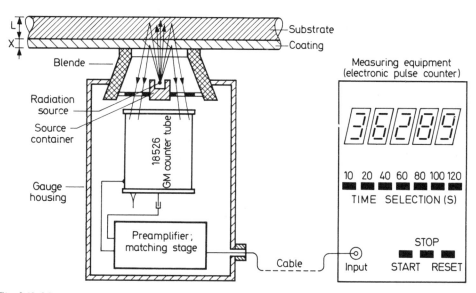

Fig. 2.69. Measurement station configuration and system arrangement for β-ray reflection coating thickness determination

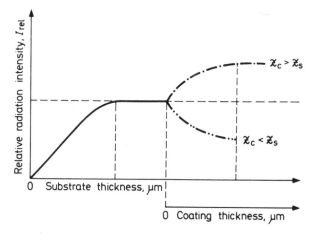

Fig. 2.70. Scattered β-ray intensity versus substrate thickness (———) and coating thickness $(-\cdot-;-\cdots-)$

the atomic number of the coating $(\mathscr{Z}_c)$ with a film thickness X is higher than that of the substrate $(\mathscr{Z}_s)$ then the intensity of the scattered radiation continues to increase with X or, in the opposite case, it decreases, according to the basic equation and to Fig. 2.68. Based on the described relations, coating thicknesses in the range of 0 to 50 μm, with accuracies of 3 to 8%, can be measured using β-ray sources, in accordance with industrial requirements.

The measuring line of β-ray reflection thickness gauges (Fig. 2.71) includes a radiation source, a source container and a detector for scattered radiation measurement. The selection of the *radiation source* is influenced by the required measurement region and the difference in atomic numbers. For thin coating, soft (low-energy) β-ray emitters are needed. If a rather wide region should be covered by a single measuring equipment, two β-ray sources, one with soft, the other with hard radiation, should be employed simultaneously. The radiation source activity varies as a function of the tested area, between 100 kBq (a few μCi) and 0.4 GBq (10 mCi). The most frequently used sources for films thickness measurements are, in the order of increasing energy, the isotopes ^{14}C, ^{147}Pm, ^{204}Tl and ^{90}Sr.

Due to the high efficiency of β-ray measurement, any *detector* type can be used. At industrial laboratories, systems using end-window GM-counters or scintillation counters are the most popular ones, though the semiconductor detector offers a more advanced solution for this problem. In the case of continuous processes, the ionization chamber is advantageous, especially in the *differential regime*. In this case, an uncoated reference sample should be applied in one of the measuring lines and a coated specimen in the other. Using current subtraction directly, the difference current is proportional to the coating thickness implying that a display unit controlled by the output of the electrometer-type measuring system can be calibrated directly in thickness values.

Industrial laboratory test equipment is used for individual measurements with random sampling. For this purpose, time-gated digital electronic counters are considered as the most favourable processing units with pulsed-type detectors. The characteristic feature of such equipment is that a gate circuit is opened up by the frequency-division signals of an internal quartz oscillator for specified periods (say, 10 to 100 s) allowing one to count all pulses detected during these intervals by the counter and to display the results at the completion of each period. The displayed numerical value stays until resetting or until it is substituted by that of the following cycle. A calibration diagram can be used for thickness readings from the displayed pulse numbers.

The most advanced nuclear gauges display directly the actual physical parameter (coating thickness) in absolute units as a consequence of a zero shift, a continuously variable scale extension and in some cases of linearization.

The provision for stable measurement geometry is absolutely important at the different measurement arrays. The β-ray reflection type monitors for coating thickness measurements are provided with precision instrument frames even for industrial plant laboratory applications to maintain a strictly constant sample-to-detector distance and angular position.

β-ray reflection thickness gauges are extensively used especially for copper, gold, silver, tin and zinc coating thickness determination, as well as being used for oxide film, paint layer and plastic coating thickness measurements, provided that the atomic number difference relative to the substrate is at least $\Delta\mathscr{Z} = 2$ to 4.

An example for thickness measurement in technological processes is shown in Fig. 2.71, illustrating the thickness monitoring of Zn-layers deposited by hot-dip coating on steel plates. Because both sides of the plates are coated with zinc, two separate detecting

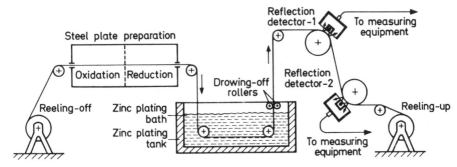

Fig. 2.71. Continuous thickness monitoring of protective layers deposited by hot-dip coating using β-ray reflection techniques

and measuring systems should be applied. If the steel plate substrate is thinner than the infinite backscattering layer, the measurement is performed at the direction-breaking steel cylinder mantles.

2.4.6. MOISTURE CONTENT MEASUREMENT

In certain industries processing moist or wetted substances, the problems associated with control requirements and quality improvement make necessary the knowledge within well-defined limits of the moisture content in processing stuffs, in starting materials or in mixtures. This problem can be solved only by continuous moisture measurement. The major requirements are rapid and continuous measurement systems with possible interconnection capabilities to control equipment.

To satisfy requirements, conventional techniques like conductivity measurement, relative permittivity measurement and mass determination before and after drying are in some respects not suitable. The industrial application of these methods is restricted by the corrosion, contamination and mechanical damage sensitivity of the electrodes, by the strict requirements concerning measurement geometries, and by the relatively small volumes of the materials measured typical of these measurement techniques. For continuous moisture determination during technological processes, nuclear techniques based on neutron stopping have proved to be valuable.

Neutron scattering and moderation can be used for moisture content determination because of the excellent neutron scattering properties of hydrogen atoms in water. Fast neutrons during collisions with hydrogen atoms lose most of their energies resulting in their slowing down. This is so due to the fact that the neutron mass and hydrogen nucleus mass are almost equal thereby leading to nearly inelastic collisions.

On the other hand, if a neutron collides with nuclei of higher atomic number elements, neutron energy moderation becomes relatively small because a practically elastic collision process is involved in that case. Therefore, if, for example, an [241]Am–Be *radiation source* with an activity range of 4 to 10 GBq (100 to 300 mCi) *emitting fast neutrons* is introduced into the media tested and *detectors sensitive to slow neutrons* are applied, then the hydrogen content or, in most cases, from the hydrogen data also the water content of the substance under investigation can be determined, using low-energy (i.e., slow) neutron intensities measured by the detector near the source. The gauge construction and its

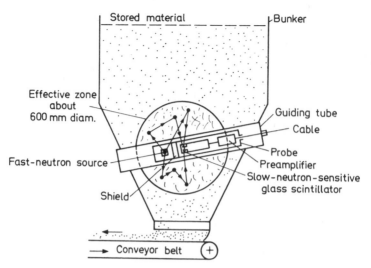

Fig. 2.72. Measurement station configuration for neutron moderation moisture content determination with intrusion probes

arrangement for measurement application is illustrated by Fig. 2.72. Apart from this mentioned, the other neutron sources used in practice are the ^{210}Po–Be, ^{226}Ra–Be and ^{239}Pu–Be systems.

Though neutron moderation is independent of temperature, pressure, *pH*-value, state of material and some other factors, moisture measurement by slowing down of neutrons does, however, have its restrictions. The reason is that, during this measurement, hydrogen content rather than free water content is obtained and, consequently, also *the crystal water and the chemically bound hydrogen in compounds* will be determined simultaneously. This means that reliable measurements can be performed only if the bound hydrogen is low or, at least, of constant composition. An additional disturbing effect is the presence of highly neutron-absorbing substance (B, Cd, etc.) and the variation in the low atomic-number elements ratio (Cl, O, S) which are possibly detectable in the media under investigation (for instance, the effect of a sulphur content higher than 1 percent is equivalent to the moisture content being higher by 0.5%.

Though the moisture content determination is related to the volume, the results, however, are false if the volume mass of the material in the test vessel (filling compactness) varies. Consequently, either the constancy of the volume mass or a continuous monitoring of it should be provided for. The *density* has to be determined if the moisture content is given in weight percentage; density can be measured by nuclear techniques (see Section 2.4.2). *Density compensated moisture measurements* can be carried out either by two distinct radiation sources (a γ-ray and a neutron source) and two corresponding detectors or by two sources again and a single detector sensitive to both radiation types. In the former option two independent measuring systems have to be applied; in the latter one a double-channel signal processing unit capable of distinguishing between γ-quanta-induced pulses and slow-neutron-induced pulses has to be applied.

The ^{241}Am–Be neutron source is used almost exclusively for moisture measurements, due to its high neutron yield and because the accompanying low-energy γ-radiation can easily be handled by shielding. The complementary density measurement is usually

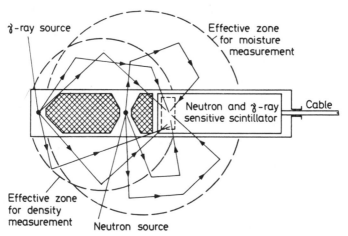

Fig. 2.73. Arrangement principle of a combined probe for density-compensated moisture measurement with a single detector and two radiation sources

performed with ^{137}Cs sources in the activity region of 0.4 to 1 GBq (10 to 30 mCi). For the single-detector option, a scintillation counter can be utilized with the two radiation components separated by a discriminator (Fig. 2.73).

Nuclear moisture gauges are applied mainly during *concrete fabrication, ore dressing, blending of ceramic materials, glass production and foundry sand conditioning* processes as well as during the production of building materials, for fast, continuous, reliable, sampling-free and contactless monitoring at various blend and aggregate concentrations. In their application for *concrete slab manufacturing* the monitors included in the blending process are used to determine the moistures of the "dry" blend components (cement, pebble, river sand) to be able to control the metering of water according to rated water ratios.

Automatic moisture content control can be realized by nuclear moisture gauges, too. Figure 2.74 shows the schematic outline of a continuous moisture measuring and control

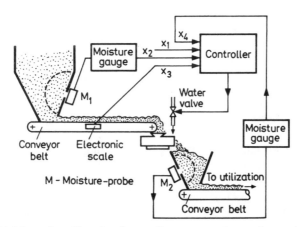

Fig. 2.74. Schematic outline of nuclear moisture measurement and control process

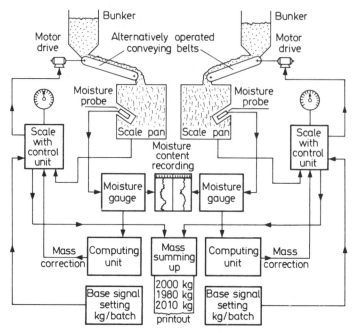

Fig. 2.75. Block scheme of an automatic coke stoker

equipment used in *foundries* or at continuous *concrete mixers*. The ability to maintain the specified water content ratings is provided by a controller with an output controlling water rationing (command signal), according to four input parameters:

— a setting signal (base signal) X_1 corresponding to the specified moisture content;

— a signal X_2 characterizing the moisture content of the basic material, provided by a nuclear measuring system M_1 at the hopper wall;

— a signal X_3, a quantitative property measured by a scale: the control unit produces from X_2 and X_3 the water dose required for adjusting the moisture content corresponding to the base signal X_1 and it sets the water valve correctly via its output signal. The supplied water is distributed uniformly by a stirring machine;

— finally, the moisture of the material to be utilized is measured by a measuring system M_2 to provide an X_4 quantity, generated for continuous self-checking of the controller from the measurement results.

In Fig. 2.75, the schematic diagram of an automatic *twin-arranged alternate-mode* coke *stoker* with a nuclear moisture gauge is presented. Its function is to provide an even *coke* supply for the blast furnace. The coke stoking in metallurgical plants and at blast furnaces should be accomplished so as to provide for identical and specified dry coke content in each batch. To meet these requirements, an automatic coke stoker has been developed using neutron moisture gauges with the probes intruding into the scale pan for moisture sensing. The moisture content detector is connected to a computing unit to obtain a mass correction for the specified base signal of the elctronic scale system by the measured moisture characteristics. The weighing corrected by moisture content ensures identical dry coke contents in each batch. An accounting machine records and sums up the real, accountable coke masses of the individual batches. The correction procedure can be extended also by nuclear techniques to include ash content fluctuation determination as well, to eliminate an effect which may lead to disturbances similar to those resulting from moisture content fluctuations.

2.5. PORTABLE NUCLEAR INSTRUMENTS

Portable nuclear instruments are used for field or for *in situ* measurements for quick surveying, requiring no sophisticated preparatory work. The most common measurement principles involved so far are:
— soil bulk density measurement with depth probes in boreholes;
— surface bulk density measurement by the reflection technique;
— surface bulk density measurement by the absorption technique;
— moisture measurement with depth probes in boreholes;
— surface moisture measurement;
— level detection (level finding by the absorption technique);
— reflection wall thickness measurement (corrosion and deposition testing).
A measurement system includes an appropriate source with radiation protection (shielding), a detector, and electronic equipment for signal evaluation:
— the *radiation source*, depending on the particular measurement problem, can be ^{137}Cs or, occasionally, ^{60}Co isotope *γ-ray sources* (100 to 200 MBq, ∼ 3 to 5 mCi) for density and wall thickness measurements; Pu–Be and Am–Be *neutron sources* for moisture determinations. The *α-ray excitation source* activity lies in the range from 1 to 3 GBq or ∼ 30 to 100 mCi;
— as *detector*, a GM-counter or a scintillator counter can be used;
— the *electronic unit* is either a portable scaler or rate meter (see Section 2.3.3).
The instrumentation for field measurements by the soil testing techniques listed above includes a single item of central equipment with a number of various optional probes. They are designed to comply with the different specified requirements and contain radiation sources in fixed geometrical configurations (Fig. 2.76).

Soil density gauges operate in the *reflection* or the *absorption* mode with γ-ray sources. The so-called density-sensing depth gauges include 0.4 GBq (∼ 10 mCi) activity ^{137}Cs sources, separating lead shields, GM-counter detectors and preamplifiers. The gauge in its container is placed on the top of the borehole and the gauge head is lowered into the hole supported by its cable. The measurement result is a value inversely proportional to the density of soil surrounding the borehole. The quantity of material participating in the interaction process can be considered as infinite in the vicinity of the gauge, implying that the number of reflected particles will be identical in any case. However, the particle number counted by the detector decreases with increasing bulk density, due to particle absorption, hence, an inverse proportionality exists between the density and the obtained intensity.

The structure compactness, for example, damming works and road construction can be determined using *surface compactness gauges* based upon the reflection measurement principle. At one end of the flat gauge is an automatic locking-device-equipped radiation source and at the other end a GM-counter detector. The probe laid on the tested structure monitors the material compactness and density at the surface by a principle similar to that used for hole density measurement with the difference of referring to the measurement hemisphere (2π sr space angle) in the present case. The locking design of the radiation source involves an automatic closure of the source on lifting the gauge head.

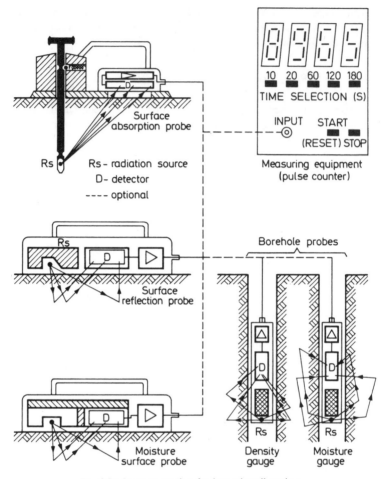

Fig. 2.76. Instrumentation for isotopic soil testing

Using an absorption flat probe, the radiation source has to be introduced into the soil tested to a specified depth. The sensitivity of the absorption measurement along a slant path is usually higher compared with reflection probes, though its drawback is that the obtained value is characteristic but to small material quantities. By introducing the radiation source at different depths, the measurement geometry and thus the soil layers tested can easily be altered.

Soil moisture gauges operate under the principle of the technological moisture gauges discussed in Section 2.4.6. Construction and application schemes of these gauges are also presented in Fig. 2.76. In the given case, a cylindrical depth gauge contains a 1 GBq (~ 30 mCi) activity *fast neutron source* and BF_3-filled slow-neutron-sensitive GM-counter tubes or scintillation detectors with lithium glass scintillators. The construction and method of application is similar to those of the soil density gauge. The moisture content in volume percent is obtained from the values determined by a portable counter (radiation gauge) using a calibration curve.

The moisture content in surface layers can be tested by neutron-source-activated flat gauges. They are used in the moisture concentration range of 0 to 50 volume percent, with a mean error of 2% in order of magnitude.

The accurate knowledge of soil and surface moisture concentration is important both in *civil engineering* for road and railway construction, for damming works and other groundworks as well as in *agriculture*.

Apart from the portable instruments listed so far there are some special instruments for related purposes such as those for absorption measurements between parallel driven boreholes, to be used also for moisture content determination through density measurement knowing when the density data of the dry measurement site is known.

The gauges of the most advanced soil testers use two radiation sources, a dual-sensitivity detector and a two-channel signal processor to perform simultaneous soil density and soil moisture measurement.

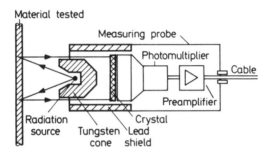

Fig. 2.77. Detector arrangement of the reflection wall thickness measurement gauge

Level finding gauges can be used, for example, for height determination when filling *steel pressure bottles* with liquefied gas. The gauge head geometry of the device is forklike, with the source in one arm and the detector in the other. The radiation source is typically a 40 MBq ($\sim$1 mCi) activity ^{137}Cs isotope, the detector is a GM-counter tube in conjunction with a rate meter. The measuring fork is arranged so that the gauge is movable in the axial direction down the bottle; on reaching the fluid level, the rate meter indicates an abrupt intensity rise. The technique is suitable for determination of the filling height with an accuracy of 1–2 cm.

Reflection thickness measurement is based on the fact that the intensity of scattered γ-rays depends on the thickness of the reflecting surface, provided that any other factors including density, geometry, source activity and radiation energy content do not change during the measurement; the observed average pulse number is higher with increasing thickness. In the detector arrangement according to Fig. 2.77, the direct radiation is shielded by a tungsten cone. The crystal of the scintillation counter senses, with high probability, only scattered radiation. In this way, using suitable equipment in appropriate configuration, *steel plate thicknesses* from 1 to 20 mm can be measured. Thickness determination from one side of the object only (such as wall thickness testing of closed containers or *corrosion* and *deposit thickness* measurement) gives an average thickness across about a 30 mm diameter circular area. The source applied is a 4 MBq ($\sim$0.1 mCi) ^{60}Co isotope or a 20 MBq ($\sim$0.5 mCi) ^{137}Cs isotope. The thickness limit of 18 mm is restricted by the penetration depth maximum, or the total absorption of particles reflected from deep-lying layers.

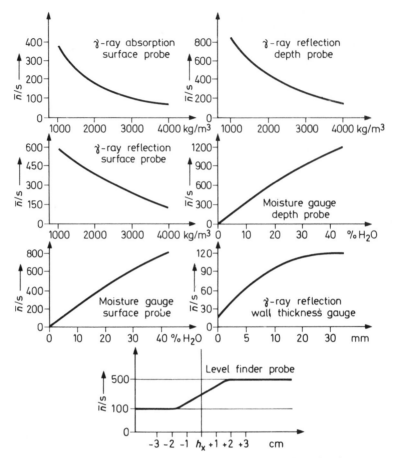

Fig. 2.78. Characteristic calibration curves for probes of portable nuclear instruments

Figure 2.78 summarizes the characteristic calibration curves for various measurement problems. In density measurement applications, the plots are descending for increasing densities even for reflection measurements. During moisture determination, the curves are rising, resulting from the growth in the number of moderated thermal neutrons with increasing hydrogen content. The rising character of the calibration curves for reflection wall thickness determination is due to reasons discussed earlier. For level finding instruments, an abrupt intensity rise is detected around the unknown level height, h_x.

Industrial or field applications of the portable nuclear instruments outlined in this Section have considerable advantages over conventional techniques. These advantages include the possibility of *in situ* observation, short measurement times, etc.

REFERENCES

[2.1] Whitehouse, W. J. and Putman, J. L., *Radioactive Isotopes.* Clarendon Press, Oxford, 1953.

[2.2] Price, W. J., *Nuclear Radiation Detection.* Mc Graw–Hill Book Company, New York, 1958.

[2.3] Kment, V. and Kuhn, A., *Technik des Messens radioaktiver Strahlung.* Akademische Verlagsgesellschaft, Leipzig, 1960.

[2.4] Snell, A. H., *Nuclear Instruments and their Uses*. John Wiley, New York, 1962.

[2.5] Hart, H., *Radioaktive Isotope in der Betriebsmesstechnik*. VEB Verlag Technik, Berlin, 1963.

[2.6] Mengelkamp, B., *Radioisotope in der Mess- und Regelungstechnik*. AEG, Berlin, 1966.

[2.7] *Radioisotope Instruments in Industry and Geophysics. I.A.E.A. Proceedings Series*, I.A.E.A., Vienna, 1966.

[2.8] Rózsa, S., (*Isotopes in Automation*) (In Hungarian). Műszaki Könyvkiadó, Budapest, 1966.

[2.9] Gardner, R. P. and Ely, R. L., *Radioisotope Measurement Applications in Engineering*. Reinhold Publishing Company, London, 1967.

[2.10] Hartmann, W., *Messverfahren unter Anwendung ionisierender Strahlung*. Akademische Verlagsgesellschaft, Leipzig, 1969.

[2.11] Kowalski, E., *Nuclear Electronics*. Springer Verlag, New York, 1970.

[2.12] Mengelkamp, B., *Radiometrie*. AEG, Berlin, 1972.

[2.13] Rózsa, S., (*Nuclear Instruments in Industrial Automation*) (In Hungarian). Műszaki Könyvkiadó, Budapest, 1972.

[2.14] Cameron, J. F. and Clayton, C. C., *Radioisotopes Instruments*. Pergamon Press, London, 1971.

[2.15] Heimann, R., *Radionuklide in der Automatisierungstechnik*. VEB Verlag Technik, Berlin, 1974, 2nd ed.

3. RADIOTRACER TECHNIQUES

3.1. GENERAL CONSIDERATIONS

Tracers can be used to label substances or objects in order to distinguish them, to follow their movement, changes of concentration, distribution between phases, etc. The tracers should allow sensitive and rapid detection, they should not change the properties of the material, i. e., the labelled entity should behave during the whole period of investigation identically with the non-labelled one.

If radioactive isotopes are used as tracers, the object or substance to be studied is labelled with a radioactive isotope *prior to the experiment.* If such a preliminary labelling is impossible owing to radiation hazard problems (e.g., in the food or cosmetics industry) then an inactive substance is added to the system which can easily be activated by subsequent neutron irradiation, due to its high activation cross-section. This process is called *subsequent or reverse labelling.*

Two groups can be distinguished within the radioactive tracer technique; it is not possible, however, to confine them within strict limits. In the case of *chemical labelling*, the radioactive isotope must follow the movement, reaction or metabolism of a particular element; in this way it is possible to draw conclusions about the chemical changes in the system. The isotope, of course, must be in the same chemical compound as the component of the system to be followed. In the case of *physical labelling* the radioactive nuclid is not an essential part of the system, therefore the chemical properties of the isotope are of minor importance; it is important only that its radiation is detectable and it is able to be attached in some way to the object or medium to be studied.

The following scheme demonstrates the essence of the *radioactive tracer technique:*

$$S(xA, yA^*) \Big\langle \begin{matrix} Z(x'A, y'A^*) \\ Z'(x''A, y''A^*) \end{matrix} \tag{3.1}$$

System S containing x atom of A type and y atom of radioactive isotope A, named A*, gets into states Z and Z'. Since isotope effects are negligible in the cases discussed here, from the physicochemical point of view isotopes of the same element behave identically. Hence, Eq. (3.2) will hold:

$$\frac{x'}{x} = \frac{y'}{y}; \quad \frac{x''}{x} = \frac{y''}{y} \tag{3.2}$$

Equal fractions of inactive and radioactive atoms will get into the intermediates or end-products of the transformation.

There are several advantages to the application of radioactive isotopes as tracers:

— their amount can be determined quantitatively with high sensitivity even in *low concentrations;* this—apart from other factors—is generally likely to result in shorter investigation times;

— *chemical labelling* is possible with radioactive isotopes only (apart from the stable isotope technique requiring very expensive instrumentation and being impossible in several cases) because—in spite of the isotope effect of low atomic number elements—the best label for every element is its own isotope;

— *the isotopes can be detected through the wall* of the tube or reactor, thus sampling or instrument installation-caused disturbance in material flow can be avoided;

— the data of field measurements can be obtained in a *computer-compatible form*, thus if the instrument has an adequate program, tables or diagrams can be constructed shortly after finishing the experiment.

When planning tracer studies, the disadvantages of radioisotopic experiments should also be considered:

— it is essential to ensure the radiation protection of both the experimenters and their environment; *health physics instructions* are complex and render the work tedious and expensive—especially when open preparations should be measured in a factory;

— *plant experiments* can be made and the samples measured and analysed only if a well-equipped isotope laboratory is available as a "background"; the necessary investment, including the measuring instruments, is added to the experimental costs;

— *special personal qualification* is required for work with radioactive isotopes.

When a tracer study is planned, the selection of the method should be preceded by a careful consideration of the advantages and disadvantages, and a rigorous estimation of the cost-effectiveness is required. On this basis it is possible to decide whether it is reasonable to use radioactive isotopes for the solution of the given problem, if there is a possibility of using another tracer adequate from the scientific or technical point of view.

The most important possible industrial applications of radioactive tracers are summarized in Table 3.1—grouped according to branches of industry as well as the nature of application.

Tracer techniques in *nuclear geophysics* is discussed in Section 5.4, *autoradiography* in Section 7.6.

3.1.1. PREPARATION OF THE INVESTIGATION

The preparation of a tracer study includes the *selection of the labelling radioisotope*, the *elaboration of the method for measurement*, i.e., the determination of the *activity* of the necessary *isotope*, as well as the planning of the *health safety* measures. These steps cannot be separated strictly, because they are not independent. The final plan must be an optimum compromise developed considering all the above factors.

3.1.1.1. SELECTION OF THE TRACER ISOTOPE

The isotope for labelling should be adapted to the nature of the process and the properties of the material to be studied. It is essential that the chemically labelled and non-labelled substance should be *chemically identical* when reaction mechanism, metabolism, physical processes depending on the intrinsic properties (solubility, adsorption, diffusion, etc.) are investigated by this method: the identity is not confined to the *atomic number*, but also to the *chemical compound* including the isotope. The selection involves here that between various isotopes of the same element. The decision should be based upon the possibilities of *measurement* as well as *health physics* problems, considering also *economy*. For example, the labelling of a sodium compound can be done by ^{22}Na or ^{24}Na: both emit γ-radiation. As far as the frequency of the emitted γ-quanta and their detectability are

Table 3.1. Main fields of application of radioactive tracers in industry

Field of application	Character of application	Effect or operation studied
Mining	I	Identification of mine cars; location of un-exploded charges
	L	Water seepage in mines; underground gasification
	V	Velocity of suspensions; air movement in mines
	MT	Mechanism of flotation
	W	Faults in pipelines; correlations between lubricants and machine part wear
	H	Homogenization of dry materials
Food industry, agriculture, forestry and fishing	I	Evapotranspiration rates; effects of fertilizers, soil composition and irrigation on crop yields; identification of wines; fish waste in rivers and seas, distribution of sewage in rivers and seas
	L	Defects in irrigation canals and pipes
	V	Flow rates in irrigation systems; water movements; water storage capacity of soils; capacity of beer and wine storage tanks; air drying in crop stores; ventilation in storage rooms and factories
	MT	Flow patterns in sugar mill settlers
	M	Washing efficiency studies; transfer of contaminants and additives from wrapping; extraction of sugar from sugar beet; water content of food pulps; air filter testing
	H	Efficiency of food mixers, distribution of vitamins, fats, additives and trace elements in foodstuffs
Construction, non-metallic mineral products	I	Distribution of cement and asphalt injections, identification of special concretes
	L	Leakage in dams; newly laid water mains
	MT	Retention, distribution, flow patterns of cement in rotary kilns; flow patterns in glass furnaces
	V	Air movement in hospitals, libraries, stores
	W	Lifetime of brick, refractory materials; wear of grinding balls in cement mills and of refractory linings in glass furnaces; corrosion of glass
	H	Mixing of concrete, asphalt and additives; distribution of additives and cement in concrete
Machinery, transport equipment	I	Efficiency of combustion in internal combustion engines
	L	Fuel and gas leaks in automobiles and aircraft
	V	Flow rates in cooling systems; flow rates of fuel; air movement in railway cars, automobiles and aircraft
	M	Study of fuel, lubricants and air filters
	W	Wear of cutting tools, gears, turbine blades, bearings, pistons, piston rings and plastic components; corrosion of metal parts, measurement of peak temperatures of machine parts in operation
	H	Mixing of liquid and solid fuels

Field of application	Character of application	Effect or operation studied
Metallurgy, casting, metal products	I	Movement of charges in blast furnaces; flow patterns in aluminium castings
	V	Gas velocity in blast furnaces; flow patterns of molten metals
	MT	Material flow in digestion equipment
	M	Solid–liquid interface phenomena; exchange of iron between slag and metal
	W	Wear of linings; corrosion of tanks, etc.
	H	Distribution of alloying components, e.g., tungsten and cobalt in steel; homogeneity studies in mixing tanks
Light industry (textile, wood, paper printing)	I	Impregnation of wood by fungicides
	V	Flow rate of water, pulp, chemicals, effluent in paper industry; waste dilution studies
	MT	Flow patterns in process vessels in textile industry; movement of chips and liquid in continuous digesters; dynamics in bleaching towers; flow patterns in paper machines
	M	Washing efficiency in textile industries, transfer of printing inks
	W	Wear of fibres; wear of wood cutting tools; abrasiveness of filler and fibrous material (e.g., asbestos)
	H	Dye and colour distribution studies; distribution of lubricants on artificial silk and nylon; distribution of wool fibres during carding; distribution of glue in laminates; penetration and distribution of printing ink
Chemicals and chemical products (rubber, petroleum)	I	Location of go-devil in pipelines
	L	Leaks in heat exhangers, double-walled process vessels, underground pipelines, pressure vessels
	V	Dilution of wastes in rivers and seas
	MT	Residence time and mixing studies in reactors
	M	Permeability of gases and liquids through plastic and vulcanized materials
	W	Type wear; corrosion of tanks
	H	Mixing of carbon black, zinc oxide, fuels and distribution of final products
Water economics	I	Silt and sand movement in harbours and rivers
	L	Tanks, water mains
	V	Flow rate of water and sewage
	M	Efficiency of sewage settlers
	W	Turbine blades
Labour safety	M	Efficiency of bacterial filters and gas masks
	H	Distribution of stack gases and dust

I = Identification of materials and following their movement (Section 3.2.1).
L = Detection of leaks and cracks (Section 3.2.2).
V = Determination of flow rates (Section 3.3).
MT = Material transport in industrial processes (Section 3.4).
M = Determination of the amount of mass (Section 3.5.1).
W = Wear studies (Section 3.5.2).
H = Study of homogeneity of mixtures (Section 3.4.4).

concerned, ^{22}Na is more advantageous, but health physics regulations allow the use of 25 times more ^{24}Na. There are essential differences between the *half-life* of the two isotopes (^{24}Na: 15h, ^{22}Na: 2.6 years) and their *price* (^{24}Na is hundred times more expensive). For these reasons, if the rate of the process to be studied permits, ^{24}Na is selected; in the case of slow processes (e.g., dissolution of sodium from enamels) the use of ^{22}Na is advisable. Several other elements provide similar possibilities, e.g., ^{54}Mn or ^{56}Mn, ^{58}Co or ^{60}Co, ^{65}Zn or ^{69m}Zn, ^{85m}Kr or ^{85}Kr, ^{110m}Ag or ^{111}Ag, ^{113}Sn or ^{121}Sn, ^{122}Sb or ^{124}Sb, ^{125}I or ^{131}I, ^{197}Hg or ^{203}Hg.

For application in an *identical chemical form*, it is often necessary to develop the method of synthesis of the labelled compound or to carry out a known synthesis.

Labelling of inorganic substances and metals is possible also by *reactor activation* of the appropriate inactive compound or metal. If the tracer study is to be performed by means of a given component of the activated compound or alloy, the other isotopes activated should be left to decay before the investigation, or the radiation must have a given energy selected by gamma-spectroscopy.

For *physical labelling*, a tracer of different chemical property can also be used. The movement, flow, mixing of liquids and gases do not depend on their chemical properties, rather on the physical ones (density, viscosity, etc.); dissolved tracers do not change these properties because of their small amounts. The tracer should fulfil the requirements discussed above even in such measurements because if the concentration ratio of the tracer and the substance to be studied are altered (e.g., due to the adsorption, condensation, precipitation of the tracer), faulty results will be obtained. In the presence of a multiple amount of an inactive carrier, the losses due to adsorption can be minimized.

Some recommended tracers for *gases, water* and *aqueous solutions, solid inorganic substances and organic systems* have been summarized in Table 3.2, together with their suitable chemical form.

For labelling *water and aqueous solutions*, water soluble salts of the isotopes (shown in Table 3.2) are advantageous. In more difficult cases, e.g., for labelling subsurface water—where minerals with considerable sorption ability may be present—strong complexes of the metals should be used, e.g., ^{51}CrNa$_2$–EDTA (sodium ethylenediaminetetraacetate acid) complex, or ^{60}Co-cyanido complexes. A further possibility is to use tritiated water, which is a good tracer, yet it has two disadvantages: its long half-life and that its detection requires a soft β-counter.

The movement and mixing of *solids* is determined by the size and shape of the granules, in addition to the density. Hence, for this purpose, the most suitable tracer is the activated sample of the material participating in the process. If direct activation is impossible, e.g., in the case of organic substances or elements having low cross-sections for (n, γ) reactions, surface labelling can be used. This can be performed by electrolysis, by adsorbing radioactive colloids on the surface of the grains, or by reducing negligible amounts of noble metals—^{110m}Ag or ^{198}Au—onto the surface.

The use of model substances, easy to activate, *imitating the physical properties of the material* is not a perfect solution, but acceptable in some cases. e.g., activated glass fibre with identical dimensions as those of the cellulose fibrils can be used in paper manufacturing, or a ^{60}Co rivet may be inserted into the wearing surface of an engine cylinder in wear studies.

The above examples demonstrate that in physical labelling, as a rule, several isotopes may give principally equivalent solutions. The selection is important for field or industrial measurements in view of the stricter regulations valid for the activity used outside the

Table 3.2. Recommended chemical composition for isotopes applied in the labelling of various substances

Isotope	Recommended chemical composition for labelling			
	gases	aqueous solutions	solid inorganic substances	organic substances
^{3}H	Gas	^{3}H$_2$O	—	Various organic compounds
^{14}C	CO$_2$	—	—	Various organic compounds
^{24}Na	—	Water soluble salt	Na$_2$CO$_3$ (polypropylene balls)	Naphthenate, salicylate
^{35}S	H$_2$S	—	—	Various organic compounds
^{38}Cl	—	—	—	Chlorobenzene
^{41}Ar	Gas	—	—	—
^{46}Sc	—	Water soluble salt	Sc$_2$O$_3$	—
^{51}Cr	—	—	Adsorbed on quartz	—
^{59}Fe	—	—	Metal	Dicyclopentadienyliron (ferrocene)
^{60}Co	—	—	Metal	Naphthenate
^{64}Cu	—	Water soluble salt	CuO	Naphthenate
^{65}Zn	—	—	ZnO	—
^{65}Ni	—	—	—	Stearate, oxalate
^{67}As	AsH$_3$	—	—	—
^{77}Ge	—	—	—	Various organic compounds
^{82}Br	CH$_3$Br	Water soluble salt	CaBr$_2$	Bromobenzene, p-dibromobenzene
^{85}Kr	Gas	—	—	—
^{86}Rb	—	Water soluble salt	—	—
^{110}Ag	—	—	Adsorbed on solid particles	—
^{124}Sb	—	—	Metal	Triphenylstibine
^{131}I	—	Water soluble salt	—	Iodokerosene, iodobenzene
^{133}Xe	Gas	—	—	—
^{140}La	—	—	La$_2$O$_3$ (polypropylene balls)	—
^{144}Ce	—	—	Ce$_2$O$_3$	—
^{182}Ta	—	—	Ta$_2$O$_3$	—
^{198}Au	—	—	AuCl$_3$ adsorbed on powder	Cyanide solution, amine salt, colloidal gold

isotope laboratories. If several—principally equivalent—isotopes are available, it is reasonable to select which has

— *lower toxicity* and—due to its half-life being shorter than 60 days—can be applied in higher concentrations;

— *γ-radiation* and can thus be measured in larger samples, eventually through the wall;

— *a higher γ-radiation intensity* (number of γ-quanta pro decay), since in this case a sample with identical activity gives more counts per unit time.

3.1.1.2. PREPARATION OF THE MEASUREMENT, CALCULATION OF THE NECESSARY AMOUNT OF ISOTOPE

Owing to the random character of the radioactive decay, increase of the specific activity of the tracer enhances not only the sensitivity of the measurement, but also its statistical accuracy: the relative standard deviation (σ) decreases. The statement is, of course, also true in the reverse sense: if the deviation of the measurement with a given isotope is fixed, then lower activity than a corresponding given value cannot be used. The most reasonable way to increase the accuracy is to use higher specific activities. This, however, is limited by health physics regulations, especially in the case of industrial and field measurements. Instead of increasing the specific activity, such an isotope may be found and such a measuring arrangement developed which will ensure the proper sensitivity even at a relatively low activity.

In plant experiments, as a rule, γ-radiation is used for tracer purposes. This involves incomparably lower losses due to absorption and self-absorption than β-radiation. This further limitation determines the measuring instrument. Since γ-quanta are detected with the best efficiency by *scintillation counters*, this is the most widespread detector type for plant or field tracer measurements.

The following equation (3.3) expresses the correlation between the *count rate.* I_{rel} (its usual unit is counts per second), and the *activity*, A (its SI unit is disintegration per second $= 1$ Bq, 1 Bq $= 2.7 \times 10^{-11}$ Ci):

$$I_{rel} = \delta \eta A \qquad (3.3)$$

where δ denotes the number of γ-quanta per disintegration, and η is the efficiency of the measurement.

The value of η is determined by several factors:

$$\eta = f \varepsilon G \qquad (3.4)$$

where f denotes the *self-absorption* rate in the source itself together with the *absorption* and *scattering* between the source and the detector; ε is the *efficiency of the counting*, i. e., the probability that a photon will produce a detectable interaction in the sensitive volume of the detector; the *geometry factor* G expresses the ratio of the radiation getting into the detector to the total amount emitted into 4π solid angle.

The following example will be of assistance in finding the correlation between I_{rel} and A. How large I_{rel} imp/min count rate can be expected if a section of a tube of V volume is measured by a scintillation detector with a counting efficiency of ε, the geometry factor of the arrangement is G, the absorption and scattering factor is f, the tracer isotope has an A activity and emits δ number of γ-quanta per disintegration? The radioactive concentration is thus $c = A/V$ MBq/dm^3. On the basis of Eqs (3.3) and (3.4),

$$I_{rel} = 6 \times 10^7 \, c \, V \delta \, f \varepsilon G \quad \text{imp/min} \tag{3.5}$$

$$I_{rel} = 6 \times 10^7 \, A \delta f \, \varepsilon G \quad \text{imp/min} \tag{3.6}$$

The attempt required is obvious: c and A should be kept at a low value and, at the same time, all other factors determining the value of I_{rel} should be selected in a way that the count rate should be as high as possible. However, this gives rise to a very complex problem:

δ, f and ε are independent: the values of all three are inseparably connected with the properties of the tracer isotope: δ is determined by the decay scheme of the isotope, f increases and ε decreased with increasing energy of the γ-quanta;

— V cannot be increased beyond certain limits because this involves the decrease of both f and G;

— the value of G can be improved by modifying the relative position of the sample to the detector, this, however, also involves a change of f.

Considering these factors, a given I_{rel} ensuring a pre-determined accuracy can be achieved partly by selecting an isotope advantageous for detection (increasing δ, ε and f), and partly by ensuring a favourable geometry (increasing the value of G). The value of V should be adjusted according to the purpose of the study and other possibilities. If these do not impose any limitation, the increase of the volume is, as a rule, advantageous for obtaining a higher count rate, in spite of the simultaneous decrease of f and G.

Even if the values of $V, \delta, f, \varepsilon$ and G are selected to be at optimum, it may occur that the minimum I_{rel} count rate necessary for the given accuracy of the measurement can only be achieved at such a high radioactive concentration, c, which does not allow safe handling and may be hazardous. In this case the tracer study has to be given up—either on the basis of preliminary calculations or preliminary laboratory experiments—and the problem should be solved in another way, e.g., by subsequent activation.

The values of δ, f and those of the absorption coefficients determining the latter factor can be found in various manuals [3.6, 3.7, 3.10].

The calculation of the geometry factor, G is illustrated below by a few simple examples.

In the case of a *point source of radiation*, let its distance from a detector window of R radius be l, and let the distance between the source and the sensitive volume be $r = l + m$. The ratio of the radiation getting into the detector to the total radiation emitted will be the ratio of the area of the corresponding spherical calotte to the total sphere (Fig. 3.1):

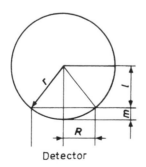

Detector

Fig. 3.1. Point source of radiation

$$G = \frac{2rm\pi}{4r^2\pi} = \frac{m}{2r} = \frac{r-l}{2r} = \frac{1}{2}\left(1 - \frac{l}{\sqrt{l^2 + R^2}}\right) \tag{3.7}$$

If a disc-shaped source with a radius of R_p is at a distance of l from the detector with a radius of R, the series of curves depicted in Fig. 3.2 give good information about G.

For *linear sources* (e.g., a thin tube filled with a liquid of homogeneous specific activity), Fig. 3.3 gives an illustration. If the length of the source is $2L$ and the number of γ-quanta emitted from unit length of the radiation source is n, the radius of the detector being R, then the number of γ-quanta arriving at the surface of the detector (n') will be:

$$n' = 2\int_0^L \frac{n\pi R^2}{4\pi r^2}\,\mathrm{d}x = \frac{nR^2}{2}\int_0^L \frac{\mathrm{d}x}{x^2 + l^2} = \frac{nR^2}{2l}\arctan\frac{L}{l} \tag{3.8}$$

and the geometric efficiency will be:

$$G = \frac{n'}{2nL} = \frac{R^2}{4Ll}\arctan\frac{L}{l} \tag{3.9}$$

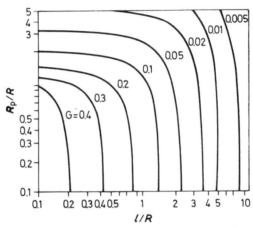

Fig. 3.2. Calculation scheme for determination of the geometry factor of a disc-shaped radiation source

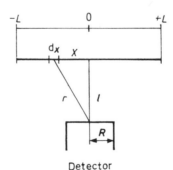

Fig. 3.3. Linear radiation source

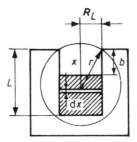

Fig. 3.4. Well-type scintillation crystal

For a *well-type scintillation crystal* (Fig. 3.4), the calculation is based on the fact that the radiation is emitted in every direction; its downwards directed part also gets through the crystal and may interact with it. The upwards part is determined by R_L and can be calculated on the basis of Eq. (3.7), although there is a fraction which is lost under any circumstances for detection.

If a cuvette with a radius of R_k, height of L, emitting n γ-quanta per cm^3 and unit time, filled with a radioactive liquid up to the line $L-b$ is placed into a hole with a radius of R_L, the number of γ-quanta passing through the crystal (n') will be:

$$n' = \int_b^L \frac{1}{2} nR_k^2\pi \, dx + \left[\frac{1}{2} nR_k^2\pi \, dx - nR_k^2\pi \, dx \frac{1}{2}\left(1 - \frac{x}{\sqrt{x^2 - R_L^2}} \right) \right] =$$

$$= \frac{R_k^2 n\pi}{2} \int_b^L \left(1 - \frac{x}{\sqrt{x^2 + R_L^2}} \right) dx = \frac{R_k^2 n\pi}{2} \left(L - b + \sqrt{L^2 + R_L^2} - \sqrt{b^2 + R_L^2} \right) \quad (3.10)$$

The geometric efficiency is as follows:

$$G = \frac{I_{rel}}{R_k^2\pi(L-b)n} = \frac{1}{2}\left(1 + \frac{\sqrt{L^2 + R_L^2} - \sqrt{b^2 + R_L^2}}{L-b} \right) \quad (3.11)$$

If a detector with a radius of R is used for detection of the γ-radiation emitted by a *tube surface* of D diameter and l length, the question can be formulated; how many γ-quanta pass through the detector if the liquid in the tube emits n γ-quanta per unit volume and time? If absorption and scattering are neglected, the following estimation can be performed:

The number of γ-quanta leaving the tube per unit surface is n'_0:

$$n'_0 = \frac{nl\pi D^2/4}{\pi Dl} = \frac{nD}{4} \quad (3.12)$$

Thus, the number of γ-quanta passing through a detector with a radius of R is:

$$n' = \pi R^2 \frac{nD}{4} \quad (3.13)$$

It is necessary to know the *counting efficiency*, ε to be able to calculate the I_{rel} count rate after having calculated the f absorption-scattering factor and the G space angle factor, using Eqs (3.3) and (3.4):

If n' γ-quanta get into a crystal with a thickness of l and an absorption coefficient of μ, the counting efficiency is:

$$\varepsilon = \frac{n'\left[1 - \exp\left(-\mu l\right)\right]}{n'} = 1 - \exp\left(-l\right) \tag{3.14}$$

This is illustrated schematically by Fig. 3.5.

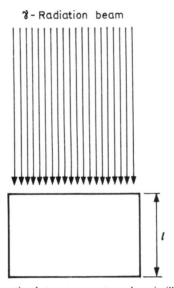

Fig. 3.5. Interaction between γ-quanta and a scintillation crystal

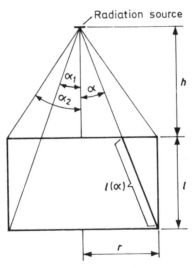

Fig. 3.6. Interaction between γ-quanta from a point source and a scintillation crystal

Figure 3.6 illustrates the interaction of a *point source* and a crystal. The value of ε is an integral form and can be calculated by numerical methods only:

$$\varepsilon = \frac{\dfrac{1}{2} \displaystyle\int_0^{\alpha_2} \{1 - \exp[-\mu l(\alpha)]\}\, \sin\alpha\, d\alpha}{\dfrac{1}{2} \displaystyle\int_0^{\alpha_2} \sin\alpha\, d\alpha} \tag{3.15}$$

Differential equations describing the interaction of an *extended source* and a scintillation crystal are even more difficult than Eq. (3.15), and are unsuitable for direct calculations. There are, however, diagrams available for the so-called total efficiency, i.e., the εG product. These have been calculated by computers and give the above product as a function of the radiation energy, considering also such parameters as the crystal dimensions, the distance between the source and crystal, etc. Such diagrams can be found in several handbooks (e.g., [3.6]); some of them will be discussed also here.

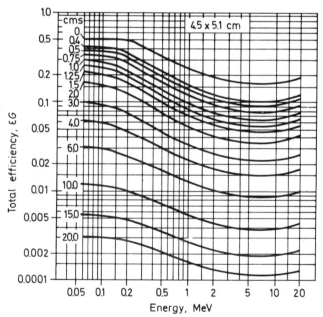

Fig. 3.7. Total efficiency values for the interaction of a point γ-source and a ϕ 4.5 cm by 5.1 cm NaI(Tl) crystal

Figure 3.7 illustrates the total efficiency values for γ-quanta emitted by a point source placed at various distances from a NaI(Tl) crystal with dimensions of 4.5 × 5.1 cm, as a function of the γ-energy. The total efficiency decreases with increasing energy and distance.

A *cylindrical vessel* with a height of 5.1 cm is filled with a radioactive solution up to a height of L. The count rate is measured with a NaI(Tl) scintillation detector at a distance of $l_0 = 1$ cm, with a height of 5 cm and diameter of 5.1 cm. The functions $\varepsilon G = \varphi(E_\gamma)$ are shown in Fig. 3.8 for different L parameters.

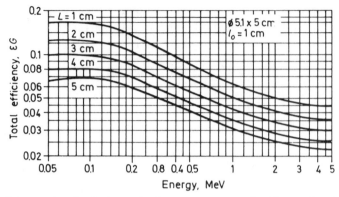

Fig. 3.8. Total efficiency values for the interaction between γ-quanta emitted by a radioactive liquid filling a
cylindrical beaker and a NaI(Tl) crystal

The *Marinelli-vessel* (Fig. 3.9) ensures a very advantageous arrangement for
measurement. The total efficiencies for this arrangement are shown in Fig. 3.10.

A *well-type scintillation crystal* is suitable to measure the activity of even very small
amounts of liquid. The total efficiencies for this device are shown in Fig. 3.11, for different
radiation energies. The data relate to liquid samples filling the hole completely.

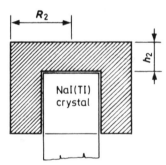

Fig. 3.9. *Marinelli*-vessel

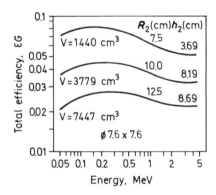

Fig. 3.10. Total efficiency values for a *Marinelli*-vessel

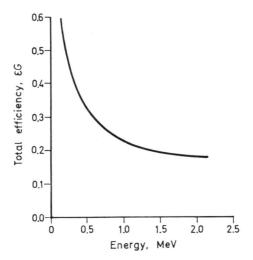

Fig. 3.11. Total efficiency values for a well-type scintillation crystal. Crystal diameter: 3.8 cm, height: 5 cm; bore diameter 1.3 cm, depth 3.8 cm

As a result of various calculations, using Eq. (3.5), if c, V, δ, f, ε, G (or εG) are known, I_{rel} can be performed, or for a fixed I_{rel}, the necessary radioactive concentration can be calculated. Plant and field experiments have demonstrated that these data provide useful information, but—owing to errors in the estimation of individual factors—they cannot be considered more than a *technological estimation*. This is particularly important and must be seriously considered if the calculation has consequences as far as health physics is concerned. In such cases, the activity values obtained by careful calculations have to be corrected by a safety factor of at least 1.5. (Chapter 8 and Section 8.5 deal with the health physics problems connected with the tracer technique.)

EXAMPLE 3.1

To demonstrate the sensitivity of the radio-tracer technique, let us calculate the m amount of radioactive substances (in gram-moles) with various $t_{1/2}$ half-lives (in minutes) necessary to ensure a count rate of $I_{rel} = 40$ imp/min, if the β-radiation is measured by means of a GM-tube and the counting efficiency is 4%.

The number of disintegrations per minute is:

$$\frac{dn}{dt} = \frac{40}{0.04} = 1000$$

The number of radioactive atoms present in the sample (n) can be calculated on the basis of the disintegration constant (λ) or the half-life ($t_{1/2}$):

$$-\frac{dn}{dt} = \lambda n = \frac{0.693}{t_{1/2}} n = 1000 \tag{3.16}$$

$$n = 1000 \frac{t_{1/2}}{0.693}$$

Table 3.3. Amounts of isotopes detectable at a count rate of 40 imp/min, in the case of the measurement of β-radiation by a 4% efficiency GM-tube

Half-life		Amount of label in the sample	
Conventional units	Minute	Number of atoms	Mole
1 hour	60	8.66×10^4	1.44×10^{-19}
1 day	1.44×10^3	2.08×10^6	3.45×10^{-18}
1 month	4.32×10^4	6.23×10^7	1.03×10^{-16}
1 year	5.26×10^5	7.58×10^8	1.26×10^{-15}
10 years	5.26×10^6	7.58×10^9	1.26×10^{-14}

The amount of tracer in gram-moles is:

$$m = \frac{n}{6.03 \times 10^{23}} = 1.658 \times 10^{-24}\, n$$

For fictitious isotopes with half-lives of $t_{1/2} = 1$ h, 1 day, 1 month, 1 year and 10 years, calculations have been carried out, the results are summarized in Table 3.3.

EXAMPLE 3.2

The tightness of tins with 5.1×5 cm dimensions are studied in an experimental pressure boiler imitating the conditions of the canning factory. The boiler contains water labelled with ^{24}NaCl and the tins contain inactive water. If the tins are not closed hermetically, ^{24}Na will appear in them in c MBq/cm^3 radioactive concentration.

Let us select the preferable method for measuring the radioactivity of water samples withdrawn from the tins, regarding that method as the more advantageous which results in a higher count rate at the same radioactive concentration.

The two measuring systems are the following:

— the count rate is $I_{rel,\, w}$ if the activity of a 15 cm^3 sample is measured by a *well-type crystal* of 55 mm O. D. and 55 mm high, with a bore of 25 mm I. D. and 42 mm in depth;

— the count rate is $I_{rel,\, cy}$ if the activity of a total amount of 100 cm^3 liquid is measured in a *cylindrical vessel* placed onto a crystal with 51 mm O. D. and 51 mm high; the diameter of the vessel is 50 mm.

According to Fig. 3.11, in the case of a *well-type crystal*, the value of εG corresponding to the most frequent 2.75 MeV γ-quantum of ^{24}Na, is about 0.17. This has to be substituted into Eq. (3.5):

$$I_{rel,\, w} = 6 \times 10^7\, c\, 15\delta f \times 0.17$$

In the case of a *cylindrical vessel*, $L \approx 5$ cm corresponds to $V = 100$ cm^3 and 50 mm O. D. As shown in Fig. 3.8, $\varepsilon G \approx 0.026$ for a γ-quantum of 2.75 MeV, thus:

$$I_{rel,\, cy} = 6 \times 10^7\, c \times 100\delta f \times 0.026$$

Dividing the two equations:

$$\frac{I_{rel,\, w}}{I_{rel,\, cy}} = \frac{15 \times 0.17}{100 \times 0.026} = 0.98$$

If the radioactive concentrations are the same, the arrangement with a bored crystal allows a less sensitive measurement because the counting rate is lower by about 2%.

3.2. IDENTIFICATION OF MATERIALS
AND DETECTION OF FAULTS

The purpose of a considerable part of industrial tracer applications is to obtain qualitative information. Such a task may be the identification of materials or objects, their location, the establishment of their presence or absence, the tracking of their movement, or the discovery of cracks and leaks.

3.2.1. IDENTIFICATION OF MATERIALS
AND TRACKING THEIR MOVEMENT

Special purpose or special quality products can be labelled by radioactive isotopes in order to facilitate their identification or recognizability. For example, the special stainless steel alloy tubes for heat exchangers of nuclear reactors can be labelled by ^{82}Ta or ^{192}Ir with a specific activity of about 40 MBq/t. Machine parts or the position of the blind splicing of wire-ropes can be labelled during their welding by a welding electrode containing ^{192}Ir.

So-called *go-devils* are used for the cleaning of petroleum and gas pipelines; if they are used only to separate two different kinds of liquids, they are called *pigs*. Especially in new pipelines, where the welding is not perfect, or after the formation of a lot of deposits go-devils may get stuck and their search is a laborious and costly procedure. If, however, a go-devil labelled with ^{60}Co isotope is started after the stuck device, its passing can be observed at detecting stations along the pipeline. The location of the stuck device can thus be limited and—examining the critical section by means of sensitive radiation detectors on the surface—its exact place can also be established. Since pipelines are, as a rule, about 1–1.5 m deep below the surface, the necessary ^{60}Co activity is about 7–20 GBq (200–500 mCi).

Figure 3.12 gives an example for the shielding of the source by lead. The radiation source s is built in the metal cylinder B. This is inserted into the bore of the lead block C during assembling. As the go-devil moves forward, the oil deforms the bellows, D, consequently the oil content E will push the cylinder, together with the source, out of the bore of the lead block. Under atmospheric conditions, the force of the spring F is sufficient to push back the cylinder into the shielding block.

The wearing of the *refractory lining of blast-furnaces* can be checked by installing, e.g., ^{60}Co sources into the lining at appropriate places, in different depths. By recognizing the disappearance of the source it can be shown when the lining becomes so thin that it threatens the safety of the plant and it is time to reconstruct the furnace. Another variant of

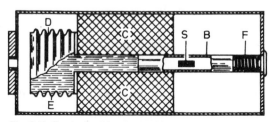

Fig. 3.12. Go-devil with a lead-shielded radiation source

the method checks the radioactivity of the cast iron or the dross; the appearance of radioactivity is an indication that the wear has reached the radiation source built into the lining. This latter method allows to draw conclusion on the extent of the wear only if different types of isotopes are built in different depths. These can be distinguished by γ-spectroscopy.

The movement of silt and ooze in rivers and sea can also be tracked by radioactive tracers. A possible tracer for this purpose is glass powder containing about 5% of $^{46}Sc_2O_3$ with a specific activity of $1,5G \times$ Bq/g. The dispersion of the tracer can be followed by a detector towed by a boat. Thus the origin of the silt can be determined.

EXAMPLE 3.3

Special purpose steel tubes made of a special alloy are labelled by a radioactive isotope in order to facilitate subsequent identification. ^{192}Ir is selected which has a half-life of $t_{1/2}$ = 74.2 days. The conditions of measurement are given. The purpose is to calculate the activity of ^{192}Ir necessary to add to a 10 t ingot in order to ensure a 99.99% probability of identification after $t = 1$ year.

The necessary activity can be calculated from Eqs (3.3) and (3.4):

$$A = \frac{I_{rel}}{\delta f \varepsilon G} \tag{3.17}$$

The count rate is determined according to the arrangement depicted in Fig. 3.3, by means of a 2.5 by 2.5 cm NaI crystal, with a background of 2000 imp/min, in 30 s runs. Equation (3.9) can be used to determine the geometry factor G of the arrangement:

$$G = \frac{R^2}{4LI} \quad \text{arc tg} \quad \frac{L}{I}$$

The data necessary for the calculation of G are as follows: $R = 1.25$ cm, $2L = 20$ cm and $I = 5$ cm.

$$G = \frac{1.563}{4 \times 10 \times 5} \quad \text{arc tan } 2 = \frac{1.563}{4 \times 10 \times 5} \times 1.1071 = 8.65 \times 10^{-3}$$

The numerical values of factors necessary for the calculation of A are the following: $\delta = 2.5$ (from the decay scheme of ^{192}Ir), $f = 1$, $\varepsilon = 0.55$ (for the given scintillation counter, on the basis of calibration), $G = 8.65 \times 10^{-3}$ (as calculated above).

The 99.99 probability means that the net count rate (after the deduction of the background) should be at least five times higher than its standard deviation. This condition can be expressed as follows for a background of 2000 imp/min and times of measurement as long as 30 s:

$$I_{rel} = 5 \sqrt{\frac{I_{rel} + 2000}{0.5} + \frac{2000}{0.5}}$$

$$I_{rel}^2 - 50 \, I_{rel} - 200\,000 = 0$$

$$I_{rel} = 473 \text{ imp/min} = 7.9 \text{ imp/s}$$

Now we have every numerical data necessary for the calculation of A:

$$A = \frac{I_{rel}}{\delta f \varepsilon G} = \frac{7.9}{2.5 \times 1 \times 0.55 \times 8.65 \times 10^{-3}} = 664 \text{ Bq} = 18.1 \text{ nCi}$$

The A activity is the total activity of a tube section of 20 cm length, with an outer diameter of 2 cm and a wall thickness of 1 mm. Knowing the density of iron (7800 kg/m³ = 7.8 g/cm³) we can calculate the necessary a specific activity of the alloy at the moment of activity measurement:

$$a = \frac{664}{20 \times 7.8(1.0^2 \times \pi \times 1.0 - 0.9^2 \times \pi \times 1.0)} = 7.13 \text{ Bq/g} \, (7.13 \text{ kBq/kg}) = 0.19 \text{ nCi/g}$$

Before 1 year, i.e. at the moment of the labelling, the a_0 specific activity is

$$a_0 = a \exp(0.693t/t_{1/2}) = 7.13 \exp(0.693 \times 365/74.2) = 215.6 \text{ kBq/kg} = 5.82 \text{ nCi/g}$$

Thus the 10 t ingot should be labelled with 2.2 GBq (60 mCi) ^{192}Ir in order to be able to identify the tubes made of it with a 99.99% probability even after a year.

3.2.2. DETECTION OF CRACKS AND LEAKS

No problem of principle arises in the *detection of cracks in gas pipelines* by means of radioactive tracers, yet, from the practical point of view, it is a very difficult task. If gaseous isotopes were introduced into a volume of several hundred or thousand cubic meters of gas, even a small leak could result in an uncontrollable escape of high amounts of radioactive gas. This fact limits the applicability of the method to low-volume systems.

A typical field of application is the detection of leaks in *pressure power cables* filled with, e.g., nitrogen. Earlier, radon produced as a decay product of radium was used as a tracer, and the activities of its solid daughter elements were measured. ^{85}Kr can also be well applied as a tracer.

Leaks in *gas-filled telephone cables* can also be detected. Underground cables are placed, as a rule, into hard, e.g. ceramic, tubes in order to protect them against corrosion and also to facilitate the removal of faulty sections. The arrangement suitable for detecting leakages is shown in Fig. 3.13. The examination can be performed, e.g., by means of ethyl bromide labelled with ^{82}Br. The air in the ceramic tube enclosing the cable is continuously sucked through an activated carbon filter. The cable to be examined is connected to a pressure vessel containing the labelled ethyl bromide. If there is a leak, the radioactive vapour gets into the ceramic tube, and further to the filter, where activated carbon adsorbs the bromide with high efficiency; its appearance can be detected here. Subsequent examination of each tube section leads to an almost certain discovery of the leak.

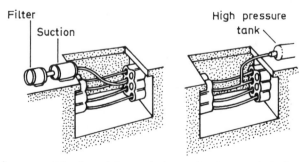

Fig. 3.13. Arrangement for determining gas leakages of underground telephone cables

Cracks and leaks in pipelines and tanks for *liquid* can be detected by labelling of the liquid, thus a part of the considerable cost of dismantling the whole system can be saved.

If concrete tanks with an inner bitumen coating suffer a crack, the liquid leaking through the bitumen coating travels a lot before appearing at the outer surface of the vessel. Therefore, the location of the surface trickle does not give a reliable information about the location of the crack on the bituminous coating. If the problem is to be solved by the tracer technique, the tank is filled with water and about 100 MBq (3 mCi) of sodium hydrogen carbonate labelled with ^{24}Na is dissolved in it. After allowing the solution to stand in the tank overnight, it is emptied and the inner walls are washed with a strong water jet in order to remove adsorbed ^{24}Na. Sensitive detectors can then trace the path of the radioactive solution under the bituminous coating, thus the crack can be located unambiguously.

When a *watermain network* is laid down it is assumed that, in the optimum case, one joint is faulty of every ten thousand. The economy of pipeline installation requires that the ditch should be filled up as soon as possible, because if the earth becomes wet in the meantime, it will not fill properly; it will sink causing harmful deformations. Before tracer application the rapid filling of the ditch was impossible because all joints had to be viewed during the pressure test.

For tracing, tablets of sodium hydrogen carbonate containing ^{24}Na are dissolved in warm dilute acetic acid diluted with water in a tank up to about 450 dm^3. When the water is flowing in the watermain from valve A toward valve B, the tracer solution is pumped from the tank into the pipeline and the activity change is monitored at B (Fig. 3.14).

After having attained a constant activity level, valve B is closed and the flow stopped. If valve A is also closed, the system will be under pressure due to the operation of the pump between the tank and the main. Maintaining the pressure for one or one and a half hour is sufficient to ensure the flow of the tracer through leaking joints into the surrounding soil. If the watermain is new, the sites of the joints are well known and they can be approached by GM-tubes inserted into holes bored near to them into the earth. Leaks are revealed by the appearance of radioactivity in the soil.

Leaking of *petroleum pipelines* must be stopped, partly because of the value lost through them, partly to protect the environment against the petroleum flow. For a tracer study, a liquid labelled with a γ-emitting radioisotope is injected into the pipeline. The tracer is rapidly mixed with the petroleum and as it proceeds, small amounts of it get into the environment through the leaks. One hour after the tracer injection a detector installed into a hydraulically driven go-devil is started to monitor the remaining activity. The signals of the battery-operated GM-detector are recorded, e.g. by a tape recorder. The detector reveals the radioactive substance in the soil as it passes before the leaks. The

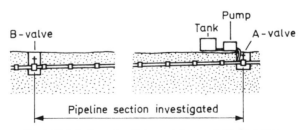

Fig. 3.14. Injection of radiotracer into a watermain system

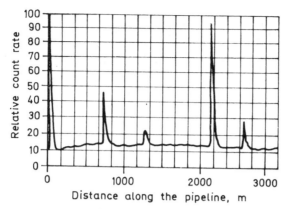

Fig. 3.15. Detection of leaks in a petroleum pipeline

intensity is proportional to the radioactivity recorded by the tape recorder. At the end of the tube, the recorder is dismounted, the tape is played back and its signals are recorded by a recorder. As a result, a diagram such as shown in Fig. 3.15 is obtained. Cracks are represented by the small peaks when the count rate is plotted as a function of the distance. Large peaks correspond to reference radiation sources with known activity placed at well-known spots. These serve to check the reliability of the device and to facilitate exact location.

EXAMPLE 3.4

A pulse-like injection of ^{133}Xe is introduced into the flow of warm nitrous gases before the heat exchanger of a nitric acid factory. The gas flow is monitored by detectors, counters and magnetic recorders installed into various spots of the system as shown in Fig. 3.16. The background is measured for 10 minutes before the injection and also for several minutes after the passing of the tracer.

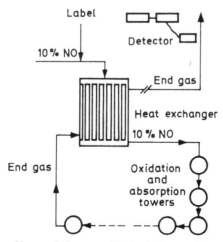

Fig. 3.16. Determination of losses of nitrogen oxides in the heat exchanger of a nitric acid plant

Table 3.4 shows the count number obtained from the magnetic recorder installed on the final gas line at the outlet of the heat exchanger. Also count differences between given time values are shown. What are the qualitative and quantitative conclusions with respect to the gas permeability of the cooler tube bundle?

Table 3.4 shows that in the t_1 time interval, background values are obtained. The flow of small amounts of active substance is recorded for 0.5 min after the injection (t_2 time interval). During the subsequent t_3 interval again background counting is observed. The bulk of the labelled gas flows before the detector in the t_4 time interval. Table 3.4

Table 3.4.Pulse counts at the measuring point shown in Fig. 3.14 and their calculated differences

Time interval	Time, min	Total counts	Count difference
t_1	− 10	0	0
	− 9	625	625
	− 8	1270	645
	− 7	1905	635
	− 6	2550	645
	− 5	3216	666
	− 4	3862	646
	− 3	4505	643
	− 2	5127	622
	− 1	5774	647
	0	6392	618
t_2	0	Injection of the tracer	
	0.5	7922	1530
t_3	1.0	8239	317
	1.5	8567	328
t_4	2.0	9177	610
	2.5	46629	37452
	3.0	76432	29803
	3.5	77340	908
	4.0	77895	555
	5.0	78657	762
t_5	6.0	79335	678
	7.0	79977	642
	8.0	80602	625
	9.0	81226	624
	10.0	81857	631

demonstrates that the passing of the gases from the site of the injection to the detection point (through the heat exchanger and the acid manufacturing towers) required 2.5–3.0 min. No active gas was detected in the t_5 time interval.

The facts indicate that the activity recorded in the t_2 period is the consequence of a short-circuit: some of the gas (with 10% NO content) entering the heat exchanger never gets to the absorption towers, but flows directly to the outlet gas through leaks and cracks of the cooler tubes or separation walls, and is lost.

The quantitative determination of the loss can be achieved by the following calculation.

The background, from the average of 15 data obtained during t_1 and t_5, is:

$$I_{rel, B} = \frac{6392 + (81857 - 78657)}{15} = 640 \text{ imp/min}$$

Considering this, the count values exceeding the background during time intervals t_2 and t_4 are:

Interval	length, min	Counts measured	Background measured	Net counts
t_2	0.5	$n_2^M = 1530$	$n_2^B = 0.5 \times 640 = 320$	$n_2^{M-B} = 1210$
t_4	3.5	$n_4^M = 75657 - 8567 = 70090$	$n_4^B = 3.5 \times 640 = 2240$	$n_4^{M-B} = 67850$

The loss in the heat exchanger due to leakage is:

$$\frac{n_2^{M-B}}{n_2^{M-B} + n_4^{M-B}} \times 100 = \frac{1210}{1210 + 67850} \times 100 = 1.75\%$$

3.3. DETERMINATION OF FLOW RATES

Radioisotope tracers are often used for the determination of flow rates of *gases, liquids* and *solids* transported pneumatically or hydraulically. Since the method is applied first of all for tracing liquid flow, our subsequent discussion, as far as quantitative considerations are concerned, will be related to the flow of labelled liquid substances, but these can be generalized also for gases and—with some modifications—also for the determination of flow rates of solid substances.

The principle of the determination of flow rate by tracers is as follows: a tracer is introduced into the liquid stream either pulse-like or continuously, and the activity changes of the liquid is measured at different distances from the site of injection. The *distance for mixing* is selected so that by that distance the tracer be completely mixed with

Table 3.5. Relative standard deviations characteristic of the extent of mixing at various mixing distances and with different ways of injection
(Reynolds number: 7.7×10^{14})

Relative standard deviation, %	Distance of mixing, L/D			
	Way of injection*			
	(a)	(b)	(c)	(d)
10	90	60	25	17
5	107	82	35	25
2	129	106	50	39
1	145	125	64	54
0.3	190	144	93	86

* According to Fig. 3.17.

the liquid, i.e., the relative standard deviation of the activity distribution should not exceed 2–5%. The distance for mixing is expressed by relative L/D ratios, where L denotes the distance between the site of injection and that of the measurement and D is the tube diameter. The distance for mixing is a function of the flow characteristics of the liquid (*Reynolds*-number) and the method of injection of the tracer. Table 3.5 summarizes

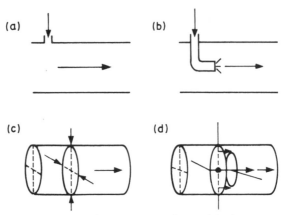

Fig. 3.17. Different methods of tracer injection

relative standard deviation values for various L/D values, for four different methods of injection (denoted by a, b, c and d in Fig. 3.17) if the *Reynolds* number is 7.7×10^4. As it is seen, the shortest distance for mixing can be achieved by method (d).

3.3.1. PULSE-LIKE TRACER INJECTION

This method involves the injection of a tracer with A activity into the flowing liquid at the moment of $t = 0$. The radioactive isotope distributed uniformly throughout the whole cross-section, flows at a rate identical with that of the liquid but, owing to longitudinal mixing, it appears in an increasing volume of the medium to be investigated. The farther the site of measurement from the site of injection, the less pulse-like will be the passing of the tracer before the detector. This is illustrated in Fig. 3.18. If several detectors are installed at various distances from the site of injection in order to monitor the radioactive concentration (c), the "pulses" will be more and more diffuse in time, as longitudinal mixing increases. If there is no activity loss, the peak areas are equal to each other, i.e., the following equation is valid for every point of detection:

$$v_V \int_0^\infty c(t)\mathrm{d}t = A, \quad \frac{\text{volume}}{\text{time}} \times \frac{\text{Bq}}{\text{volume}} \times \text{time} = \text{Bq} \qquad (3.18)$$

Rearranging this equation, we obtain the formula for the *volume flow rate* (v_V):

$$v_V = \frac{A}{\int_0^\infty c(t)\mathrm{d}t} = \frac{I_{\text{rel}}}{\int_0^\infty i(t)\mathrm{d}t} \qquad (3.19)$$

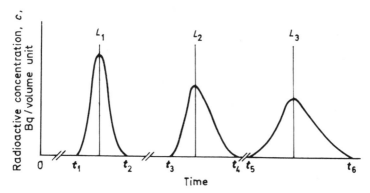

Fig. 3.18. Radioactive concentrations measured at increasing distances from the point of injection as a function of time

According to Eq. (3.3), A and $c(t)$ can be substituted by I_{rel} (imp/min) and i (imp/min. cm^3) values, respectively, if both are determined with the same η efficiency.

In practice, the initial (t_1) and final (t_2) times of the passing of the radioactive tracer are selected as the limits of integration.

Several methods are known for determining the value of $\int_{t_1}^{t_2} i(t)dt$; the methods based upon pulse-like injection will be classified according to these:

Sampling technique: Several consecutive samples are taken at the monitoring point at short intervals and, on the basis of their radioactive concentrations, the $i(t)$ functions are constructed. The function is integrated by a graphical method. I_{rel} is determined from an aliquot part of the tracer, under identical conditions of measurement as in the determination of $i(t)$. The *advantage* of the process is that the sample volume and the period of counting can be chosen freely; its *disadvantage* is that high flow rates cannot be monitored by sampling.

Full sample technique: During the flow of the radioactive tracer the system is tapped at a constant flow rate of $v_{V_m} \ll v_V$; the liquid sample is collected and its radioactive

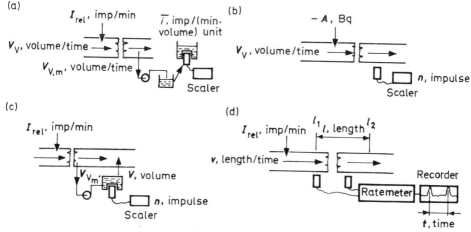

Fig. 3.19. Methods based on pulse-like introduction of a tracer for determining flow rates

concentration is determined (expressed as imp/(min cm^3)): see Fig. 3.19(a). The value obtained is an integral average, $\bar{i}$:

$$\bar{i}(t_2 - t_1) = \int_{t_1}^{t_2} i(t)\,dt \tag{3.20}$$

Thus, it is true that:

$$v_V = \frac{I_{rel}}{\bar{i}(t_2 - t_1)} \tag{3.21}$$

where t_1 and t_2 are the initial and final moments of the tapping, respectively. I_{rel} is determined as mentioned for the sample series technique.

Total count technique: If the medium flows through a pipeline, a scaler connected to a detector mounted on the pipeline can also serve as an integrator (Fig. 3.19(b)). The recorded n count number is higher if the injected total activity is larger and if the flow rate is smaller:

$$\frac{A}{v_V} Q = n, \left[\frac{Bq}{volume/time} \times \frac{imp/time}{Bq/volume} = counts \right] \tag{3.22}$$

where Q is a conversion factor: the count rate given by the detector if the tube section around it is filled with a liquid with a radioactive concentration of 1 Bq/volume unit.

If n is measured and Q is determined by means of calibration using a model with identical diameter and wall thickness as the tube to be investigated, v_V can be calculated with known A value:

$$v_V = \frac{AQ}{n} \tag{3.23}$$

If a bypass is applied, the integration can also be performed by a scaler (Fig. 3.19(c)). A fraction of the liquid flow is led continuously through a measuring vessel of V volume; the flow rate of the side circuit is constant and equal to v_{V_m}. A detector and a scaler is attached to the measuring vessel. The total count number measured by the scaler will be (after deduction of the background):

$$n = V \int_0^{\infty} i(t)\,dt \left[volume \frac{counts}{volume \times time} time\text{-}counts \right] \tag{3.24}$$

.It follows then that:

$$v_V = \frac{I_{rel}}{\int_0^{\infty} i(t)\,dt} = \frac{I_{rel} V}{n} \tag{3.25}$$

I_{rel} denotes the counts per unit time caused by the total amount of the tracer solution in the given measuring vessel of V volume, measured by the given detector; it can be determined from the aliquot fraction of the tracer, after diluting it several times.

Measurement on the basis of the velocity of the pulse. The linear flow rate of the liquid can also be determined by measuring the speed of the advancing of the injected tracer. The detectors are installed at distances L_1 and L_2 with respect to the injection point (Fig. 3.19(d)). The time interval $t_2 - t_1 = t$ can be determined and the *linear flow rate of the liquid* is as follows:

$$v = \frac{L_2 - L_1}{t_2 - t_1} = \frac{L}{t} \tag{3.26}$$

If we know the cross-section (q) of the tube, also the volume rate (v_V) can be calculated:

$$v_V = v.q = \frac{L.q}{t} = \frac{V}{t} \qquad (3.27)$$

If the measurement for determining the linear flow rate (v) on the basis of the pulse rate is combined with one of the methods of determining the volume flow rate (v_V), the "actual flow-through cross-section" of the tube can also be measured:

$$q = \frac{v_V}{v} \qquad (3.28)$$

EXAMPLE 3.5

The rate of a water flow is measured by means of pulse-like injection of $^{24}NaCl$, using the full sample technique (Fig. 3.19(c), Eqs (3.21) and (3.23)). Let us calculate the v_V volume flow rate on the basis of the given data.

Using the *full sample technique*, about 10 dm^3 of sample is collected 600 m away from the site of injection of ^{24}Na, from $t_1 = 5$ min to $t_2 = 25$ min after injection. A $500 = cm^3$ portion of the sample is poured into a Marinelli flask surrounding a 50 by 50 mm NaI(Tl) crystal. A count rate of 4000 imp/min is measured. On this basis:

$$\bar{i}(t_2 - t_1) = 4000 \times 2 \times 20 \times 60 = 9.6 \times 10^6 \text{ imp s/(min dm}^3)$$

150 cm^3 of $^{24}NaCl$ solution is used for labelling. Its count rate is determined from 10 cm^3 samples. In a dilution of 50 000 times, the same Marinelli flask gives 27 900 counts, on an average:

$$I_{rel} = 27\,900 \frac{150}{10} 50\,000 = 2.09 \times 10^{10} \text{ imp/min}$$

Hence the volume rate, v_V is:

$$v_v = \frac{I_{rel}}{\bar{i}(t_2 - t_1)} = \frac{2.09 \times 10^{10}}{9.6 \times 10^6} = 2.18 \times 10^3 \text{ dm}^3/\text{s} = 2.18 \text{ m}^3/\text{s}$$

Using the *bypass total count technique*, water is pumped through a 390 cm^3 measuring vessel at a constant flow rate during the total period of the experiment. The count rate is measured continuously by means of a scaler. The net count number (corrected for the background) is $n = 37\,940$.

150 cm^3 of ^{24}Na solution is used for labelling. The count rate of the tracer solution is determined in the same vessel (of 390 cm^3), from a 10 cm^3 sample, at a dilution of 50 000 times. The average count rate is 17 280 imp/min.

$$I_{rel} = 17\,280 \times \frac{150}{10} \times 50\,000 = 12.96 \times 10^9 \text{ imp/min}$$

Thus:

$$\frac{I_{rel} V}{n} = \frac{12.96 \times 10^9 \times 0.390}{3.794 \times 10^4} = 1.332 \times 10^5 \text{ dm}^3/\text{min} = 2.22 \text{ m}^3/\text{s}$$

11*

3.3.2. CONTINUOUS TRACER INTRODUCTION

This method involves the continuous introduction of a tracer with a radioactive concentration of c_j and at a volume rate of $v_{V,j}$ into the liquid flow to be measured having a volume flow rate of v_V and a radioactive concentration of c_0, Bq per unit volume. The c_V radioactive concentration is determined at a distance necessary for complete mixing (Fig. 3.20). The following mass balance can be written:

$$v_{V,j} c_j + v_V c_0 = (v_V + v_{V,j}) c_V \qquad (3.29)$$

Rearranging the equation we have:

$$v_V = v_{V,j} \frac{c_j - c_V}{c_V - c_0} \qquad (3.30)$$

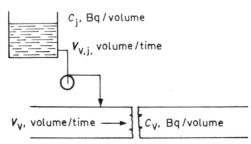

c_j, Bq / volume

$v_{V,j}$, volume / time

v_V, volume/time $\longrightarrow$ c_V, Bq / volume

Fig. 3.20. Methods based on continuous introduction of a tracer for determining flow rates

The value of c_0 is, as a rule, zero, except for systems with recirculation.
On the other hand, $c_j \gg c_V$. Considering this:

$$v_V = v_{V,j} \left(\frac{c_j}{c_V} - 1 \right) \approx v_{V,j} \frac{c_j}{c_V} \qquad (3.31)$$

The values of radioactive concentrations (expressed in Bq per unit volume) can be substituted by count rate per volume values (expressed, e.g., as imp/(min cm^3)) if the measurement is carried out with the same detector and in the same arrangement:

$$v_V = v_{V,j} \frac{i_j}{i_V} \qquad (3.32)$$

The tracer should be diluted several times for determination of the value of i_j.

EXAMPLE 3.6

In a water labelling study, the purpose is to determine a flow rate of $v_V \approx 5 \text{ m}^3/\text{s}$ with a relative standard deviation of 1%, NH$_4$^{82}Br produced in a reactor is used as a tracer.
What is the necessary activity (A) of ^{82}Br to carry out this relatively accurate measurement?
Considering industrial conditions and other aspects, a *continuous tracer introduction method* will be applied to measure the flow rate [Eqs (3.31) and (3.32), Fig. 3.20]. The tracer ^{82}Br with a radioactive concentration of c_j is introduced for a period of $t = 5$ min at a rate

of $v_{V,j}$ into the liquid flow having a rate of v_V. The value of c_V is determined by scintillation counting of a 10 dm^3 sample in an arrangement having a sensitivity to ^{82}Br as high as 405.4 imp/(s. $MBq.m^3$) [11.0 imp/(s nCi m^3)]. The samples and the background can be measured for 10 min, the background count rate is 100 imp/s.

The accuracy of measurement of v_V will depend on the error in the values of $v_{V,j}$, c_j and c_V. Since $v_{V,j}$ can be regulated with an accuracy of 0.5% ($\Delta v_{V,j}/v_{V,j} \times 100 \leq 0.5\%$) and the radioactive concentration (c_j) of the tracer can also be adjusted very accurately, it is obvious that the error of v_V will be determined by the deviations in the measurement of c_V (i.e., I_{rel}).

The following correlation can be written between the net count rate I_{rel} necessary for achieving the required 1% relative deviation and the background and the period of counting:

$$\sigma = \pm \frac{\sqrt{\dfrac{I_{rel} + 100}{10 \times 60} + \dfrac{100}{10 \times 60}}}{I_{rel}} = 0.01$$

On this basis:

$$0.06\, I_{rel}^2 - I_{rel} - 200 = 0$$

$$I_{rel} = 66.7 \text{ imp/s}$$

Considering the sensitivity of the measuring system, the count rate of 66.7 imp/s corresponds to a radioactive concentration of $c_V = 0.1645$ MBq/m^3 (4.5 $\mu Ci/m^3$).

The activity necessary for measurement is:

$$A = v_{V,j} c_j t = v_V c_V t = 5 \times 0.1645 \times 5 \times 60 = 246.75 \text{ MBq (6.7 mCi)}$$

3.3.3. EXAMPLES OF APPLICATION

Radiotracer methods are often used for the *calibration of flow meters*, both of liquids and gases. This is necessary when the arrangement of the pipelines cannot comply with the prescriptions for installation of the measuring elements, consequently the calculated values should be checked.

The models and equations discussed above hold good also in measuring and calculating flow rates of gases, just as well as for liquids: e.g., for a study of *ventilation*, for the determination of the flow rates of air and heating gases in *Siemens–Martin* steel production, methyl iodide labelled with ^{131}I can be applied [3.7].

The radiotracer technique is especially suitable for the measurement of the flow rate of liquids, as it is proved by the widespread application of the method. The cooling water requirement of *petroleum refineries* can be determined by the total count technique. Water requirements of *water turbines*, the cooling water consumption of *thermoelectric power station* can be investigated by the continuous tracer technique. Checking of the *sewage outlet* of factories requires continuous tracer technique in some cases, combined with the measurement by the pulse velocity method; in other cases the full sample technique is advantageous. The methods are not restricted to aqueous solutions: they are well suited for determination of flow rates of *petroleum* or other *organic liquids* if a tracer with appropriate solubility is available.

If the liquid does not fill the whole cross section of the tube, or if the flow occurs in an open system where the flow-cross section fluctuates between wide limits, radiotracer technique is about the only method permitting the determination of the flow rate and velocity. This is the case in measurements of *rivers, brooks, cavern waters,* and *sewage or irrigation water flowing in surface canals.*

Radioactive isotopes have become competitors of tracer dyestuffs or salt solutions used in studies of surface or *subsurface water systems;* the advantage is the extremely high sensitivity of their detection. The selection from among the radioactive isotopes listed in Table 3.2 should be done in consideration of several factors characteristic of the given task. Owing to the rapid development of the measuring technique of soft β-emitters, tritium has become more and more widespread in water labelling studies.

The flow rate of solid substances can be measured by hardly any other method than using radioactive tracers. Pulse-like injection and measurement of the pulse flow rate (Fig. 3.19(d)) is the most advantageous technique here.

The mass flow rate can be determined in this manner in *rotary kilns of cement factories* (Fig. 3.21); in general, ^{140}La isotope is used. In a given study it was found by means of measuring the tracer with detectors along the kiln that the mass flow was very slow in the first 30 m of the oven, presumably owing to a slowdown caused by the chains installed here. Granulation is characteristic of this section; as a result of this process, the flow rate is doubled along the further kiln section (about 90 m). In the final section of the kiln, the mass flow is again slower; this can be attributed to softening, aggregation, due to the formation of clinker.

In a continuous *paper digesting tower* ^{64}Cu fibres are used to study the flow rate of chips. It was found that the flow rate decreased gradually along the height of the tower (Fig. 3.22). From the technological point of view it is essential to prove that the chips flow through the tower under identical time, independently of their point of entry into the tower.

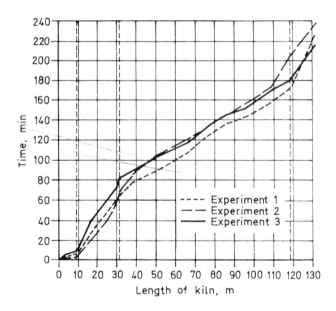

Fig. 3.21. Investigation of material transport in a rotary kiln of a cement factory

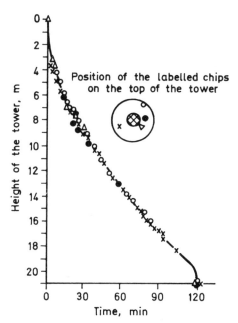

Fig. 3.22. Determination of the flow rate of ground wood in a digestion tower of a paper factory

3.4. STUDY OF MATERIAL FLOW IN INDUSTRIAL PROCESSES

Various types of material flow are required in different continuous systems. For example, the ventilation of a room is ideal if fresh air flows through the full cross-section, i.e., there are no unventilated dead volumes and, further, if the fresh air is practically not mixed with the vitiated one but it displaces the latter. In other cases the purpose is to produce a homogeneous composition; in such cases the substances introduced should be mixed with each other and with the substances being already in the system as completely as possible.

3.4.1. BASIC PRINCIPLES OF UNIT OPERATIONS

Mixing during flow determines the residence time distribution of the flowing substance in the system. Hence the concept of residence time distribution function can be used to characterize the mixing during flow, and thereby the flow itself.

The residence time is the time *t* during which a volume element of a given material can be found within an operation unit (tank, reactor, tube, etc.), i.e., the time between the entry and exit of the volume element.

The age of the volume element denotes the time elapsed since the time of its entry into the system.

In a batch type unit the period of time between the entry and exit of each volume element is identical with the time between the charging and emptying of the reactor; therefore, if the durations of the charging and emptying operations are relatively negligible, the residence time and the age of each element are equal.

Continuous units are simultaneously and continually charged and emptied. One of the characteristics of the residence of material in these units is the mean residence time $\bar{t}$, which can be calculated as the ratio of the full volume of the unit (V) to the volume rate of the feed (v_V):

$$\bar{t} = \frac{V}{v_V} \tag{3.33}$$

The mean residence time is, however, characteristic of the total fluid flow only, because it would be an ideal limiting case if every portion of the flow spent identical time in the apparatus. In this case no volume element of the feed would get ahead of another, so the volume elements would leave the unit in the order of their entry.

This ideal case is called *plug or piston flow* and this is similar to the movement of a plunger in a cylinder. Such a flow is characteristic of *ideal tube reactors*.

Perfect mixing would be characteristic of *stirred tanks*, where the entering volume elements would be divided instantaneously over the full volume of the unit. Neither of these limiting cases can be realized in practice.

The character of fluid flow in industrial units is *between these two ideal cases* because there is a more or less complete mixing in them. Longitudinal mixing causes some volume elements to be fast, others to be late. Therefore the flow leaving a continuous operation unit consists of *volume elements with various residence times*.

In order to investigate processes in industrial operation units precisely, it is necessary to know the functions describing the residence time distributions and their interrelationships. These distribution functions are age distributions. They will be discussed in a dimensionless system, where $t_A = t/\bar{t}$.

Internal residence time distribution, i.e., the *age of the material* in the reactor can be characterized by the $I(t_A)$ distribution function. The fraction of V volume in the unit with an age between t_A and dt_A is $I(t_A)\,dt_A$.
Since

$$\int_0^\infty I(t_A)\,dt_A = 1 \tag{3.34}$$

the fraction of material in the unit with an age lower than $\bar{t}$ is:

$$\int_0^1 I(t_A)\,dt_A \tag{3.35}$$

and similarly, the fraction with an age higher than $\bar{t}$ is:

$$\int_1^\infty I(t_A)\,dt_A = 1 - \int_0^1 I(t_A)\,dt_A \tag{3.36}$$

The mean age of the material in the reactor is:

$$\bar{t}_A = \int_0^\infty t_A\, I(t_A)\,dt_A \tag{3.37}$$

The shapes of the $I(t_A)$ functions are shown in Fig. 3.23 (a) for plug flow, in Fig. 3.23 (b) for perfect mixing and in Fig. 3.23 (c) for a practical unit.

The external residence time distribution, i.e., the distribution of the age of the material leaving the unit is shown in Fig. 3.24 (a) for plug flow, in Fig. 3.24 (b) for perfect mixing and for a practical case in Fig. 3.24 (c).

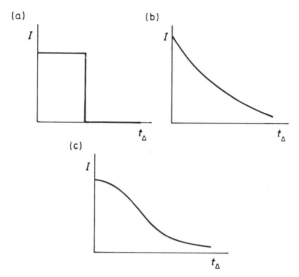

Fig. 3.23. Internal residence time distributions: (a) for plug flow; (b) for perfect mixing; (c) in actual practice

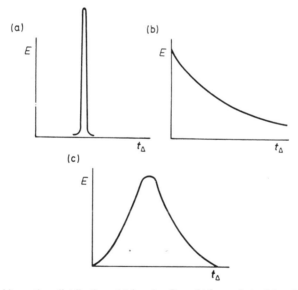

Fig. 3.24. External residence time distributions: (a) for plug flow; (b) for perfect mixing; (c) in a practical case

According to the definition of the $E(t_A)$ function, the fraction of the total outflow from the unit the age of which is between t_A and $t_A + dt_A$ is $E(t_A) dt_A$. Since the total amount is between $t_A = 0$ and $t_A = \infty$,

$$\int_0^\infty E(t_A)\, dt_A = 1 \tag{3.38}$$

The fractions having lower and higher ages than $\bar{t}$, respectively, are:

$$\int_0^1 E(t_A)\,dt_A \text{ and } \int_1^\infty E(t_A)\,dt_A = 1 - \int_0^1 E(t_A)\,dt_A \tag{3.39}$$

The mean age of the material leaving the reactor is:

$$\bar{t}_{A,E} = \int_0^\infty t_A\, E(t_A)\,dt_A \tag{3.40}$$

The following correlation exists between $I(t_A)$ and $E(t_A)$:

$$\int_0^1 E(t_A)\,dt_A = 1 - I(t_A) \tag{3.41}$$

The distribution function of the residence time $F(t_A)$ gives the relative occurrence of ages lower than $\bar{t}$:

$$F(t_A) = \int_0^1 E(t_A)\,dt_A = 1 - I(t_A) \tag{3.42}$$

Figure 3.25 shows the shapes of the function $F(t_A)$ for the basic types of flow.

Various factors have been proposed to *characterize the extent of deviation from the ideal cases of plug flow and complete mixing*. These factors assume the knowledge of the residence time distribution functions of the given system.

If there are dead volumes (V_d) in the V volume of the apparatus, the actual volume participating in the mass flow is $V_a = V - V_d$. In this case, the actual value of residence time, $\bar{t}_a$, is different from the value calculated from the volume rate v_V by means of the formula $\bar{t} = V/v_V$:

$$\bar{t}_a = \frac{V - V_d}{v_V} < \bar{t} = \frac{V}{v_V} \tag{3.43}$$

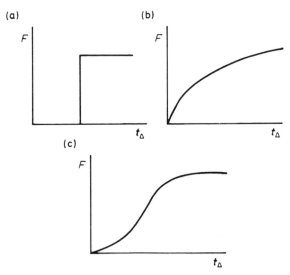

Fig. 3.25. Distribution functions of the residence time: (a) for plug flow; (b) for perfect mixing; (c) in actual practice

The relative size of the dead volume is:

$$\frac{V_d}{V} = 1 - \frac{\bar{t}_a}{\bar{t}}$$ (3.44)

The *H hold-back* expresses the fraction of the original filling of the reactor being still in it after the flow of V volume, i.e., at $t_A = 1$, if the mass composition is changed at the initial moment. H expresses the fraction of material older than $t_A = 1$:

$$H = 1 - \int_0^1 I(t_A)dt_A = 1 - \left[\int_0^1 [1 - F(t_A)]\,dt_A\right] = \int_0^1 F(t_A)dt_A$$ (3.45)

In the case of *plug flow* $H = 0$, i.e., the value of hold-back reflects the deviation from this ideal case, in other words, the extent of mixing. Since—as it can be derived—for complete mixing:

$$F(t_A) = 1 - \exp(-t_A)$$ (3.46)

the hold-back in this case is:

$$H = \int_0^1 [1 - \exp(-t_A)]\,dt_A = \frac{1}{e} = 0.368$$ (3.47)

The mixing of the material (equalization of concentration) often depends on factors other than turbulence, recirculation, etc. in the tank. The actual process may involve irregularities in the fluid flow: short circuits, or conversely, dead or partly dead volumes, where the material stagnates or flows very slowly. The two types of effect can be distinguished on the basis of the intensity function $\lambda(t_A)$:

$$\lambda(t_A) = \frac{E(t_A)}{1 - F(t_A)}$$ (3.48)

If the $\lambda(t_A)$ function has a maximum or decreases monotonically, this indicates stagnation of the fluid flow. In the case of regular mixing characteristic of ideal stirred tanks or tanks and plug flow volumes linked in series, $\lambda(t_A)$ is constant or increases monotonically.

3.4.2. DETERMINATION OF RESIDENCE TIME DISTRIBUTION FUNCTIONS

In order to obtain information about the flow or mixing in a given unit, some property of the entering material should be changed. This can be done according to an arbitrary time function. The shape of this signal is modified by mixing and flow in the unit: *the output signal has changed as compared with the input*. The shape of the output signal is a function of the flow and mixing phenomena occurring in the reactor, thus, with a given input signal, its change at the output can give sufficient information to calculate the distribution functions of the residence time of the material.

Radioactive labelling is a concentration labelling as far as its essence is concerned; it is generally applied according to two types of time functions:

Pulse-like tracing means the injection of a tracer of A Bq activity during an infinitesimally short dt time period. The radioactive concentration (c) of the input material changes according to the following time functions:

$$c(t) = \begin{cases} 0, & t<0 \\ \infty, & t=0 \\ 0, & t>0 \end{cases} \quad \text{and} \quad \int_{-\infty}^{+\infty} c(t)dt = A \qquad (3.49)$$

Figure 3.26 (a) shows the shape of the *input signal*. The *output signal* is the radioactive concentration of the material leaving the unit, as a function of time: $[c'(t)]$. Let us investigate—with a given input signal—the correlation between $c'(t)$ and the frequency function of the external *residence time distribution* characteristic of the system. When doing so, it should be considered that $c'(t)$ is the radioactive concentration of the material leaving the system between t and $t+dt$ (expressed in Bq per unit volume).

If the flow rate is v_V, the activity leaving the system between t and $t+dt$ is $v_V dt c'(t)$ Bq. The same activity can also be expressed by the frequency function of the external residence time distribution: $AE(t)dt$ activity leaves the system at t time, of the total input activity of A.

Comparing the two expressions:

$$AE(t)dt = v_V dt c'(t) \qquad (3.50)$$

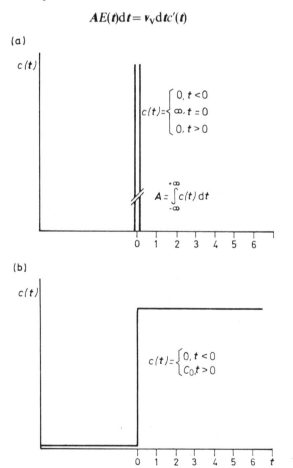

Fig. 3.26. Introduction of a radiotracer: (a) pulse-like injection; (b) continuous introduction

Obviously, until $t = \infty$, the total activity leaves the unit, thus:

$$A = v_V \int_0^\infty c'(t)dt, \quad \text{since} \quad \int_0^\infty E(t)dt = 1 \tag{3.51}$$

From Eqs (3.50) and (3.51) we have:

$$E(t) = \frac{c'(t)}{\int_0^\infty c'(t)dt} \tag{3.52}$$

The $c'(t)$ radioactive concentration can be substituted in Eq. (3.52) by the specific count rate, $i(t)$, expressed in imp/min.

The above discussion shows the method of determination of the $E(t)$ function: a tracer pulse is injected into the fluid flow entering the operation unit, and the count rate of samples taken from the output flow, or the count rate measured at a section of the outlet tube, is used to construct an i vs. t diagram. The i value corresponding to any t time divided by the area under the curve gives the value of $E(t)$.

Continuous tracing involves the dosing of a radioactive tracer, at a constant input rate from $t = 0$ onwards, into the input stream. Thus, the radioactive concentration in the fluid will increase by a sudden jump up to c_0 at $t = 0$, and remain constant at this value. Accordingly, the input signal (Fig. 3.26 (b)) will be as follows:

$$c(t) = \begin{cases} 0, & t < 0 \\ c_0, & t > 0 \end{cases} \tag{3.53}$$

As a consequence of labelling, the output stream becomes radioactive; its radioactive concentration, i.e., the output signal increases from $c'(t) = 0$ up to c_0.

According to the definition of the $F(t)$ distribution function of the residence time (Eq. (3.42)), the correlation existing between the input and output signals is the following:

$$F(t) = \frac{c'(t)}{c_0} \tag{3.54}$$

The value of $c'(t)$ can be substituted by the specific count rate.

Even the shape of a residence time distribution function determined by radiotracers may give useful information in a qualitative sense, but it can also be quantified if the above-discussed characteristics, the real mean residence time, the dead volume, the hold-back and the intensity function are calculated.

In the knowledge of the distribution functions and the above-mentioned characteristics, the modelling of the system to be investigated can be attempted. This is justified if the evaluation points to considerable differences between the real system and either of the ideal limiting cases. Modelling involves the construction of a system resulting in a residence time distribution similar to the one to be investigated. The construction can be done by *coupling ideal stirred tanks* (creating a so-called *cascade* (tanks-in-series)), or by so-called dispersion spaces or by combining various flow spaces.

EXAMPLE 3.7

In a sugar factory the liquor to be purified passes through a subsider where the sludge precipitates. The residence time of the material, t, should be determined in the tank of a volume $V = 3260$ dm^3, at a volumetric flow rate of $v_V = 61$ m^3/h. The task is solved by a pulse-like injection of 200 MBq of ^{82}Br into the tank. The radioactivity of the outflow is measured.

If a well-soluble material is added into a tank into which a liquid is flowing at a constant rate and outlet stream is leaving it at the same rate, then, assuming ideal mixing, the following differential equation will describe the concentration change in the tank and in the outflow, respectively, as a function of the time:

$$-\frac{dc}{dt} = \frac{v_V}{V} c \tag{3.55}$$

where c denotes the concentration of the material at t time; v_V is the flow rate of the liquid and V the volume of the tank.

The solution is the following (with the initial condition: if $t=0$, $c=c_0$):

$$c = c_0 \exp(-v_V t/V) \tag{3.56}$$

Let us introduce the half-life for characterizing the process; accordingly, if $t=t_{1/2}$ then $c=c_0/2$:

$$t_{1/2} = \frac{V}{v_V} 0.693 \tag{3.57}$$

The residence time is characterized in our example by the half-life according to Eq. (3.57) as follows:

$$t_{1/2} = \frac{3260}{61\,000} 0.693 \times 60 = 2.2 \text{ min}$$

Figure 3.27 shows the actual concentration change: the decrease is initially faster, afterwards slower than expected on the basis of the theoretical calculation. The experimental half-life is 2.9 min, and this is rather close to the calculated value.

The result indicates that the mixing in the settler approaches ideal conditions fairly well.

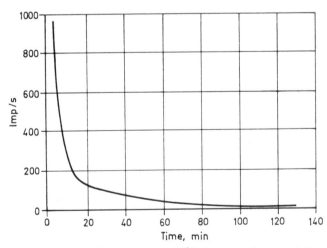

Fig. 3.27. Changes of the count rate of ^{82}Br in a sugar factory subsider

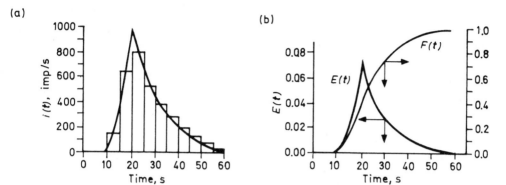

Fig. 3.28. Tracer study of a tube reactor: (a) the count rate, with a pulse-like tracer injection, at the outlet of the reactor as a function of time; (b) distribution functions $E(t)$ and $F(t)$ calculated from the above data

Table 3.6. Data and results to Example 3.8

i	t, S	$i(t)$	$i(t)\Delta t$	$E(t$	$E(t)\Delta t$	$F(t$	$tE(t)\Delta t$
(0)	(1)	(2)	(3)	(4)	(5)	(6)	(7)
0	7.5	0	0	0	0	0	0
1	12.5	150	750	0.0092	0.046	0.046	0.575
2	17.5	650	3250	0.0400	0.200	0.246	3.500
3	22.5	815	4074	0.0502	0.251	0.497	5.650
4	27.5	535	2675	0.0329	0.164	0.661	4.510
5	32.5	385	1925	0.0237	0.118	0.779	3.840
6	37.5	285	1425	0.0175	0.088	0.867	3.300
7	42.5	200	1000	0.0123	0.062	0.929	2.640
8	47.5	130	650	0.0080	0.040	0.969	1.900
9	52.5	75	375	0.0046	0.023	0.992	1.205
10	57.5	25	125	0.0015	0.008	1.000	0.460
11	62.5	0	0	0	0	1.000	
$\sum\limits_{i=0}^{n}$	—	—	16249	—	1.000	—	27.58

Remarks:
(1) Taken from Fig. 3.28 (a)
(2) Taken from Fig. 3.28 (a)
(3) $\Delta t = 5$ s

(4)
$$E(t) = \frac{i(t)}{\int\limits_0^\infty i(t)\,dt} \approx \frac{i(t)}{\sum\limits_{i=0}^\infty i(t)\Delta(t)} \qquad (3.52)$$

(5)
$$E(t)\Delta t = 5E(t)$$

(6)
$$F(t) = \int\limits_0^t E(t)\,dt \approx \sum\limits_{i=0}^\infty E(t)\Delta t \qquad (3.42)$$

(7)
$$\bar{t} = \int\limits_0^\infty tE(t)\,dt \approx \sum\limits_{i=0}^\infty tE(t)\Delta t \qquad (3.40)$$

EXAMPLE 3.8

A liquid flow entering a tube reactor is labelled in a pulse-like way and the count rate is measured at the outlet as a function of time, i.e., the $i(t)$ function (Fig. 3.28 (a)).

Data have been summarized in Table 3.6. Let us calculate the following characteristics:

— the frequency function of the external residence time distribution $E(t)$ according to Eq. (3.52);

— the distribution function of the residence time, $F(t)$, according to Eqs (3.52) and (3.42);

— the mean residence time, $\bar{t}$, according to Eq. (3.40);

— the percentage of outlet material with $t = 20$ s residence time (Eq. (3.42));

— the percentage of the material in the reactor, the residence time of which is between $15 \leq t \leq 20$ s according to Eq. (3.42): $I(t_A) = 1 - F(t_A)$, but since $t_A = t/\bar{t}$, $I(t) = 1/\bar{t}[1 - F(t)]$.

The solution of the first three parts of the example can be read from the Table: functions $E(t)$ and $F(t)$ have been plotted in Fig. 3.28 (b); $\bar{t} = 27.58$ s.

As Fig. 3.28 (b) shows, $F(20) = 0.4$, i.e., 40% of the material leaves the reactor with a residence time $t \leq 20$ s.

The fraction of the material with a residence time between t and $t + \Delta t$ corresponds to $I(t)\Delta t$. We look for $I(15)$:

$$I(15) = \frac{1}{27.58} [1 - F(15)] = \frac{1}{27.58} (1 - 0.12) = 0.032$$

$$I(t)\Delta t = I(15) \times 5 = 0.032 \times 5 = 0.16$$

Thus, 16% of the material in the reactor has a residence time t between 15 and 20 s.

3.4.3. INDUSTRIAL EXAMPLES OF MATERIAL FLOW STUDIES

In the *production of alumina*, the slurry flow was studied in two continuous bauxite digesting autoclave series by means of a bauxite labelled with ^{59}Fe. Analysis of distribution functions of the residence time revealed the presence of considerable dead volumes in the autoclave battery stirred with direct steam blow, and smaller dead volumes in the battery equipped with mechanical stirrer. The study indicates that a change in the way of stirring may be advisable, as this could lead to an increase of the useful volume of the autoclave battery.

The flow of red mud and that of the slurry in Dorr thickeners of alumina factories can be investigated by red mud labelled with ^{59}Fe and *aluminate liquor* labelled with ^{24}Na. The study allows the construction of the hydrodynamical model of the thickener.

In the paper industry, tracer studies can detect dead volumes in the bleaching process of cellulose fibres. The real residence times may be considerably lower than those calculated from the volume of the system and the flow rate. This can be attributed to the gradual breaking of Raschig rings and channelling in the tower packing. The fluid will pass through these channels instead of being distributed over the entire cross-section. Glass fibres activated by neutron irradiation (^{24}Na) can be used as tracers.

In petroleum pipelines, e.g., ^{140}La-naphthenate may serve as a tracer. In this way, the mixing of various types of petroleum with each other, as well as that of gasoline and Diesel fuel during flow can be studied. The plots of activity *vs.* time give almost regular *Gaussian* curves at various detection sites, their maxima decreasing and the half-width-valves

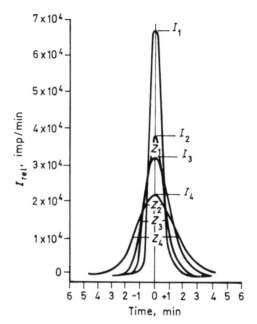

Fig. 3.29. Investigation of mixing processes in a petroleum pipeline. I_1, I_2, I_3, I_4 denote the height of the curves, Z_1, Z_2, Z_3, Z_4 their half-width values

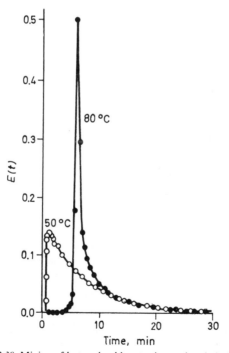

Fig. 3.30. Mixing of hot and cold water in an electric boiler

increasing as the distance from the injection point increases (Fig. 3.29). The purpose of such studies is to determine the amount of mixtures formed when different liquids are pumped after each other, eventually requiring repeated refining. A mathematical analysis of the distribution curves allows the solution of the problem.

In electric boilers, a flow study with $Na^{131}I$ tracer demonstrated that the mixing of warm and cold water displacing the former showed hardly any dependence on the structure and shape of the installed heater in some constructions, but there was a close correlation with the temperature of the hot water found in the boiler. If the hot water temperature was 80°C, its displacement by cold water occurred almost like a plug flow, whereas a very intense mixing with the inlet cold water occurred when the water temperature was 50°C. This is unambiguously demonstrated by the shape of the distribution functions of residence time, shown in Fig. 3.30.

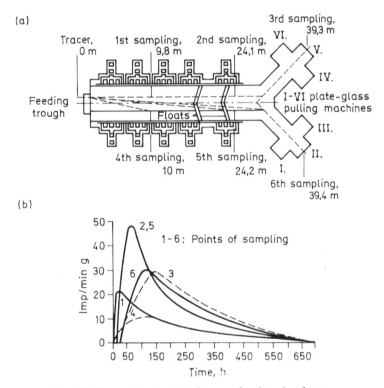

Fig. 3.31. Flow studies in a tank furnace of a plate-glass factory

The flow in a 720 t continuous tank furnace of a *plate-glass factory* was studied by a glass containing ^{51}Cr tracer. The experimental arrangement and the activity *vs.* time plots at given sampling points are shown in Figs. 3.31 (a) and (b), respectively. The analysis of the curves shows that the bulk of the material flows in the middle of the first quarter of the furnace. Later on, the flow is distributed throughout the whole cross-section, but the last third of the furnace responds very sensitively to the loading changes of the continuous plate-glass pulling machines. Axial dispersion characteristic of the flow is rather high and recirculation flows also contribute to it.

3.4.4. STUDY OF THE HOMOGENEITY OF MIXTURES

Homogeneity in a chemical sense means the *uniformity of the chemical composition*. In *multiphase disperse systems*, the individual chemical properties are bound to well-defined particles, e.g., in solid mixtures, to macroscopic granules of the components. Obviously, the uniformity of chemical composition of such systems depends on the spatial distribution of the granules.

Two extremes can be distinguished:

— the particles of each component are found in *separate groups*; this is the state of maximum inhomogeneity;

— the particles fill the space in a *random way*; this is the state of maximum mixing or random homogeneity.

These two extreme cases are the initial and final states of a *batch-like mixing* (homogenizing) process; during mixing, the former state transforms into the latter continuously.

The *concentration* of a given j component related to the whole system (i.e., its average concentration, $\bar{c}$) remains unchanged, but the spread of this average concentration to samples taken from various points of the system gradually decreases.

The measure of the homogeneity of the system at time t is the standard deviation (σ) of the differences between the concentrations measured in samples taken from various places (c_t) and the mean concentration ($\bar{c}$).

If the value of c_t is determined at time t, in sample n and $\bar{c}$ is calculated, the following values are obtained for the standard deviation (σ) and the relative standard deviation (σ_{rel}), respectively:

$$\sigma = \pm \sqrt{\frac{\sum_{i=1}^{n} (c_{t,i} - \bar{c})^2}{n-1}} \tag{3.58}$$

$$\sigma_{rel} = \pm \sqrt{\frac{\sum_{i=1}^{n} (c_{t,i} - \bar{c})^2}{\bar{c}^2 (n-1)}} \tag{3.59}$$

The relative standard deviation is a more advantageous characteristic of the homogeneity of the system, because its value is independent of the concentration unit (mole fraction, mass (weight) fraction, rel.%, etc.). Its value remains unchanged even if the concentration is expressed in specific activity (a [Bq per mass]), radioactive concentration (c [Bq per volume]) or specific count rate (i, e.g., imp/(min g)).

The tracer study of granulated materials in batchwise mixing or homogenizing processes can be carried out by admixing a radioactive tracer before starting the mixing. The most advantageous tracing may consist in the neutron activation of a small sample of one of the components, most suitably that of the important agent of the blend. Several samples should be taken at various time intervals during mixing and the relative standard deviation of their specific activity should be determined, according to Eq. (3.59). The progress of homogenization can be checked by the gradual decrease of the relative standard deviation of the specific activity with time. When the state of random homogeneity is reached, the relative standard deviation does not change any more. This indicates the time necessary for homogenization with given materials and a given apparatus.

This technique is applied for the study of mixing of *glass, refractory materials, starting materials for artificial carbon, synthetic fertilizers, agents and carriers for herbicides, petrol plus ethyl fluid, PVC powder and ingredients for plastics production*. The results can be used for characterizing of the homogeneity of the products, too, as well as for *classifying mixing machines of various design*.

EXAMPLE 3.9

50 kg cobalt and tungsten carbide 450 kg powder are homogenized in a drum mixer for hard metal production. Determine the time necessary to reach homogeneity!

Tungsten carbide containing [187]W produced by neutron irradiation is used as a tracer. One hundred gram of this tracer is added to the mixture. Samples of 10 g each, 10 at the same time, are withdrawn and their relative count rates determined (*i* values are shown in Table 3.7, expressed in imp/(min g)).

Table 3.7. Specific count rates (imp/(min g)) of samples taken at various moments of mixing and their relative standard deviations

Number of the sample	Specific count rate, i, imp/min g			
	$t_1 = 2.5$ min	$t_2 = 10$ min	$t_3 = 20$ min	$t_4 = 30$ min
1	1547	1292	1255	1262
2	1429	1191	1274	1226
3	1313	1305	1290	1268
4	1741	1272	1327	1230
5	1264	1130	1285	1244
6	1484	1242	1288	1221
7	1058	1342	1262	1264
8	918	1187	1260	1294
9	1248	1370	1313	1233
10	998	1269	1246	1258
$\bar{i}$	1300	1260	1280	1250
σ	260	74	25.9	23.1
σ_{rel}	0.200	0.0587	0.0202	0.0185

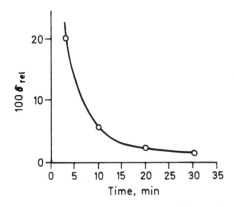

Fig. 3.32. Homogenization of tungsten carbide powder

The mean value of the specific count rates (i) is determined, then the relative standard deviation (σ_{rel}) of the measurement series is calculated. The results of the calculation are also given in Table 3.7. Figure 3.32 shows the values of σ_{rel} as a function of time.

The figure indicates that the value of σ_{rel} does not decrease any more after mixing for 20 minutes; the given blend has become randomly homogeneous, its uniformity cannot be increased by further mixing.

3.5. MEASUREMENT OF MASSES OR VOLUMES AND MASS TRANSFER

3.5.1. DETERMINATION OF MASSES OR VOLUMES

The method is based on the same principles which are used in isotope dilution or reverse isotope dilution analysis (see Section 4.3.1), independently of the fact whether a simple mass measurement or measurements repeated several times in connection with mass transfer processes are made. The method can only be applied in systems where the homogeneous mixing of the isotope is ensured. Homogeneity should be checked separately.

If a tracer with a volume of V_j and with a radioactive concentration of c_j is dispersed in a material of V volume and c_0 radioactive concentration, and the radioactive concentration after homogenization is c_i, the following mass balance can be written:

$$Vc_0 + V_j c_j = (V + V_j) c_i \qquad (3.60)$$

The equation can be solved for either V or V_j; the former is the basic equation for *isotope dilution analysis*, whereas the latter is the same for *reverse isotope dilution analysis*:

$$V = V_j \frac{c_j - c_i}{c_i - c_0} \qquad (3.61a)$$

$$V_j = V \frac{c_i - c_0}{c_j - c_i} \qquad (3.61b)$$

If $c_0 = 0$ and $c_j \gg c_i$, we have:

$$V = V_j \frac{c_j}{c_i} = \frac{A_j}{c_i} \qquad (3.62a)$$

$$V_j = V \frac{c_i}{c_j} = \frac{A_i}{c_j} \qquad (3.62b)$$

If specific activities (a_0, a_j, a_i) are used (expressed in Bq/g) instead of radioactive concentrations, it is possible to determine the mass of the material (m, m_j) instead of their volume:

$$m = m_j \frac{a_j}{a_i} = \frac{A_j}{a_i} \qquad (3.63a)$$

$$m_j = m \frac{a_i}{a_j} = \frac{A_i}{a_j} \qquad (3.63b)$$

Absolute activities (A_j, A_i), radioactive concentrations (c_j, c_i) and specific activities (a_j, a_i) can be substituted in Eqs (3.62) and (3.63) by the count rates $(I_{rel,j}, I_{rel,i}, imp/min)$ and count rates per volume unit $(i_{j,c}$ and $i_{i,c}, imp/(min\ cm^3); i_{j,a}$ and $i_{i,a}\ imp/(min\ g))$ determined in the same arrangement. The determination of $I_{rel,j}$ and i_j should be carried out after a manifold dilution.

In the case of the application of the *isotope dilution method*, the material to be investigated is labelled with a tracer of known absolute and specific radioactivity; after its homogeneous dispersion the resulting specific activity is determined, and the quantity of material is calculated by Eqs (3.61a), (3.62a) and (3.63a).

For determination, of *the amount of dross in blast* or *cupola furnaces*, a rare earth metal oxide (e.g., $^{140}La_2O_3$) is used as a tracer, because it is not reduced to metal and certainly remains in the dross.

In electric furnaces, the amount of the molten metal should be known with certainty in order to be able to calculate the dosage of expensive alloying metals. The isotope dilution method can be used with advantage to solve this problem in metallurgy. Procedures using short half-life ^{65}Ni isotope for this purpose are, as a rule, successful.

From time to time, the "stock-taking" of mercury in alkali chloride electrolysis cells should be performed. If a ^{203}Hg tracer is used, switching off the cell and emptying the mercury are unnecessary. The accuracy of the tracer method is higher than that of weighing, because graphite elements of the cell sometimes retain as much as 100–150 kg of mercury. The accuracy of this dilution method is 0.8 rel.%.

A reverse isotope dilution method is applied when a small fraction of a large amount of material is admixed to a known amount of another substance and the determination by chemical methods is—owing to the low concentration—uncertain. Equations (3.61b), (3.62b) and (3.63b) should be used here.

The method was used to *detect leaking* of the cooling water into tins after the sterilization process. Such a leaking may cause spoilage of closed cans or jars. The cooling water was labelled with ^{24}Na in an experimental autoclave. In order to avoid adsorption of ^{24}Na, the tins and jars to be investigated were filled with brine. It was possible to detect the leaking of such a low amount of cooling water as 0.2 μl. By means of this examination various methods of sealing could be compared to establish their efficiency (see Example 3.2).

Material loss in manufacturing processes is an interesting and important technological characteristic. For example, the mercury stock in certain *alkali chloride electrolysis cells* should be labelled with ca. 10 GBq (0.1–1 Ci) ^{203}Hg per cell. This allows the determination of the Hg content in the Cl_2 and H_2 gases, as well as in the NaOH and NaCl solutions even if the mercury concentration is as low as 0.1–1 μg/cm^3. Thus, in the knowledge of the mass balance of the manufacturing, the sources of the mercury loss can be discovered.

In nitric acid factories, gases leaving the ammonia oxidation plant and containing about 10% nitrogen oxides are cooled by the cold end gases leaving the absorption system. Labelling the hot gas with ^{133}Xe, it can be traced if even 1–2% of it is mixed to the end gases through the leaks of the cooler, avoiding thus the absorption system. This fraction of the gas is lost (see Example 3.4).

Quantitative data can be obtained as far as the losses of the *platinum net catalyst* in nitric acid production are concerned. The method gives information about the sites of concentration of platinum powder carried over by the gas stream. The process is traced by ^{192}Ir; the net with a given mass has a known specific activity with respect to the radioactive isotope.

In petroleum refinery, entrainment in vacuum distillation towers is determined by reverse isotope dilution technique. From lower trays, not only vapour gets to the upper ones but also liquid drops entrained by the vapour. Consequently, low-boiling distillates taken from the upper trays are contaminated with high-boiling fractions. If the liquid on the lower tray is labelled by a nonvolatile substance (^{60}Co-naphthenate, triphenyl antimonate labelled with ^{121}Sb, etc.), then any radioactivity appearing on the upper trays is the indication of entrainment.

Corrosion studies are carried out according to the reverse dilution method. The advantage of the procedure is that corrosion losses are determined by measuring the activity of the dissolved metal instead of following the loss of activity of the solid plates. The latter technique would ensure much less sensitivity. The examinations can be extended to the study of corrosion of nonmetallic coatings: e.g., the dissolution of enamels can be studied by enamels containing ^{22}Na tracer. The same method may be used for the determination of chemical resistivity or the solubility of glasses.

The study of *action mechanism* of *detergents* is an important field of application in mass transfer research. Appropriate tracers enable us to check the efficiency of *de-greasing before electroplating* or to study the efficiency of *washing ingredients* or that of *flotation* as a function of the used additives (wetting agents or modifiers).

Mass transfer studies have been carried out also in *welding*. The application of a ^{51}Cr tracer may help to decide the origin of chromium in arc welding of stainless steels. This may be important in the reduction of chromium losses. By labelling either the substance to be welded, or the electrode, or the welding rod, it can be decided whether the chromium in the weld originates from the welding rod. The optimum chromium content can be controlled by a reasonable increase of the chromium content of the welding rod.

3.5.2. WEAR STUDIES

The importance of tracer investigations of mass transfer processes increases as longer and longer lifetimes in machine industry are required. The determination of smaller and smaller changes in dimensions means, from the point of view of metrology, that negligibly small amounts have to be measured in comparison with the original dimensions of a wearing machine part.

Traditional wear studies—such as the measurement of mass or dimensional changes of the machine part in question, the application of arched scratches or *Vickers*-traces, as well as the chemical analysis of wear products accumulated in the lubricant—have limited sensitivities, require lengthy experiments and are expensive. Only the last method on the above list does not require periodic dismantling of the unit containing the machine part to be investigated. Hence, the wear plot is determined by most methods in a discontinuous way, involving also the disadvantage of repeated running-in after reassembling. A further disadvantage of traditional methods is that the factors causing wear should be maintained at a constant level and this requirement can be fulfilled with difficulties in practice. The above-mentioned principial and practical difficulties have been eliminated to a large extent by radioactive wear studies.

The *machine part* to be investigated is labelled with a radioactive isotope. The active part is built in the appropriate machine unit and operated properly. The activity of worn products getting into the lubricant and suspended there is measured by a detector. The extent of wear is indicated by the count rate being proportional to the amount of worn products. The proportionality factor is determined by appropriate calibration; in this way it is possible to obtain information not only about the changes of the nature of the wear curve, but also about the absolute amounts worn off.

Several possibilities are possible for *labelling*. It is possible to label the metal *during casting*, in the molten state, before preparing the labelled object. This method requires the use of a relatively large amount of radioactive isotope and this is undesirable from the point of view of radiation protection. In the *diffusion* process, the tracer is made to diffuse into the surface. In this case it should be considered, however, that the specific activity will decrease toward the bulk of the solid from its surface. *Galvanic coating* of the surface by a tracer alter the surface parameters and falsifies thus the results of the wear studies, except for the case when the worn surface is electroplated anyway (e.g., chromium plated piston rings). The "forced wear" principle involves inserting radioactive rivets into holes bored into the surface whose wear is to be studied. This measurement gives, however, very uncertain information only, as far as the behaviour of the whole surface is concerned. The most suitable method is the *neutron activation* of the whole object to be investigated—if this is possible—although inhomogeneities in the neutron flux may cause specific activity differences in larger objects. The chemical analysis of worn products can also be performed by *subsequent neutron activation*—when justified—this is, as a matter of fact, an activation analysis (see Section 4.2).

Isotopic wear studies are successfully used in the *motor vehicle industry*. The wear of *piston rings, cylinders, bearings, fuel pumps* of engines, etc. can be determined by detecting the radiation of radioactive worn products getting into the lubricating oil or fuel. Wear of piston rings is studied not only as a function of the engine parameters (number of revolution, loading, etc.), but one can also draw conclusions about the effect of outer factors, because the ring wear is influenced by the quality of the fuel and lubricant, as well as by the efficiency of the oil and air filtration.

Figure 3.33 shows an experimental arrangement for the determination of radioactive piston ring wear of a *diesel engine* built in an autobus. Figure 3.34 shows a scheme for a better understanding of the operation of an apparatus suitable for the measurement of wear of piston rings in an air compressor. Compressors *1*, driven by electric motors *13*, compress air into the air tank *2* (with a blow-off valve, oil separator *12* and a common pressure regulator *3*), from where it goes off into the atmosphere through an adjustable valve *11*. Two independent lubricating circuits serve for oiling the compressors. An exactly known amount of lubricating oil is filled into the oil tanks *4*. The oil is thermostated by the water circuit of a thermostat *8* (e.g., at 80 °C) and is pumped by centrifugal pumps immersed into the tanks toward the measuring elements *5*, from where it gets into the

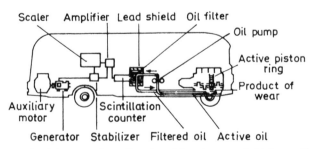

Fig. 3.33. Study of the wear of a radioactive piston ring mounted into a vehicle engine

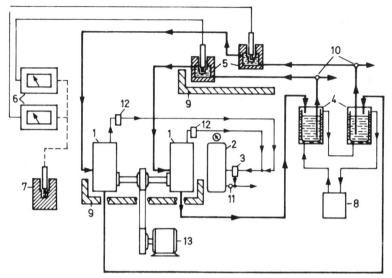

Fig. 3.34. Laboratory device for wear studies of piston rings in an air compressor

compressors. Scintillation counters are immersed into the measuring elements shielded with lead (*7*). The pulses are recorded by the counters *6*. The measuring of standard preparations from time to time serves for checking the proper operation of the counters and measuring elements. Three-way stopcocks *10* render it possible to make samples from the oil at regular intervals. The oil leaving the oil outlet orifices of the compressors is led back to the oil tanks. Lead shields *9* protect both the measuring elements and the experimenters from the radiation of the radioactive piston rings.

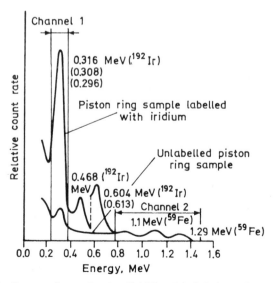

Fig. 3.35. γ-Spectra of normal and radioiridium-labelled piston ring samples

γ-spectrometry opens up the possibility to perform wear studies in multicomponent systems, i.e., by simultaneous labelling of two or more parts. Different parts worn in the same process are labelled by different isotopes emitting γ-quanta of different energies. Abrasion products are analysed by single- or multichannel analysers. Figure 3.35 shows the γ-spectra of piston rings labelled with ^{192}Ir and ^{59}Fe, worn in the apparatus shown in Fig. 3.34. The simultaneous wear of the two types of ring can be determined by the measurement in the two channels of the spectrometer (Fig. 3.36).

Motor car tyres can also be labelled to study their wear: the running surface can be prepared of rubber labelled with triphenyl phosphate containing ^{32}P. The outer 4 mm thick layer of the tyre should contain an activity as high as about 7 GBq (0.2 Ci). The dispersion of worn off rubber in air is monitored by a detector installed in the car; the worn substance adhering to the road surface can be determined by autoradiography.

The kryptonate technique has brought interesting results in the field of both wear studies and investigations of other surface changes of solid objects. Since ^{85}Kr is a noble gas, it belongs to the least dangerous radioisotopes from the biological point of view.

Diffusion or the *ion* implantation *bombardment*, technique is used for incorporation of ^{85}Kr of high radioactive concentration into the surface layer of the solid to be examined. A pressure of 30–40 MPa (300–400 bar) is necessary for the diffusion method. For ion bombardment, a discharge tube is used; the "kryptonation" process involves the ionization of krypton gas by electrons accelerated by high voltage, followed by the impact of the positive krypton ions into the target switched as a negative electrode. The impact causes incorporation of the ions (Fig. 3.37) into a depth of 0.1–10 μm. This fact explains why the desorption of "sorbed" noble gas atoms is influenced very strongly by any interaction involving the surface. This makes the kryptonate technique suitable for studies in the field of reaction kinetics, analysis, solid state physics, wear, corrosion and surface deformation.

When measuring *peak operation temperatures*, the metal part to be studied is kryptonated by, e.g., ion bombardment, then mounted into place and let to operate. The extent of desorption of krypton will depend on the maximum operation temperature. After dismantling, the part is heated gradually, starting from room temperature, and the count rate proportional to the ^{85}Kr content of the surface is measured at the various temperatures. There is a "memory effect" indicating the maximum operation temperature;

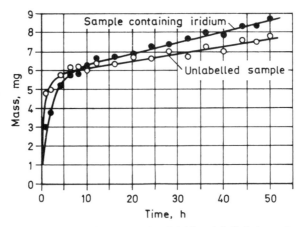

Fig. 3.36. Wear plots of normal and radioiridium-labelled piston rings

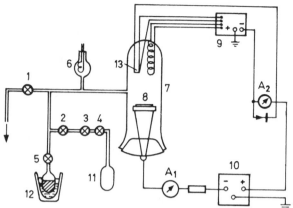

Fig. 3.37. Device for kryptonation by ion bombardment. 1, 2, 3, 4, 5 — vacuum stopcocks; 6 — vacuum gauge; 7 — discharge tube; 8 — target; 9, 10 — voltage source; 11 — krypton gas cylinder; 12 — activated charcoal; 13 — electron gun

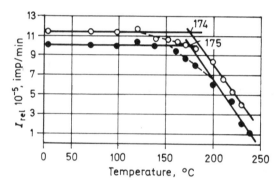

Fig. 3.38. Peak temperature measurement by kryptonation technique

owing to this, the radioactivity of the part begins to decrease only at temperatures exceeding the value which had been attained before. Figure 3.38 shows a typical plot indicating the count rate of kryptonated piston rings as a function of the temperature.

EXAMPLE 3.10

The wear of balls in a ball bearing is studied, by labelling the balls by neutron activation. The composition of the material is as follows: 1.5% C, 0.3% Mn and 98% Fe. The wear rate under the conditions studied is about 1 µg/h. The thermal neutron flux available in the reactor is $\Phi = 8 \times 10^7$ neutron/(m$^2 \cdot$ s). The wear study of the greased bearings can be carried out 48 h after completing the irradiation, and the bearing is operated for 1 h. The worn particles are quantitatively separated from the grease by treatment with a solvent, and their activity is measured.

Table 3.8 gives information about the radioactive nuclei formed and their properties.

The radiation of ^{56}Mn cannot be detected 48 h after irradiation, owing to its short half-life. ^{55}Fe captures K-electrons, its determination is impracticable. The experiment should, therefore, be based on the β- and γ-radiation of ^{59}Fe.

Table 3.8. Radioactive nuclei formed during thermal neutron irradiation in a specified ball bearing material

Element	Radionuclide	Half-life	Radiation emitted	Saturation activity, Bq	
				for 1 kg element	for 1 kg bearing material
Carbon	–	–	–	–	–
Manganese	^{56}Mn	2.58 h	β, γ	1.16×10^{14}	3.49×10^{11}
Iron	^{55}Fe	2.7 y	K	1.41×10^{12}	1.37×10^{12}
	^{59}Fe	44.6 d	β, γ	2.89×10^{10}	2.81×10^{10}

Two detectors are available in the laboratory. The questions are: which of them allows a more sensitive measurement, what are the specific activities obtainable under the given circumstances, and how long should be the irradiation time?

The background of the given GM-tube is $I_{rel, B} = 10$ imp/min. We wish a 2% relative error in the net count rate; then, the following count times and count rates should be selected:

$$t_M = 30 \text{ min}; \quad n_M = 3300 \text{ counts}; \quad I_{rel, M} = 110 \text{ imp/min};$$

$$\underline{t_B = 10 \text{ min}; \quad n_B = 100 \text{ counts}; \quad I_{rel, B} = 10 \text{ imp/min};}$$

$$I_{rel, M-B} = 100 \text{ imp/min}$$

Then:

$$\sigma_{M-B} = \sqrt{\frac{n_M}{t_M^2} + \frac{n_B}{t_B^2}} = \sqrt{\frac{3300}{30^2} + \frac{100}{10^2}} \approx 2$$

The self-absorption of the sample is neglected, and it is assumed that the β-particles of ^{59}Fe, with maximum energies of 0.27 and 0.46 MeV, pass without energy loss through the GM-tube having a mica window of 10–20 g/m^2. Equation (3.3) is reduced then to a form of $I_{rel} = GA$.

Equation (3.7) serves for calculation of the G geometry factor; $l = 0.3$ cm and $R = 1.3$ cm:

$$G = \frac{1}{2}\left(1 - \frac{l}{\sqrt{l^2 + R^2}}\right) = \frac{1}{2}\left(1 - \frac{0.3}{\sqrt{0.09 + 1.69}}\right) \approx 0.4$$

100 imp/min divided by 0.4 is equal to 4.17 Bq (0.11 nCi). After an operation period of 1 h, the sample contains about 1 μg of abrasion product, i.e., its specific activity is 4.17 Bq/μg = 4.17 GBq/kg (0.11 Ci/kg). The 48 h elapsed from the completion of the irradiation can be neglected as compared with the 45 days which is the half-life of ^{59}Fe, thus let us study how long the irradiation time (t) should be to ensure the 4.17 GBq/kg (0.11 Ci/kg) ^{59}Fe activity in the ball material.

$4.17 = 28.12[1 - \exp(-0.693t/t_{1/2})]$; hence, we have:

$$t \approx 10 \text{ days}$$

The other measuring device consists of a scintillation well-type crystal (O.D. 3.8 cm, height: 5 cm, hole diameter: 1.3 cm, hole depth: 3.8 cm). Its background is $I_{rel, B} = 2000$

imp/min. We wish a 2% relative error in the net count rate ($I_{rel, M-B}$); then the following count times and count rates should be selected:

$$t_M = 30 \text{ min}; \quad n_M = 85\,800 \text{ counts}; \quad I_{rel, M} = 2860 \text{ imp/min}$$

$$t_B = 10 \text{ min}; \quad n_B = 20\,000 \text{ counts}; \quad \underline{I_{rel, B} = 2000 \text{ imp/min}}$$

$$I_{rel, M-B} = 860 \text{ imp/min}$$

Then:

$$\sigma_{M-B} = \sqrt{\frac{85\,800}{30^2} + \frac{20\,000}{10^2}} = 17.2$$

$$\frac{17.2 \times 100}{860} = 2\%$$

^{59}Fe emits $\delta \approx 1$ γ-quantum per disintegration. The efficiency of the well-type crystal with respect to the 1.1 and 1.29 MeV γ-quanta of ^{59}Fe is 0.22 (see Fig. 3.11). A relative count rate of $I_{rel, M-B} = 860$ corresponds to the following activity: $860/0.22 = 3909$ disintegration/min $= 65.15$ Bq (1.76 Ci). According to the above reasoning, this means a 65.15 GBq/kg (1.76 Ci/kg) ^{59}Fe activity in the bearing material.

Since the thermal neutron flux available is not sufficient to produce the activity of 65.15 GBq/kg (1.76 Ci/kg) required for the scintillation measurement, it is evident that the GM-tube arrangement should be selected. In this case—as it has been shown—the necessary time of irradiation is $t \approx 10$ days.

3.6. TRACER STUDIES IN PHYSICAL AND CHEMICAL RESEARCH

The first radiotracer studies—carried out by the Hungarian-born Nobel prize winner G. Hevesy as early as in 1912—aimed at solving scientific problems. Radiotracer methods have remained useful for scientific research up till now. Such applications are to be mentioned also in this book, because both the principles and the methods are the same as in industrial tracer studies. The systems investigated are often of practical interest and, in addition, the results may be useful for industrial applications.

Applications in selected fields of physics and chemistry will be illustrated by examples which are thought to be typical. It should be noted that *stable isotopes* as tracers partly as one of the *double tracers* are gaining more and more importance due to the development of increasingly sophisticated techniques. The simplicity of radioactivity detection represents, however, such an advantage that radiotracers have not and expectedly will never become obsolete.

3.6.1. TRACERS IN PHYSICAL OPERATIONS

Radiotracer methods used in scientific research are usually connected with *mass transfer studies*. This chapter will deal with those applications which do not involve chemical reactions.

3.6.1.1. TRACERS IN SEPARATION PROCESSES

The distribution of a microcomponent between a solution and the crystals of a macrocomponent precipitating from the solution can be expressed by the following equation:

$$\frac{x}{y} = D\frac{a-x}{b-y} \tag{3.64}$$

where a and b denote the initial amounts of the micro- and macrocomponents (g), respectively, x and y their precipitated amounts (g) and D is the recrystallization or fractionation parameter. D is related to the K partition coefficient by the following equation:

$$D = \frac{Kc}{\rho} \tag{3.65}$$

where c is the concentration of the macrocomponent in the saturated solution (kg/m^3) and ρ is the density of the crystals (kg/m^3). The above equation expresses an equilibrium situation. In rapid crystallization, under nonequilibrium conditions, the following equation will hold:

$$\log\frac{a}{a-x} = \lambda\log\frac{b}{b-y} \tag{3.66}$$

where λ denotes a constant.

D is a thermodynamical and λ is a kinetic constant; in an ideal limiting case, $D = \lambda$. The assumption of a transition layer on the surface may help to resolve the apparent contradiction.

These theoretical studies have gained particular interest in separation methods of radioisotopes. *Together with other partition methods* (ion exchange, chromatography), *coprecipitation methods* are suitable for the separation and enrichment of very dilute isotopes.

It is also necessary to take into consideration similar *adsorption processes* in the electrolytic separation of radioisotopes. Exchange processes may be rate determining during *electrolysis*, too. The application of an anode with very small surface area promotes these exchange processes to a very high extent.

The efficiency of aerosol filters is investigated by test aerosols. These aerosols should be as monodisperse as possible to avoid the separation effect according to particle sizes.

A very monodisperse ^{192}Ir aerosol of spherical particles can be produced by a low intensity, high voltage "Tesla" arc discharge. A constant air stream blows the aerosol through a buffer tank and through the filter to be investigated. After the tested filter, a so-called "absolute" filter is switched into the gas stream. This, in principle, retains all aerosol particles present. The radioactivity of both filters is measured in a specially constructed high cross-section lead shielding.

The relative activity measured in the filter $I_{rel,f}$ is proportional to the amount of aerosol retained by it. If we relate this value to the amount of aerosol retained by the absolute filter ($I_{rel,u}$), we get the *retention* (R) of aerosol by the filter in question:

$$R = \frac{I_{rel,f} - I_{rel,u}}{I_{rel,f}} = 1 - \frac{I_{rel,u}}{I_{rel,f}} \tag{3.67}$$

The penetration (D) is defined as follows:

$$D = 1 - R = I_{rel, u}/I_{rel, f} \tag{3.68}$$

Typical results obtained with a ^{192}Ir aerosol (most frequent particle size: about 0.3 μm) are shown in Fig. 3.39. Of the subsequent cellulose fibre filter layers, for the sake of clarity, only retention by layers Nos 1, 4 and 8 are shown. The loss through the absolute filter was less than 0.0025% and could be neglected.

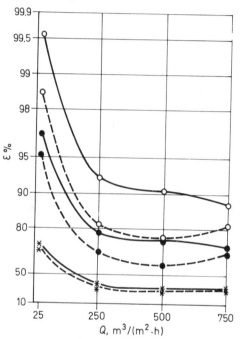

Fig. 3.39. Filtration efficiency (ε) of fibrous viscose filters of different compactness as a function of filter loading (Q). Solid line: loose filter; dotted line: filter compacted in wet state and dried; * first layer (surface mass: 1.5 kg/m²); ● fourth layer (surface mass: 6 kg/m²); ○ eighth layer (surface mass: 12 kg/m²). Reproduced by permission of Hirling [3.23]

As the filter loading increases, the retention decreases to a fairly constant value. The logarithmic scale shows that the filtration process can well be followed even in the range of almost complete aerosol retention (above 99%). A loose filter has a higher efficiency than more compact layers, because of the higher residence time of the air in it.

With a ^{24}NaCl test aerosol it is possible to study the behaviour of cubic particles.

3.6.1.2. DIFFUSION

Radiotracers can be applied successfully for the determination of the very slow diffusion and self-diffusion processes taking place in solids; this is the only method for following the latter phenomenon.

Two main measurement techniques can be distinguished for the study of *self-diffusion processes in solids:*

(a) A thin labelled layer is applied over a thick metal plate, e.g., by electrolysis. After keeping the system under the conditions to be investigated (at a given temperature), thin layers are cut or dissolved parallel to the surface. The radioactivity of subsequent fractions indicates the extent of diffusion. The dependence of the activity on the depth permits one to calculate the diffusion coefficient.

(b) A thin layer of an α- or β-radiating isotope is applied to the surface. The intensity measured by an outer detector decreases with time, thus the rate of disappearance of radioactive surface species is proportional to the rate of diffusion.

The so-called capillary technique serves for studying *diffusion processes in liquids.* A capillary with one open end is filled with a radioactive liquid. The capillary is immersed into the inactive liquid which is stirred intensively. The change of radioactive concentration in the capillary as a function of time is used to calculate the diffusion coefficient.

3.6.1.3. TRACER STUDIES IN SORPTION
AND SURFACE CHEMISTRY

The measurement of *radioisotopes attached to a solid surface* is possible by means of a device the principle of which is shown in Fig. 3.40 (a). Counter *2* monitors the radioactivity in the gas phase plus that adsorbed on the surface *1*. Gas phase counts can be subtracted either by removing *1* from the device or by applying a second "blank" chamber with the same geometry and flow conditions, but without the adsorbent. This latter solution has also been described with thin-window GM-tubes. The catalyst sample can be moved by means of a magnet to and from a furnace. Another, more recent version permits *in situ* heating (Fig. 3.40 (b)). Similar devices can be used for detecting radiotracers adsorbed on electrodes from liquid electrolytes.

The partition of an isotope between a solution and a crystalline substance may give information about the surface of the solid.

Krypton adsorbate is often used for sensitive *surface area determination* of solids by the BET method. If radioactive ^{85}Kr is added to the adsorbate, pressure measurement can be substituted by radioactivity measurement. Surfaces as small as 100 cm^2 have been measured by this method. The application of ^{133}Xe as adsorbate permitted the measurement of a surface area as small as 1 cm^2. Similar sensitivities can be attained if the radiation of ^{85}Kr actually adsorbed on the solid surface is directly measured as an indicator of the adsorbed amount.

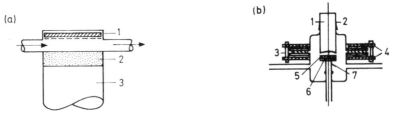

Fig. 3.40. (a) The principle of measurement of adsorbed radioactivity. 1 — catalyst or electrode; 2 — scintillator; 3 photomultiplier. Reproduced by permission of Mink *et al.* [3.21]. (b) High temperature chamber for the continuous monitoring of radiotracers on a catalyst surface. 1 — GM-tube; 2 — rubber seal; 3 — Teflon gasket; 4 brass clamps; 5 — catalyst; 6 — heating coil; 7 — thermocouple. After Norval and Thomson [3.35], reproduced by permission of The Chemical Society, London

Aluminate hydrate adsorbs readily chemical *impurities* from the aluminate liquor during decomposition. The radioactivity of alumina precipitated from a liquor labelled with ^{65}Zn (measured after drying) reaches a saturation value as a function of time of contact with the labelled solution according to a *Langmuir* isotherm. The adsorption capacity of alumina was found to be about 10 μg Zn/g hydrate at 293 K. Experiments carried out with ^{59}Fe-labelled solutions have shown that iron behaves similarly to zinc in the aluminate liquor.

In corrosion studies (see also Section 3.5.1), it has been demonstrated by the use of labelled *corrosion protective agents* (such as [^{14}C]-labelled sodium benzoate and sodium cinnamate) that a coverage as little as 7–10% of a monolayer can ensure corrosion resistance. Obviously, the agent occupies the most active surface sites where in its absence also corrosion would commence. Autoradiographies (Section 7.6) have confirmed that the distribution of the protective substance is not uniform on the surface.

Surface heterogeneity in chemisorption (Example 3.11) or *the lifetime of molecules* adsorbed or reacting on a *catalyst surface* (Example 3.12) can also be followed by radiotracers.

EXAMPLE 3.11

Radioactive benzene is adsorbed over a nickel catalyst at 423 K and displaced first by an inactive benzene (423 K) flow and then by hydrogen flow (from 423 up to 653 K). The substance removed from the catalyst surface is trapped in a 1:1 cyclohexane–benzene mixture. The cyclohexane and benzene are separated by column chromatography and their radioactivities determined. The following results are obtained:

Removing agent	Relative radioactivity, impulses	
	in cyclohexane	in benzene
Benzene	260	4350
Hydrogen	2360	320

Let us determine the ratio of reversibly and irreversibly adsorbed benzene. The total count is 7290. Of this, the ratio of reversibly adsorbed benzene is

$$\text{Reversibly adsorbed} = \frac{260+4350}{7290} \times 100 = 63.2\%$$

The amount of cyclohexane within this value is 5.6%. Thus, 5.6% of the reversibly adsorbed benzene *transformed* under the conditions applied, i.e., at least this fraction of the reversible adsorption, took place by chemisorption.

The amount of irreversibly adsorbed benzene is 36.8% of the total adsorption, this is certainly chemisorbed by dissociation of one or more C—H bonds [3.26].

EXAMPLE 3.12

Chlorobenzene is passed in a hydrogen stream over a Pd/SiO_2 catalyst promoting dechlorination (practically without any other reaction) [3.35]:

$$C_6H_5Cl + H_2 \longrightarrow C_6H_6 + HCl \tag{3.69}$$

13 Földiák

Using the device shown in Fig. 3.40 (b), let us determine the mean lifetime of the molecules on the surface of the catalyst and the SiO_2 support alone. Let us determine the ratio of molecules leaving the Pd/SiO_2 catalyst with and without reaction.

The mass of the catalyst is $m = 250$ mg; the flow rate: $F = 22.0 \times 10^{16}$ molecule/s (over the Pd/SiO_2) and 15.1×10^{16} molecule/s (over the SiO_2), respectively.

Let us regard the catalyst surface as a "chamber" with a capacity of C_c molecules. Let us inject a radiotracer with a total activity of A Bq into the flow. The number of total counts n is obtained by integrating the time-dependent count rates (I_{rel}):

$$n = \int_0^\infty (I_{rel}) dt = A\tau_s \tag{3.70}$$

where $\tau_s = C_c/F$ is the "occupancy", i.e., the mean time of sojourn of a molecule on the surface. The C_c capacity can be calculated as follows (n and I_{rel} being the difference of the values measured in the catalyst chamber and the "blank" chamber):

$$C_c = \frac{F}{A} \int_0^\infty (I_{rel}) dt = \frac{F \cdot n}{A} \tag{3.71}$$

If a rectangular pulse of $[^{14}C]$-chlorobenzene is passed through the system, during its desorption the count rate I_{rel} will vary according to $-dI_{rel}/dt = I_{rel}/\tau_s$. On integration we have:

$$\ln I_{rel} = -t/\tau_s - \ln I_{rel,o} \tag{3.72}$$

The logarithm of count rates as a function of time are shown in Fig. 3.41. The mean lifetime of molecules on the catalyst, determined graphically, is $\tau_s = 0.81 \times 10^3$ s, and on the support: $\tau_s = 3.04 \times 10^3$ s. This shows that the support serves as a reservoir for the chlorobenzene molecules; the catalytic reaction ensures an enhanced opportunity for them to leave the surface after reacting.

The "turnover number" (λ_s), i.e., the probability of desorption of any molecule in unit time, is given by the reciprocal value of the mean lifetime: $\lambda_s = 1/\tau_s$. This consists of two parts: the probability of desorption after reaction (λ_r, 1/s) and that without reaction (λ_e, 1/s); the turnover number for reaction λ_r, 1/s is the probability that a molecule will leave

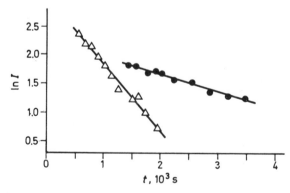

Fig. 3.41. Plot of the logarithm of the count rate, I, 1/s, as a function of the time t during the desorption of ^{14}C-chlorobenzene. Δ — Experiment with Pd/SiO_2; X — Experiment with SiO_2 support alone. After Norval and Thomson [3.35], reproduced by permission of The Chemical Society, London

the surface as benzene in unit time. Its value is given by w/C_c where w is the reaction rate and C_c the capacity of catalyst surface. The latter is calculated from Eq. (3.71) and its value is $C_c = 4.64 \times 10^{17}$ molecule/(mg catalyst).

$\lambda_e = \lambda_s - \lambda_r$ gives the probability that a molecule will desorb without reacting.

The reaction rate in a flow system is (with a measured conversion of $x = 0.15$):

$$w = \frac{Fx}{m} = \frac{22 \times 10^{16} \times 0.15}{0.25} = 1.32 \times 10^{17} \text{ molecule/(s g catalyst)}$$

$$\lambda_r = \frac{w}{C_c} = \frac{1.32 \times 10^{17}}{4.64 \times 10^{20}} = 2.84 \times 10^{-4}, \, 1/s$$

Since $\lambda_s = 1/\tau_s = 1/0.81 \times 10^3 = 12.3 \times 10^{-4}$ s, it can be seen that 23% of the molecules react on the surface. The turnover number of molecules not reacting $\lambda_e = (12.3 - 2.84) \times 10^{-4} = 9.46 \times 10^{-4}$, and this corresponds to 77% of the total flux.

3.6.2. STUDY OF REACTION MECHANISMS AND KINETICS

Radiotracers may be very useful in elucidating reaction pathways, kinetic parameters and mechanisms. Apart from earlier reviews [3.14, 3.15], a recent excellent book is cited here dealing with the applications of isotopes (stable and radioactive) in heterogeneous catalysis [3.36].

3.6.2.1. ISOTOPIC EXCHANGE

Isotopic exchange is essentially a "nonreaction", because the chemical nature of the starting material and end product is the same, only their isotopic compositions change. This reaction cannot, in principle, be studied without (radioactive or stable) tracers. Stable isotopes (using mass spectrometric analysis) ensure the additional advantage that also the *isotope distribution* can be determined.

Exchange reactions give information about the character and strength of chemical bonds and the mechanism of the reactions accompanying the exchange. The possibility of exchange should always be considered in chemical investigations and care must be taken to ensure that exchange processes do not falsify the results. Applications of exchange reactions in chemical analysis are shown in Section 4.3.6.

If metals are immersed in the solution of their ions, a rapid *heterogeneous exchange* occurs up to several monoatomic layers. For example, about 400 monoatomic layers can be exchanged within 30 minutes between Mn and a radioactive 0.5 M $MnSO_4$ solution; in systems like $Pb/Pb(NO_3)_2$ and $Bi/Bi(NO_3)_2$, the extent of exchange can be nearly 1000 layers. The rate of the exchange is influenced by the nature of the anion and increase with higher acid concentration. Although inert metals undergo exchange to a lesser extent, there is no direct correlation with the electrode potential.

Such investigations may serve as a basis for *corrosion studies* (see Section 3.5.1). If inactive nickel is electroplated over an iron plate containing ^{59}Fe, radiactive iron "clusters" can be observed in the coating. A rapid exchange must have taken place between the galvanizing bath (nickel solution) and solid iron, as confirmed by the appearance of radioactivity in the bath. Surface iron deposits are the starting points of later corrosion, as indicated by autoradiographs.

Catalytic exchange reactions between organic molecules and radioisotopes have been studied extensively (e.g. [3.37]).

3.6.2.2. THE INVESTIGATION OF INTERMEDIATES AND REACTION PATHWAYS

Radiotracer methods may help to elucidate whether a given intermediate is formed or not in a reaction.

If the starting substance is labelled and its mixture with inactive moiety of the assumed intermediate is made to react, the *molar* (or *specific*) *radioactivities* of the individual components after the reaction can give useful information about the reaction pathway. If the specific activity of the intermediate is as high as that of the end product, its intermediate character is verified.

In heterogeneous catalysis, the desorption of the intermediate may be hindered. Then it may not appear in the final product; the intermediate is "masked". This difficulty may be overcome if the intermediate in question is produced in the reaction system. If the difference in the reactivities of the two components is very high, one of the components may be consumed entirely. This will falsify the results, because the radioactive starting material will thereafter react as if it had been fed alone.

Both features are illustrated in Fig. 3.42. Here the role of cyclohexene was studied as a possible intermediate in benzene *hydrogenation*. The starting mixture consisted of

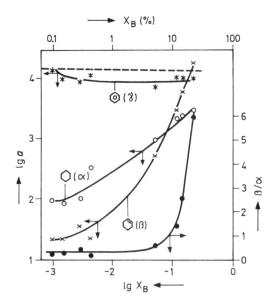

Fig. 3.42. Molar relative radioactivities *a* of the individual fractions of benzene hydrogenation (cyclohexane: α; cyclohexene: β; benzene: γ) and the value of β/α as a function of the degree of benzene conversion over various amounts of Ni catalyst. Starting mixture: ^{14}C-benzene + inactive cyclohexadiene. After Derbentsev *et al.* [3.28], reproduced by permission of Akademische Verlagsgesellschaft, Wiesbaden

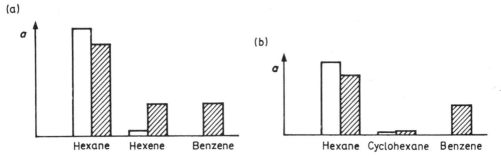

Fig. 3.43. Relative molar radioactivity values (a) in the aromatization of a mixture of radioactive hexane plus inactive hexene (a) and radioactive hexane plus inactive cyclohexane (b) on Pt-black. (Empty columns: starting mixtures, shaded columns: after reaction.) After Tétényi *et al.* [3.29]

[^{14}C]-benzene plus inactive cyclohexadiene; thus *any* cyclohexene found in the product was necessarily formed on the catalyst surface. Its radioactivity increases with increasing conversion (as cyclohexadiene is consumed).

Instead of the molar (or specific) activities, another, relative calculation method can also be used. If the distribution of the individual components is calculated from both the mass and radioactivity analyses (e.g., from the chromatograms), two compositions are obtained, consisting of the mass—or molar—($m\%$) and radioactivity ($r\%$) percentages. If the molar (or specific) radioactivities of all components are equal, dividing the two percentage series gives unity for each peak ($r\%/m\% = 1$). If, however, radioactivity is distributed nonuniformly between the components, the $r\%/m\%$ values will vary. It is just the extent of nonuniformity which gives the answer required by the tracer study. Instead of the molar or specific radioactivities of the components, the $r\%/m\%$ ("relative molar radioactivity") values can be used to characterize radioactivity distributions (Example 3.13).

Radiotracer studies have shown that cyclohexane is not an intermediate in the platinum- and nickel-catalysed *dehydrocyclization* of *n*-hexane, but 1-hexene is (Fig. 3.43). Hexadienes and hexatriene-1,3,5 have also been found to be intermediates of the dehydrocyclization. This has rendered it possible to develop the so-called "stepwise" or "hexatriene" mechanism of dehydrocyclization. The scheme shows the reactions proved (thick arrows) and disproved (crossed arrows) by the radiotracer method, including also the triangular scheme proven for the hydrogenation–dehydrogenation of cyclohexane–cyclohexene–benzene:

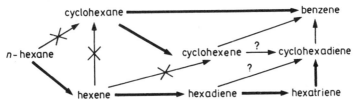

The manufacture of hydrocarbons or methanol from synthesis gas is gaining more and more importance. The mechanism of the *Fischer–Tropsch synthesis* was studied by adding ^{14}C-labelled ketene (0.25–2%) to a mixture of CO + H$_2$ reacting over a Co and Fe catalyst.

Ketene (CH_2=C=O) undergoes dissociation under these conditions. If the label was in the methylene group, the molar radioactivity of ethane produced was equal to that of the added ketene, in the presence of iron catalyst. The molar radioactivities of higher hydrocarbons increased in proportion to their carbon number. With cobalt catalyst, the molar radioactivity was lower and independent of the chain length.

If the C=O group of the ketene was labelled, the molar radioactivity of ethane was very low, but that of higher hydrocarbons increased proportionally with the chain length, in the presence of either both catalyst.

These results point to the role of active surface =CH_2 groups in initiating the chain growth which, afterwards, takes place via incorporating CO units.

Methanol synthesis may, in principle, take place by the following two reactions:

$$CO + 2 H_2 \longrightarrow CH_3OH \tag{3.73}$$

$$CO_2 + 3 H_2 \longrightarrow CH_3OH + H_2O \tag{3.74}$$

Tracer studies (Table 3.9) point to the occurrence of reaction (3.74). Carbon monoxide reacting with water dilutes the radioactive CO_2 in the final mixture. Iron catalyst promotes direct hydrogenation of CO to methanol.

Table 3.9. Tracer studies of methanol synthesis from CO, CO_2 and H_2 over $CuO/ZnO/Al_2O_3$ catalyst

| Component | Specific relative radioactivity,* impulse (s mg) | | | |
| | Mixture 1 | | Mixture 2 | |
	initial	final	initial	final
CO	400	190	0	490
CO_2	0	9	5900	790
CH_3OH	–	5	–	1450

* Expressed per mg $BaCO_3$ after processing the samples. Reproduced after Kagan et al. [3.38].

3.6.2.3. REACTION PATHWAY STUDIES BY THE DETERMINATION OF THE POSITION OF THE LABEL

Sometimes the mere detection of radioactivity in any given product does not provide the necessary information; in this case the product of the reaction should be isolated and the position of the label determined by its chemical processing. This involves, as a rule, laborious and sophisticated processes (Example 3.14). Therefore, they are used only if no other experimental technique would give the necessary information. Sometimes stable isotope studies (applying, e.g. ^2H, ^{13}C, etc.) are more useful for this purpose, expecially if sophisticated instrumentation (microwave or nuclear magnetic resonance spectroscopy) is available for analyis. These methods generally allow the direct determination of the position of the label, but they require much higher amounts of isotopic species than radiotracer methods do.

Ethylcyclohexane *isomerizes* catalytically to give 1,2-, 1,3- and 1,4-dimethylcyclo-hexanes. Distinction between the theoretically possible two mechanisms (A: side methyl transfer; B: ring concentration–expansion) is possible by means of radioactive tracers.

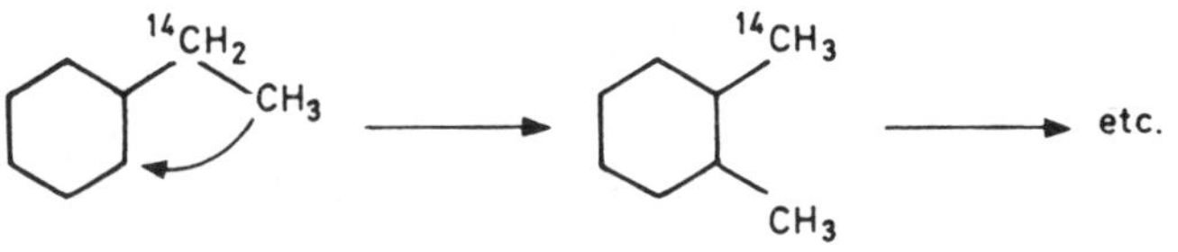

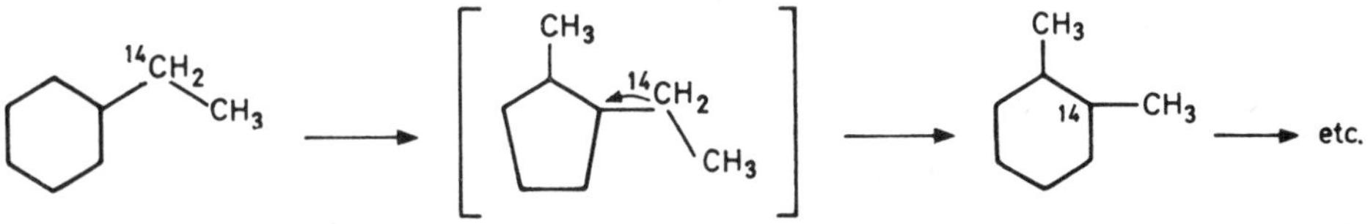

^{14}C-labelled ethylcyclohexanes were passed over a nickel–silica–alumina catalyst. The product was chromatographed over a silica column, dehydrogenated, then the aromatic fraction was oxidized with potassium permanganate and the resulting benzoic, phthalic, isophthalic and terephthalic acids were separated. These acids originated from ethylcyclohexane, 1,2-, 1,3-, and 1,4-dimethylcyclohexanes, respectively, as illustrated on the example of isophthalic acid:

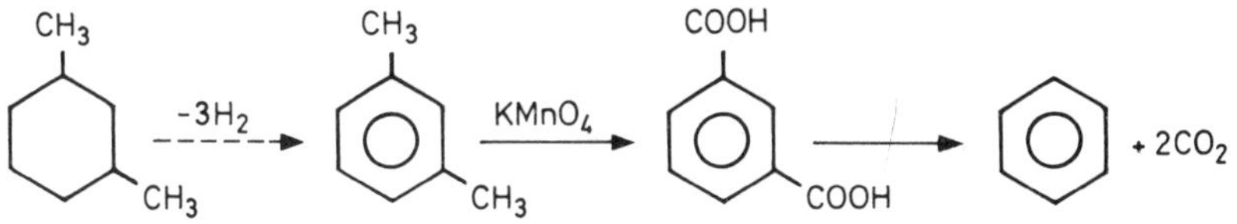

Specific relative activity values (a_{rel}) from typical runs are shown in Table 3.10. The decarboxylation of the acid gives CO_2 plus benzene. If the a_{rel} values of these fragments are related to that of the original acid, the fraction of label in the side chain and the ring, respectively, is obtained.

The considerable ring activity of 1,3-dimethylcyclohexane points to the validity of Mechanism B and points to several subsequent ring contraction—expansion steps during one run. Other isomers gave analogous results.

Table 3.10. Specific radioactivity values of isophthalic acid and its degradation products obtained in the isomerization of labelled ethylcyclohexanes

Feed	Product	Specific relative activity*	Percent activity distribution
Ethyl-1-[^{14}C]-cyclohexane	Acid	6728	100
	Benzene	3486	Ring: 52
	Carbon dioxide	3222	Side: 48
Ethyl-2-[^{14}C]-cyclohexane	Acid	1648	100
	Benzene	1272	Ring: 77
	Carbon dioxide	336	Side: 20

* Expressed as impulses/[min (mass unit $BaCO_3$)] and corrected for the number of unlabelled carbon atoms in the parent molecule. After Pineş and Shaw [3.30].

3.6.2.4. THE KINETIC ISOTOPE METHOD

This method has been developed for following complex reaction sequences by tracers. The theory and practice of the method have been described in detail in Refs [3.16] and [3.17]; the basic principles will here be illustrated by a few examples.

(a) Let us consider a reaction sequence:

$$A \to B \xrightarrow{\ w_1\ } X \xrightarrow{\ w_2\ } Y \xrightarrow{\ w_3\ } C \to \dots$$

If a small portion of labelled X (with a total activity of I_X) is introduced into the system, the *specific activity* of X denoted by α, and Y denoted by β can be characterized by the following equations:

$$\frac{d\alpha}{dt} = -\frac{\alpha w_1}{[X]} \tag{3.75}$$

$$\frac{d\beta}{dt} = \frac{(\alpha - \beta)w_2}{[Y]} \tag{3.76}$$

At $t=0$, $\beta=0$. As time elapses, the specific radioactivity of Y (i.e., β) increases. At maximum, $d\beta/dt=0$, and $\alpha=\beta$. Thus, the value of β as a function of the time gives a *maximum curve* which intersects the plot of α *vs.* time at the maximum of the former curve.

(b) In the case of another reaction sequence,

$$A \longrightarrow B \longrightarrow X \longrightarrow Y \longrightarrow D \longrightarrow \dots$$
$$K \nearrow$$

a *similar maximum curve* is obtained, but the value of β will never reach that of α; there is no intersection of the two curves.

(c) In a third case:

$$A \to B \to X \to C \to Y \to D \to \dots$$

the intersection of the specific activities of X and Y is *not at the maximum* of the specific activity of Y. A similar picture is obtained with a *triangular* reaction scheme, too.

The reaction of xylene *isomerization* was studied over a technical Pt/Al$_2$O$_3$ catalyst. The feed is *p*-xylene with 10% of labelled *m*-xylene and 1% of *o*-xylene added. The specific radioactivities of the individual fractions obtained in the presence of hydrogen are depicted in Fig. 3.44.

The specific radioactivity curve of *o*-xylene intersects at its maximum with that of *m*-xylene. A similar curve is obtained for *p*-xylene when labelled *m*-xylene is added to an *o*-xylene feed. Therefore, case *(a)* is valid here, and the reaction can be characterized by the scheme:

$$o\text{-xylene} \rightleftarrows m\text{-xylene} \rightleftarrows p\text{-xylene}$$

The isomerization of xylene with the same starting conditions as before was also studied in nitrogen atmosphere. The results are shown in Fig. 3.45.

The specific activity plot of *m*-xylene intersects here with that of *o*-xylene at the left of the maximum of the latter. This points to case *(c)*. At the same time, another curve

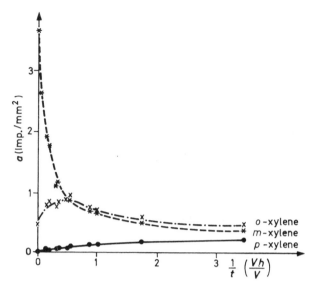

Fig. 3.44. Isomerization of *p*-xylene in hydrogen flow. The specific relative activities (a_{rel}) of the individual xylene isomers are plotted as a function of the time of residence (*t*). Reproduced by permission of Dermietzel *et al.* [3.31]

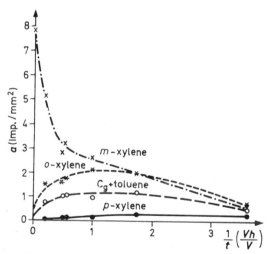

Fig. 3.45. Isomerization of *p*-xylene in nitrogen flow. The specific relative activities (a_{rel}) are plotted as a function of the time of residence (*t*). Reproduced by permission of Dermietzel *et al.* [3.31]

appears characterizing the specific radioactivity of C_9-aromatics and toluene (present in equimolar concentrations). This maximum curve does not intersect with that of the *m*-xylene, i.e., for these substances case *(b)* is valid.

Therefore, the reaction pathway in nitrogen involves a triangular scheme. Trimethyl-benzenes (first of all, 1,2,4-trimethylbenzene) are the intermediates of the pathway of transformation of *o*- and *p*-xylenes into each other. The following scheme can be written:

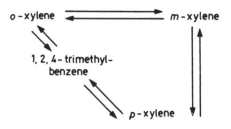

3.6.2.5. DOUBLE LABELLING

The simultaneous use of two isotopic species permits one to carry out more refined kinetic or mechanism studies. The apparatus required is, however, more complicated. If two radioactive isotopes are used (e.g., ^{14}C and ^{3}H), energy-selective detection is necessary by using at least a *two-channel instrument*. If one of the isotopes is stable (e.g., deuterium), a mass spectrometer is required in addition to the radioactivity measurement.

Acetylene hydrogenation on a catalyst surface occurs according to the following mechanism:

$$\begin{array}{ccc} (C_2H_2)_{gas} & (C_2H_4)_{gas} & (C_2H_6)_{gas} \\ \uparrow\downarrow & \uparrow\downarrow & \uparrow\downarrow \\ (C_2H_2)_{surf.} \rightleftarrows & (C_2H_4)_{surf.} \rightleftarrows & (C_2H_6)_{surf} \end{array}$$

If ^{14}C-labelled acetylene is hydrogenated by means of *deuterium*, any ^{14}CHD$_2$—CHD$_2$ produced (especially near to zero conversion) would allow not only the measurement of the formal rate of "direct" hydrogenation of acetylene to ethane, but it would also show the existence of a concerted mechanism of ethane formation. The use of ^{14}C-ethylene may, however, give more information, since even if the *adsorption/desorption* equilibrium for ethylene is not established, the rate of decrease of its specific radioactivity is a measure of the ethylene formation by hydrogenation of its nonradioactive surface precursor.

If a mixture of acetylene plus [^{14}C]-ethylene is hydrogenated over a palladium catalyst, the specific radioactivity of ethane extrapolated to zero conversion is zero. Hence, ethane must have been formed—at least initially—from acetylene, *via* a *concerted* pathway. The ethylene radioactivity decreases simultaneously owing to the hydrogenation of inactive acetylene. At the same time, considerable amounts of $C_2H_2D_4$ appear, the quantity of which decreases with the time elapsed. Careful analysis of the distribution of the deuterated products also confirms the existence of a concerted mechanism of acetylene hydrogenation to ethane [3.39].

EXAMPLE 3.13

Inactive ethylene reacts with tritium "retained" over a platinum catalyst. Product analysis is carried out by gas chromatography; radioactivity counts are monitored by a scaler. The results of measurements are shown in Table 3.11. Let us calculate the relative molar radioactivity values in terms of $r\%/m\%$.

Table 3.11. Input data and results for the calculation of the relative molar radioactivities in Example 3.13

Data		Peak no.		
		1 (methane)	2 (ethane)	3 (ethylene)
Peak height, mm	h_i	17	58	135
Peak half-width, mm	w_i	2.5	4.5	8.7
Sensitivity factor	S_i	10	10	20
Time, s	t	140 (0–I)	140 (I–II)	140 (II–III)
Pulse counts	n	3900 (at I)	12350 (at II)	1915 (at III)
Background, imp/s	i_b	6	6	6
Results:				
$m\%$		1.60	9.84	88.56
$r\%$		18.40	45.76	35.84
$r\%/m\%$		11.50	4.65	0.40

After Paál and Thomson [3.32]

The molar and radioactivity compositions, respectively, are calculated from Eqs (3.77) and (3.78) (with 3 peaks):

$$m_i\% = \frac{h_i w_i S_i}{\sum_{i=1}^{3} h_i w_i S_i} \times 100 \qquad (3.77)$$

$$r_i\% = \frac{n_{i+1} - n_i - (i_b t_i)}{n_3 - n_0 - (i_b t_{0-\mathrm{III}})} \times 100 \qquad (3.78)$$

The results are also tabulated in Table 3.11.

Comparison of the relative molar radioactivity values shows that the tritium uptake of methane is about twice as high as that of ethane. Thus, the hydrogenolysis of ethylene must involve more dissociated surface species than its hydrogenation. Ethylene underwent hydrogen isotope exchange to some extent, and it was the least radioactive [3.32].

EXAMPLE 3.14

The n-Heptane-[^{14}C] is aromatized over a chromia–alumina catalyst. The position of the label is determined in the product toluene according to the scheme shown in Figure 3.46. This degradation procedure allows one to determine all label positions except for Position 4 with respect to the methyl group [3.33]. Let us calculate the relative importance of the individual dehydrocyclization pathways.

The following data were obtained in a typical run (counts per 10 min per 100 mg BaCO$_3$) [3.33]: quinolinic acid decarboxylation 128; nicotinic acid decarboxylation 2660; nicotinic acid combustion 488; benzoic acid decarboxylation 92; benzoic acid combustion 874.

The methyl activity is

$$\frac{92}{7 \times 874} \times 100 = 1.5\%$$

since there are six inactive carbon atoms in the benzoic acid for each radioactive carbon).

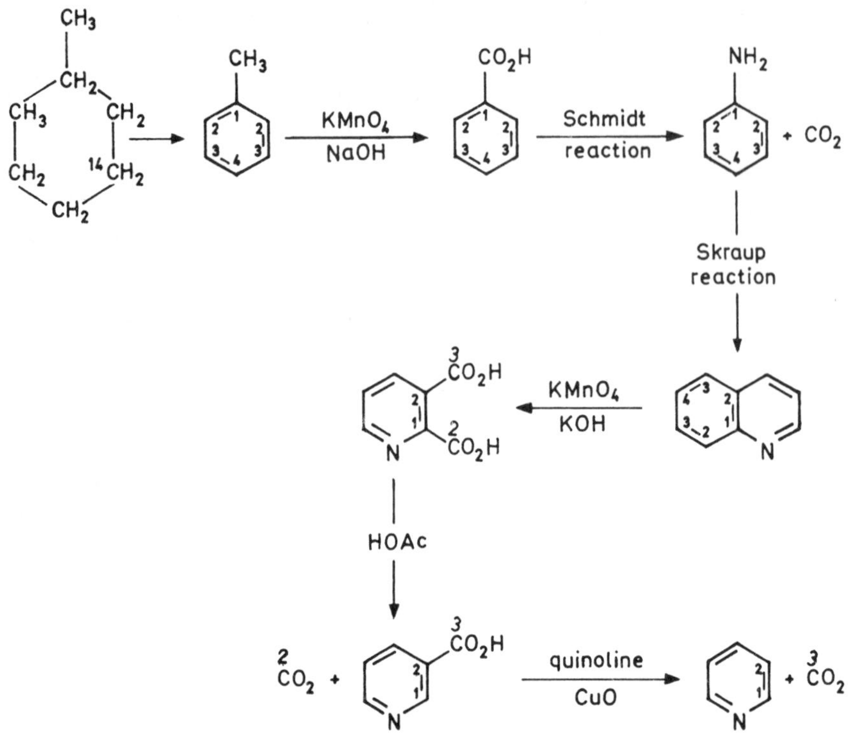

Fig. 3.46. Reaction scheme of the degradation of the dehydrocyclization product of n-heptane-4-[^{14}C] for clarification of the position of the label in the product. After Feighan and Davis [3.33], reproduced by permission of Academic Press, New York–London

The relative activity of the total combustion of nicotinic acid contains half of the activity of C-3 and C-2 and the total activity in C-1 (because half of the original activity in the C-2 and C-3 positions was lost when the quinoline was oxidized to quinolinic acid). The relative activity of nicotinic acid must be multiplied by 6, because there are five inactive carbon atoms in the molecule for each active carbon. The situation is represented in the following scheme:

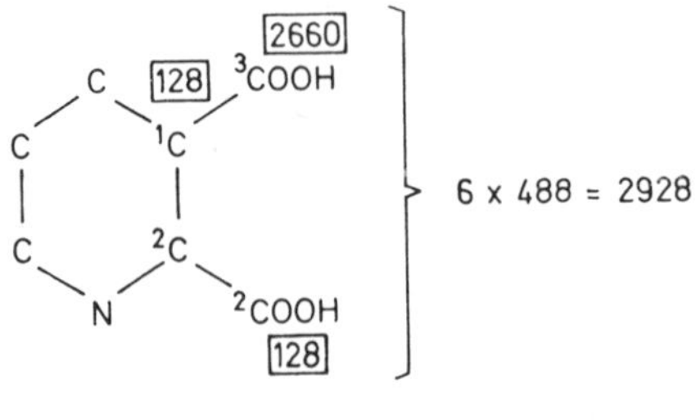

The activity in Position 1 will then be $2928 - (2660 + 128) = 140$ counts.
The total activity in the ring is:

$$A_{rel} = 140 + (2 \times 128) + (2 \times 2660) = 5716$$

The percent activity values will be (considering also the 1.5% methyl radioactivity):

Methyl	C-1	C-2	C-3
1.5%	2.4%	4.4%	91.7%

The 91.7% C-3 activity corresponds to C_6-ring closure without isomerization; nearly 10% of the heptane underwent some kind of skeletal rearrangement during aromatization.

3.7. RADIOTRACERS IN STUDIES
OF BIOLOGICALLY ACTIVE CHEMICALS

Commercially available isotopes such as ^3H, ^{14}C, ^{32}P and ^{35}S and the development of sensitive instruments make possible to apply the radiotracer technique for testing the fate and behaviour of chemicals in biological systems. The method has an outstanding importance in *pharmacokinetic and metabolic studies of drugs*, as well as in the evaluation of *agrochemicals*.

3.7.1. PHARMACOKINETIC
AND METABOLIC STUDIES OF DRUGS

Pharmaceutical industry producing compounds for the treatment of sick human beings, animals or plants is interested in learning the mechanism of the biochemical action of these agents and their fate within the organism.

In the last two decades, *new drugs* are being submitted to stricter and stricter safety testing prior to their commercial production and marketing. At the same time, also the scope of testing methods has become wider and wider. The strictest criteria are applied when a decision is made about the acceptance or application of a drug. The substance must meet several biological requirements and more and more profound safety testing is required to detect potentially harmful (toxic) side-effects. These tests are called the pre-clinical safety evaluation of a new drug.

In addition to investigations such as that of *chronic toxicity* (over 30 days), *teratology* (whether there is a harmful effect on the development of embryos), *carcinogenesis* (whether the compound may cause cancer or not) and *mutagenesis* (a study of possible effects on the chromosomes), the pharmacokinetic and metabolic study of pharmaceuticals belongs also to the compulsory pre-clinical safety evaluation. Pharmacokinetic studies support, explain and complete other pharmacodynamic data.

3.7.1.1. PHARMACOKINETIC STUDIES

The scope of pharmacokinetics is to find quantitative correlations between the living organism and a potential drug taken by it, in order to attempt to answer the following fundamental questions of determining whether the compound may become a drug or not:

— is the compound *absorbed* or not from the gastrointestinal tract of laboratory animals?

— how is the pharmacon *distributed* between the individual organs and in the aqueous phases of the organism?

— is it *bound by proteins* in the serum or in the tissues?

— does it show a *specific accumulation* in any of the organs?

— what is the extent and pathway of its *elimination*, i.e., excretion?

High resolution and sensitive methods are required to answer these questions, because the concentration of the compounds in question in the blood and tissues is usually $1 \, \mu g/cm^3$ or less. Using ^{14}C-labelled compounds, the concentration of the parent compound and/or metabolite(s) can be determined with high accuracy, even if they are present in the biological samples in amounts as low as a few ng.

The *distribution of the drug in the tissues* and its *accumulation* in separate organs can also be studied by radiotracer methods provided that a radionuclide with appropriate molar (specific) activity is built into a suitable position of the molecule. Pharmaceuticals labelled with ^{14}C isotope have considerably broadened the spectrum of pharmacological, toxicological and teratological research.

The *metabolism of normal body-constituents* and the fate of foreign substances within the organism (decomposition, excretion, etc.) can equally be studied by labelled compounds; these substances get into the organism together with the food or are just pharamaceuticals.

Radioactive tracers must be considered *potentially hazardous* to individual organs or tissues. No results have been reported, however, that soft β-emitters (such as ^3H and ^{14}C) in amounts as low as 30–50 kBq (ca 1 µCi) would actually cause even the slightest damage in laboratory animals. Yet, because of potential genetical radiation damage, the kinetics and metabolism of labelled pharmaceuticals is studied mainly in animals. On the other hand, histological data are available that the harder (higher energy) radiation of ^{32}P caused damage in the brain tissue of young rats when applied with specific activities higher than 400 kBq/kg (10 µCi/kg). The radiation dose considered safe in a single application of the isotope also depends on the nature of the carrier molecule.

In the case of *intravenous studies* (i.e., when the compound containing an isotope is injected intravenously and is detected subsequently from consecutive blood samples), the combination of several single exponential equations are used to describe the concentration decline of the compound. This latter is caused by the distribution of the substance between various liquid spaces—compartments—of the organism. If, however, the different compartments have already been equilibrated, the concentration changes of the compound in the blood can be further described by a single equation:

$$\frac{dy}{dt} = -k_2 y \tag{3.79}$$

where dy/dt is the rate of elimination of the substance; k_2 is the elimination coefficient and y is the actual concentration of the compound in the blood. The elimination coefficient can be obtained by integrating Eq. (3.79):

$$k_2 = \frac{1}{t_2 - t_1} \ln \frac{y_1}{y_2} = \frac{1}{t_2 - t_1} \ln \frac{a_1}{a_2} \tag{3.80}$$

where t_2 and t_1 denote the times of blood-taking; y_2 and y_1 are the appropriate concentrations, a_2 and a_1 the specific activity values.

In the case of intravenous application of labelled compounds, the *specific activity* is determined instead of the concentration of the compound in the blood. The "blood level" of a compound determined by measurement of its radioactivity represents, therefore, not only the elimination of the parent compound, but also the rate of disappearance of its radioactive metabolites.

In elimination studies, radioactive substances present in the blood are separated (e.g., by thin-layer chromatography) and their activity is determined (e.g., by liquid scintillation counting). *Elimination coefficients* characteristic of the original substance and the metabolites are calculated from the metabolite concentrations belonging to different time values.

If the application is not intravenous, the *invasion* (absorption, distribution within the organism) *precedes elimination*. In this case the rate of the increase of drug concentration in the blood is proportional to the amount of unresorbed drug.

An equation disregarding *elimination*, describing *invasion* only, is as follows:

$$\frac{dy}{dt} = k_1(a-y) \tag{3.81}$$

where dy/dt is the rate of invasion; k_1 is the *invasion coefficient*, and $a-y$ denotes the concentration of the unresorbed drug.

The invasion coefficient can be expressed by integrating Eq. (3.81):

$$k_1 = -\frac{\ln(1-y/a)}{t} \tag{3.82}$$

where a is the saturation concentration; t the time of the investigation.

In the case of oral administration, *invasion and elimination take place simultaneously.* Considering both processes, the actual blood concentration pertaining to a given time is as follows:

$$y = \frac{ak}{k_1-k_2}[\exp(-k_1 t) - \exp(-k_2 t)] \tag{3.83}$$

where y is the "blood concentration" belonging to the t time of the measurement (*Bateman*-function). This is valid for any natural process involving two processes of opposite direction both of them characterized by exponential functions.

By measuring the time of the maximum "blood concentration" (t_{max}) and by a preliminary calculation of the elimination coefficient, k_1—i.e., the coefficient characteristic of the invasion of the given compound—can be calculated by Eq. (3.84):

$$t_{max} = \frac{1}{k_1-k_2} \ln \frac{k_1}{k_2} \tag{3.84}$$

If the $\alpha = k_1/k_2$ substitution is introduced:

$$k_2 t_{max} = \frac{1}{\alpha-1} \ln \alpha \tag{3.85}$$

This equation cannot be solved for α but, knowing the values of $k_2 t_{max}$, the value of α can be taken from appropriate tables (Table 3.12). Knowing α and k_2, the k_1 invasion coefficient can be calculated.

As far as the type of the applied technique is concerned, pharmacokinetical methods for studying the fate of labelled drugs in the organism can be divided into two groups:

— whole body autoradiography;

— liquid scintillation counting of organs, tissues, body liquors, excreta.

RADIOTRACER TECHNIQUES

Table 3.12. Data table for the function $k_2 t_{max} = \dfrac{1}{\alpha-1} \ln \alpha$

α	$k_2 t_{max}$	α	$k_2 t_{max}$
0	∞	8.0	0.297
0.01	4.652	9.0	0.275
0.02	3.992	10.0	0.256
0.03	3.615	12.0	0.226
0.04	3.353	14.0	0.203
0.05	3.153	16.0	0.184
0.07	2.859	18.0	0.169
0.1	2.558	20.0	0.157
0.2	2.012	25.0	0.134
0.4	1.526	30.0	0.117
0.6	1.276	40.0	0.049
0.8	1.115	60.0	0.069
1.0	1.000	80.0	0.055
1.5	0.811	100.0	0.047
2.0	0.695	200.0	0.027
2.5	0.610	300.0	0.019
3.0	0.549	400.0	0.015
4.0	0.462	600.0	0.011
5.0	0.402	800.0	0.008
6.0	0.358	1000.0	0.006
7.0	0.324	∞	0.000

In the case of *whole body autoradiography* (see also Section 7.6), the laboratory animal (e.g., mouse, rat, rabbit, monkey) receives the labelled compound in the most suitable dose as indicated by preliminary pharmacological investigations. The animals are narcotized by ether each at different times after the dosing and placed into a dry ice–hexane mixture ($t = -78\ °C$). The frozen and solidified animals are embedded in carboxymethylcellulose and sliced at temperatures between -20 and $-15\ °C$ by means of a microtome-cryostat. The slices are placed on a transparent tape, lyophilized and autoradiographed (Fig. 3.47). Dark spots on the film indicate the accumulation of the labelled substance in various parts of the animal.

Whole body autoradiography—though elegant and documentative—gives only semi-quantitative information about the fate of a compound within the organism.

Liquid scintillation methods are, as a rule, more complicated but allow a quantitative evaluation.

For "blood concentration" studies, the labelled substance is administered to the laboratory animals in various amounts and blood samples are taken at predetermined intervals. The blood is dissolved in a tissue solvent (e.g., in TEH = tetraethylammonium hydroxide or Soluene of the firm Packard, or Hyamine 10x of the firm FLUKA), discoloured by hydrogen peroxide and added to an appropriate liquid scintillator. About 50 mg blood is generally used for an activity determination. The samples are placed into a liquid scintillation counter and measured.

Figure 3.48 shows the concentration of a labelled compound in blood, as a function of the time.

The treatment involved an intravenous dose of 3 µg/g and an oral dose of 100 µg/g. It can be seen that oral administration is followed by a rapid and complete absorption through the intestinal tract. Fifteen minutes after the treatment the concentration in the

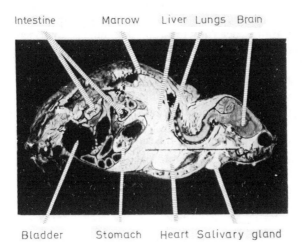

Fig. 3.47. Autoradiography on the distribution of Probon-[^{14}C] (made by Chinoin, Budapest) in whole body, 30 s after intravenous injection

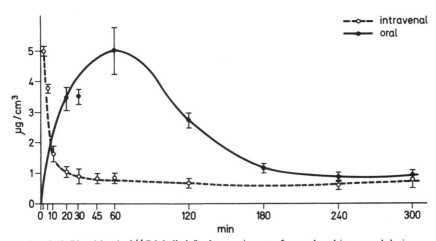

Fig. 3.48. Blood level of ^{14}C-labelled β-adrenerg in rats after oral and intravenal dosing

blood is as high as 3 µg/cm^3. Elimination starts rapidly and, after 3 h, the residual blood concentration is hardly detectable. In the case of intravenous dosing, a blood concentration profile characteristic of intravenous drugs can be observed.

When the binding to proteins is studied, the blood of animals treated with labelled compounds is centrifuged, then the aliquot part of the serum is precipitated by trichloroacetic acid. The sample is centrifuged again, and the activity of the supernatant and the precipitate is measured. The radioactivity in the precipitate will be characteristic of the drug present in free state in the serum.

The strength of binding to the protein can be determined by dialysis. The serum is dialysed against physiological salt solution for 24 h and the radioactivity in the dialysate and the dialysing liquid is determined.

In the case of a *chromatographic analysis of the serum*, after having carried out the "blood concentration" determination, the serum is analysed by an appropriate radiochromatographic method (Section 4.3.5) in order to determine whether the radioactivity in the blood originates from the original compounds or its radioactive metabolites. The serum is extracted by an organic solvent after precipitation of the protein; the extract is then analysed by thin-layer chromatography. The qualitative and quantitative analysis of the chromatograms is carried out either radiometrically (Section 4.3.5.6), or, after autoradiography, densitometrically (Section 7.6.3).

The purpose of *"blood concentration" investigations after subacute treatments* is to determine whether the labelled compound or its metabolites are accumulated in the blood or not. Rats are treated with the labelled compound for 10 days. Blood is taken prior to the treatment and after it—at the moment of the maximum of blood concentration caused by the acute, i.e., the first treatment—and the activities of the samples are measured.

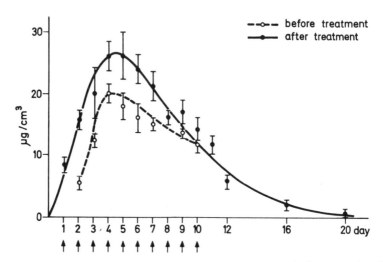

Fig. 3.49. Blood level of a ^{14}C-labelled compound after 10 days of subacute treatment

Figure 3.49 shows the blood concentration of a labelled antihyperlipidic compound during a 10-day subacute treatment. It can be seen that the daily oral doses led to the increase of the blood concentration for 3 days by a value of 8 μg/(cm^3 day); on the fourth day, however, the maximum level was reached. On proceeding with the treatment, the blood concentration decreased by a value of about 3 μg/(cm^3 day). After stopping the dosing, the rate of elimination of the radioactivity accelerated. The phenomenon can be explained by the enzyme inductive effect of the compound, proved also by pharmacological and toxicological investigations.

Distribution studies should be carried out at different times. Experiments performed in the early post-treatment period are aimed at the study of the initial distribution of the compound; subsequent investigations may supply information about the eventual organ-specific accumulation of the compound.

The organs to be studied in these experiments are selected according to a preliminary whole-body autoradiography. The organs are homogenized and suspended in distilled water. Aliquot amounts of the suspension with known organ concentration are measured into a liquid scintillation cuvette, discoloured and measured as described for blood samples.

The method used in *elimination studies* depends on the main route of elimination of the compound: by the exhaled air, urine, faeces or bile. In the first case, the animal is treated with the labelled compound then placed in a metabolic cage (Fig. 3.50). The metabolic cage is constructed so that urine and faeces can be collected separately. It is possible to pass air through it by means of a vacuum pump; a gas washer is installed into the vacuum line. The washer is filled with a 25% solution of tetraethylammonium hydroxide absorbing the radioactive carbon dioxide exhaled by the animal. If the radioactive concentration of the alkaline liquid is measured, the amount of the exhaled radioactive carbon dioxide can be established.

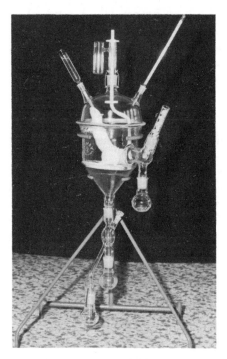

Fig. 3.50. Metabolic cage

Radioactivity eliminated *via urine* and *faeces* can be determined by the liquid scintillation method. If the volume of the urine and the amount of the faeces are known, the percentage of the introduced substance eliminated in the urine and faeces can be calculated. This allows also to calculate the *biological half-life* of the substance.

The measurement of *excretion through bile* is carried out on rats narcotized with urethane. *The ductus choledochus* is cannulated, i.e., bile is made obtainable, through a plastic tube. The labelled compound is then introduced, and the amount and radioactivity of the bile is measured in every hour. The amount of the radioactive material eliminated through the bile can be calculated on the basis of these two values.

Compounds eliminated through the bile get into the small intestines and can be reabsorbed. Blood takes them into the liver from where they can be eliminated again either in unchanged form or after having suffered some chemical transformation. This process is called *enterohepatic circulation* and can be followed by several methods. The most frequently applied procedure is as follows: the *ductus choledochus* of rats previously starved for 18 h is cannulated, and the bile is passed through a plastic tube into the duodenum of another animal. The biliary canal of the second animal is also canuled and the bile is passed into the duodenum of a third rat. This method permits one to determine the activity of the excreted bile, as well as the frequency and extent by which the compound is circulated.

Studies on pregnant animals can determine whether the compound or its radioactive metabolites can get into the rat *embryo*; important information is thus obtained for subsequent teratological experiments. In these studies, rats in the 18th day of their *gravidity* are treated with labelled compounds and, at the moment of the maximum acute blood concentration, an operation is carried out in ether narcosis. Several embryos are taken out of the animal together with the amnion. The radioactivities of the embryo, the placenta and the amniotic fluid are determined.

3.7.1.2. METABOLIC STUDIES

Drugs in the human and animal organisms are treated as *foreign substances* and are, therefore, eliminated as soon as possible. This, however, takes place only if the compound is polar enough to be water-soluble, because the kidney cannot excrete lipoid-soluble apolar substances. The enzyme system of the organism attempts to transform any foreign substance into a polar compound which can be rapidly eliminated from the organism. These transformation processes in the organism are called *metabolism*.

Metabolic studies are primarily designed to detect transformation processes and to decide whether a compound can be administered safely. Additional information, however, can be obtained about the fact whether the original compound or one/some of its metabolites is/are responsible for the pharmaceutical action. The importance of such investigations is continually increasing.

In the case of an advantageous metabolism, sometimes a pharmacologically more active substance is obtained than the one introduced originally, or the metabolite with the desired action may not possess any harmful side-effects. Thus, metabolism studies can promote and render more expedient the research and synthesis of biologically active substances.

For rapid and successful metabolism studies, as a rule, ^{14}C-labelled substances are applied. This is due to the fact that (as shown in Section 3.7.1.1) compounds introduced into the animal or plant organism are there distributed, transformed and the resulting concentration of the metabolites in the body fluid is so low that common chemical or physical methods are not sufficiently sensitive to detect them properly. Amounts as low as a few ng can be detected by radiometry or contact autoradiography, provided that preparations with proper specific radioactivities are applied. Of course, some metabolic studies can also be carried out by classical analysis; in this case, however, the time requirement of the experiments increases, identification becomes uncertain, and several very expensive instrumental analyses may be necessary.

Metabolic studies can be divided into four groups:
— determination of the location of the metabolism;
— isolation of metabolites;
— elucidation of the structures of metabolites;
— determination of the concentration of metabolites.

Various methods are used to determine the *location of metabolism*. Metabolism may commence during the passage of the drug through the intestinal tract, but it takes place in most cases in the liver or in the kidney. The changes of the compound introduced into the isolated stomach or intestine can be monitored by radiochromatography (Section 4.3.5); the same method can be applied for bile secreted by the liver. The metabolizing role of the liver can be proved by "liver perfusion" studies carried out on the isolated liver. In most cases, however, metabolites are identified from human or animal urine.

The steps of *metabolite isolation* are as follows:
— removal of endogenous substances;
— extraction;
— separation;
— identification.

The biggest problem is represented by the total or partial removal of *endogenous substances* from the urine. Organic solvents dissolve not only the drug or its metabolites, but they may also extract some endogenous physiological substances. The presence of such substances makes the isolation of metabolites considerably difficult.

Apart from various salts and colouring substances, the most important disturbing material is urea. The first task is, therefore, the removal of urea from the urine. Urine is, as a rule, supersaturated with respect to urea, thus some of it precipitates spontaneously and can be removed by sedimentation. The removal can be enhanced by keeping the fresh urine between 0 and $+45\,°C$ for 24 h, and decanting the liquid after the sedimentation of urea. The quantitative removal of urea is only possible by digesting with an urease enzyme.

The removal of colouring substances and other impurities can be attempted by treatment with Florisil. Care should be taken, however, to avoid the loss of radioactivity during this process.

After having removed the endogenous materials, the metabolites can be extracted from the urine. They are present either in the free state or as various conjugates (sulphates, glucuronides, etc.).

After extraction, the metabolies are separated by means of *various separation methods*. Preparative thin-layer chromatography is the method employed most frequently. Detection is carried out by contact autoradiography (Section 4.3.5). The preparation of the layer requires special care because its material may contain several impurities. They can be removed by multiple pre-running.

The first step of identification is autoradiography; the spots detected this way are eluted and the structures of the compounds are determined by gas chromatography and/or mass spectrometry.

Application of radioactive substances and their detection by autoradiography ensures such advantages in thin-layer chromatography which make this process unique. No other separation methods possess such advantages. Metabolites produced from the introduced compound can be detected in most cases only on the basis of their radioactivity because, owing to their unknown chemical structure, no chemical investigations are possible.

3.7.2. RADIOTRACER TECHNIQUE IN PESTICIDE RESEARCH

Agrochemicals include fertilizers and pesticides. *Pesticides* are compounds of specific toxicity and their combinations can exterminate certain harmful or pathogenic species (as total killers) while they are harmless to human organism. Modern pesticides, such as organophosphoric esters, have extremely high toxicity but, having low persistency (chemical stability), their human toxicity is very low.

The following classes of agents belong to the group of pesticides:
— *herbicides* (against weeds);
— *fungicides* (against fungi);
— *insecticides* (against insects).

In addition, special pesticide groups are: *regulators* (plant growth substances), *bactericides* (preventing the reproduction of bacteria), *rodenticides* (killing rodents), etc.

Pesticides are, in some respect, similar to human or veterinary drugs (see Section 3.7.1). The most significant differences are found in the amount and quality of the chemicals used, as well as in the specificity of action:
— whereas human or veterinary drugs are applied in doses as low as a few mg or g, the usual dose of pesticides is 10–100 kg per hectare;
— whereas drugs act with sparing one or a few species (human or domestic animal), pesticides should spare several plant species and, at the same time, kill as many species as possible of harmful fungi, weeds and insects, respectively.

Research in the field of pesticides has accelerated considerably during the last decade, therefore the metabolic studies of these agents has to be accelerated, too. Acute and subacute toxicological and kinetic studies, applied generally in earlier years, have been completed with other studies such as mutagenesis, teratology, as well as the investigation of possible changes in the organisms under the effect of chronic exposure (very low doses for very long time). Comparison of the biological activities of decomposition products of a parent pesticide permits one to recognize the active ingredient on a molecular level, i.e., to interpret the correlation between chemical structure and biological activity.

Environmental protection requires long duration studies as far as potential damage to human individuals and the whole population is concerned. In addition to the protection of human genetics, the potential effects on the soil, on surface waters and subsurface water reservoirs, on different categories of plant and animal species are also to be investigated, together with the *protection of nature* (wild-life and aesthetics of the surroundings). For all these purposes, *rapid and sensitive analytical methods* have to be developed and applied; among these, radiotracer technique and radioanalytics (see Section 4.3) occupy an important position.

Of toxicity studies, the investigation of *acute toxicity* takes the first place. First of all, the value of *lethal dose* (*LD*) should be determined. As a rule, the dose causing 50% lethality (LD_{50}) is established, mainly on rats, and also on mice and rabbits. If several laboratory animals are used in these experiments, the toxicological characterization of lethality becomes also possible, namely by comparing sublethal doses. The determination of LD_{50} values is necessary mainly with respect to *labour* nutrition hygienics; the characterization of lethality in subacute doses is important for the same reasons. Directly connected with these studies is the research in the field of first medical aid and antidotes; this latter is aimed at the elimination of toxicosis caused by the pesticide. In addition to LD values, eye irritation hazard, skin sensitivity and allergic effects are also to be determined.

In the case of more detailed studies, the dose causing 90% lethality (LD_{90}) is also determined.

When chronic toxicity is studied, subacute doses are administered for longer periods; thus the accumulation of pesticide can be detected in certain special organ(s). Accumulation is usually accompanied by cellular damage. Tracer methods permit one in such cases to verify directly certain cellular localizations and to study the biological effects of pesticides and their metabolites on cellular structures.

The mutagenic action of a pesticide is tested by means of cultures of special microorganisms, as a rule, bacteria. This follows directly acute toxicological studies. It is the *protection of human genetics* which peremptorily necessitates the widening of the scope of mutagenetic studies as well as the qualifying of general biological comparisons.

The testing of *teratogenic activity* of pesticides is also of basic importance. The tests can be performed by treating one or several generations of pregnant rats. In addition to the determination of the number of progeny and the birth weight, measurement of the absolute and relative masses of endocrine glands gives also important information, first of all, when *histological and radiopharmacological methods* are used in combination. Fundamental information can be obtained in many cases by chronic toxicological studies of one or several generation(s) on the basis of profound histological and organological studies of new-born offsprings in the early stages of their life, because sometimes this is the only stage when chronic toxicological defects appear.

Biological effectiveness as far as *carcinogenesis* is concerned, requires particular experiments, either with tissue cultures or with laboratory animals. Tissue cultures used in the study of carcinogenesis consist usually of leukocytes (*lymphocytes*); thus the specific hindrance and/or stimulation of mitosis can be evaluated also in a quantitative way. Investigation in other types of cultures is more clumsy and the evaluation of results is also more difficult, hence the generalizations are valid in a smaller circle. Tracer technique or cellular localization carried out with leukocytes is often used for studying *leukocytosis*. These latter studies in tissue cultures are, however, not regarded as full value carcinogenetic evaluations; a carcinogenic induction is only stated when abnormal cell (tissue) growth can be observed after treatment of living laboratory animals. In cases of verified carcinogenesis, it is necessary to apply also radiopharmacological methods.

From the point of view of the development and selection of pesticides, it is very important to learn as much as possible about the reaction mechanism and the kinetics of the processes. One of the most important data is the knowledge of *uptake (absorption)* of the model substance or the pesticide developed to a ready preparation. This is important from the point of view of application of active ingredients. These latter substances get into the living organism through lipoidic membranes, thus lipoid solubility influences directly or indirectly the absorption of these agents into the tissues exposed. The study of the uptake by means of labelled pesticides supplies well evaluable data for both intact organisms and isolated organs.

It is very important to study the *distribution* of pesticides. Active ingredients are transferred from absorbing cells to other ones *(translocation)* or special transporting tissues *(transport)* thus they are distributed in the organism. The distribution is, however, almost *never uniform*: absorbed active ingredients are generally accumulated in certain organs or tissues (e.g., DDT is accumulated in fat tissues of insects—apart from nervous and brain tissues—due to its favourable solubility in lipoids). Agents well absorbed through roots and distributing favourably (systemic pesticides) are accumulated predominantly in dividing (meristematic) tissues, thus uniform distribution is more and more distorted. Continuous growth makes the situation even more complicated.

The effect of active ingredients on the *metabolism of host organisms* is studied by means of radiotracer-labelled active ingredients in large series and by several tests. Such tests on plants are: the study of intensity of $^{14}CO_2$ fixation in photosynthesis, the study of protein synthesis by means of incorporation of amino acids labelled with radiocarbon or tritium, the investigation of the extent of evaporation using HTO, etc.

The determination of *stability* and *persistency* of an active ingredient in the treated organisms, in products to be utilized, in soil, in waters, etc. is, by far, not a negligible task. There is a close positive correlation between the effectivity and persistency of pesticides.

Metabolism and the decomposition mechanism of pesticides are analogous processes: active ingredients are generally oxidized or fragmented *via* enzymatic or abiotic (e.g., photochemical) transformations. If active ingredients are introduced in which the label is in a predetermined position, radiochemical identification of the unchanged agent and its partially fragmented metabolites gives precise information about the mechanism of decomposition. The identification of parent and secondary compounds in host organisms or in laboratory animals is necessary not only in order to determine the direction and extent of the decomposition, but also because decomposition may produce toxic products. An example is shown in Fig. 3.51 depicting the metabolism of 2,4,5-trichlorophenoxy-ethanol (TCPE) applied to maize root. It can be seen that the concentration of the parent pesticide decreases as a function of time, that of its metabolic products (e.g., 2,4,5-trichlorophenoxyacetic acid = 2,4,5-T) increases. The constancy of the overall radioactivity indicates a negligible translocation.

The figure shows that some of the metabolites could not be identified, although their amounts and concentrations may not be negligible. Therefore, potential derivatives, which are the most important from the point of view of labour- and food-hygienic effects and hazards, are to be studied separately: a comparison with synthetic reference substances of known composition may clarify whether various properties of the unknown

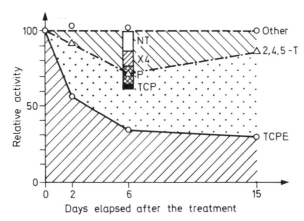

Fig. 3.51. Changes in the quantity of metabolites after the treatment of maize root; — x — TCPE; — · — · 2,4,5-T; —O— other metabolites (the 6-day value of the latter in breakdown is shown in a separate column; TCP: trichlorophenol; P: chlorinated polyphenols; X4: unidentified metabolite present in highest amount; NT: total amount of all other unidentified metabolites)

metabolite are the same as those of a hazardous key compound. If a radiochromatographic analysis points to the opposite, it can be regarded as proven that the given hazardous substance is not produced during the metabolism of the active ingredient.

Since the purity of pesticides is, as a rule, of technical grade, a *detailed study of contaminations (accompanying substances)* is also justified: e.g., 2,4,5-trichlorophenoxyacetic acid (2,4,5-T) is usually contaminated with dioxin (2,3,7,8-tetrachlorodibenzodioxin, TCDD) which has a chronic toxicity, causing teratogenesis.

Localization of labelled pesticides in organs, tissues and cells can be directly studied by autoradiography (Section 7.6). In addition to the specific and extensive *accumulation* of the original substance and its decomposition products, the mechanism of their *elimination (excretion)* can be studied directly and reasonably by tracer technique only.

When the effects of *plant growth substances* are studied, the label is not introduced into the biologically active substance itself, but into those *metabolic intermediates* which directly participate in the metabolism. Thus, stimulation of the accumulation of dry constituents can be measured by the extent of stimulation of the photosynthetic fixation of $^{14}CO_2$. Alternatively, the effectiveness of bioactive substances stimulating protein synthesis can be tested by following the incorporation of amino acids labelled with radiotracers (^3H, ^{14}C, ^{35}S) into the proteins. The increase of radioactivity can be correlated with the extent of stimulation.

The determination of the level of pesticide residues in products of plant or animal origin represents a fundamental task from the point of view of environmental protection. For this purpose, it is advantageous to label the product with a radioisotope, followed by radiochromatographic separation and quantitative evaluation (see Section 4.3.). The application of this method permits not only the radiochemical identification of the original pesticide but also the quantitative determination of its metabolites.

It is possible to determine the optimal concentration of a nutritive solution and the optimal dose (kg/ha) of a pesticide; isotopic tracers also allow to decide whether multiple washing (watering) will result in the complete removal of the pesticide from the plant, i.e., its translocation into the soil.

Radiotracer studies may answer the question whether the application of a biologically active agent *through* the *root* or through the *foliage* is more advantageous or more economical.

3.8. RADIOISOTOPIC AGE DETERMINATION

In technology it may be important to check the age of *rubber and plastic articles* liable to aging. Double labelling (see also Section 3.6.2.5) renders this possible; one of the isotopes should have a long half-life (serving as the basis of comparison), the other a short half-life. If the two radioisotopes are introduced in an appropriate ratio of activity, the age of the object can be determined. Suitable pairs of isotopes are, e.g., ^{36}Cl–^{54}Mn, ^{36}Cl–^{95}Nb and ^{99}Tc–^{95}Nb.

Natural radioactive elements are from the beginning "multiply labelled". *Mother elements* decay, as a rule, very slowly (the half-life of ^{238}U is, e.g., 4.5×10^9 years, that of ^{232}Th 1.4×10^{10} years). The half-lives of the *daughter elements* derived from the long-living heavy elements are much shorter. Daughter elements decay further—according to the well-known decay schemes—until an end-product, such as a stable isotope of lead, is obtained.

The half-life of every daughter element of the ^{238}U family is incomparably shorter than that of the mother element, except for ^{234}U with a half-life of 2.5×10^5 years. Thus, an isolated sample of uranium represents a decay series in equilibrium within itself. If it can be assumed that no member of the decay series is lost, the rate of formation of the end product, ^{206}Pb, will truly correspond to the decay of ^{238}U.

In minerals containing uranium, the products are trapped with high probability, hence comparison of the *lead content* of the sample with its uranium content gives information about the age of the mineral. The value obtained is the time which has elapsed since the *solidification of the mineral*: when the "trap" for the products was formed.

Eight members of the ^{238}U family emit α-particles, i.e., each ^{238}U atom in the ore generates eight α-particles. Their absorption results in the formation of helium, which cannot escape from many types of ores. Consequently, the amount of *helium* produced from uranium may also give useful information about the age of a mineral. Since helium is a noble gas, it can be separated by simple dissolution of the ore: 1 g of ^{238}U gives 0.12 cm^3 NTP of helium during 1 million years.

The natural radioactive isotope of *potassium*, ^{40}K, gives rise to argon by K-electron capture. If the amount of argon in a lava is related to that of potassium, it is possible to estimate the time elapsed since the solidification of the lava, i.e., since the *eruption of a volcano*.

Cosmic rays induce free neutrons by colliding with the atoms of the earth surface and the atmosphere; therefore the whole earth is exposed to a low-level but more or less uniform, neutron bombardment inducing an extremely low level of radioactivity.

The neutron capture cross-section of *chlorine* is high, thus the natural ^{35}Cl isotope gives relatively considerable amounts of radioactive ^{36}Cl under the effect of this background neutron irradiation. The measurement of the activity of radioactive chlorine isotopes permits one to determine the lifetime of surface minerals, if this is lower than one million years.

Radiocarbon is produced under the effect of cosmic rays from the nitrogen of the atmosphere. The half-life of ^{14}C is 5730 years, i.e., by several orders of magnitude lower than that of isotopes discussed above. This half-life corresponds to the order of magnitudes of *historical*, and not geological, periods of time (2000–30 000 years). ^{14}C transforms, as a rule, into carbon dioxide, which is assimilated by green living plants by photosynthesis. General experience has shown that the ^{14}C isotope concentration of living organisms is the same all over the world. If an organism dies, the accumulation of ^{14}C stops, and decay will occur. Hence, the specific radioactivity of the carbon decreases during the time between the death of a specimen of plant or animal origin, and the date of measurement. This method is suitable for determining the age e.g., of peat, wood (charcoal, wooden objects), leaves, hair, leather, charred bones, clothes, paper. Petroleum and coal had been formed so long ago that its radiocarbon content cannot be measured any more and their formation cannot be dated by this method.

Tritium with a half-life as short as 12 years, is also produced continuously in the atmosphere under the effect of cosmic rays. Hence it can always be found in atmospheric water. Its activity is too low for direct measurement, therefore it should be enriched, e.g., by electrolysis. The radioactivity of water samples from *subsurface water layers* is much lower than that of *surface waters*; the extent of the difference is characteristic of the age of the water below the surface, i.e., it indicates the date when the interaction with the cosmic rays ceased. This way the age of a water layer can be determined (between 1 and 30 years) and even the amount of subsurface water supply can be estimated.

EXAMPLE 3.15

Let us calculate the age of an uranium ore with a mass ratio of ^{238}U to ^{206}Pb, $B = 30$. Let us assume that every ^{238}U atom which disintegrated up to the time of measurement gave ^{206}Pb.

If the mineral contains n_U ^{238}U atoms at the moment of the measurement, and it contained n_U^0 atoms when it solidified, the correlation between these two values is given by the following expression:

$$n_U = n_U^0 \exp(-\lambda t) \tag{3.86}$$

where t denotes the age of the ore.

The amount of ^{206}Pb atoms will be

$$n_{Pb} = n_U^0 - n_U^0 \exp(-\lambda t) \tag{3.87}$$

and

$$K = \frac{n_U}{n_{Pb}} = \frac{\exp(-\lambda t)}{1 - \exp(-\lambda t)} \tag{3.88}$$

Thus

$$\exp(\lambda t) = \frac{K+1}{K} \tag{3.89}$$

and

$$\lambda t = \ln \frac{K+1}{K} \tag{3.90}$$

Considering that

$$\lambda = \frac{\ln 2}{t_{1/2}} \tag{3.91}$$

the t age of the mineral can be calculated:

$$t = \frac{1}{\lambda} \ln \frac{K+1}{K} = t_{1/2} \frac{\ln \dfrac{K+1}{K}}{\ln 2} \tag{3.92}$$

where $t_{1/2}$ is the half-life of ^{238}U, i.e., 4.5×10^9 years.

The atomic ratio can be obtained from the B mass ratio by multiplying it by the ratio of mass numbers ($\mathscr{A}$):

$$K = B \frac{\mathscr{A}_{Pb}}{\mathscr{A}_U} = 30 \frac{206}{238} = 26 \qquad \frac{K+1}{K} = \frac{27}{26} = 1.04$$

The age of the uranium ore investigated is:

$$t = 4.5 \times 10^9 \times \frac{3.92 \times 10^{-2}}{0.693} = 2.54 \times 10^8 \text{ years}$$

EXAMPLE 3.16

Wooden residues have been found which originated assumedly from glacial drift. The specific relative radioactivity of CO_2 formed when burning these wooden pieces was 3.8 disintegration/(min g). The ^{14}C "content" of living organisms is at present 15.3 disintegration/(min g). How old is the finding? Its age could be dated back to the end of Pleistocene glaciation of the earth.

$$n = n_0 \exp(-\lambda t) \tag{3.93}$$

where t is the time that elapsed between the decay of the plant in the glacial period and the date of measurement.

The half-life of radiocarbon ^{14}C is 5730 years, i.e.,

$$\lambda = \frac{\ln 2}{5730} = \frac{0.693}{5730} = 1.21 \times 10^{-4}$$

The ratio of the ^{14}C concentration of living organisms (no) to the concentration of the found sample (n_0) is:

$$\frac{n_0}{n} = \frac{15.3}{3.8} = 4 = \exp(1.21 \times 10^{-4} \cdot t)$$

thus the age of the finding is:

$$t = \frac{\ln 4}{1.21 \times 10^{-4}} \cong 11457 \text{ years}$$

Radioactive dating on the basis of ^{14}C determination is a measurement demanding care. The low-energy β-radiation of ^{14}C can be determined by special detectors, e.g., liquid scintillators. The low absolute difference of the count rates is another problem: in the example, 15.3 imp/(min g) was compared with 3.8 imp/(min g).

REFERENCES

[3.1] Erwall, L. G., Forsberg, H. G. and Ljunggren, K., Radioaktive Isotope in der Technik. Friedrich Vieweg & Sohn, Braunschweig, 1965.

[3.2] Fodor, J., (Radioisotope Tracer Technique in Industry) (In Hungarian). Műszaki Könyvkiadó, Budapest, 1962.

[3.3] Földiák, G., (Production and Application of Radioactive Isotopes) (In Hungarian). Tankönyvkiadó, Budapest, 1970.

[3.4] Földiák, G., Balla, B., Hirling, J. and Rózsa, S., (Industrial Applications of Isotopes) (In Hungarian). Tankönyvkiadó, Budapest, 1970.

[3.5] (Research on Isotope Chemistry) Yearbook of the Institute of Isotopes of the Hungarian Academy of Sciences, (In Hungarian). Budapest, 1969.

[3.6] Keszthelyi, L., (Scintillation caunters) (In Hungarian). Műszaki Könyvkiadó, Budapest, 1964.

[3.7] Lengyel, T. and Jász, Á., (Handbook for Isotope Laboratories) (In Hungarian). Műszaki Könyvkiadó, Budapest, 1966.

[3.8] Molnár, J., (Isotope Applications) (In Hungarian). Vols. I–II. Tankönyvkiadó, Budapest, 1970.

[3.9] Nagy, L. Gy., (Radiochemistry and Isotope Technique) (In Hungarian). Tankönyvkiadó, Budapest, 1983.

[3.10] Nagy, L. Gy. and Szokolyi, L., (Neutron Activation Studies) (In Hungarian). Műszaki Könyvkiadó, Budapest, 1966.

[3.11] Radioactive Isotopes in Scientific Research. A symposium. Budapest, 12–13 November, 1969.

[3.12] Radioisotope Tracers in Industry and Geophysics. IAEA, Vienna, 1967.

[3.13] Kiss, I. and Vértes, A., (Nuclear Chemistry) (In Hungarian). Akadémiai Kiadó, Budapest, 1979.

[3.14] Broda, E. and Schönfeld, T., Die technischen Anwendungen der Radioaktivität. Vol. I, VEB Deutscher Verlag für Grundstoffindustrie, Leipzig, 1962.

[3.15] Haissinski, M., Nuclear Chemistry and its Applications. Addison-Wesley, Reading–Palo Alto–London 1964.

[3.16] Neiman, M. B. and Gál, D., The Kinetic Isotope Method and Applications. Akadémiai Kiadó, Budapest–Elsevier, Amsterdam, 1971.

[3.17] Gál, D., Danóczy, E., Nemes, I., Vidóczy, T. and Hajdu, P., Ann. N.Y. Acad. Sci., 213, 51 (1973).

[3.18] Derbentsev, Yu. I. and Isagulyants, G. V., Uspekhi Khimii, 38, 1597 (1969).

[3.19] Campbell, K. C. and Thomson, S. J., Progress in Surface and Membrane Science, 9, 163 (1975).

[3.20] Pines, H. and Goetschel, C. T., J. Org. Chem., 30, 3530 (1965).

[3.21] Mink, Gy., Móger, D. and Nagy, F., *Magyar Kém. Folyóirat*, **76**, 411 (1970).
[3.22] Reid, J. U., Thomson, S. J. and Webb, G., *J. Catal.*, **29**, 421 (1973).
[3.23] Hirling, J., *Izotóptechnika*, **20**, 125 (1977).
[3.24] Bujdosó, E., Miskei, M. and Ormos, Gy., *Kohászati Lapok*, **97**, 146 (1964).
[3.25] Bujdosó, E. and Tóth, P., *Proc. Res. Inst. Non-Ferrous Metals (Budapest)*, **9**, 117 (1971).
[3.26] Tétényi, P. and Babernics, L., *J. Catal.*, **8**, 215 (1967).
[3.27] Tétényi, P., Babernics, L. and Thomson, S. J., *Acta Chim. Acad. Sci. Hung.*, **34**, 335 (1963); Tétényi, P., Paál, Z. and Dobrovolszky, M., *Z. phys. Chem. (Frankfurt)*, **102**, 267 (1976).
[3.28] Derbentsev, Yu. I., Paál, Z. and Tétényi, P., *Z. Phys. Chem. (N.F.)*, **80**, 51 (1972); Tétényi, P. and Paál, Z., *Z. Phys. Chem. (N.F.)*, **80**, 63 (1972).
[3.29] Tétényi, P., Guczi, L., Paál, Z. and Babernics, L., *(Metal Catalyzed Hydrocarbon Reactions)* (In Hungarian). Akadémiai Kiadó, Budapest, 1974; Paál, Z., *Advances in Catalysis*. Vol. 29, Academic Press, New York, 1980, p. 273.
    Kiadó, Budapest, 1974; Paál, Z., *Advances in Catalysis*. Vol. 29, Academic Press, New York, 1980, p. 273.
[3.30] Pines, H. and Shaw, A. W., *J. Am. Chem. Soc.*, **79**, 1474 (1957).
[3.31] Dermietzel, J., Bauer, F., Rösseler, M., Jockisch, W., Franke, H., Klempin, J. and Barz, H.J., *Isotopenpraxis*, **12**, 57 (1976).
[3.32] Paál, Z. and Thomson, S. J., *Kémiai Közlemények*, **43**, 463 (1975).
[3.33] Feighan, J. A. and Davis, B. H., *J. Catal.*, **4**, 594 (1965).
[3.34] Guczi, L., Sharan, K. M. and Tétényi, P., *Monatsh.* **102**, 187 (1971).
[3.35] Norval, S. V. and Thomson, S. J., *J. Chem. Soc., Faraday Trans.*, *I*, **75**, 1798 (1979).
[3.36] Ozaki, A., *Isotopic Studies of Heterogeneous Catalysis*. Kodansha, Tokyo–Academic Press, New York, 1977.
[3.37] Blyholder, G. and Emmett, P. H., *J. Phys. Chem.*, **63**, 962 (1959); **64**, 470 (1960).
[3.38] Kagan, Yu. B., Rozovskii, A. Ya., Liberov, L. G., Slivinskii, E. V., Lin, G. I., Loktev, S. M. and Bashkirov, A. N., *Dokl. Akad. Nauk SSSR*, **221**, 1093 (1975); **224**, 1081 (1975); Rozovskii, A. Ya., *Kinet. Katal.*, **21**, 97 (1980).
[3.39] Guczi, L., LaPierre, R. B., Weiss, A. H. and Biron, E., *J. Catal.*, **60**, 83 (1979).

4. RADIOANALYTICAL METHODS

The analysis of trace elements represents an up-to-date field of analytical chemistry: i
has an outstanding importance in *nuclear energetics*, in *electronics* and in compute
technology with regard to the investigation of high purity materials; it is also o
outstanding importance in *biology* (trace elements of vital importance and toxic ones
environment pollution). This area was first exposed by the requirements for extremely
pure materials used in nuclear reactors (beryllium, thorium, uranium, etc.); here the
amount of impurities must not exceed 10^{-6}–$10^{-4}\%$, especially of elements with high
neutron capture cross-section, such as boron, europium, gadolinium, cadmium, etc.
Purity requirements are also very strict for some metals and alloys (e.g., cobalt, copper
molybdenum, nickel, titanium, tungsten, vanadium), as well as for semiconductor
(germanium, silicon, etc.); e.g., contamination with antimony, arsenic, bismuth, cadmium
lead, tin should not exceed 10^{-5}–$10^{-4}\%$. This means that these elements must b
determined with high sensitivity (10^{-9}–$10^{-6}\%$) not only in the end products, but also i
raw materials.

Apart from "classical" chemical *methods* and optical spectroscopy and spec
trophotometry, the most suitable ways of trace analysis are atomic absorption
radioanalytical and mass spectrographic methods. The combination of various methods i
often advantageous. The scope of application of some microanalytical methods is show
in Table 4.1.

Table 4.1. Application ranges of microanalytical methods

Analytical method	Number of elements detected in various ranges			
	Total	900–$10 \cdot 10^{-7}\%$	10–$0.1 \cdot 10^{-7}\%$	$<0.1 \cdot 10^{-7}\%$
Atomic absorption	49	27	22	
X-Ray fluorescence	23		13	10
Emission spectroscopy	61	36	25	
Fluorescence	26	17	9	
Isotopic dilution	37		10	27
Kinetic methods	25	10	9	6
Flame photometry	66	32	31	3
Neutron activation	67	16	37	14
Spectrophotometry	57	39	18	
Mass spectrometry	83		47	36

Several radioanalytical methods are also suitable for *rapid (express) chemical analysis*
The purpose of such measurements is generally the determination of chemica
components present in considerable amounts; in field work or in the production proces
such measurements should be carried out, if possible, without any loss of time. Thes
methods may also be used for controlling automatics.

Geophysical rapid methods will be discussed in Sections 5.2 and 5.3.

Many of the nuclear analytical methods developed in the last decade are only applie
for scientific purposes, therefore they do not belong to the scope of this book and here w
refer only to some excellent monographs [4.24–4.26].

4.1. ANALYTICAL METHODS BASED ON PHYSICAL INTERACTIONS WITH RADIATION

Phenomena involving *absorption* or *scattering* of radiation, as well as *excitation* or *ionization* of the electron orbitals, represent the simplest forms of interactions between radioactive radiation and material. The analytical applications of interactions between radiation and matter are based on the differences between individual elements. Because the energy of radioactive radiation generally exceeds by several orders of magnitude that of chemical bonds, it is *only possible to determine elementary composition*. Only refined measurement techniques allow sometimes to detect differences between compounds and to apply these differences (e.g., Mössbauer effect).

4.1.1. GAS ANALYSIS BASED ON IONIZATION

If a radioactive radiation passes through a gas, it will ionize and excite gas atoms and molecules; if an electric field is applied, electrons and positive ions move towards the respective electrodes and are neutralized there. This is detected as an "ionization current". Electrons moving towards the positive electrode can be captured by atoms, molecules and even aerosol particles; therefore the speed of their migration decreases. Hence the probability of their collision and recombination with positive ions increases by several orders of magnitude. The outward manifestation of this phenomenon will be the decrease of the ionization current. The energy of excited noble gas atoms—produced by ionizing radiation or even by ion collisions under the effect of a sufficiently strong electric field—is high enough to ionize other (generally organic) molecules if they collide with them. The presence of low amounts of easily ionizable substances, under otherwise identical conditions, leads to the considerable increase of the basic ionization current—up to several orders of magnitude.

It follows that *ionization gas detectors* are such *ionization chambers* (Fig. 4.1, see also Section 2.1.1.1.) in which ions are produced by radioactive irradiation and *the changes of the current are used to draw conclusions about the composition of the gas*. Depending on the fact which of the basic phenomena is favoured by detector geometry, by the applied electric field and by the carrier gas used, there are cross-section, electron capture and noble gas (e.g., argon) ionization detectors.

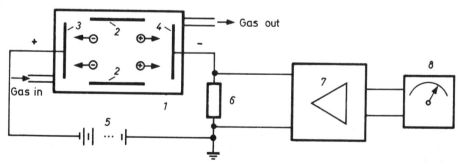

Fig. 4.1. Simple scheme of a gas ionization detector. *1* — Detector; *2* — Radiation source; *3* — Anode; *4* — Cathode; *5* — Direct current supply; *6* — Input resistance; *7* — Amplifier; *8* — Reading instrument

The sensitivity of these detectors is determined mainly by the number of primary ions, hence suitable *radiation sources* are α- and β-radiating isotopes with high specific ionization ability, such as ^{210}Po, ^{241}Am, ^{226}Ra, ^{241}Am and ^{3}H, ^{63}Ni, ^{85}Kr, ^{90}Sr–^{90}Y, ^{99}Tc and ^{147}Pm, respectively. The main factors in selecting the isotope are the half-life and the danger from the point of view of radiation protection.

The ionization currents are in the order of magnitude of 10^{-14}–10^{-7} A, i.e., appropriate amplification is indispensable (see Section 2.1).

Ionization detectors are extensively used in *gas chromatography* and in *industrial gas analysis*.

Cross-sectional detectors (CSD) are generally much smaller than the maximum action radius of the ionizing radiation in the carrier gas of the chromatograph; in such cases the *ion current* can be written as follows:

$$i = \frac{\Phi p}{RT} \Sigma \, c_i Q_i \qquad (4.1)$$

where Φ is a constant depending on the geometry, p and T are the pressure and the temperature (K) of the gas in the detector, R is the universal gas constant, c_i is the molar fraction of the i-th component and Q_i is the relative ionization cross-section of the i-th component (as a rule, that of the carrier gas is taken as unity).

The ionization cross-sections of the elements increase generally with increasing atomic number, therefore He or H_2 can be used as carrier gases. The ionization cross-section of molecules can be calculated from elementary constants by simple methods. These constants can be found in tabulated form in handbooks (e.g., [10]).

Isotopes emitting β-radiation are preferred to α-radiating ones for the purposes of *gas chromatography*, because the same ion current is produced in the latter case by fewer α-particles, therefore the statistical fluctuation and thus the noise background of the detector is higher.

The great advantage of CSD is its sturdiness; it is simple to select its feed voltage (100–200 V), and it is linear from $10^{-6}\%$ up to 100%. Its selectivity is small (Example 4.1).

Noble gas detectors (argon or helium detectors, ArD, HeD) are based on the fact that the energy of the metastable excited state of noble gases is very high (Ar: 11.5–11.7, He: 19.8, 20.6 eV), exceeding the ionization energy of most substances. The voltage of the chamber is generally chosen to be high (1000–2000 V) in order to render it possible that also electrons from the secondary ionization may participate in the production of the metastable excited state.

The detector has an extremely high sensitivity (10^{-14} g/s), but is very sensitive to impurities as well as to pressure and temperature changes.

Electron capture detectors (ECD) are extensively used in *gas chromatography* and also in other fields, e.g., as *aerosol detectors*. Their operation is based on electron capture and recombination following primary ionization. The probability of the electron capture can be characterized by the electron affinity of the atom, molecule or aerosol particle. This is equal to the energy necessary to remove the captured electron.

The electron affinity coefficient K can be calculated by an exponential function from the electron affinity energies and extends over seven orders of magnitude for organic substances; if the value for benzene is taken equal to unity, that for carbon tetrachloride is 10^7 [4.10]. This principle ensures very high selectivity, because the coefficient affects the ion current exponentially:

$$i = i_0 \exp\left(-Kc\Phi\right) \qquad (4.2)$$

and if $c \to 0$, then

$$i - i_0 = i_0 K c \Phi \qquad (4.3)$$

where i and i_0 denote the ion current intensity in the presence and absence, respectively, of a substance with c concentration and K coefficient; Φ is a constant representing a geometric and dimensional factor.

Using nitrogen as a *carrier gas*, the sensitivity can be increased up to 10^{-15} g/s for substances having strong electron affinity, such as halogenated hydrocarbons. ${}^3\text{H}$ and ${}^{63}\text{Ni}$ isotopes are the best radiation sources.

Aerosol detectors are special types of the electron capture detectors, which have found industrial application is some fields, and are thought to become more wide-spread mainly for *air pollution measurements* in *environmental protection*. There are two new problems in connection with the measurement of aerosol concentration by means of electron capture detectors:

— the measurement is, of necessity, to be carried out in the presence of a *strong electron captive substance* (O_2 of air); this disturbing effect can be decreased by increasing the electric field, because the electron capture by O_2 decreases if the electrons have higher energies (0.1–1 eV);

— the electron captive properties of aerosol particles *increase in proportion to their diameter*, therefore reproducible concentration measurement can only be expected with homodisperse aerosols. If, however, the concentration is expressed gravimetrically (in mg/m³), this type of aerosol detector gives a higher sensitivity for submicrometer particles, whose detection is, in general, impossible by traditional methods of measurement and separation, but which are more hazardous for health.

Smoke detectors (Fig. 4.2) are based on the same principle as aerosol detectors and contain a *double ionization chamber*. One of these is closed; the other is in contact with the air space to be monitored. In the case of smoke appearance, the equilibrium of the chambers is disturbed and a warning signal is generated.

For the measurement of concentration of an air pollutant, the high sensitivity of aerosol detectors can also be utilized in such cases when the pollutant itself gives no signal, but can be transformed into an aerosol by means of a *suitable chemical reaction*. For instance, sulphur-containing pollutants (H_2S, SO_2, organic sulphur compounds) can be measured

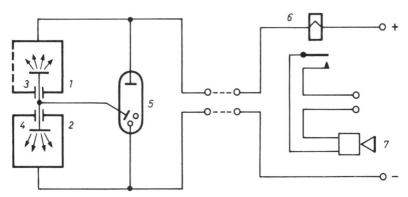

Fig. 4.2. Smoke detector. *1* — Open measuring chamber; *2* — Closed reference chamber; *3, 4* — Radiation sources mounted on electrodes; *5* — Thyratron; *6* — Relay; *7* — Alarming hooter

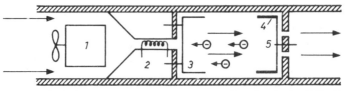

Fig. 4.3. Detector of a Sulphodet instrument. *1* — Pump; *2* — Reactor; *3* — Anode; *4* — ^{14}C radiation source; *5* — Measuring electrode (cathode)

after their oxidation over a Pt-catalyst to SO_3, as in a portable device (Fig. 4.3), whose sensitivity is in the range of mg/m^3. Owing to requirements in environmental protection, the detector has a sealed ^{14}C radiation source of 200 MBq ($\sim$5 mCi) activity. The same device can measure the concentration of *dust* and *smoke*; with a glowing tungsten filament, it is suitable for the measurement of aerosols generated by thermal decomposition, such as metal carbonyls and organometallic compounds, e.g., gasoline lead additives. By applying triethylamine vapours, the measurement of acidic pollutants can be carried out in a selective way.

EXAMPLE 4.1

Let us calculate the percentage change of the ion current in the ionization cross-sectional detector, if 0.01 volume-% ethane (C_2H_6) appears in the H_2 carrier gas. If we apply a ^{90}Sr–^{90}Y source of 400 MBq ($\sim$10 mCi) activity, can this concentration be detected? Basic data are as follows [4.10]: $Q_H = 1$, $Q_C = 4.16$, the time constant of the electronic device is $\tau = 1$ s.
Calculation:

$$Q_{H_2} = 2 \times 1 = 2 \text{ (because } H_2 \text{ contains two hydrogen atoms)}$$

$$Q_{C_2H_6} = 2 \times 4.16 + 6 \times 1 = 14.32$$

From Eq. (4.1):

$$\frac{\Delta i}{i_0} \times 100\% = \frac{(0.01 \times 14.32 + 99.99 \times 2) - 100 \times 2}{2} = \frac{0.01 \times 12.32}{2} = 0.062\%$$

At the same time, based on Eq. (2.4), the statistical fluctuation of the ion current is as follows:

$$\frac{\Delta i}{i_0}\% = \frac{100}{\sqrt{n \times \tau}} = \frac{100}{\sqrt{1 \times 4 \times 10^8}} = +5.0 \times 10^{-3}\%$$

It follows that 0.01 vol.-% of ethane can be detected with a high certainty, because the fluctuation of the basic ion current is below $5.0 \times 10^{-3}\%$ in 68% of the cases, according to the definition of standard deviation, τ calculated from the above equation and the signal of ethane in the given concentration does exceed this value more than ten times.

EXAMPLE 4.2

The sensitivity of an electron capture detector for a given pesticide is 40 A/(g s). Let us calculate how large concentration of the given pesticide will result in a change of the ion current of 3 nA by 30% (which is still in the linear range). Let us calculate the concentration of the pesticide in a sample of 1 mm^3 (1 μl), if the peak of the pesticide is about 1 min long. The flow rate of the carrier gas (w) is 3 dm^3/s.

$$i = 0.3 \times 3 \times 10^{-9}\,\text{A} = 0.9 \times 10^{-9}\,\text{A} = \frac{40\,\text{A}}{\text{g s}}\, c\,(\text{g/dm}^3) \times (\text{dm}^3/\text{s})$$

$$i = 40\,\text{A}\, c\, \frac{3}{3600}$$

$$c = \frac{0.9 \times 10^{-9} \times 3.6 \times 10^3}{40 \times 3} = 2.7 \times 10^{-8}\ \text{g/dm}^3$$

This is the pesticide concentration in the carrier gas causing a 30% ion current decrease.

Since 3 dm^3/h = 0.05 dm^3/min, this pesticide is dispersed in 50 cm^3 of carrier gas. (The volume of the detector is neglected here.) The Gaussian distribution of the concentration can be approximated by an isosceles triangle; the area of the triangle gives the mass of the pesticide:

$$g_{\text{pesticide}} = \frac{2.7 \times 10^{-8}\ \text{g/dm}^3 \times 0.05\ \text{dm}^3}{2} = 6.75 \times 10^{-10}\ \text{g}$$

1μl ≈ 10^{-3} g, so the pesticide concentration is:

$$\frac{6.75 \times 10^{-10} \times 10^2}{10^{-3}} = 6.75 \times 10^{-5}\%$$

4.1.2. ANALYSIS ON THE BASIS OF β-ABSORPTION AND SCATTERING

The rules of elementary analysis based on the absorption and scattering (β-reflection) of β-rays have been studied in detail, but there is still no consistent theory permitting one to predict the experimental results on a theoretical basis. Distinction between absorption and reflection has a technical meaning only: the method consists in the studying of the same interaction from two different aspects.

The regularities of β-absorption can be approximated by Eq. (4.4) in the range of a layer thickness of about 20–80% of the maximum action radius:

$$\ln \frac{I_0}{I} \approx \rho \times l \cdot \mu_{\text{m},i} \cdot c_2 \qquad (4.4)$$

where I_0 and I are the radiation intensities before and after passing through a layer of ρ density and l thickness, $\mu_{\text{m},i}$ is the mass absorption coefficient of the i-th element and c_i is the mass fraction of the i-th element.

The mass absorption coefficient for a given element is roughly *proportional to the specific electron density*, but it also depends on the geometry of the measuring device and even on the nature of the detector (ionization chamber or GM-tube) and on the maximum energy of the radiation source. The mass absorption coefficient *of hydrogen is about twice as high* as that of other low atomic number elements and, further on, from atomic number $\mathscr{Z} = 12$ it increases slowly with increasing atomic number (the exponent of the atomic number is about 0.3). The increase is not independent of the energy of β-radiation.

Hydrogen determination is the only measurement of analytical importance, because other, more sensitive methods are available for the other elements. Some of these are also nuclear methods, such as γ-absorption, X-ray fluorescence.

The measurement of the hydrogen content is important in the analysis of liquid hydrocarbons (petroleum products); here the material to be analysed is regarded as a two-component system consisting of carbon and hydrogen only. The other elements are regarded as impurities the concentration of which has to be known from independent measurements and taken into consideration. If the layer thickness is constant, the density must be known with great accuracy (± 0.1 kg/m^3). It is possible also to measure a constant amount of sample into a measuring cell with constant cross-section; in this case, the density measurement can be omitted because the condition $\rho \cdot l =$ constant is fulfilled.

In practical measurements, the absorptivity of unknown samples is related to one or more well-known *standards* (e.g., pure hydrocarbons); in this case, the determination of I_0 can be omitted. The advantage of using several standard materials is that the deviation from Eq. (4.4) can also be taken into consideration, and an error not exceeding $\pm 0.02\%$ (w/w) hydrogen can be ensured.

The low limit of experimental error involves very strict requirements with regard to the measuring technique, too. *Two cuvettes* can be found in the absorption section of the device shown in Fig. 4.4; one contains the liquid to be analysed, the other a standard sample. *Two opposite beams* of the radiation source (isotope: ^{90}Sr–^{90}Y; activity: 700 MBq $= 19$ mCi) irradiate the cuvettes; the *difference of the intensities* is measured by two ionization chambers, switched in opposite way to each other. The difference of ion currents brings about a potential difference in the input resistance of an amplifier; a two-decade *compensator* decreases it to a value to be recorded; the uncompensated voltage can be read from the recorder.

In order to eliminate any temperature effects, the ionization chamber, the amplifier, the input resistance and the cuvettes themselves are thermostated. It is advisable to carry out the measurement in the order of increasing densities and to insert at least three pure hydrocarbons (*n*-heptane, cyclohexane, benzene, decaline, tetraline, etc.) between the unknown samples.

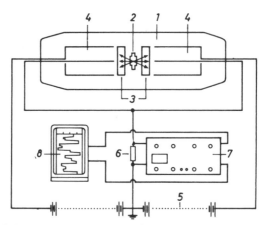

Fig. 4.4. A ^{90}Sr–^{90}Y radiation absorption instrument for determination of the hydrogen content of liquid petroleum products. *1* — ^{90}Sr–^{90}Y radiation absorption instrument; *2* — Radiation source; *3* — Cuvettes; *4* — Ionization chambers; *5* — High voltage supply; *6* — Input resistance; *7* — Amplifier; *8* — Line recorder

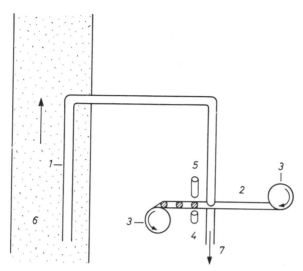

Fig. 4.5. Dust measurement by β-absorption. *1* — Sampling tube; *2* — Filter paper strip; *3* — Strip forwarder; *4* — ^{14}C-radiation source; *5* — GM-tube; *6* — Gas line; *7* — Suction

Several authors have dealt with the development of a *continuous method* of measurement, but there is no practicable method as yet, since the small differences in the mass absorption coefficients raise so strict requirements with regard to the device which cannot be satisfied with mechanically moving structures (as opposed to γ-absorption analysers) (Example 4.3).

The continuous measurement of immission or emission *dust concentrations* is also possible by means of a β-absorption method (Fig. 4.5). The air to be examined is sucked through a strip of filter. A given section of the filter is placed by the programmer between a ^{14}C-source (activity: 4 MBq ≈ 100 μCi) and a thin-window GM-tube, and the corresponding radiation intensities are measured.

Owing to the low range of the β-radiation of ^{14}C (140 g/m^2) and to the presence of other absorbing media (filter, GM-tube window, etc.), the loading of the filter may not be higher than 20–30 g/m^2. Nevertheless, by proper selection of the suction rate and time, the range of the measurement can be between a few times 10 μg/m^3 and about 5 mg/m^3. The accuracy of the measurement is sufficient for practical purposes if the disturbing effect of air humidity is eliminated by heating the air prior to filtration up to 90 °C. Measurement of, and correction for, the accidental radioactivity of the dust is also possible.

Figure 4.6 depicts the principle of the *β-reflection technique*. Arrangement (b) is more advantageous, because the detector is shielded from the primary radiation; at the same time, the intensity of the radiation towards the sample considerably increases due to reflection on the shielding.

The main regularities of β-reflection and the consequent conclusions are as follows:

— (a) With increasing layer thickness, the intensity of the reflected beam increases first proportionally, then, at a thickness equal to about 1/3 of the maximum range of β-particles, a constant saturation value is reached. This principle renders the determination of *layer thickness* possible.

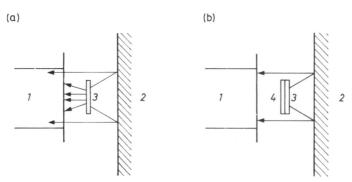

Fig. 4.6. Principal scheme of a β-reflection measurement. *1* — Detector; *2* — Sample to be analyzed; *3* — Radiation source; *4* — Shielding

— (b) The intensity belonging to the saturation layer thickness of various materials is proportional approximately to the square root of the atomic number. The only exception is hydrogen, having an extremely low (negative) value; these values may depend on the geometry and β-energy. This correlation renders the measurement of *coating thickness* possible (e.g., Sn-coating on iron).

Some authors believe that the *G* reflection factor expressing the fraction of the reflected primary particles, can be calculated as follows:

$$G = 1 - \exp\left(-\frac{\mathscr{Z}}{40}\right) \tag{4.5}$$

Other researchers propose its approximation by straight lines, the slopes of which decrease as the elements are in higher and higher periods of the periodic table. Practice shows that the function is dependent on the apparatus, thus the reflection factors should be determined experimentally for each arrangement.

— (c) The β-reflection of compounds or mixtures consisting of several elements can be calculated from the mass fraction, considering the elementary values. On this basis it is possible to determine the composition (or rather the average atomic number) of *two-component systems*. Within the linear section, the layer thickness should be constant; above the saturation value, the thickness is indifferent.

— (d) The average energy of radiation reflected from higher atomic number elements is higher than in the case when the reflecting elements have lower atomic number. This phenomenon makes possible to *sensitize the measurement to the higher atomic number component* by placing a thin absorber in the way of the reflected radiation. It has been reported that this phenomenon is valid only for separate layers and not for homogeneous two- or multicomponent systems.

— (e) As opposed to β-absorption, β-reflection shows a minor dependence on the density, consequently, it may be better suited—from this respect—e.g., for hydrocarbon analysis.

Several sources of error originating from the change of geometry, surface impurities, bubble formation, etc., are serious disadvantages of the reflection method.

The measurement of *ash content of coals* is based on the fact that the components of the coal (H, C, N, O) have lower atomic numbers than those of the ash (Mg, Al, Si, S, Ca, Fe). In this respect, coal can be regarded as a two-component system, i.e., the intensity of the reflected beam—with the same ash composition—is an unambiguous function of the ash content. This simple measuring technique permits one to eliminate the time-consuming and expensive combustion (calorimetric) method. The rapidity of the measurement is a great advantage, because it allows adjustment of the operation parameters to the current ash content, whereby the operation of a power plant becomes more economical.

At the same time, the sources of error are also to be considered (sensitivity to the changes in the humidity and ash composition, as well as to geometric factors); they can be decreased by drying, milling, layer uniformization, etc. In this way an accuracy of ± 0.4–1% ash content can be achieved.

Several methods of determining compositions have also been recommended; examples are determination of the *salt content of aqueous solutions*, the concentration of As, Bi, Br, Cu, Fe, I and Zn in medicines, of W in steels, as well as the measurement of the elementary composition of hydrocarbons.

EXAMPLE 4.3

Let us calculate the necessary accuracy of density determination in a β-absorption hydrogen analysis where the error should not exceed ± 0.02 mass-$\%$ hydrogen. The mass absorption coefficients of hydrogen and carbon are [4.4]:

$$\mu_{m, H} = 1.52 \text{ m}^2/\text{kg}, \quad \mu_{m, C} = 0.74 \text{ m}^2/\text{kg}$$

If we substitute the relationship $\Sigma c_i = 1$ into Eq. (4.4), express c_H and form a total differential, we obtain:

$$dc_H = \frac{1}{\mu_{m, H} - \mu_{m, C}} \left[\frac{d \ln \frac{I_0}{I}}{\rho \cdot I} - \frac{\ln \frac{I_0}{I} d(\rho \cdot I)}{(\rho \cdot I)^2} - \sum_{i=3}^{n} dc_i (\mu_{m, i} - \mu_{m, C}) \right]$$

The first term of the expression in brackets denotes the effect of intensity measurement, the second, that of the inaccuracy of the surface mass, the third, that of foreign elements on the hydrogen determination. Let us assume 750 kg/m^3 for density, 4 mm for thickness (as a usual value) and 14% for hydrogen content; the following expression is then obtained on the basis of Eq. (4.4):

$$\ln \frac{I_0}{I} = 750 \times 0.004 [0.14 \times 1.52 + 0.86 \times 0.74] = 2.55$$

$$0.0002 = \frac{1}{1.52 - 0.74} \left[-\frac{2.55 \times \Delta\rho \times 0.004}{(750 \times 0.004)^2} \right]$$

$$\Delta\rho = -\frac{0.0002(1.52 - 0.74)(750 \times 0.004)^2}{2.55 \times 0.004} = \pm 14 \text{ kg/m}^3$$

EXAMPLE 4.4

Let us estimate the error of a β-reflection ash content measurement around 40 mass-$\%$ ash content.

Let us assume that the coal consists of carbon with the atomic number $\mathscr{Z} = 6$, and of 40% SiO_2:

$$\mathscr{Z}_{SiO_2} = \frac{28 \times 14}{60} + \frac{2 \times 16 \times 16}{60} = 15.1$$

Thus, the mean atomic number is:

$$\bar{\mathscr{Z}} = 0.6 \times 6 + 0.4 \times 15.1 = 9.64$$

In the range between $\bar{\mathscr{Z}} = 2\text{--}10$ it holds true for the correlation between the reflection factor G and the atomic number [17] that:

$$G \times 100\% = 1.2311 \mathscr{Z} - 2.157$$

The correlation between the atomic number and the ash content will be as follows:

$$\mathscr{Z} = 6 + (15.1 - 6) = 6 + 9.1 \text{ (ash mass fraction)}$$

Thus:

$$G \times 100\% = 1.2311 \times (6 + 9.1 \text{ ash mass fraction}) - 2.157$$

and

$$\Delta G \times 100\% = 11.2 \times \text{(ash mass fraction change)}.$$

Let us assume that we count 10^6 pulses; its uncertainty is from Eq. (2.4):

$$\Delta G \times 100\% = \frac{100}{\sqrt{10^6}} \% = 0.1\%$$

This corresponds to

$$\frac{0.1}{11.2} = 0.009, \text{ i.e. } \pm 0.9 \text{ mass-}\% \text{ ash content.}$$

Practical measurements give $\pm 0.7\%$, the deviation is not significant.

4.1.3. ANALYSIS BY ABSORPTION AND SCATTERING OF X- AND γ-RAYS

In radioanalysis it is justified to discuss X-rays and γ-radiation under a common heading; sometimes these two radiations are called quantum radiation, disregarding the fact that the first of them originates from the electron shell, the later the nucleus.

Of the three main mechanisms of the interaction of quantum radiation and matter—photoelectric effect, Compton scattering and pair formation—only the former two have analytical importance.

Equation (4.4) holds good also for the *absorption of radiation*—disregarding the way of interaction—if the beam is sufficiently collimated, "monochromatic", and the absorber is not too thick.

Since the sum of the mass (weight) fractions is equal to unity, for the total elementary analysis of a multicomponent system, the number of the necessary absorption measurements is equal to the number of components minus one; in practice, however, only two-component (or practically two-component) systems can be analyzed.

The I_0 intensity passing through the empty cuvette is practically so high that it cannot be measured exactly, owing to the limited resolving power of the detectors. Therefore, the measurement of the radiation intensity is related either to a basic material with a concentration of $c_1 = 0$, or to a standard solution with known $c_{1\,std}$ concentration. Hence Eq. (4.4) is modified as follows:

$$\ln \frac{I}{I'} = (\mu_{m,1} - \mu_{m,0}) c_1 \rho l \qquad (4.6)$$

$$\ln \frac{I}{I_{std}} = (\mu_{m,1} - \mu_{m,0})(c_1 - c_{1,std})\rho l \qquad (4.7)$$

where I denotes the intensity belonging to the substance of c_1 concentration, I' that belonging to the solution of $c_1 = 0$ concentration, and I_{std} is the intensity belonging to $c_{1,std}$ concentration (the mass absorption coefficient of the solvent being $\mu_{m,0}$).

The values of the mass absorption coefficients depend considerably on the predominating way of interaction.

In the case of photoelectric effect—predominating, e.g., for Al under 50 keV, for Pb under 500 keV—the mass absorption coefficient increases for a given energy with the fourth power of the atomic number; or else, for a given element it increases considerably with decreasing energies, except for the position of absorption maxima where there is a sharp decrease (Fig. 4.7). This strong dependence on the atomic number of the photoelectric effect makes possible the determination of any high atomic number components in a medium with low mean atomic number (e.g., hydrocarbons, water), by measuring the absorption intensities.

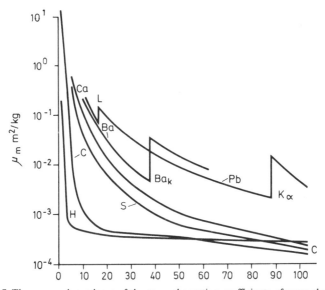

Fig. 4.7. The energy dependence of the mass absorption coefficients of some elements

There are relatively few isotopes suitable for radiation absorption measurements, owing to the strict requirements concerning the radiation source (long half-life, "monochromatic" radiation, low price, high specific activity, etc.). Such isotopes are: ^{55}Fe (5.9 keV), ^{109}Cd (22 and 87 keV), which are K-radiating; ^{3}H/Ti, ^{3}H/Zr, ^{147}Pm/Al (with an average bremsstrahlung of 5.8 and 22 keV); ^{241}Am and ^{170}Tm radiation sources are often used for the determination of the ash content of coals.

Figure 4.7 suggests that sources such as ^{55}Fe and ^{3}H/Zr (and occasionally ^{3}H/Ti) can be used for the determination of elements between $\mathscr{Z} = 6$ and 24 in a hydrocarbon medium; sources like ^{109}Cd and ^{147}Pm/Al are suitable for elements above $\mathscr{Z} = 20$, and the 87 keV radiation of ^{109}Cd is useful for heavy elements.

Table 4.2. Main features of laboratory analysis methods based on radiation absorption

Isotopic radiation source (symbol, radiation type, energy, half-life)	^{55}Fe K 5.9 keV 2.6 y	^{3}H/Zr-target 8 keV 12.3 y	^{109}Cd K 22 keV 1.3 y
Recommended surface mass, kg/m^2	3.8	7.2	43
The error of analysis (σ) in absolute % of the element studied	±0.16 O ±0.013 S ±0.012 Cl	±0.35 O ±0.029 S ±0.026 Cl	±0.057 S ±0.048 Cl ±0.011 Co ±0.004 Pb ±0.01 Ba and Zn
Permissible error of density measurement, kg/m^3	±2	±3	—
Time requirement of one measurement, min	5	5	5

In the analysis of liquid hydrocarbons, the most advantageous surface density depends on the energy of radiation used for absorption measurement, and even on the activity of the source and on its radioactive impurities. These also influence the accuracy of the analysis. Table 4.2 summarizes the main parameters of the determination of individual elements in a hydrocarbon medium by means of ^{55}Fe, ^{3}H/Zr and ^{109}Cd (22 keV). These calculated data have been verified by practice.

Figures 4.8 and 4.9 give information about the constructional details of the devices. The ^{55}Fe radiation absorption device operates with *beryllium window* cuvettes and the liquid layer thickness is constant. Consequently, the density of the liquid sample to be analysed has to be known with an accuracy of ±2 kg/m^3. The device can be used mainly in the petroleum industry, for the determination of the *sulphur content* in Diesel fuels and fuel oils. As it can be seen from Table 4.2, the ^{3}H/Zr target can be used practically for the same purposes as a ^{55}Fe source, with the difference that the higher half-life involves some advantages; at the same time, the continuous energy distribution of the radiation results in a nonlinear calibration equation (see Example 4.5).

For determination of the concentration of *motor oil additives*, measurement with a ^{109}Cd radiation source is advantageous, especially if the additive contains Ba, Ca, or Zn. The method substitutes combustion analysis requiring several hours by a measurement taking only a few minutes, and the nuclear method is more accurate. Density measurement

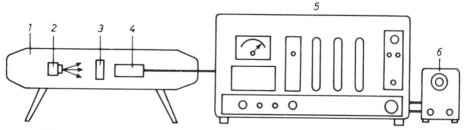

Fig. 4.8. A ^{55}Fe radiation absorption instrument. *1* — ^{55}Fe radiation absorption instrument; *2* — Radiation source; *3* — Cuvette; *4* — GM-tube; *5* — Pulse counter; *6* — Time switch

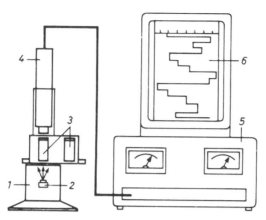

Fig. 4.9. A ^{109}Cd radiation absorption instrument. *1* — ^{109}Cd radiation absorption device; *2* — Radiation source; *3* — Measuring vessels; *4* — Scintillation counter; *5* — Ratemeter; *6* — Voltage compensator and recorder

means no problem here because the mass of the sample to be poured into the constant cross-section cuvettes should be identical. If the 87 keV energy of ^{109}Cd is used for the determination of Ba, the presence of sulphur and phosphorus in the sample disturbs only insignificantly.

Continuous analysis methods can be developed, because the use of mechanical density compensators permits one to cut free the measurement from the changes of the density (and, consequently, from those of temperature, too). Thus, the radiation intensity depends only on the elementary composition. Figure 4.10 depicts a device for the continuous determination of the *sulphur content in gas oil*. The apparatus shown in Fig. 4.11 is used for the monitoring of *lead additive in motor gasoline*. The former device has to be adjusted once or twice a week on the basis of data obtained from combustion measurements; at the same time, the lead measuring device can be zeroed to lead-free gasoline and the decay of the isotope can be corrected by adjustment of a potentiometer.

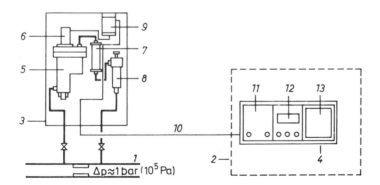

Fig. 4.10. The scheme of a continuous sulphur analyzer. *1* — Gas oil pipeline; *2* — Control room; *3* — Measuring section; *4* — Instrumental section; *5* — Radiation absorption device; *6* — Ionization chamber; *7* — Flow meter; *8* — Control valve and filtration equipment; *9* — Preamplifier; *10* — Electric cable (max. 120 m); *11* — Power supply; *12* — Electrometer; *13* — Recorder

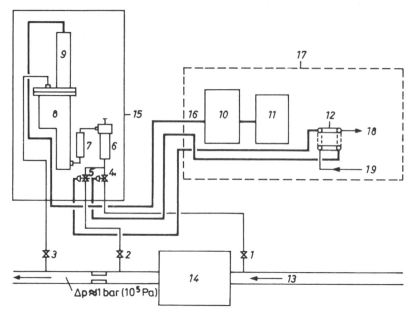

Fig. 4.11. A continuous equipment for measuring the lead content of motor gasoline. 1, 2, 3 — Valves; 4, 5 — Pneumatic valves; 6 — Filter and control valve; 7 — Rotameter; 8 — Radiation absorption device; 9 — Scintillation counter; 10 — Industrial ratemeter; 11 — Recorder; 12 — Adjusting valve for the pneumatic valves; 13 — Gasoline mains; 14 — Additive mixer; 15 — Measuring section; 16 — Five-core electric cable; 17 — Control room; 18 — Blow-down; 19 — Compressed air

Compton scattering is brought about by the interaction of the quanta with the electrons of the absorbing matter. Consequently—as it was seen also for β-absorption (see Section 4.1.2)—the effect of the specific electron density of the matter can be seen in the mass absorption coefficient.

In the range of energies between 100 keV and 1 MeV—which is characteristic for Compton scattering—the *hydrogen content* and *moisture content* of the substances can be measured. The 662 keV radiation of ^{137}Cs can be used well for this purpose. The surface mass (i.e., the layer thickness and the density) of the liquid should be known here. The principle of the measurement is not very practicable because of the considerable layer thicknesses to be used here (the mass absorption coefficients are equal to 60 cm^2/kg, except for hydrogen, in which case it is twice as high), further owing to the considerable activities necessary in view of collimation, the necessity of measurement or adjustment of the surface mass and the inconvenience of the large shielding against hard γ-radiation.

The measurement of *reflection of quantum radiation* requires such a geometry in which the detector is placed on the same side of the sample as the radiation source, because the radiation intensity deflected by 90–180° from the direction of the original beam is measured here. The layer thickness (or more correctly, the surface mass) of the medium is "infinite"—or at least, constant—here. The regularities of reflection of quantum radiation are rather complex, due to the complexity of the scattering processes:

— (a) with a given primary energy and scattering angle, the scattering abilities of the elements decrease exponentially with increasing *atomic numbers*; this is the basis of the analytical applications;

— *(b)* using higher energies, the extent of the above decrease is smaller, consequently, *lower energies* are more advantageous for analytical purposes.

The intensity of the scattered radiation greatly depends on the *apparatus* (on the geometry, on the quality of the detector, etc.), therefore calibration is necessary.

The measurement of the *ash content of coals* has a great practical importance. Mining itself requires a rapid qualifying method suitable for monitoring the stopping. Apart from this, rapid control measurements are necessary for coal separation. Direct information about the actual ash content, calorific value of a solid fuel used in power plants is similarly valuable. The necessity of development and introduction of appropriate radioanalytical methods is obvious in view of the laborious sampling for combustion ash content and calorific value determinations, as well as the time requirement of this measurement lasting several hours. The above-discussed regularities of absorption and reflection of quantum radiations—especially the strong dependence of the reflection and photoelectric absorption on the atomic number—permit one to apply a rapid nuclear measurement for determination of the ash content—consequently that of the calorific value—of coals, regarding them as two-component systems, the two components having different mean atomic numbers (see Section 4.1.2).

Coal analysis *based on the absorption* of quantum radiation in the energy range between 20 and 100 keV can be applied both in laboratories and under plant conditions, if the identical surface density of the irradiated layers is ensured (Figs 4.12a and 4.12b). Targets containing ^{90}Sr–^{90}Y, ^{206}Tl, ^{147}Pm β-radiating isotopes as well as ^{170}Tm isotope were used for this purpose initially; recently ^{241}Am isotope has been applied which is more suitable in view of both its energy (59.8 keV) and half-life (458 years).

The disadvantage of this method is that the sample preparation (grinding, drying and weighing) is time-consuming and expensive; occasional changes in high atomic number ash components (e.g. Fe) can cause a considerable error. The advantage is that the method gives information on a relatively large sample (about 1 kg).

Two-beam measurement represents a considerable improvement in the method; this applies a second radiation with an energy higher than 100 keV (here the dependence on the atomic number is not significant); this gives information about the surface mass of the coal layer measured. Thus, by applying a scintillation detector and two-channel energy selection, the measurement can be rendered independent of the changes in the surface density. At the same time, the grinding and drying of the sample remain necessary in order to ensure a higher accuracy.

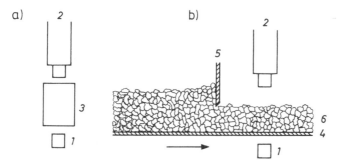

Fig. 4.12. Ash content measurement by γ-absorption. (a) — Laboratory device; (b) — Industrial device. *1* — Radiation source; *2* — Detector; *3* — Sample; *4* — Conveyor belt; *5* — Layer thickness adjustment; *6* — Coal layer

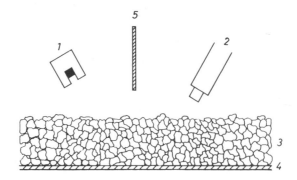

Fig. 4.13. Ash content measurement by γ-reflection. *1* — Radiation source; *2* — Detector; *3* — Coal layer; *4* Conveyor belt; *5* — Shielding

The error of ash content measurements based on radiation absorption is between 0 and 2% (absolute).

The ash content measurement *based on reflection* (Fig. 4.13) has a few advantages a compared with the absorption measurement:

— *(a)* as a result of compact geometry, the radiation source and the detector a placed on the same side of the coal layer;

— *(b)* if the layer thickness exceeds a given value (12–15 cm), the surface mass of th sample need not be known;

— *(c)* by ensuring a high space angle geometry, information can be obtained about high mass of coal (25–30 kg).

The accuracy of the measurement is influenced by the uniformity of the coal layer (i degree of coarseness), by the moisture content and by the changes in the ash compositio (e.g., iron content). The latter effect can be eliminated by electronic correction, on the bas of measurement of the intensity of the Fe K_α-radiation excited by a ^{3}H/Zr source.

Both *portable and fixed devices* can be used for reflection ash content determinatio Portable devices are especially suited for the rapid monitoring of coal contents of min cars. The qualification can be carried out by a three-minutes measurement, with an err of ±2–5% (absolute).

Fixed devices are suited for the monitoring of coal quality *on conveyor belts.* Th radiation source is, as a rule, ^{241}Am; scintillation counters, and also GM-tubes switched i parallel can be used to advantage. The signal of the detector is led to an industri ratemeter which can be rendered more sensitive by shifting its zero.

The ash content can be read from an instrument or from a recorder, after appropria calibration. At the same time, the output signal can be used for operating automat controls of the power plant system.

EXAMPLE 4.5

Let us calculate the percentage intensity decrease of the 5.9 keV radiation of ^{55}Fe whe passing through a 0.4 cm thick layer of gas oil with 1% sulfur content ($\rho = 820$ kg/m^3

The mass absorption coefficients of the elements present are [4.4]:

$$\mu_{m,H} = 0.05 \text{ m}^2/\text{kg}; \qquad \mu_{m,C} = 1.04 \text{ m}^2/\text{kg};$$

$$\mu_{m,S} = 20.4 \text{ m}^2/\text{kg}; \qquad c_{m,H} = 13.0 \text{ mass-\%}$$

From Eq. (4.4):

$$\ln \frac{I_0}{I} = \rho l \Sigma \mu_{m,i} c_i = 820 \times 0.004\,(0.13 \times 0.05 +$$

$$+ 0.86 \times 1.04 + 0.01 \times 20.4) = 3.28 \times (0.0065 + 0.894 + 0.204) = 3.623$$

Thus:

$$\frac{I_0}{I} = 37.4$$

and

$$I = \frac{100}{37.4} = 2.7\%$$

The intensity of the ^{55}Fe radiation decreases to 2.7% of the original value.

EXAMPLE 4.6

Let us calculate the percentage transmittance of the above gas oil with respect to cyclohexane.

The density of cyclohexane is 779 kg/m^3, its hydrogen content is 14.4 mass-%.
From Eq. (4.4):

$$\ln \frac{I_0}{I} = 779 \times 0.004\,(0.144 \times 0.05 + 0.856 \times 1.04) = 3.117\,(0.0072 + 0.890) = 2.797$$

$$\frac{I_0}{I} = 16.4; \qquad I = 6.1\%$$

Using the results of Example 4.5:

$$\frac{2.7 \times 100}{6.1} = 44.2\%$$

The transmittance of the gas oil with respect to cyclohexane is 45.9%.

4.1.4. ISOTOPIC X-RAY FLUORESCENCE ANALYSIS
(IXRF)

Discovery of the principle of X-ray fluorescence analysis (XRF) can be dated back to 1913 when Moseley stated that the anticathode of an X-ray tube emits X-rays whose lines are characteristic of the element present in the anticathode and this X-ray spectrum is an unambiguous function of the atomic number. It was also Moseley who drew conclusions from the line intensities of brass anodes on their copper and zinc contents, laying thus the foundations of XRF. XRF spectrometers have found increasing use from the 1930s; they diffract the X-rays according to their energy by a crystal dispersion technique, corresponding to Bragg's law.

The beginning of IXRF can be dated to the 1950s, when radioisotopes became more widespread. These measurements were based on energy dispersion, i.e., on the energy sensitive response of various detectors (proportional and scintillation counters).

As far as its analytical performance is concerned, IXRF can be compared with other branches of modern instrumental analysis (atomic absorption, neutron activation, gas chromatography, etc.), yet such a comparison would lead too far; a comparison, however, with its closest competitor, crystal diffraction X-ray analysis, can be very enlightening.

The *advantages* of IXRF are as follows:

— *(a) its source–sample–detector assembly is simple and compact*, the device does not contain any moving, i.e., wearing parts, it does not require mains current, water, vacuum, therefore its volume is small and the device is portable; the *price* is lower than that of a comparable crystal dispersion apparatus;

— *(b) its operation is stable, and the handling is easy:* therefore it can be readily applied in an on-stream and automatic on-line way of operation;

— *(c) it has better resolving power*, if a semiconductor detector is applied in the 20–100 keV energy range, than a crystal dispersion instrument of medium resolving power has; by applying a generally monochromatic and high-energy exciting radioisotope radiation source, the disturbing effect of other elements (matrix effect) will be less;

— *(d) it simultaneously measures* the intensity of the characteristic radiation of each component; this is especially advantageous with multicomponent samples, encountered, e.g., in environmental control studies.

The *disadvantages* of IXRF are:

— *(a) it has a poorer resolving power* when samples with low and medium atomic numbers are analysed;

— *(b) the relative error of the measurement is larger* due to the lower radiation intensity; this results in a worse detection limit.

4.1.4.1. THEORETICAL BASIS

Let us outline the part-processes as a result of which X-rays, emitted by the substance to be analysed under the exciting effect of the nuclear radiation and reaching the detector, carry all information permitting one to obtain qualitative and quantitative data characterizing the sample under investigation. The transformation of the radiation into electric pulses and their further processing will be discussed in the next section.

The radiation of a given radioisotope falls on the sample to be studied and the radiation emitted by the latter is detected by the detector (Fig. 4.14). The following main *part-processes* of the interaction between radiation and matter should be considered here:

— *(a)* the exciting radiation is absorbed and scattered in the sample (primary absorption);

— *(b)* as a result of the absorption, the atoms of the sample are excited;

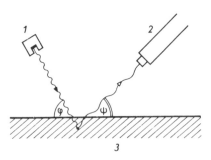

Fig. 4.14. Excitation of X-ray fluorescence. *1* — Radiation source; *2* — Detector; *3* — Sample

— (c) the excited state collapses into the ground state and a characteristic X-radiation or Auger electrons are emitted;

— (d) the characteristic X-rays of various elements interact with one another; as a result, secondary absorption and secondary excitation may occur;

— (e) since the part-processes listed above take place in space, and the radiation source, the sample and the detector have finite dimensions, and their distance from one another should also be considered, the geometry of IXRF is also important, especially as far as absolute intensities are concerned.

The regularities of radiation absorption, especially those of the photoelectric effect, which is of primary importance in XRF, have already been summarized in Sections 4.1–4.3; here it is sufficient to point out some consequences:

— (a) a small layer thickness of the sample has a role only in determining the intensity relations of the X-ray fluorescence, because the intensity of the exciting radiation decreases exponentially with increasing layer thickness and the characteristic radiation emitted has always a lower energy than the exciting one, thus its absorption is greater. It follows that it is advisable to select an exciting energy just above the characteristic X-ray energy, because excitation with high energy radiation becomes ineffective in far away layers owing to secondary absorption;

— (b) the surface roughness, grain size, inhomogeneity of the sample are of considerable importance, especially in the analysis of low atomic number elements; this is due also to absorption phenomena (see Section 4.1.4.2);

— (c) matrix effect is also mainly due to secondary absorption; as a result, various elements influence the intensity conditions of each other (decrease or increase by secondary excitation).

Scattering is, as a rule, negligible as compared to the absorption of the exciting electromagnetic radiation; the intensity of the scattered radiation can, on the other hand, be used for consideration of the matrix effect (Section 4.1.4.1).

The fundamental precondition of the generation of *characteristic X-rays* is that the exciting radiation should transfer its energy to one of the electrons of the atom and the electron with excess energy should leave the electron orbital. The orbital can achieve stabilization either by emitting an Auger electron, or by electron transition from an outer shell. The ratio of electron transitions resulting in X-ray fluorescence to the total number of excited states is called *fluorescence efficiency*; this efficiency starts from zero with low atomic number elements and reaches the value of unity at high atomic numbers by an S-shaped curve (Fig. 4.14). Depending on the fact whether the original excitation of the atom occurs on the K, L, M, N, etc. shells, the fluorescence X-ray is called K, L, M, N, etc. radiation; the fact that electron transition occurs from the neighbouring shell or from a farther one is denoted by indices α, β, etc. Within this, various energy levels are denoted by numbers. Thus, we can speak of $K_{\alpha 1}$, $K_{\alpha 2}$, $L_{\alpha 1}$, $L_{\alpha 2}$, $L_{\beta 1}$, ... $L_{\beta 10}$ radiation, provided that the atomic number and consequently the structure of the electron shells permits such a variety. Handbooks contain energy values for the characteristic X-rays of various elements [4.13].

The following correlation can be written between the frequency (v) and atomic number ($\mathscr{Z}$) of various elements, as far as the K- and L-series are concerned (after Moseley): for the K-series:

$$\sqrt{v} = 0.498 \times 10^8 (\mathscr{Z} - 1) \tag{4.8}$$

for the L-series:

$$\sqrt{v} = c(\mathscr{Z} - 7.4) \tag{4.9}$$

where c is a constant.

The relative intensities of the individual series are the following:

$$K_{\alpha 1} : K_{\alpha 2} : L_{\beta 1} \approx 4 : 2 : 1$$

$$L_{\alpha 1} : L_{\alpha 2} : L_{\beta 1} : L_{\beta 2} : L_{\beta 3} \approx 10 : 1 : 6 : 2 : 1$$

The above correlations can be used for corrections for overlapping energies.

Qualitative analysis by means of IXRF presents no particular problem, especially in tl case of higher concentrations of the components. The identification of trace components a more complicated task, especially if they are present together with other elements wit close atomic numbers. The solution of this problem is related to quantitative analysis. Tl energy calibration of the device can be performed by means of available elements (compounds, or by using well-known radiation energies. The signal of an energy-sensitiv detector increases, as a rule, proportionally to the energy, still, extrapolation requiri caution. Some detectors are extremely sensitive to the changes of operation parametei (voltage, temperature, gas pressure, etc.); other instruments exhibit energy shifts as function of the intensity.

Qualitative identification can be disturbed by scattering phenomena occurring in th detector and in the surroundings, as well as by fluorescence of the detector material an the appearance of an escape peak. Qualitative analysis is the more reliable, the better th resolving power of the detector and the lower the background due to scattered radiatioi When identifying individual lines, the simultaneous appearance of several lines of the sam element should also be considered.

The measurement of K_α-radiation intensity serves generally as a basis for *quantitativ analysis*. This is the most intensive radiation and the least sensitive to disturbing effect (matrix, grain size, etc.). The intensity of K_α-radiation of a sample consisting of even *single element* depends on several factors:

— *(a)* the nature of the radiation source, the intensity of the exciting radiation;

— *(b)* the thickness, surface roughness, particle size of the sample, the fluorescenc efficiency of the element;

— *(c)* the efficiency of the detector;

— *(d)* the absorptivities of the radiation source, the sample, the detector window, a well as of the medium between them (air, or He, eventually H_2 with low energies) absorptivity depends on the nature of the material and its thickness;

— *(e)* the geometry (the surface and distance of the radiation source, the sample anc the detector) and the arrangement of them.

Measurement of the *thickness of coatings* is based on the phenomenon that with a giver radiation source, detector, geometry, instrument, etc., the I_A intensity of X-ray fluorescence first increases in proportion to the l layer thickness to reach a saturation value (see also Section 4.1.4.3). The intensity of an "infinitely" thick layer (as a rule, a few mm) is denoted by $I_{A, 100}$. If the amount of the sample allows it, the layer thickness should be selected high enough to exceed the saturation value for all the elements present with certainty; in this case, the dependence of the intensity on the thickness can be eliminated and the measurement becomes simpler.

Generally, if it is valid that

$$\rho_A l \bar{\mu}_{m, A} \leqq 0.1 \tag{4.10}$$

we are in the proportional range, and if

$$\rho_A l \bar{\mu}_{m, A} \geqq 1 \tag{4.11}$$

we are in the saturation range of the intensity *vs.* layer thickness function. Here ρ_A is the density, kg/m^3, l is the layer thickness, m, and $\bar{\mu}_{m,A}$ denotes the combined mass absorption coefficient, m^2/kg:

$$\bar{\mu}_{m,A} = \frac{\mu_{m,AE_1}}{\sin\varphi} + \frac{\mu_{m,AE_2}}{\sin\psi} \qquad (4.12)$$

where φ is the incidence angle of the exciting radiation of E_1 energy, ψ is the emission angle of the X-ray fluorescence radiation with E_2 energy, μ_{m,AE_1} and μ_{m,AE_2} are the mass absorption coefficients of element A with respect to E_1 and E_2 energy (Fig. 4.15). The linear section may extend up to 0.3–5 μm with Ni, Fe, Cr, and up to 20 μm with Sn, depending on the exciting energy and on geometrical factors (see Example 4.7).

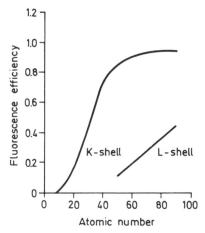

Fig. 4.15. Dependence of fluorescence efficiency on the atomic number

In the case of *two- and multicomponent samples*, the quantitative analysis in the linear range has several advantages as compared to the saturation range. This is utilized especially in analyses for environmental protection measurements where the amount of the available sample is generally small. The advantages are as follows:
— *(a)* the intensity of the characteristic X-rays is proportional to the surface mass of the elements to be analyzed (kg/m^2);
— *(b)* interactions between elements are negligible;
— *(c)* the ratio of fluorescence to scattered radiation increases, thereby the sensitivity becomes higher;
— *(d)* correction for the dependence on particle size is easier.
On the other hand, the relative accuracy is lower in the linear range.
Further on, the regularities of analyses with "infinite" thick layers will only be discussed.
One can use the *method of basic parameters* (based on the parameters of the material: mass absorption coefficients, excitation efficiencies, etc.) for converting measured radiation intensities into concentration values; this, however, requires computer-processing of several parameters. Therefore, more wide-spread are the methods based on intensity measurements of samples of known compositions. These *empirical methods* will be dealt with in detail below.

The presence of a second (B) component may increase or decrease the X-ray intensity emitted by the element to be investigated, depending on the fact whether absorption (due to accompanying elements with lower atomic number) or secondary excitation (in the case of accompanying elements with higher atomic number) is predominating.

The consideration of the absorption interaction can only be ensured by the introduction of the regression coefficient (r_{AB}); the form of the correlation $I_A = f(c_A)$ (radiation intensity *vs.* concentration) will then be as follows:

$$I_A = \frac{c_A I_{A,100}}{c_A + (1 - c_A) r_{AB}} \tag{4.13}$$

where c_A denotes the concentration in mass fraction, $c_A + c_B = 1$; the r_{AB} regression coefficient can be calculated from the mass absorption coefficients of components A and B with respect to the characteristic radiation of component A, as calculated by means of Eq. (4.12):

$$r_{AB} = \frac{\bar{\mu}_{m,B}}{\bar{\mu}_{m,A}} \tag{4.14}$$

$I_{A,100}$ is the intensity belonging to the pure component A.

The value of r_{AB} is positive, lower or higher than unity; if $r_{AB} = 1$, the above equation is linear for both components. Since r_{AB} is constant, in principle it is possible to determine the functions $I = f(c)$ for both components on the basis of a sufficiently accurate analysis of a single sample with known composition.

It follows from the definition of the regression coefficient according to Eq. (4.14) that it also depends on the energy of the exciting radiation (see Example 4.8).

At low concentrations, the form of Eq. (4.13) will be as follows:

$$I_A = \frac{c_A I_{A,100}}{r_{AB}} \tag{4.15}$$

The intensity of the X-ray fluorescence of substance A being present in trace concentration in a matrix M of constant composition, is inversely proportional to the mass absorption coefficient ($\mu_{mM E_1}$) of the matrix with respect to the exciting radiation (E_1)

$$I_A = \text{prop} \frac{c_A}{\mu_{mM E_1}} \tag{4.16}$$

At the same time, it can be shown that the intensity of the diffuse scattered radiation with respect to the E_1 excitation energy (U_{E_1}) is also inversely proportional to the mass absorption coefficient of the matrix M; consequently, measurement of the intensity of scattered radiation permits one, in principle, to calculate the concentration according to Eq. (4.16), especially if the proportionality factor is known in a standard matrix.

The calculation according to Eq. (4.13) is rather complex with three- or multicomponent samples, and is also inaccurate because the equation does not take into consideration mutual secondary excitation effects of the components. At present, the most accurate equation, suitable also for computer data processing is as follows [4.16]:

$$\frac{c_i I_{i,100}}{I_i} = 1 + \sum_{ki} A_{ik} c_k + \sum_{k \neq i} \frac{B_{ik} c_k}{1 + c_i} \tag{4.17}$$

where A_{ik} is a constant expressing the effect due to the absorption of k element, exerted on the concentration of the i-th element, B_{ik} is the same, but with respect to the excitation effect.

In principle it is possible but it has never been observed that one of the two constants be not equal to zero; thus the calculation of the constants from calibration data is considerably simplified.

The determination of the calibration constants is based on the intensity measurement of samples with known composition. The authors propose an approximation method for the calculation of concentrations in samples with unknown composition. This can be *easily performed by a computer*; as a rule, the second approximation gives already sufficiently accurate results (Example 4.9).

In industrial practice, the following equation is useful, especially within a narrow concentration range:

$$c_i = aI_i + b \tag{4.18}$$

where a and b are empirical factors and can be found in tables for various alloys [4.23].

The statistical error of the concentration measurement of a given element depends on several factors—like absolutely intensity itself—but here only the problems connected with intensity measurement will be considered.

If n pulses are counted for the calculation of a concentration value, the relative error of the concentration measurement $\left(\dfrac{100\Delta c}{c}\%\right)$ cannot be lower than permitted by Eq. (4.19):

$$\frac{100\Delta c}{c} = a\frac{100}{\sqrt{n}} \tag{4.19}$$

This correlation is an accuracy limiting factor of paramount importance, especially with multichannel analysers. If the value of n is increased by two orders of magnitude, the relative error decreases by one order of magnitude only (see Section 4.1.4.2). Practically this decrease is still lower because the n net count is calculated as a difference of the overall count and the background. The calculation of this difference also results in the propagation of the errors; it should also be considered that the background count cannot be measured directly and the extrapolation or interpolation has its own error. This latter is the larger, the smaller the resolving power of the detector used.

Consequently, a relative error of 0.3% can be ensured only exceptionally; 1–2% can be regarded as a good accuracy.

The detection limit is also very important in view of the analysis of components present in trace amounts. This c_{DL} value is defined as the lowest concentration or amount of an element which can be detected with certainty above the background value, i.e., the count rate of which is higher, with 99% certainty, than the lowest standard deviation of the background, $\sqrt{n}$:

$$c_{DL} \geqq 2.3\sqrt{n}\left(\frac{c}{n}\right)_{St} \tag{4.20}$$

where n denotes the net count rate obtained for a c standard concentration; 2.3 is the factor corresponding to the 99% certainty.

The detection limit is a function of the excitation efficiency and the matrix: for medium and high atomic number elements in a matrix consisting of low atomic number elements its value may be as low as a few thousandth percent; in other cases a value between 0.01–0.1% is acceptable. Lowering of the detection limit is especially important in environmental protection measurements.

If a polarized exciting radiation is applied, the intensity of the background radiation can be decreased by a factor of three; this improves the detectability, although the intensity of useful radiation also decreases by a factor of two, and this is disadvantageous for the sensitivity.

Several possibilities are available for decreasing the instrumental errors. The increase of the resolving power of the detector, the use of selective filters, sample preparation (grinding, melting, etc.), the use of various background correction methods, the application of internal or external standards—they all help to decrease the experimental error. Ample literature is available concerning this problem [4.13].

4.1.4.2. INSTRUMENTS

Figure 4.16 shows the scheme of an X-ray fluorescence apparatus working with isotopic excitation. The exciting radiation originating from the radiation source excites the X-rays of the sample; these X-rays reach the detector, eventually through filters, and its pulses, proportional to the radiation are transformed in the preamplifier to voltage pulses. The energy-selective separation of the pulses occurs in an analyzer. This is followed by data processing and data recording, or displaying. As a rule, a separate high-voltage power supply is necessary for the detector.

Isotopes emitting α-, β-, γ- and K-radiation can be used as *radiation sources. Targets emitting secondary radiation* may serve as an alternative solution. No generally applicable radiation sources exist; even for the same purpose more than one isotope can be used, as a function of the type, energy, activity of the emitted radiation, of the half-life of the isotope, of its specific activity, price, class of danger, etc.

α-*Radiating isotopes* (e.g., ^{210}Po) are recommended for exciting low atomic number elements, owing to their good fluorescence efficiency and weaker background radiation; the increased radiation hazard is a disadvantage.

β-*Radiating isotopes* are not used directly for excitation because the characteristic X-rays of the sample are accompanied by a considerable background, due to bremsstrahlung. The application of targets (e.g. ^{3}H/Ti, ^{3}H/Zr, ^{147}Pm/Al, ^{90}Sr–^{90}Y/U) is

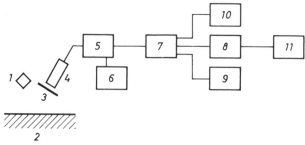

Fig. 4.16. Scheme of X-ray fluorescence analysis. *1* — Radiation source; *2* — Sample; *3* — Filter; *4* — Detector; *5* — Preamplifier; *6* — High-voltage power supply; *7* — Analyzer; *8* — Recorder; *9* — Printer (puncher, tape recorder); *10* — Screen; *11* — Computer

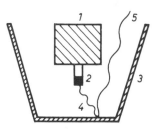

Fig. 4.17. A secondary radiation target. *1* — Shielding; *2* — Radiation source; *3* — Target; *4* — Primary radiation;
5 — Fluorescent radiation

much more widespread, they are necessary for excitation of low atomic number elements. Owing to secondary excitation, the necessary activity is higher by 2–3 orders of magnitude than otherwise (see Example 4.10).

γ- and K-radiating isotopes predominate among radiation sources: ^{55}Fe, ^{109}Cd, ^{125}I, ^{210}Pb, ^{238}Pu and ^{241}Am emit in the range of low energies (4–60 keV), ^{57}Co and ^{153}Gd in the range of medium energies (60–150 keV) and ^{192}Ir, ^{137}Cs and ^{60}Co in the range of high energies (above 150 keV).

The development of secondary radiation targets (Fig. 4.17) has been promoted by the attempts to bring near to each other the energy of the exciting radiation and that of the characteristic X-ray emitted, in order to obtain as good an excitation efficiency as possible. The primary radiation source is fixed inside a cavitated target having the shape of a truncated cone, and the primary radiation is shielded in the direction of the emission; thus practically only the X-rays emitted by the target material reach the sample. A disadvantage is that such targets are not pinhole-like radiation sources and their use ensures a relatively low intensity below $\mathscr{L} = 30$. ^{109}Cd, ^{147}Pm/Al and ^{241}Am are suitable as primary radiation sources.

The sample may be solid or liquid, seldom gaseous. The analysis of liquids and gases presents no particular problems. In several cases it is possible to immerse the detector and the radiation source into the medium to be analyzed; this is an advantage especially with continuous analysis. The main requirements with respect to the window of the sample holder are low radiation absorptivity and the stability of its shape; these are of importance especially with low energies. Beryllium, polypropylene and Melinex (polyethylene–terephthalate) films are the most suitable materials for window preparation.

The intensity of X-rays is also influenced by the *surface roughness* of the solid samples; this disturbing effect is more serious with components of low atomic numbers. For example, a surface roughness of 50 μm can cause an intensity decrease of 1% with Cu, but 17% with Al. The same effect is found in the case of powdered samples; another inaccuracy may originate from the different wear rates of various particles in ore and mineral samples, together with the tendency of separation according to density or size. The intensity correction of powdered samples can be carried out by applying a factor of $(1+ba)^2$ [4.6] where a denotes the particle size, μm, b is a factor, 1/μm, depending on the material; this latter value can be found in tables of this Section.

In order to eliminate problems arising from particle size effect as well as from the influence of the matrix, the *melting of the sample with borax or lithium borate* is widely used. This increases the time and labour requirement of the measurement, but the results become more reproducible and more accurate. The sample added to the borax is, as a rule, between 1 and 10%; five minutes at 1000°C is sufficient time for homogenization.

Ionization chambers (Section 2.1.1.1), proportional counters (Section 2.1.1.1), scintillation counters (Section 2.1.1.2) and semiconductor detectors (Section 2.1.1.3) can be used as *detectors*. Each type has its particular advantage and disadvantage in X-ray analysis; the applicability for a given purpose should be considered in the knowledge of these.

The great advantage of *ionization chambers* used in the mode of operation for current measurement is their insensitivity to disturbing effects (e.g., changes in the voltage and temperature); their disadvantage is that they can be applied only if the intensity of one single component is to be measured (e.g., measurement of layer thickness). The application of filters increases the selectivity.

The resolution, the efficiency and the simplicity of operation are the factors which should be considered when the other three detector types are compared. As far as resolution is concerned, scintillation counters are the poorest, proportional counters are a little bit better (recently the resolving power has been increased to 8%) and the *semiconductor detector* is by far the best. In respect of the measuring efficiency, acceptable values are obtained with proportional counters, from the lowest energies (e.g., the K radiation of carbon) up to about 20 keV; with Si(Li) detectors, between 3 and 30 keV; with Ge(Li) detectors, above 3 keV. The appropriate selection of the so-called counting gas in the proportional counter (argon, neon, xenon, etc.) can ensure a kind of selectivity, due to the different efficiencies; at the same time, X-rays originating from the gases may appear in the spectrum and interfere with the measurement of some components.

It is a difficulty that Si(Li) and Ge(Li) *semiconductor detectors* require cooling with liquid nitrogen not only during measurement, but also during storage, although pure Ge detectors can be stored at room temperature without any damage. *Scintillation detectors* are less sensitive to the changes of the voltage supply, but industrial conditions accompanied by higher temperatures do them more damage than to proportional counters.

Filters should be placed between the samples and the detector; their task is to improve the discriminating power of the detector. Simple filters such as aluminium foils can be used for filtration of low energies if the energy differences are not higher than a few keV. Their effect is based on the difference of the mass absorption coefficients. The sharp changes of the mass absorption coefficients at the absorption edges (see Section 4.1.3, Fig. 4.7) can also be used for selective energy filtration. For example, a 10 µm thick Al filter absorbs the K_α radiation of Si almost completely, at the same time, that of aluminium only to 50%.

In the atomic number range between 20 and 30 (where technologically important metals such as Co, Cr, Cu, Fe, Mn, Ni, Ti, V and Zn are found), the energies of the K-absorption edges of elements differ from each other by about 8% each. This serves as a basis for the differential filtration method of Ross applying a *pair of filters made of two elements for the selective measurement* of fluorescent X-rays of a given element. One of the filters has an absorption edge just below, the other just above the energy to be investigated. In the range outside the absorption edges both filters have identical absorptivities. For example, a filter pair made of Mn and Fe (K-absorption edges: 6.5 and 7.1 keV, respectively) can be used for the measurement of the K-radiation of Co (6.9 keV). Mn has a mass absorption coefficient about eight times as high as that of Fe at the given energy, yet hardly any difference can be observed in other energy ranges, and even this small difference can be eliminated by the appropriate selection of the surface masses. By applying filter pairs, the resolving power may attain the efficiency of semiconductor detectors; by virtue of this, the simultaneous analysis of several components can be carried out if suitably selected filters are used for each component.

The geometry of the system source–sample–detector (Fig. 4.14) has an important role as far as sensitivity and reproducibility are concerned, as well as the easy handling of the apparatus. The great advantage of energy-dispersive X-ray analysis is that the radiation source and the detector can be built together, into *one single compact unit*, which can be immersed into the liquid to be analysed or fitted to the solid sample surface. The optimum geometry is determined mainly by the dimensions of the source, the sample and the detector, its determination should be carried out experimentally.

The instruments can be divided into three main groups:

Portable devices are operating with proportional counters or scintillation counters and can be *battery-operated*. They are used in geology, ore mining and for the identification of metal products in metallurgy. Several devices apply filter pairs. Recently, devices with digital instruments are preferred, but integrating analysers can also be found together with those using traditional ratemeter switching. Miniaturization is a general trend: a device as light as 2.5 kg can be found on the market. Some items can be used for multipurpose analyses, by using exchangeable sources, detectors, filters.

The supply is very abundant as far as *laboratory instruments* are concerned. Their characteristic detector is a semiconductor cooled by liquid nitrogen; they usually apply a built-in multichannel analyser (10^6 counts per channel), they display the spectrum on a (coloured) screen during measurement (continuously), the evaluation (background correction, resolving of overlapping lines, calculation and integration of intensity data, eventually concentration calculation) and recording (by typewriter or on a punched type, on an X–Y recorder or magnetic tape) are usually automatic. The time requirement of the analysis is determined, in general, by the lowest concentration element and the relative error demanded. For example, if a component present at a concentration of $10^{-3}\%$ has to be analysed with an error of $\pm 10^{-4}\%$, with an X-ray intensity of 2×10^4/s, the apparatus should receive 10^7 photons: theoretically this requires 500 s, but practically the time requirement is 2–3 times greater, due to the operation conditions of multichannel analysers. Built-in programs facilitate the work to a great extent, but they often introduce instrumental errors which are difficult to control.

Industrial on-line or off-line instruments are used for both the monitoring and control of production. X-ray analysis is especially suitable for this purpose because it is rapid, accurate, reproducible and is based on purely physical phenomena. The main characteristic of such instruments is high-level automatization extending to sampling, preparation, weighing, data processing and data output.

The apparatus grinds the solid average sample, taken with special care, to the fineness required, then a tablet of standard compactness is pressed or melted from the sample. The X-ray intensity of the tablet is then measured for a predetermined period. Some devices are capable of the *on-stream measurement* of material flow, e.g., by measurement of the thickness of coatings, by the analysis of slurries or powders. Metal samples are fed generally by hand into the device; the sample preparation includes surface etching and polishing only. Some more will be written about these devices in Section 4.1.4.3.

As a rule, *microcomputers* (with a capacity of 10, occasionally 20 Kb) are applied in the on-line way of operation. Of this capacity, 4 K generally controls the operation, the rest carries out data processing including intensity corrections and the calculation of composition. A great advantage of such systems is rapidity whereby the analysis data can be used for direct production control. A further advantage is that the computer checks directly the work of the measuring device and detects any error instantaneously.

If the X-ray intensities of various elements overlap owing to the limited resolving power of the apparatus, it is necessary to use big computers (with a capacity of 50–100, or even 700 K) in the off-line way of operation. Background correction should precede line processing (the correction uses generally linear interpolation and gives the net spectrum). After resolving the lines, matrix corrections can be introduced, thus the composition data can also be calculated.

Since a big computer can serve several instruments, its application for IXRF purposes may be cheaper than the use of a small computer in the on-line way of operation for each instrument. Off-line operation is accompanied, however, by the loss or decrease of the advantages of rapidity and direct operation control.

4.1.4.3. PRACTICAL APPLICATIONS

The measurement of thickness of coatings is a separate group within IXRF analyses; affording higher accuracy, it has, in several fields, replaced reflection measurements which were in general use earlier (Section 2.3.3). The predominant majority of applications involves the measurement of Sn and Zn coatings on iron. The analysis can be carried out, in principle, by both direct and indirect methods with both coatings; with Sn, the X-rays of the coating, with Zn, the absorption of X-rays by the iron base is measured.

The measurement of *tin coatings* (thickness: between 0.1–2.3 µm) can be performed either directly or indirectly. The excitation of the K_α-radiation of Sn (25.2 keV) can be best effected by a ^{241}Am source or by a ^{241}Am/Cs target; proportional and scintillation counters are equally suited for detection. With a time constant of 1.5 s, the relative error of the measurement is between ± 5–1%, decreasing with the thicker layers. If the analysis is to be carried out by the measurement of the absorption of the K_α-radiation of iron, excitation can be carried out by a ^{3}H/Zr target, and the best detector is an ionization chamber. An aluminium filter can easily eliminate the disturbing effect of the L-radiation of Sn (3.4 keV). A current of about 10^{-11} A corresponds to uncoated iron; the relative error is $\pm 0.2\%$ with a time constant of 7.5 s.

A zinc coating can be prepared by electroplating or by an immersion technique; the typical thickness is 5 µm in the former, and 80 µm in the latter case. The thickness of the Zn-coating can be measured by the indirect method up to 5 µm (by exciting the 8.6 keV K_α-radiation of Fe by a ^{3}H/Zr target and measuring its intensity). The filtration of the Zn K_α-radiation can be effected by means of a Ni filter; the detector is an ionization chamber. Only the direct method can be applied for coatings prepared by the immersion technique (Fig. 4.18); use of the combination of a ^{241}Am source and an ionization chamber is the best, but interference from Fe should be eliminated. This can be done by a simple Al filter, or by using a double ionization chamber in a differential mode of operation, one of them provided with a Ni, and the other with a Cu filter. Thus the differential current of the two chambers is a function of the thickness of the Zn coating only.

Reliable measurement of *the silver content in photographic films and papers* is possible by means of a ^{241}Am/Sb target and a scintillation counter. The range is between 1–40 g Ag/m^2, the error being ± 0.2–0.5 g Ag/m^2.

The thickness of *nickel-iron films evaporated onto glass* (between 20 and 200 nm) can be measured by the decrease of the intensity of the K_α-radiation of Si or Ca.

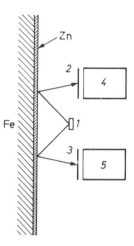

Fig. 4.18. Zinc-coating thickness measurement. *1* — ^{241}Am radiation source; *2* — Ni filter; *3* — Cu filter; *4, 5* — Ionization chambers

The application of X-ray analysis in *iron and steel manufacturing* is very widespread. Several elements should be analysed here: Co, Cr, Cu, Fe, Mn, Mo, Nb, Ni, Si, Ta, Ti, V, W, Zr, etc. owing to this complex composition, the necessary resolving power can be ensured by crystal dispersion devices only; however, strong interactions can often be decreased by dissolution in acids, by melting with borax, by adding BaO_2, etc. (Section 4.1.4.2).

IXRF is very suitable for measurement of the *iron content of ores*. The radiation source used almost exclusively for excitation of the K_α-radiation of Fe is a ^{3}H/Zr target. If a ^{170}Tm/Cd target is used, changes in the mean atomic number are corrected on the basis of the β-reflection of a ^{90}Sr–^{90}Y source. Reproducible geometry is a precondition of continuous measurement; this is the critical factor determining the error being, as a rule, $\pm 0.5\%$ Fe.

A method has been developed for the measurement of *titanium and zirconium content in ores*, using a ^{55}Fe radiation source.

In slag analysis, the determination of its Fe and Mn contents is important together with that of alkalinity, as calculated from the $CaO : SiO_2 : P_2O_5$ ratio; the problems of the measurement of the latter elements are practically the same as those in cement analysis (see there).

Portable IXRF devices proved to be very useful in the *classification of finished and semi-finished goods*. The identification is based on the measurement of a few key components. The instruments are characterized by easily replaceable radiation sources (^{55}Fe, ^{109}Cd, ^{210}Pb, ^{239}Pu, ^{241}Am) and by the application of selective filter pairs. The measurement of the concentration of each component lasts about 1 minute; a planar surface with a diameter of 2 cm is sufficient for the measurement.

In the production of *tool steels*, the task is generally the determination of Co, Cr, Mn, Mo, Si, V, W, etc.

IXRF devices are very efficient in the analysis of *metals and their ores*. A considerable demand is already evident in *mining*. Unfortunately, the accuracy is very limited, partly because the required planar sample surface can seldom be ensured, and partly because of the inhomogeneity of ores. Yet, semi-quantitative data provided by such devices may be very valuable in mining.

By analysis of *drilling cores* the requirement of reliable geometry can be more or less fulfilled. Heterogeneity, however, still remains a problem. Rapid information and sample selection are the advantages offered here. The analysis of *drilling holes* is also possible, but development has not reached the level of industrial application.

For ore analysis, mainly laboratory methods have been developed; the measurement itself is carried out with powdered samples in most cases. Ag, Co, Cu, Mo, Ni, Pb, Sn and Zn are analysed most frequently. Various radiation sources are used. Because of the poor resolving power of scintillation detectors used almost exclusively, the use of filter pairs is general.

The application of IXRF is very important in *slurry analysis in ore enrichment,* because it makes possible an on-stream measurement. It is widely used for the analysis of both concentrates and refuses. Intensity data with respect to the elements to be measured are still insufficient for the unambiguous calculation of the concentrations, therefore the density of the slurry is measured separately by a radiation absorption technique (Section 2.3.3) and used for correction (Fig. 4.19). Industrial analyses of Ba, Cu, Mo, Nb, Pb, Sn

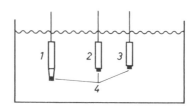

Fig. 4.19. Slurry analysis in a flotation plant. *1* — Density measurement; *2* — Zinc analysis; *3* — Lead analysis; *4* — Radiation sources

have been introduced in practice. The detector is, as a rule, a scintillation counter provided with a filter, but in some cases, proportional counters filled with xenon or argon and even ionization chambers are used. The relative error has been given to be between 1 and 10%; this is satisfactory for industrial purposes.

The traditional X-ray technique is generally used for the analysis of *light metals and alloys*; the applicability of IXRF is still unexplored, although aluminium production is an important branch of industry. Besides Al, elements such as Cr, Cu, Fe, Ge, Mg, Mn, Si, Ti, V, Zn, etc. occur in aluminium, in its raw materials and products of processing.

The application of IXRF in *silicate industry* is very wide-spread. Though crystal dispersion devices are used mostly in *cement industry,* the IXRF technique (Fig. 4.20) will be able to substitute a considerable number of traditional devices in the near future, first of all because of its lower investment costs. The problems of applications of IXRF in this field are due to the low energies to be detected and to the low excitation efficiencies, the main components of cement being Al, Ca, Mg and Si, i.e., elements with low atomic numbers. Apart from these elements, the properties of cement can be influenced by the presence of Fe, and traces of Mn, P, S, Sr and Ti may also be present.

α-Radiating ^{210}Po with an activity between 150–400 MBq ($\sim$4–10 mCi) is a good exciting source; good efficiencies can also be obtained with ^{3}H/Ti and ^{3}H/Zr targets using an activity higher by three orders of magnitude. The use of windows with low absorptivities is very important; absorption conditions are improved by purging the device with helium or by evacuation. Proportional counters filled with 90% Ar + 10% CH$_4$ are used most frequently as detectors. The electronics is conventional

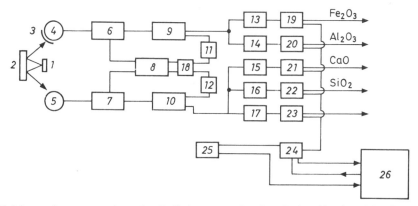

Fig. 4.20. Scheme of a cement analyzer. *1* — Radiation source; *2* — Sample; *3* — Absorber; *4, 5* — Proportional counters; *6, 7* — Preamplifiers; *8* — High-voltage power supply; *9, 10* — Linear amplifiers; *11–17* — Amplitude analyzers; *18* — Double amplitude stabilizer; *19–23* — Scalers; *24* — Timer; *25* — Control unit; *26* — Computer

preamplifier, amplifier, amplitude discriminator, timed scaler, computer. Grounding is very important in order to decrease the noise level; the stabilization of the amplification can be ensured by a special pulse-height stabilizer of the high-voltage part.

Direct analysis of powder samples is not possible, due to variations in particle size and density, therefore the sample is melted together with $Li_2B_4O_7$ or with borax. Typical analysis data and errors for the individual components are as follows: $Fe_2O_3 : 2 \pm 0.05$ mass-%, CaO : 44 ± 0.16 mass-%, $SiO_2 : 14 \pm 0.18$ mass-% and $Al_2O_3 : 3 \pm 0.09$ mass-%. The analysis lasts for 20 minutes; of this, sample preparation requires 10 minutes.

The analysis of $CaCO_3$, being the most important raw material of cement manufacturing, is very important, because its concentration has to be known in the 5–85% range with an error of $\pm 1\%$, with 95% certainty. The continuous analyser works with a ^{55}Fe radiation source of an activity of about 100 MBq (~ 3 mCi) and with an argon-filled proportional counter. If the particle size of the sample is smaller than 0 μm, the standard deviation of the analysis is $\pm 0.35\%$ $CaCO_3$.

Besides the atomic absorption method, also crystal dispersion X-ray analysis finds increasing application in the analysis of *silicates and glasses.* In plate glass manufacturing, the main analytical problem is the control of the prescribed quality of the product, in case of container glasses (bottles, etc.), however, dangerous elements (Cd, Pb, etc.) eluted by their liquid contents, are to be determined. International standards (the so-called EC-series of the European Co-operation) are available for checking the various analytical methods.

The analysis of *minerals* presents similar problems as discussed in the case of cement, but the composition expected is much more complicated (Al_2O_3, CaO, Fe_2O_3, K_2O, MgO, MnO, Na_2O, P_2O_5, SiO_2, TiO_2, etc.). Therefore, the use of traditional X-ray devices is general in this field.

The application of X-ray analysis in *petroleum industry* is not new. This is attributed to the fact that the elements to be determined are in a hydrocarbon matrix and usually in liquid medium, thus their mutual interference and the sample preparation present no particular problem. Above a few times $10^{-4}\%$ concentration, elements can generally be determined directly; below this value combustion in the presence of sulphuric acid can be recommended; in the latter case, Co or Cr is added as an internal standard.

The determination of Cl, S and V is important in *petroleum products. Sulphur determination* is possible by means of a commercial instrument in the range between 0.005–4%. The experimental error is ±0.0015% in the lowest concentration range. The exciting source is [55]Fe, the detector a neon-filled proportional counter. The device should be purged by He or N_2, because the argon content of the air interferes with the measurement of the K_α-radiation of sulphur.

The determination of *chlorine content* is important because this element poisons some catalysts; the measurement is similar to the procedure, described for sulphur.

The lead content of motor gasolines can be determined generally with the excitation of the L_α-radiation of Pb. The excitation of the K_α-radiation would require the use of [153]Gd

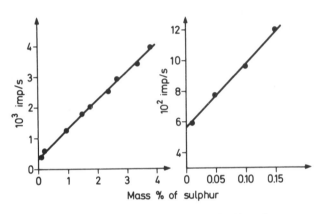

Fig. 4.21. Sulphur determination in petroleum products by X-ray fluorescence analysis

of high energy; this source has, however, several disadvantages: relatively short half-life, the requirement of a considerable radiation shielding and large samples. Both [238]Pu and [109]Cd are suitable for exciting the L_α-lines. When using the former one, a correction should be applied for the bromine content of the additive and the error is still twice as high as with [109]Cd; the 86 years' half-life is, however, an advantage. The half-life of [109]Cd is 1. years only, but the error can be as low as ±70 kg/m[3], provided that Ge and Ga filter pair and xenon counting gas are applied.

Additives of lubricating oils improving their quality contain Ba, Ca, Cl, P, S and Zn. The measurement of their concentrations is, as a rule, directly possible, moreover their surface concentrations on engine parts can also be studied (between 0 and 750 mg/m[2]

Deposits, corrosion products in engines and oil furnaces can also be analysed by IXRF The determination of Cu, Fe, Ni, Pt and V in *catalysts* may also be important; Co is recommended as an internal standard for this purpose.

In coal ash analysis, the Fe content gives rise to an unacceptable error in the isotopic measurement; its correction can be carried out by an IXRF measurement.

Purely analytical-chemical IXRF applications are not very significant. The determination of Br, Cl, I, P, S and metals in organic substances can be carried out with the advantage that usually no sample preparation is necessary. In some cases the products of certain compounds or functional groups with selective reagents (e.g., those of olefins with bromine, NH_3 with Hg-containing Nessler reagent) can be utilized for IXRF purposes Determinations of Al, Cl, Co, Fe, Mn, Sb, Ti and Zn in *plastics* have been reported in the literature; the analysis of *pharmaceutical products* as far as their As, Cu, Hg, Ni, Pb and Se content is concerned, can also be mentioned.

The XRF method is very useful for the analysis of *elements which are difficult to separate* (such as Hf in Sc, Zr and Y in rare earth metals, actinides, the parallel determination of Mo and W, etc.); most methods have been developed, however, for crystal dispersive spectrometers.

IXRF has some advantages over all other analytical methods in the field of *environmental protection*. This can easily be understood if it is considered that by minimal sample preparation and by the use of maximum 3 or 4 radiation sources, it is possible to determine simultaneously all elements between Na and U with a generally satisfactory accuracy. At the same time, the noise level is only a fraction of that observed with X-ray tube excitation—owing to the monochromatic excitation—and the matrix effect is usually negligible. There are also problems, of course. With elements of low atomic number, the criterium of proper thin layers (i.e., the linear relationship between layer thickness and intensity—see Section 4.1.4.1) can hardly be fulfilled; corrections should be applied in the case of fine powder samples, due to multilayer deposition, and the calibration problem of quantitative evaluation is not entirely solved. Best calibration standards are thin layers of elements or compounds prepared by evaporation onto the sample holder.

IXRF is used for the analysis of both molecular and serosol particles in the examination of *air pollution*. A known amount (a few cubic meters) of air is sucked through an appropriate filter in both cases. There are very strict requirements concerning the filter material; they can be fulfilled in practice only partially: it should consist of low atomic number elements, its very thin layer should ensure a 100% filtration efficiency, and it must not be hygroscopic. Cellulose-based filters are the best suited to these requirements. For the selective measurement of *air pollutants in molecular dispersion*, the filter is treated with a reagent: for example the Cl_2 content of air can be measured in the range between $0.3-13$ g/m^3 using a filter treated with *o*-toluidine; for the analysis of H_2S and SO_2, the appropriate reagents are $AgNO_3$ and $NaOH$, respectively.

Particle collection is the main problem in the analysis of *aerosol particles*. It has been established that Br, Ni, Pb and Zn are bound to small, V to medium, Ca, Fe, Mn and Ti to large particles. Therefore, instead of filters, devices ensuring separation according to particle size are often used. They work generally on the basis of kinetic principles. Some authors assign various elements to definite pollution sources: Fe, Ga, Mn and Ti come from the soil, Br, Fe, K and Mn from coal, Br and Pb from petrol, Ni and V from fuel oil, S from coal or petroleum, Fe from steel manufacturing, Al and Si from other branches of industry.

Water analysis (Fig. 4.22) shows analogous features with air pollution measurements. Dissolved and suspended pollutants can be distinguished in the phase of sample preparation. Direct measurement of *dissolved impurities* is possible only with concentrated sources of contamination, such as sewage from factories; dilute solutions require *preconcentration*. A possible way of doing this is precipitation or co-precipitation (e.g., the concentrating of Pb and Zn by means of iron hydroxide). Extraction can also be used; sodium diethyl-dithiocarbamate and ammonium 1-pyrrolydine-dithiocarbamate with low selectivity are suitable for this purpose. They permit one to measure metal traces in concentrations as low as a few times $10^{-7}\%$. Ion exchange resins are extensively used for binding cations; either the resin or the solution displaced from the resin, sometimes both, are measured by IXRF. Finally, the simplest, but most time-consuming concentrating operation, i.e., evaporation should also be considered. Literary data are available

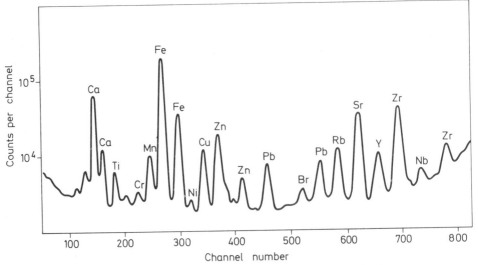

Fig. 4.22. X-ray fluorescence spectrum of the evaporation residue of a river-water

mainly with respect to the analyses of As, Ba, Ca, Cd, Co, Cu, Fe, Mn, Ni, Pb, Sr and Zn in water.

Filtration is generally used for determination of *colloidal* or suspended contaminants; another required property of the filter is that it should be wettable by water.

The investigation of materials in the *food chain* can also be carried out by IXRF analysis. Trace contaminations in the fodder can be studied after concentration by drying and powdering, occasionally by combustion. Gelatine standards are used for trace analysis of contaminants in meats and canned food. The lead content of a drop of blood on filter paper can already be determined with sufficient accuracy; about $10^{-4}\%$ of As, Br, Cr, Cs, Hg, Sr and W can be determined in dried blood. Urine samples are concentrated usually by ion exchange in order to meassure traces of Br, Cu, Pb, Sr and Zn.

EXAMPLE 4.7

A Cr coating on iron is excited by a ^{55}Fe radiation source. Let us calculate the layer thickness up to which the K_α-intensity of Cr increases linearly, and determine the thickness where its value becomes constant.

The mass absorption coefficient of Cr with respect to the Mn K_α-radiation emitted by ^{55}Fe is 50 m^2/kg, with respect to its own radiation it is 6 m^2/kg. The density of Cr is $\rho = 7100$ kg/m^3.

For the linear section:

$$\rho_{Cr}\, l\, \bar{\mu}_{m,Cr} \leq 0.1$$

$$\bar{\mu}_{m,Cr} = 50 + 6 = 56 \text{ m}^2/\text{kg}$$

Assuming that the surface is perpendicular to both the incident and emitted radiations, from Eq. (4.10) we have:

$$7100 \times l \times 56 \leq 0.1, \quad \text{and} \quad l \leq 0.25 \text{ μm}$$

Thus, the K_α-radiation intensity of Cr increases linearly up to a layer thickness of 0.25 μm. Saturation thickness begins at 2.5 μm.

If a ^{109}Cd radiation source is used for excitation (22 keV), then

$$\mu_{m, Cr, 22\,keV} = 1.8 \ m^2/kg, \qquad \mu_{m, Cr} = 7.8$$

and the linear section extends up to

$$l = \frac{0.1}{7100 \times 7.8} \ m = 1.8 \ \mu m$$

The extent of the linear section depends strongly on the exciting energy.

EXAMPLE 4.8

A mixture of Nb_2O_5–Ta_2O_5 is excited by the 84 keV energy radiation of ^{170}Tm. Let us calculate the regression coefficients determining the correlation between the concentration and intensity on the basis of the following intensity values:

If $c_{Nb_2O_5} = 0.5$, the count rate is 24 600 pulses and

if $c_{Nb_2O_5} = 1.0$, the count rate is 57 220 pulses.

$$r_{Nb/Ta} = \frac{0.5}{1 - 0.5} \times \frac{57\,220 - 24\,600}{24\,600} = \frac{32\,620}{24\,600} = 1.33$$

From the definition of the regression coefficient according to Eq. (4.14) it follows that

$$r_{Ta/Nb} = \frac{1}{1.33} = 0.75$$

Thus, for the K_α-radiation of Nb, a negative, for that of Ta, a positive deviation will be experienced from linearity.

EXAMPLE 4.9

The following relative intensities $\left(\dfrac{I}{I_{100}}\right)$ are measured in a Fe—Ni—Cr alloy of unknown composition: Fe 0.4659, Ni 0.0684, Cr 0.2839. Let us calculate the composition of the alloy!

By exciting samples of known composition, the following constants of Eq. (4.17) have been determined experimentally for the K_α-radiations of the components:

	A		B	
	Fe	Cr	Fe	Ni
Fe K_α	0	2.099	0	−0.456
Ni K_α	1.711	1.226	0	0
Cr K_α	0	0	−0.419	−0.234

The calculation can be carried out by an iteration technique. As a first approximation, let us neglect the interactions (A_{ik} and B_{ik} are zero) and let us calculate the concentrations which are then equal to the relative intensities. Then, substituting these into Eq. (4.17) we get:

$$c_{Fe} = 0.4649 \, (1 + 2.099 \times 0.2839 - 0.465 \times 0.0684) = 0.727$$

$$c_{Ni} = 0.0684 \, (1 + 1.711 \times 0.4649 + 1.226 \times 0.2839) = 0.147$$

$$c_{Cr} = 0.2839 \, (1 - 0.419 \times 0.4649 - 0.234 \times 0.0684) = 0.224$$

$$\overline{1.098}$$

Since the sum of the concentrations is equal to 1.098, the calculation should be continued. Therefore, the previous concentration values are normed to unity $\left(\text{multiplied by } \dfrac{1}{1.098} \right)$ and the calculation is carried on with these concentration values until the results of two subsequent calculations are equal to each other within ± 0.001, and their sum is equal to unity.

The final result is

$$c_{Fe} = 0.628$$

$$c_{Ni} = 0.151$$

$$c_{Cr} = 0.209$$

These data are equal to those obtained by chemical analysis within a few tenths of percent.

EXAMPLE 4.10

Let us estimate the necessary activity (A) of a ^{3}H/Zr target when the required K_α intensity (I) of S is 10^2/s cm^2 for a gas oil with 1% sulphur content.

The efficiency of generation of bremsstrahlung (ε) by the 0.018 MeV β-radiation (E_{max}) of ^{3}H in an element of atomic number $\mathscr{Z}$ is calculated from the following equation [4.8]:

$$\varepsilon = 5 \times 10^{-4} \mathscr{Z} E_{max} = 5 \times 10^{-4} \times 40 \times 0.018 = 3.6 \times 10^{-4}$$

The fluorescence efficiency (ω) of S (Fig. 4.15) is 0.1.

Geometry gives a multiplication factor (θ) of about 10^{-2} if the distances between the source–sample–detector are taken equal to 1 cm each, and the corresponding space angles are also considered. Two orders of magnitude arise from the 1% concentration ($c = 0.01$) and a decrease of one order of magnitude may arise from the absorption ($\tau = 0.1$) of the K_α-radiation of S.

Summarizing:

$$A = \frac{I}{\varepsilon \times \omega \times \theta \times c_s \times \tau} =$$

$$= \frac{10^2}{3.6 \times 10^{-4} \times 10^{-1} \times 10^{-2} \times 10^{-2} \times 10^{-1}} = 2.8 \times 10^{11} \text{ Bq} = 7.5 \text{ Ci}$$

4.1.5. ANALYTICAL METHODS BASED ON NEUTRON ABSORPTION, SCATTERING AND MODERATION

Neutrons interact with the nuclei of the medium only. The interactions include *elastic and inelastic scattering, absorption and neutron formation (nuclear fission)*. The interactions are discussed in the present section from the side of the neutrons; processes and applications connected with the measurement of the activity of a substance (activation analysis, see Section 4.2), as well as those involving fission or neturon emission will not be discussed.

At the same time, humidity measurements by neutrons belong to the determinations of purely physical parameters (Section 2.3). Owing to their special character, a separate section (Section 5.3) will deal with applications belonging to nuclear geophysics of deep drilling. The considerations discussed here, however, serve as a basis of those applications, too.

High energy neutrons from a neutron source *collide elastically* with atoms of light elements, thus they lose their energy gradually (they are "moderated"); at the same time the path of their travel is altered: they are scattered. Energy loss in elastic collision can be described by purely mechanical laws. The collision number is used to characterize the slow-down ability of various substances; this is the number of collisions necessary for decreasing the neutron energy from a level of 2 MeV to the thermal level, i.e., to 0.025 eV. This number is 18 for hydrogen, 114 for carbon, 150 for oxygen and 2150 for uranium. At the same time this number itself is not sufficient to characterize the moderating ability which is an outstandingly high value for hydrogen: if the number of atoms per unit volume is also included, the value of *linear slow-down ability* is obtained; its value for water is 153, for graphite, 6.4/m. The neutron decelerating ability of hydrogen and water serves as a basis of *moisture determination* (Section 2.3) and of the measurement of *hydrogen contents* in general (e.g., Section 5.3.1.1). The measurement itself requires the use of such instruments and detectors which can distinguish between high energy and thermal neutrons.

The concentration distribution of *thermalized neutrons* can be described on the basis of the *gas diffusion theory* ([4.97] and Section 5.3.1.1). Accordingly, the concentration of neutrons is the highest near to the source, and exponentially decreases in radial direction. Thus, the detector has to be placed as close to the source as possible; only with extremely low (hydrogen) concentrations may modify this situation because in this case the neutrons have to travel far until they are thermalized.

Inelastic collisions of neutrons occur parallel with elastic collisions; this is considerable with medium or heavy nuclei and above 1 MeV neutron energies. The energy loss of neutrons due to inelastic collisions is higher than calculated on the basis of the mechanical laws of collision; the energy difference is expended on the excitation of the nucleus.

Neutron absorption takes place parallel with the collision processes; it is characterized by the cross-section value of absorption and can be given in m^2. This is not necessarily identical with the cross-section of activation (Section 4.2.1). *The absorption cross-section* of some elements—or rather some isotopes—may be higher by several orders of magnitude than that of other elements. This is the basis of the determination of these elements by means of high sensitivity neutron absorption analysis. Some nuclei absorb neutrons having a given discrete energy with high cross-section; this phenomenon is called *resonance absorption*. Its analytical application would require, however, sophisticated and expensive instruments.

The following *analytical applications* have to be mentioned:

The determination of the hydrogen content of liquid hydrocarbons is possible on the basis of both *neutron scattering* and neutron absorption. Both methods are suited for continuous industrial measurements. The scattering technique requires a minimum sample volume of $5\,cm^3$, the absorption method, $10\,cm^3$. These measurements are practically insensitive to the changes of the density; the experimental error of the scattering method is ± 0.02, that of the absorption technique, ± 0.05–0.1% hydrogen. Owing mainly to difficulties accompanying the application of a neutron source, these methods have not spread in practice. For laboratory purpose only the absorption determination can be utilized; one measurement lasts about 10 min.

Neutron absorption analysis is advantageous in such cases when the absorption cross-section of some elements is much higher than that of other elements being present simultaneously; at the same time, nuclear reactions which do not result in induced radioactivity are preferred because of practical reasons. Neutron absorption can be described approximately by the following correlation:

$$\ln \frac{I_0}{I} \approx l \Sigma n_i \sigma_i \tag{4.21}$$

where I_0 and I are the neutron intensities before and after passing an absorbing layer of l m thickness, n_i is the number of atoms of the i-th element in $1\,m^3$ sample, and σ_i denotes the absorption cross-section of the i-th element. (In the case of deviations from the natural isotopic composition, every isotope has to be regarded as a separate component.) Mainly at higher concentrations, the validity of Eq. (4.21) ceases (the correlation becomes nonlinear).

The importance of the *concentration measurement of boric acid* is that it is applied in concentrations of 0–$20\,kg/m^3$ in the cooling water of nuclear power plants, for stabilizing the neutron flux. The apparatus uses a Pu–Be neutron source and a BF_3 detector; is operates continuously, and shows the H_3BO_3 concentration in a digital form. The error is ± 0.02–$0.1\,kg/m^3$.

In addition to boron (1%), Cd (3.2%), Li (6.4%)—e.g., in castings—Gd (0.2%), Hg (36%), In (42%) and Cl (71%) can be determined by neutron absorption (the percentage values written after the symbol of the elements denote the concentrations belonging to identical neutron absorption).

4.1.6. ANALYSIS BASED ON THE MÖSSBAUER EFFECT

The Mössbauer effect, besides its applications in nuclear physics and solid state physics, is a sensitive and effective tool for solving chemical—in particular, analytical—problems.

The Mössbauer effect is based on *resonance absorption*: if strict physical and chemical conditions are fulfilled, the nuclei of a given isotope with a high cross-section absorb exactly the same energy which is emitted by the nuclei of the same isotope. The detectability of this resonance absorption requires the simultaneous fulfilment of several conditions as far as the radiation source and the absorber are concerned.

In the case of a constant E_0 energy nuclear radiation of the *radiation source*, the *natural spread of energy* has a band width of Γ; for a given E_0 energy, Γ can be calculated from

Heisenberg's principle:

$$\Gamma, eV = 6.58 \times 10^{-16}/\tau \qquad (4.22)$$

where t is the average lifetime of the excited state in seconds. If the value of τ is about 10^{-8} s, the natural spread of the emitted radiation is very low: 6×10^{-8} eV. The *recoil energy* taken up by a free atom in the disintegration process should be subtracted from E_0; this recoil energy is, however, higher by about six orders of magnitude than the above value.

The disordered movement of *gaseous* radiating isotopes due to *Doppler's principle* may also exceed the natural energy spread of the emitted quantum by about six orders of magnitude. In this case (and likewise in *liquids*) the number of photons having just the energy which can be taken up by the absorbing nuclei is undetectably small.

In solid substances, the conditions of recoil-free energy loss are not fulfilled above the so-called Debye temperature which is characteristic of the substances. The Debye temperature for various materials is between 80 and 440 K; those values which cannot be found in tables can be estimated on the basis of the molar heat: the Debye temperature is that temperature where the molar heat reaches the value of 23.630 kJ/(mol K).

Below the Debye temperature the solid can take up energy quanta only. If the energy of the quantum is much higher than the recoil energy of the atom, energy loss due to recoil will have a low probability, thus the quantum emission will occur with the neutral band width, because the recoil energy will be taken up by the *whole crystal lattice*.

The energy uptake *via* resonance absorption by the *absorber* also has several criteria. Like for the emitting atom, the condition is valid for the absorbing atom, too, that quantum capture should occur *recoil-free*, i.e., below the Debye temperature, in the solid phase. At the same time, the energy necessary for excitation is influenced—although to a minor extent—by the changes

— in the structure of the electron shell *(chemical bond)* and
— in the surroundings of the atoms *(the presence and distance of other atoms)*.

These may result in the necessity of changes of the photon energy with natural band width in order to obtain a resonance absorption.

The change of quantum energies can be achieved by moving the absorber and the radiation source with respect to each other (by changing their relative speed), i.e., on the basis of the *Doppler-effect*. The sensitivity of the method is demonstrated by the fact that a few hundredth mm/s speed change may cause in several cases considerable absorption changes. The scattering geometry may be advantageous in some cases (surface analysis, thick sample, high internal conversion, etc.); here the intensity of secondary radiations accompanying absorption is measured, instead of the weakening of the primary radiation. Absorbing nuclei either emit another γ-photon or transfer their energy to the electrons on the s orbital (internal conversion), which results in the emission of characteristic X-rays and electrons with discrete energy. The two sorts of geometry are depicted in Fig. 4.23.

Figure 4.24 shows an example of *Mössbauer spectra*. The ordinate shows the radiation intensity, the abscissa, the speed of the radiation source (or that of the absorber). The most characteristic feature of the spectrum is the positions of the absorption (or scattering) maxima.

Changes in the electron density of the atom are accompanied by *isomeric shifts* in the spectrum. Such changes are induced by different chemical binding, different oxidation state, by exchanging the neighbouring atoms in the compound, etc. *The quadrupole split* in the spectrum is the result of an electron orbital of nonspherical symmetry and permits one

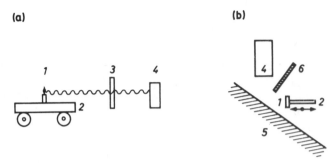

Fig. 4.23. Recording of Mössbauer spectra. (a) Absorption geometry; (b) scattering geometry. *1* — Radiation source; *2* — Moving device; *3* — Absorber; *4* — Detector; *5* — Sample; *6* — Shielding

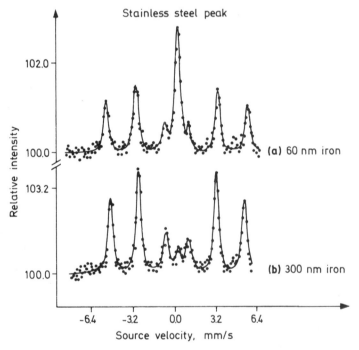

Fig. 4.24. Mössbauer spectrum of iron, based on the measurement of conversion electrons; two layers of different thickness evaporated onto stainless steel

to study the character of the chemical bond and the intramolecular symmetry. An external magnetic field, as well as the magnetic field of the electrons of the atom itself, may cause a *magnetic split* in the spectrum, which may solve problems concerning the electron spin.

The application of the Mössbauer effect is the most important in the chemistry of *complex and organometallic compounds.* The method is mainly helpful in the study of Fe-, Np-, Ru-, Sb-, and Sn-complexes, together with organometallic compounds containing Fe and Sn. Information can be obtained, among others, about the oxidation state, the covalent character (the strength) of the chemical bond, the coordination number, ligand symmetry, as well as about the high and low spin number transformations of the electron structure of complexes containing Mössbauer atoms. Molecular bonds, symmetry and the structure of organometallic compounds can be studied by Mössbauer investigations.

The number and importance of *metallurgical* applications is also very great, because natural *iron*—although the concentration of the Mössbauer-active ^{57}Fe is only 2.19% in it—is suitable for resonance absorption studies without any enrichment. The following investigations can be mentioned:

— qualitative and quantitative *studies of different phases;* and

— the measurement of *properties of individual phases.*

A deeper understanding of the *mechanism of hardening* is promoted by the Mössbauer studies carried out with *iron samples containing carbon.* The carbon atom is built in the lattice of γ-Fe (austenite) as detected by the change of the lattice constants. The next step is the rearrangement of the lattice: the carbon–austenite system becomes martensite.

In multiphase systems, the quantitative evaluation of the complex Mössbauer spectra can be facilitated by the stripping technique, where the evaluation is carried out by a computer using reference spectra (see Section 4.2.3.2).

4.2. ACTIVATION ANALYSIS

4.2.1. THEORETICAL BASIS

4.2.1.1. NEUTRON ACTIVATION

Activation analysis is a method of nuclear elementary analysis giving information by means of a *nuclear reaction* about the qualitative and quantitative elementary composition of the substance to be investigated. The radiation induced by nuclear reaction, e.g., γ-radiation, neutron, charged particle (proton, deuteron, triton, ^{3}He, etc.), transforms a fraction of the nuclei in the substance to be investigated. Study of the *characteristic properties of the products* of transformation permits one to draw conclusions about the total or partial elementary composition of the substance and about the amount of the elements identified qualitatively. In practice, the majority of activation analysis is still based on neutron activation, but one comes across charged particle and high energy γ-photon activation more and more frequently, because complex accelerators—first of all, cyclotrons—are installed at several places where activation analysis laboratories are organized.

On the basis of radiation inducing nuclear reaction, three *types* of activation analysis can distinguished:

— *neutron activation analysis* (NAA);

— *charged particle activation analysis* (CPAA);

— *photon activation analysis* (PAA).

The study of the irradiated substance may be carried out by a *consecutive (delayed)* or a *prompt (parallel)* measurement. The delayed method may be *destructive* (D) or *nondestructive* (ND).

A characteristic up-to-date trend is the widening of the circle of nondestructive *prompt methods (activation spectroscopy)* in activation analysis.

Activation analysis (reliably applicable also in the 10^{-7}–10^{-4}% concentration range) takes a *special position* among high sensitivity instrumental analytical methods due to its complexity and seemingly high costs. High performance radiation sources are necessary to achieve an appropriate high sensitivity—e.g., a nuclear reactor, or a cyclotron—and sample processing requires, in most cases, well-equipped radioanalytical laboratories. All this is true, but it is the apparent surface only, because the radiation sources are installed in most cases for other than analytical purposes and an additional activation analysis application improves the degree of utilization, while its contribution to the total operation costs is small. Radioanalytical laboratories are suitable for several simultaneous works and they operate in several places.

Table 4.3 compares the detection limits for some elements using NAA and other analytical methods.

Table 4.3. Absolute detection limits of some elements, ng

Element	Spectro-photometry	Emission spectroscopy	Atomic absorption	Flame photometry	Neutron activation analysis	Mass spectroscopy
Ag	5	0.5	5	3	0.01	0.2
Au	5	40	20	50	0.005	0.2
B	50	0.5	6000	5		0.01
Ca	100	5	2	0.2	70	0.03
Cd	3	20	5	20	3	0.3
Cr	7	1	5	0.5	0.5	0.05
Cu	2	0.5	5	1	0.06	0.08
Dy		40	400	4	0.0008	0.5
Eu		2	200	1	0.0006	0.2
Fe	200	3	5	3	100	0.05
In	50	10	50	1	0.002	0.1
K	200	20	5	0.01	2	0.03
Mg	2	10	0.5	0.8	20	0.03
Mn	5	0.3	3	0.5	0.01	0.05
Na		10	5	0.01	0.2	0.02
Pb	6	10	10	0.5	2000	0.3
Rh	50	200	30	10	0.02	0.09
Sb	4	10	200	30	0.1	0.03
W	30	10	3000	200	0.05	0.5
Zn	100	10	2	80	10	0.1

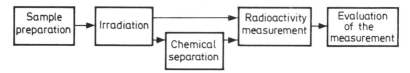

Fig. 4.25. Main steps of the consecutive activation analysis

The following are the most important *steps of the consecutive method* (Fig. 4.25):
— sample preparation;
— production of radioactive isotopes by irradiation;
— total or partial chemical separation of the individual radioactive elements;
— measurement of the radioactivity induced;
— evaluation of the measured data.

The nuclear reactions applied are as follows:

The (n, γ) process is the most common nuclear reaction, e.g.:

$$^{75}_{33}As + ^{1}_{0}n \rightarrow ^{76}_{33}As + \gamma$$

$$^{133}_{55}Cs + ^{1}_{0}n \rightarrow ^{134}_{55}Cs + \gamma$$

or, by a simpler denotation

$$^{75}As\,(n, \gamma)\ ^{76}As$$

$$^{133}Cs\,(n, \gamma)^{134}Cs$$

If the target is placed into a neutron beam, the nuclear reaction between the neutrons and the target nucleus takes place in *two steps: first* the target nucleus absorbs the bombarding particle and *an excited transition nucleus* (complex nucleus) *is formed, then this decomposes and emits* (as a rule, promptly) *photons* or nucleons and stable or radioactive product nuclei are formed. For example, the following processes take place with irradiated aluminium:

$$^{27}_{13}Al + ^{1}_{0}n \rightarrow ^{28}_{13}Al \begin{cases} ^{28}_{13}Al + \gamma & (n, \gamma) \\ ^{27}_{13}Al + ^{1}_{0}n' & (n, n') \\ ^{27}_{12}Mg + ^{1}_{1}H & (n, p) \\ ^{24}_{11}Na + ^{4}_{2}He & (n, \alpha) \end{cases}$$

The probabilities of the four simultaneous nuclear reactions at a given neutron energy are very different. *With thermal neutrons* ($E_n \approx 0.025$ eV), practically the (n, γ) *process is the* only reaction. The target atom captures neutrons, it emits one or more γ-quanta (prompt γ); the atomic number remains unchanged and the radioactive element stays in the target which is chemically identical with it.

In (n, n') *reactions* it may occur that the analysis can be carried out by the investigation of the nucleus remaining in excited state, for example in the case of the

$$^{115}In\,(n, n')\ ^{115m}In$$

reaction. The cross-section for the (n, n') nuclear reactions can reach the value of 10^{-29} m^2, because of the low threshold energy (0.2–1 MeV).

For effecting (n, p) *reactions* neutrons above 1 MeV are usually necessary. The atomic number decreases here by one unit, the target atom is transformed into another element:

$$^{27}_{13}Al\,(n, p)\,^{27}_{12}Mg$$

The (n, α) *reactions* similarly to the (n, p) reactions, take place under the effect of *high energy neutrons*. The atomic number of the target atom decreases by two units:

$$^{27}_{13}Al\,(n, \alpha)\,^{24}_{11}Na$$

In the case of nuclear reactions accompanied by the emission of *charged particles*, (e.g., p- or α-particles), the energy of the bombarding neutron should be above a given value *(threshold energy)* in order to help the emitted positive particle to overcome the potential barrier (Coulomb barrier) in the close vicinity of the nucleus. The values of such threshold energies are, e.g.:

$$E_t = 4.7 \text{ MeV} \qquad \text{for the reaction} \qquad {}^{52}\text{Cr}(n, p)^{52}\text{V}$$

$$E_t = 8.3 \text{ MeV} \qquad \text{for the reaction} \qquad {}^{55}\text{Mn}(n, \alpha)^{52}\text{V}$$

Using appropriate detectors, it is possible to identify the characteristic radiations and half-lives of the components of the activated sample.

The rate of formation of a radioactive isotope produced by a given nuclear reaction from a given isotope is:

$$\frac{dn^*}{dt_i} = n\sigma\Phi - \lambda n^* \tag{4.23}$$

where n^* denotes the number of radioactive nuclei, n the number of target nuclei to be transformed, σ the cross-section in m^2, Φ the flux $1/(\text{m}^2 \text{ s})$, t_i the irradiation time, s, λ the disintegration constant of the radioactive isotope, $1/\text{s}$.

Assuming that n, σ and Φ are constant, the activity of the given radioactive isotope after an activation period of t_i can be calculated by integration:

$$A = A_\infty [1 - \exp(-\lambda t_i)] \tag{4.24}$$

or

$$A = A_\infty \left[1 - \exp\left(-0.693 \frac{t_i}{t_{1/2}}\right) \right] \tag{4.25}$$

where A denotes the activity, Bq, $t_{1/2}$ the half-life, s, and $A_\infty = n\sigma\Phi$ is the so-called saturation activity.

The value of n can be calculated by the following formula:

$$n = \frac{N_A m\theta}{\mathcal{A}} \tag{4.26}$$

where N_A is the Avogadro number, 6×10^{23} $1/\text{mol}$; m is the weighed mass of the element in question, g, θ denotes the mass fraction of the isotopic abundance and $\mathcal{A}$ the atomic mass of the element.

The factor $[1 - \exp(-\lambda t_i)]$ in Eq. (4.24) is called the *saturation factor* and is denoted by S. If the time of irradiation is relatively high as compared to the half-life of the nucleus, i.e., $t_i \gg t_{1/2}$, then the saturation factor is about unity and the activity is

$$A \approx n\sigma\phi \tag{4.27}$$

After activation, the activity of the radioactive isotope decreases, the sample is "cooling". With a simple decay, the A_c activity of the isotope after a t_c cooling time will be

$$A_c = A \exp(-\lambda t_c) \tag{4.28}$$

Natural isotopic abundance is very important with respect to the calculation of the sensitivity and also for determining the nature of radiation measured when the given element is determined. The different stable isotopes of the elements do not behave identically when irradiated. For example only one natural gold isotope exists, ^{197}Au, giving by an (n, γ) reaction ^{198}Au, but there are four stable iron isotopes: ^{54}Fe (5.82%), ^{56}Fe (91.66%), ^{57}Fe (2.19%) and ^{58}Fe (0.33%). Since only ^{54}Fe and ^{58}Fe give active products in (n, γ) reaction, 6.15% of the iron can be activated.

The cross-section (see Section 1.2) is characteristic of the extent of interaction between the target nucleus and the bombarding particle, but its value is also a function of the energy of the latter. If, e.g., the bombardment is carried out by thermal neutrons, the thermal cross section should be considered when the induced activity is calculated. The cross-section is sometimes given for individual nuclei (isotopes), in other cases for the natural isotopic composition. It is very important that these two versions should be distinguished when the calculations are carried out.

By increasing the neutron flux, the number of atoms activated during irradiation can be increased.

The activity is a function of the irradiation time, too.

If the flux, the cross section, and the activity of the radioactive isotope formed from the given element are known, the amount of the element in the sample can be *calculated* by using Eqs (4.24)–(4.27).

The relative method is, however, used much more often than the calculation because it does not require the measurement of absolute activity, and the errors from flux fluctuations are eliminated. In this case, the unknown sample and the standard pure material (containing the same elements as the sample and having a known mass) are irradiated at the same place of the reactor for the same time and with the same flux; under real conditions, the specific activity should be the same in the standard and the unknown sample. If the measurement of both samples is carried out under identical conditions, i.e., the efficiency of the measurement is the same, then

$$m := \frac{m_s A}{A_s} \qquad (4.29)$$

$$m = \frac{m_s n}{n_s} \qquad (4.30)$$

where m and m_s are the masses of the unknown and the standard, respectively, A and A_s their activities, n and n_s their pulse counts measured under identical conditions.

The comparator method is the most up-to-date procedure in neutron activation analysis by reactors. The above-mentioned relative method is laborious, time-consuming and has several possibilities of error when a standard of each element should be irradiated and measured in the case of the analysis of several (20 or more) elements. The comparator method has retained the advantages of the classical relative method, its precision and accuracy is practically the same, but in serial analyses it saves a lot of time and it has the advantage that it allows the determination of one (or more) unexpected element(s) quantitatively.

The essence of the comparator method is that the ratio of the specific count rates (I_{sp}) of the analytical γ-lines arising from the products of the (n, γ) reactions of the elements to be analysed as well as from those of the comparator elements, the so-called k-factors, are determined experimentally (the asterisk denotes the comparator):

$$k = \frac{I_{sp}}{I_{sp}^*} = \frac{\mathscr{A}^* f_\gamma \theta_i \varepsilon (\phi_{th} \sigma_{th} + \phi_e I_0)}{\mathscr{A} f_\gamma^* \theta_i^* \varepsilon^* (\phi_{th} \sigma_{th}^* + \phi_e I_0)} \qquad (4.31)$$

where

$$I_{sp} = \frac{N}{SDCm} \qquad (4.32)$$

and $\mathscr{A}$ is the atomic mass, θ_i the isotopic abundance of the target nucleus, f_γ the absolute frequency of the measured γ-radiation, ε the efficiency of the detector at the γ-energy in question, ϕ_{th} the conventional thermal neutron flux, σ_{th} the cross-section for the capture of thermal neutrons, and ϕ_e the epithermal neutron flux per unit $\ln E$ neutron energy interval;

$$I_0 = \int_{E_{Cd}}^{\infty} \sigma(E) \frac{dE}{E}$$

the resonance integral together with the so-called "$1/v$" contribution (where v is the velocity of neutrons), $E_{Cd} = 0.55$ eV, the cadmium cut-off energy, N the area of the peak of the total energy of the measured γ-radiation; S denotes the saturation factor, according to Eq. (4.24), $D = \exp(-\lambda t_c)$ the so-called "cooling" factor, $C = \dfrac{1 - \exp(-\lambda t_m)}{\lambda}$ the so-called correction factor of the measurement, correcting the measured peak area for unit time and for disintegration per the t_m measuring interval, m is the mass of the given element.

Since k is the function of the neutron energy spectrum of the radiation source (as a matter of fact, of the ratio of the ϕ_{th}/ϕ_e thermal/epithermal fluxes) the k factors obtained from Eqs (4.31)–(4.32) are valid for a given site of irradiation. De Corte et al. have made the method generally applicable by showing that the k_{ref} factors determined with a so-called reference irradiation channel can be converted for any other so-called analytical site of irradiation by means of Eq. (4.33):

$$k_{anal} = k_{ref} \frac{\left[\left(\dfrac{\phi_{th}}{\phi_e}\right)_{anal} + \left(\dfrac{I_0}{\sigma_{th}}\right)\right]\left[\left(\dfrac{\phi_{th}}{\phi_e}\right)_{ref} + \left(\dfrac{I_0}{\sigma_{th}}\right)^*\right]}{\left[\left(\dfrac{\phi_{th}}{\phi_e}\right)_{anal} + \left(\dfrac{I_0}{\sigma_{th}}\right)^*\right]\left[\left(\dfrac{\phi_{th}}{\phi_e}\right)_{ref} + \left(\dfrac{I_0}{\sigma_{th}}\right)\right]} \tag{4.33}$$

The application of Eq. (4.33) requires the knowledge of the ϕ_{th}/ϕ_e flux ratios. This can be determined on the basis of the measurement of the so-called *cadmium ratio*, using Eq. (4.34):

$$R_{Cd} = \frac{I_{sp}}{(I_{sp})_{Cd}} = \frac{\phi_{th}}{\phi_e} \frac{\sigma_{th}}{I_0} + 1 \tag{4.34}$$

where $(I_{sp})_{Cd}$ is the specific count rate of the peak of total energy of the sample irradiated in a cadmium holder (with usually 1 mm wall thickness). Mostly [197]Au isotope is used for this measurement, having an I_0/σ_{th} ratio of 15.7, which is a generally accepted reference value.

The methods recommended for *determination* of the $(\phi_{th}/\phi_e)_{anal}$ in Eq. (4.33) have developed parallel to the comparator method itself:

— in the *two-foil method* two isotopes with considerably different I_0/σ_{th} ratio are irradiated together with the unknown sample (e.g., [197]Au and [59]Co);

— in the *multiisotopic foil method* such an element is used as a flux ratio monitor (and if possible at the same time as a comparator, too) which forms several radioisotopes. A necessary precondition is here, too, that the I_0/σ_{th} ratio of these isotopes should be as far from each other as possible. Such elements are ruthenium, tin, zirconium, etc.

The introduction of so-called k_0 factors further simplifies and widens the method. The k_0 factors are to be determined experimentally, by measuring relative intensities. They can be regarded as combined nuclear constants according to Eq. (4.35), and their application eliminates the necessity of determination of the k_{ref} factors. Apart from the known value of I_0/σ_{th} and the experimental value of $(\phi_{th}/\phi_e)_{anal}$, the only necessary function is the relative efficiency curve of the semiconductor detector used for measurement (see Eq. 4.36).

$$k_0 = \frac{\mathscr{A}^*\theta_i f_\gamma \sigma_{th}}{\mathscr{A}\theta_i^* f_\gamma^* \sigma_{th}^*} = \frac{\varepsilon^* I_{sp}\left[\left(\frac{\phi_{th}}{\phi_e}\right)_{ref} + \left(\frac{I_0}{\sigma_{th}}\right)^*\right]}{\varepsilon I_{sp}^*\left[\left(\frac{\phi_{th}}{\phi_e}\right)_{ref} + \left(\frac{I_0}{\sigma_{th}}\right)\right]}$$ (4.35)

and

$$k_{anal} = k_0 \frac{\left(\frac{\phi_{th}}{\phi_e}\right)_{anal} + \left(\frac{I_0}{\sigma_{th}}\right)}{\left(\frac{\phi_{th}}{\phi_e}\right)_{anal} + \left(\frac{I_0}{\sigma_{th}}\right)^*} \times \frac{\varepsilon}{\varepsilon^*}$$ (4.36)

One or the other version of the above-outlined method can be used for the analysis of trace elements in most various samples, e.g., *pure metals*, *geological samples* and *biological substances*.

The minimum detection limit *of a substance* (m_{min}) depends on the specific activity of the isotope formed and on the minimal activity detectable by the instrument:

$$m_{min} = \frac{A_{min}}{a}$$ (4.37)

The minimal detectable activity is determined by the background measured by the scaler, the confidence level and by the time of measurement:

$$A_{min} = k\sqrt{\frac{2I_{rel,B}}{t_m}}$$ (4.38)

where k is a constant, $I_{rel,B}$ the background, imp/(s µg), and t_m is the measuring time.

The specific count rate of the element to be measured with respect to 1µg irradiated substance is:

$$i = \frac{I_{rel}}{m} \times 10^{-6} = \frac{N_A \sigma \theta \phi}{\mathscr{A}} SD\varepsilon\eta \times 10^{-6} \text{ imp/(s µg)}$$ (4.39)

where m is the mass of the irradiated sample, N_A the Avogadro number $= 6 \times 10^{23}$ 1/mole, σ the effective microscopic cross-section, m^2, θ the natural isotopic ratio, ϕ the irradiation flux, 1/(m^2 s), $\mathscr{A}$ the atomic mass of the atom investigated; $S = 1 - \exp\left(-\frac{\ln 2}{t_{1/2}} t_i\right)$ the saturation factor, $D = \exp(-\lambda t_c)$ the "cooling" factor, ε the absolute efficiency of the detector in the given system, η the ratio of the measured radiation intensity to the whole radiation intensity, t_i the time of irradiation, s, and t_c the "cooling" time, s.

EXAMPLE 4.11

Copper has two stable isotopes; upon irradiation, both give isotopes with well-measurable γ-radiation. Calculating the sensitivities from Eq. (4.39) for both isotopes, let us state the measurement of which isotope can be performed with higher sensitivity.

For the nuclear reaction $^{63}Cu(n, \gamma)^{64}Cu$:

$$\theta = 69.1\%; \qquad \sigma = 4.1 \times 10^{-28} \, m^2 \; (4.1 \; barn); \qquad t_{1/2} = 12.8 \, h$$

$$E_\gamma = 0.511 \; MeV \; annihilation \; radiation; \qquad \eta = 2.19\%;$$

$$\varepsilon = 10\%; \qquad t_i = 10 \, h; \qquad t_c = 1 \, h; \qquad \phi = 2 \times 10^{17} \; 1/(m^2 \, s)$$

$$t_m = 1000 \, s$$

The count rate per mass unit (i) will be:

$$i = \frac{6 \times 10^{23} \times 4.1 \times 10^{-28} \times 0.691 \times 2 \times 10^{17}}{63.54} \times \left[1 - \exp\left(-\frac{0.693 \times 10}{12.8} \right) \right] \times$$

$$\times \exp\left(-\frac{0.693 \times 1}{12.8} \right) \times 0.1 \times 0.0219 \times 10^{-6} \; imp/(s \, \mu g) = 467.53 \; imp/(s \, \mu g)$$

When the efficiency of $\varepsilon = 10\%$ was given, it was assumed that the detector is a 75×75 mm NaI(Tl) scintillation counter. In order to eliminate background radiation, the detector is usually placed into a $80 \times 80 \times 80$ cm inner dimension iron cage having a 10 cm wall thickness. The measured background with the given system is in the 200 keV–2 MeV range

$$I_{rel, B, total} \approx 150 \; imp/min$$

Of the background determined in the $2000—200 = 1800$ keV energy range, a fraction corresponding to the double of the Γ half-width value is the share of the peak to be measured. Assuming a detector with a resolving power of 9%, the range to be evaluated for the 662 keV line of ^{137}Cs is $2 \times 60 = 120$ keV, thus the average background under the peak is

$$I_{rel, B} = \frac{150}{1800} \, 120 = 10 \; imp/min = 0.16 \; imp/s$$

The energy dependence of the background should also be considered for more accurate calculations.

From Eq. (4.38), on the basis of the data already calculated, we have:

$$I_{rel, min} = 2 \frac{2 \times 0.16}{1000} = 3.6 \times 10^{-2} \; imp/s$$

since, with a 95% confidence, $k = 2$.

$$m_{min} = \frac{I_{rel, min}}{i} = \frac{3.6 \times 10^{-2}}{467.53} = 77.1 \times 10^{-12} \, g = 77.1 \; pg$$

The course of calculation is the same for the $^{65}Cu(n, \gamma)^{66}Cu$ reaction. The initial data are: $\theta = 30.91\%; \sigma = 2 \times 10^{-28} \, m^2$ (2.0 barn); $t_{1/2} = 5.1$ min; $E_\gamma = 1.04$ MeV; $\eta = 9\%; \varepsilon = 5\%;$ $t_i = t_c = 5$ min; $\phi = 2 \times 10^{17} \; 1/(m^2 \, s)$.

On the basis of Eq. (4.39):

$$i = \frac{6 \times 10^{23} \times 2 \times 10^{-28}}{63.54} \times 0.31 \times 2 \times 10^{17} \times$$

$$\times \left[1 - \exp\left(-\frac{0.693 \times 300}{5.1 \times 60} \right) \right] \exp\left(-\frac{0.693 \times 300}{5.1 \times 60} \right) \times$$

$$\times 0.05 \times 0.09 \times 10^{-6} = 5.85 \times 10^{-7} \times 2 \times 10^{17} \times 0.49 \times$$

$$\times 0.51 \times 0.05 \times 0.09 \times 10^{-6} = 131.32 \; imp/(s \; \mu g)$$

$$m_{min} = \frac{I_{rel,\,min}}{i} = \frac{3.6 \times 10^{-2}}{131.32} = 2.74 \times 10^{-4} \; \mu g$$

To sum up, with the same neutron flux and measuring system, by irradiating copper isotopes nearly to their half-life, the sensitivity of the determination is higher if the $^{63}Cu(n, \gamma)^{64}Cu$ reaction is used.

Disturbing nuclear reactions can increase the error of activation analytical determinations considerably, or decrease their sensitivity.

In the case of *primary disturbing reactions*, contaminations give rise to the same radioactive isotope as the element to be determined, e.g.:

$$^{27}Al(n, p)^{27}Mg, \qquad and \qquad ^{26}Mg(n, \gamma)^{27}Mg$$

$$^{23}Na(n, \gamma)^{24}Na, \qquad and \qquad ^{24}Mg(n, p)^{24}Na$$

$$^{28}Si(n, p)^{28}Al, \qquad and \qquad ^{27}Al(n, \gamma)^{28}Al$$

$$^{65}Cu(n, \gamma)^{66}Cu, \qquad and \qquad ^{69}Ge(n, \alpha)^{66}Cu$$

Secondary disturbing reactions are those in which the radioactive disintegration following the nuclear reaction gives such a stable isotope whose further activation produces a disturbing isotope; e.g., the nuclear reaction $^{75}As(n, \gamma)^{76}As$ is disturbed by ^{76}As produced by the activation of germanium:

$$^{74}Ge(n, \gamma)^{75}Ge \xrightarrow[82 \, min]{\beta^-} {}^{75}As(n, \gamma)^{76}As$$

The two origins of ^{76}As cannot be distinguished here, but it can be calculated how much ^{76}As is produced in the secondary disturbing reaction.

Another example:

$$^{30}Si(n, \gamma)^{31}Si \xrightarrow{\beta^-} {}^{31}P(n, \gamma)^{32}P; \qquad and \qquad ^{31}P(n, \gamma)^{32}P$$

Errors arising from the *self-absorption of neutrons* can be attributed to the presence of elements with high capture cross-section in the sample.

Consideration of the disturbing effect of *fission products* is required when uranium and thorium compounds are investigated. Owing to the high number of fission products, several elements must be considered. For example, the determination of molybdenum by the nuclear reaction $^{98}Mo(n, \gamma)^{99}Mo$ is disturbed by the reaction $^{235}U(n, f)^{99}Mo$.

EXAMPLE 4.12

Neutron-induced nuclear reactions give rise to the formation of ^{28}Al from both aluminium and silicon. Activities originating from each element are determined by the so-called *double irradiation technique*. The sample to be analyzed is irradiated by *fluxes of different neutron energy distribution and the activities are measured*. The aluminium and silicium content of the sample can thus be determined separately.

In a simple case, the different energy distribution is effected by irradiating the sample without a *cadmium casing*, then, after complete decay of the induced activity, once again in a cadmium casing. The following system of equations helps to determine the aluminium and silicon content of the sample:

$$a_{Al}m_{Al} + a_{Si}m_{Si} = A - A_B \tag{4.40}$$

$$a_{Al, epit}m_{Al} + a_{Si, epit}m_{Si} = A_{epit} - A_{B, epit} \tag{4.41}$$

Here a_{Al}, a_{Si}, and $a_{Al, epit}$ and $a_{Si, epit}$ denote the measured activity of the 1 µg aluminium and silicon standard irradiated without and with cadmium casing, respectively; m_{Al} and m_{Si} the aluminium and silicon content of the sample, mg; A and A_{epit} the activities of the samples irradiated without and with cadmium casing, respectively, A_B and $A_{B, epit}$ the respective background activities.

Solving the system of equations for m_{Si}:

$$m_{Si} = \frac{(A_{epit} - A_{B, epit})a_{Al} - (A - A_B)a_{Al, epit}}{a_{Al}a_{Si, epit} - a_{Al, epit}a_{Si}} \tag{4.42}$$

The minimum detectable amount of silicon cannot be calculated in this case from the usual Eq. (4.25), because the sensitivity of silicon determination is limited by the reaction $^{27}Al(n, \gamma)^{28}Al$ leading to the same radionuclide, i.e., by the aluminium content of the sample.

The error of m_{Si} calculated by Eq. (4.42) can be determined on the basis of Gauss' law on the quadratic error distribution. If the statistical error of the standards is neglected, and it is assumed that the minimum detectable amounts are

$$m_{Al, min} = \alpha \sigma m_{Al} \qquad \text{and} \qquad m_{Si, min} = \alpha \sigma m_{Si}$$

the value of, e.g., $m_{Si, min}$ is given by the positive solution of a quadratic equation; σ is the standard deviation and α the reciprocal value of the desired relative standard deviation (Currie).

$$m_{Si, min}^2(a_{Al, epit}a_{Si} - a_{Al}a_{Si, epit})^2 - \alpha m_{Si, min}(a_{Si}a_{Al, epit}^2 + a_{Si, epit}a_{Al}^2) -$$
$$- \alpha^2[m_{Al}a_{Al}a_{Al, epit}(a_{Al} + a_{Al, epit}) + 2a_{Al, epit}^2a_B + 2a_{Al}^2a_{B, epit}] = 0 \tag{4.43}$$

Knowing the parameters to be measured, the values of $m_{Si, min}$ can be calculated as a function of the values of m_{Al}.

Let us assume that the following pulse counts were measured in a given experimental arrangement:

$$i_{Al} = 2460 \text{ imp/(µg 100 s)}$$

$$i_{Al, epit} = 720 \text{ imp/(µg 100 s)}$$

$$i_{Si} = 5.3 \text{ imp/(µg 100 s)}$$

$$i_{Si, epit} = 210 \text{ imp/(µg 100 s)}$$

$$i_H = 200 \text{ imp/(µg 100 s)}$$

$$i_{H, epit} = 300 \text{ imp/(µg 100 s)}$$

$$\alpha = 10$$

Let us solve the quadratic Eq. (4.42) in order to determine the detection limit of Si:

$$2.66 \times 10^{11} m_{Si, min}^2 - 400 \times 1.28 \times 10^9 m_{Si, min} -$$

$$- 400 \times (5.65 \times 10^9 m_{Al} + 1.92 \times 10^9) = 0$$

$$m_{Si, min} = \frac{5.14 + \sqrt{106.9 + 240 m_{Al}}}{5.32}$$

If the sample contains no aluminium, i.e., $m_{Al} = 0$, then

$$m_{Si, min} = 2.9 \ \mu g$$

Analogously, the detection limit of Si in the presence of 10 μg Al is 10.4 μg, with 100 μg Al, 30 μg. Obviously, the increase of the concentration of the disturbing element is disadvantageous for the detection limit.

A similar calculation can be carried out for the minimum detection limit of aluminium as a function of the silicon concentration.

4.2.1.2. γ-ACTIVATION

The technique of activation analysis carried out by *high energy γ-photons* is similar to that of the neutron activation. The most frequent nuclear reactions induced by high energy photons are (γ, n) processes. The maximum cross-section with light elements is at 20–25 MeV, with other elements, at about 15 MeV. The method is especially well-suited for the relatively sensitive determination of low atomic number elements. It has the advantage that by changing the photon energy, given nuclear reactions *can be induced selectively*, i.e., the determination is selective, if the threshold energies are selected appropriately. Although investigations with high energy γ-photons are often very sensitive, the measurement of most elements can be carried out with a lower sensitivity than with neutron activation.

In some cases, the determination of elements difficult to determine by other methods is possible. Such elements are fluorine, oxygen, carbon (see also in Section 5.2.3). It is a disadvantage that usually positron-emitting isotopes are produced, always giving rise to an annihilation γ-radiation of 0.51 MeV energy. This can be detected easily, but if more than one β^+-radiating isotope is produced in the sample, the determination of the half-life is also necessary for identification. The determination of impurities in elements of high neutron absorption cross-section is more advantageous by this method than by neutron activation analysis, because the self-absorption of the matrix should not be considered here.

One of the variants of activation analysis based on (γ, n) nuclear reactions is the method where the *neutrons produced during nuclear reactions are measured* instead of determining the radioactivity of the isotope formed (e.g., deuterium measurement by means of a ^{24}Na or ^{228}Th γ-radiation source).

High activity γ-radiation sources with *lower energy (about 1 MeV)* can also be used for analytical purposes. The theoretical basis of these methods is that in the process of inelastic scattering of γ-photons on the nucleus, the latter transforms into its excited isomeric state according to the equation: A(γ, γ')A. Such an effect was shown in the determination of about 25 elements (e.g. Ag, Br, Se, Sr by means of the γ-radiation of ^{60}Co, ^{116}In and ^{182}Tl). The advantage of the method is its favourable selectivity and, as a rule, the lack of disturbing reactions; the low sensitivity of the measurement is a disadvantage. By increasing the activity of the γ-radiation sources and the dose rate of the radiation sometimes a sensitivity as low as in the range of a few µg-s can be attained.

4.2.1.3. DETERMINATIONS BY MEANS
OF CHARGED PARTICLES

Activation by charged particles is one of the rapidly developing fields of activation analysis.

An ion beam produced by an *accelerator* is used for the determination. The beam collides after the necessary focussing into the target and induces a reaction of A(a, b)B type where A is the nucleus of the sample, B that of the product, (a) the particle of the beam, (b) the particle(s) produced promptly on the reaction. The bombarding particle may be a proton (p), deuteron (d), triton (t), $^3He^{2+}$ or $^4He^{2+}$ (α-particle).

The threshold energy is an important characteristic of nuclear reaction induced by charged particles; the amount of nuclear reactions is a function also of the flux *density* of the beam (in practice, the ion current is usually given in µA which can be measured more easily), of its energy, the irradiation time, of the thickness and density of the matrix therefore the specific activity decreases toward the bulk of the sample. The integral cross section of the nuclear reaction, as well as the specific energy loss of the bombarding ion in the substance of the target should also be taken into consideration. Since the mathematical calculation of the specific energy loss of the bombarding particles and that of the consequent changes of cross-section is difficult, the actual reaction rates are determined empirically.

The requirement from the *accelerators* is that the particle beam should be stable, well reproducible, occasionally of variable energy (between 0 and 50 MeV); it should be well focussed, have an identical flux rate within a well-defined cross-section with sharp limits and even the type of the particles should be changeable according to the purpose. Analysis by accelerators puts forward strict requirements concerning the physical properties of the sample to be investigated, too; thermal and electrical conductivity and smooth surface are important. If an average concentration with respect to the whole sample has to be given, it is important to ensure that the impurities have a homogeneous distribution within the sample, or measurement of the inhomogeneity within the sample should be possible. Since the corresponding facilities are at present very expensive, their application in activation analysis has a secondary importance as yet.

The detection of the reaction products can be realized by means of the measurement of the annihilation radiation, because the products are in most cases β^+-radiating isotopes. The simultaneous determination of several β^+-emitting isotopes is possible by decomposing the combined decay curves graphically, or using a computer provided that the half-lives are sufficiently different (see Section 4.2.3.2). In cases when the isotope to be determined is stable or has a very long half-life, the detection of prompt particles is more suitable: the sensitivity of the measurement can be as low as 1 g/m^2 in the determination of surface oxygen and carbon. Prompt protons can be detected by a semiconductor silicon detector, neutrons e.g., by a BF_3 counter.

The range of application is wide; the sensitivity of determination of light elements, such as beryllium, boron, fluorine, sulphur, nitrogen, oxygen and carbon is below about $10^{-6}\%$, that of phosphorus, calcium, chlorine, nitrogen and silicon, about below $10^{-5}\%$, if the energy of the beam is as high as 20 MeV and the ion current is at least 50 µA. The determination of surface layer thickness is a frequent task.

It is possible to produce charged particles in a *nuclear reactor* by means of secondary nuclear reactions, induced by high-energy charged particles produced in neutron-induced primary processes (see Section 4.2.1.1). The particle flux produced in this way is generally lower by 3–5 orders of magnitude than that of accelerators. This causes a considerable decrease of the sensitivity of the analysis. The most frequent neutron reaction used to produce charged particles is the process $^6Li(n, \, ^3H^+)^4He^{2+}$ giving rise to α- and $^3H^+$-particles, as well as the reaction $^{10}B(n, \, \alpha)^7Li$ producing also α-particles. Triton produced from ^6Li (2.75 MeV) is often used in the determination of oxygen. The powdered sample has to be mixed with LiF powder, because the range of triton is low in the sample (about 30 μm). To determine the surface oxygen in GaAs, LiF powder is suspended in polystyrene swollen by benzene and applied as a layer; after irradiation it is removed by dissolving it in benzene. The process has the advantage that the oxygen of Li_2CO_3 always present in considerable amounts in LiF does not interfere with the analysis.

4.2.2. RADIATION SOURCES USED FOR ACTIVATION ANALYSIS

The main characteristics of neutron sources used for activation analysis are shown in Table 4.4.

Table 4.4. Neutron sources for activation analysis

Type of the source	Character of radiation	Average intensity of radiation
Nuclear reactor		
Experimental nuclear reactor	Slow and fast (about 2 MeV) neutrons	10^{16}–10^{18} 1/(m² s)
Training reactor	Slow and fast neutrons	10^{15}–10^{16} 1/(m² s)
Pulse reactor	Slow and fast neutrons	10^{20}–10^{22} 1/(m² s)
Neutron generators		
Ng-2 neutron generator	Fast (14 MeV) and slow neutrons	10^{10} 1/s 10^8 1/s
Closed tube generator (Soviet NG-2)	Fast (14 MeV) and slow neutrons	10^8 1/s 10^7 1/s
Isotopic neutron sources		
^{210}Po–Be; ^{241}Am–Be; ^{239}Pu–Be	Slow and fast (1–6 MeV) neutrons	10^5–10^8 1/s
Neutron multipliers	Slow and fast (below cadmium) neutrons	10^5–10^8 1/s
^{252}Cf (1 mg)	Fission spectrum	10^8–10^9 1/s
Accelerators		
Linear accelerator	p, d, He, γ-radiation	10^{11} 1/s
Betatron (30 MeV)	Fast neutron, γ-radiation	20 mGy/s 0.3 mGy/s
Microtron (30 MeV)	Fast neutron γ-radiation	10^{11} 1/s, 300 mGy/s 40 mGy/s
Van de Graaf accelerator (5 MeV)	Proton, deutron, helium	10^{14} 1/s 10^{13} 1/s
Cyclotron (25 MeV)	He, p, deuteron	10^{14}–10^{15} 1/s

Nuclear reactors. It has been shown above (see e.g., Eqs (4.23), (4.24)) that the sensitivity of activation analysis is directly proportional to the flux *density* of the bombarding particles; highest fluxes can be obtained in the core of nuclear reactors. Reactors are used most often as radiation sources for activation analysis. The thermal neutron flux produced by reactors is about 10^{17} $1/(m^2 s)$ in most cases, but some special reactors may reach the value of 10^{19} $1/(m^2 s)$. Table 4.5 shows the sensitivity of neutron activation analysis if the flux is as high as 10^{17} $1/(m^2 s)$.

In addition to the overall value of the *neutron flux*, its *energy* and *distribution in space and time* are also important. Nuclear fission produces fast neutrons that are slowed down by the moderator; therefore we can distinguish in, or near to, the active zone of any reactor

Table 4.5. Detection limits for some elements by a thermal reactor
$\Phi_{th} = 10^{17}$ $1/(m^2 s)$

Atomic number	Element	Nuclear reaction	Half-life	Sensitivity, μg
1	Hydrogen	$^2H(n,\gamma)^3H$	12.26 y	—
3	Lithium	$^7Li(n,\gamma)^8Li$	0.84 s	—
4	Beryllium	$^9Be(n,\gamma)^{10}Be$	2.5×10^6 y	—
5	Boron	$^{11}B(n,\gamma)^{12}B$	0.027 s	—
6	Carbon	$^{13}C(n,\gamma)^{14}C$	5730 y	—
7	Nitrogen	$^{15}N(n,\gamma)^{16}N$	7.35 s	—
8	Oxygen	$^{18}O(n,\gamma)^{19}O$	29.4 s	—
9	Fluorine	$^{19}F(n,\gamma)^{20}F$	11 s	3×10^{-3}
10	Neon	$^{22}Ne(n,\gamma)^{23}Ne$	20.4 s	9×10^{-2}
11	Sodium	$^{23}Na(n,\gamma)^{24}Na$	15.05 h	8×10^{-6}
12	Magnesium	$^{26}Mg(n,\gamma)^{27}Mg$	9.5 min	3×10^{-3}
13	Aluminium	$^{27}Al(n,\gamma)^{28}Al$	2.3 min	3×10^{-5}
14	Silicon	$^{30}Si(n,\gamma)^{31}Si$	2.6 h	4.0
15	Phosphorus	$^{31}P(n,\gamma)^{32}P$	14.28 d	6×10^{-5}
16	Sulphur	$^{34}S(n,\gamma)^{35}S$	87.9 d	2×10^{-3}
17	Chlorine	$^{37}Cl(n,\gamma)^{38}Cl$	37.5 min	2×10^{-4}
18	Argon	$^{40}Ar(n,\gamma)^{41}Ar$	1.8 h	2×10^{-5}
19	Potassium	$^{41}K(n,\gamma)^{42}K$	12.36 h	3×10^{-4}
20	Calcium	$^{48}Ca(n,\gamma)^{49}Ca$	8.8 min	4×10^{-3}
21	Scandium	$^{45}Sc(n,\gamma)^{46}Sc$	83.9 d	5×10^{-7}
22	Titanium	$^{50}Ti(n,\gamma)^{51}Ti$	5.8 min	2×10^{-3}
23	Vanadium	$^{51}V(n,\gamma)^{52}V$	3.75 min	3×10^{-6}
24	Chromium	$^{50}Cr(n,\gamma)^{51}Cr$	27.8 d	2×10^{-4}
25	Manganese	$^{55}Mn(n,\gamma)^{56}Mn$	2.58 h	3×10^{-6}
26	Iron	$^{58}Fe(n,\gamma)^{59}Fe$	45.6 d	8×10^{-3}
27	Cobalt	$^{59}Co(n,\gamma)^{60m}Co$	10.5 min	3×10^{-5}
28	Nickel	$^{64}Ni(n,\gamma)^{65}Ni$	2.56 h	3×10^{-3}
29	Copper	$^{63}Cu(n,\gamma)^{64}Cu$	12.8 h	1×10^{-5}
30	Zinc	$^{68}Zn(n,\gamma)^{69m}Zn$	13.8 h	9×10^{-4}
31	Gallium	$^{71}Ga(n,\gamma)^{72}Ga$	14.1 h	1×10^{-6}
32	Germanium	$^{74}Ge(n,\gamma)^{75}Ge$	1.33 h	6×10^{-4}
33	Arsenic	$^{75}As(n,\gamma)^{76}As$	26.5 h	5×10^{-5}
34	Selenium	$^{80}Se(n,\gamma)^{81m}Se$	57 min	6×10^{-6}
35	Bromine	$^{81}Br(n,\gamma)^{82}Br$	35.9 h	2×10^{-5}
36	Krypton	$^{82}Kr(n,\gamma)^{83m}Kr$	1.88 h	4×10^{-5}
37	Rubidium	$^{87}Rb(n,\gamma)^{88}Rb$	17.8 min	2×10^{-3}
38	Strontium	$^{86}Sr(n,\gamma)^{87m}Sr$	2.8 h	2×10^{-4}

Atomic number	Element	Nuclear reaction	Half-life	Sensitivity, µg
39	Yttrium	$^{89}Y(n, \gamma)^{90}Y$	64 h	3×10^{-5}
40	Zirconium	$^{96}Zr(n, \gamma)^{97}Zr$	17 h	1×10^{-2}
41	Niobium	$^{93}Nb(n, \gamma)^{94m}Nb$	6.6 min	6×10^{-3}
42	Molybdenum	$^{100}Mo(n, \gamma)^{101}Mo$	14.6 min	4×10^{-4}
44	Ruthenium	$^{104}Ru(n, \gamma)^{105}Ru$	4.5 h	1×10^{-4}
45	Rhodium	$^{103}Rh(n, \gamma)^{104m}Rh$	4.5 min	3×10^{-6}
46	Palladium	$^{108}Pd(n, \gamma)^{109}Pd$	13.5 h	7×10^{-5}
47	Silver	$^{109}Ag(n, \gamma)^{110m}Ag$	255 d	2×10^{-5}
48	Cadmium	$^{114}Cd(n, \gamma)^{115}Cd$	53 h	3×10^{-4}
49	Indium	$^{115}In(n, \gamma)^{116}In$	54 min	3×10^{-7}
50	Tin	$^{122}Sn(n, \gamma)^{123}Sn$	40 min	3×10^{-3}
51	Antimony	$^{121}Sb(n, \gamma)^{122}Sb$	2.8 d	1×10^{-5}
52	Tellurium	$^{130}Te(n, \gamma)^{131}Te$	25 min	5×10^{-4}
53	Iodine	$^{127}I(n, \gamma)^{128}I$	25 min	3×10^{-5}
54	Xenon	$^{132}Xe(n, \gamma)^{133}Xe$	2.3 d	3×10^{-4}
55	Cesium	$^{133}Cs(n, \gamma)^{134m}Cs$	3.2 h	2×10^{-4}
56	Barium	$^{138}Ba(n, \gamma)^{139}Ba$	1.42 h	7×10^{-5}
57	Lanthanum	$^{139}La(n, \gamma)^{140}La$	40.2 h	4×10^{-6}
58	Cerium	$^{140}Ce(n, \gamma)^{141}Ce$	33.1 d	1×10^{-4}
59	Praseodymium	$^{141}Pr(n, \gamma)^{142}Pr$	19.2 h	8×10^{-5}
60	Neodymium	$^{148}Nd(n, \gamma)^{149}Nd$	1.8 h	5×10^{-5}
62	Samarium	$^{152}Sm(n, \gamma)^{153}Sm$	47 h	1×10^{-6}
63	Europium	$^{151}Eu(n, \gamma)^{152}Eu$	9.2 h	6×10^{-7}
64	Gadolinium	$^{158}Gd(n, \gamma)^{159}Gd$	18 h	2×10^{-4}
65	Terbium	$^{159}Tb(n, \gamma)^{160}Tb$	72.1 d	5×10^{-6}
66	Dysprosium	$^{164}Dy(n, \gamma)^{165}Dy$	2.32 h	4×10^{-6}
67	Holmium	$^{165}Ho(n, \gamma)^{166}Ho$	27.3 h	8×10^{-5}
68	Erbium	$^{170}Er(n, \gamma)^{171}Er$	7.5 h	3×10^{-5}
69	Thulium	$^{169}Tm(n, \gamma)^{170}Tm$	134 d	4×10^{-6}
70	Ytterbium	$^{174}Yb(n, \gamma)^{175}Yb$	4.2 d	2×10^{-5}
71	Lutetium	$^{176}Lu(n, \gamma)^{177}Lu$	6.8 d	4×10^{-6}
72	Hafnium	$^{179}Hf(n, \gamma)^{180m}Hf$	5.5 h	8×10^{-6}
73	Tantalum	$^{181}Ta(n, \gamma)^{182}Ta$	16.5 min	5×10^{-3}
74	Tungsten	$^{186}W(n, \gamma)^{187}W$	24 h	1×10^{-5}
75	Rhenium	$^{185}Re(n, \gamma)^{186}Re$	3.8 d	5×10^{-6}
76	Osmium	$^{192}Os(n, \gamma)^{193}Os$	32 h	4×10^{-4}
77	Iridium	$^{193}Ir(n, \gamma)^{194}Ir$	19 h	8×10^{-6}
78	Platinum	$^{196}Pt(n, \gamma)^{197}Pt$	18 h	7×10^{-3}
79	Gold	$^{197}Au(n, \gamma)^{198}Au$	2.7 d	5×10^{-7}
80	Mercury	$^{196}Hg(n, \gamma)^{197m}Hg$	24 h	1×10^{-4}
81	Thallium	$^{205}Tl(n, \gamma)^{206}Tl$	4.2 min	1×10^{-3}
82	Lead	$^{206}Pb(n, \gamma)^{207m}Pb$	0.8 s	—
83	Bismuth	$^{209}Bi(n, \gamma)^{210}Bi$	5.01 d	2×10^{-2}
90	Thorium	$^{232}Th(n, \gamma)^{233}Th$	27 d	1×10^{-5}
92	Uranium	$^{238}U(n, \gamma)^{239}U$	23.5 min	1×10^{-5}

Comments:

1. Nuclear reactions induced by thermal neutrons with the maximum yields and minimum disturbing effect are shown; all processes are (n, γ) reactions. In some cases (such as B, Be, C, Fe, N, O, P, Si) the reactions given are unfavourable or unpracticable. In these cases it is advisable to use the fast neutron spectrum of the reactor.

2. Isotopes with very short half-life are not shown except for B, Li and Pb.

3. Sensitivity data (detection limits) are disturbance-free values extrapolated from actual measurements; detector: 75×75 mm NaI(Tl) crystal.

the following classes: *fast* neutron flux (above 0.1 MeV), *epithermal* flux (between 0.2 eV and 0.1 MeV) and *thermal neutron flux* (below 0.2 eV). Figure 4.26 shows the energy distribution of the neutron flux of a research reactor. Since quantitative activation analysis used in most cases a relative method, i.e., the activity of the sample is compared with the activity of a standard irradiated under identical conditions, accurate and reproducible results can only be obtained if the distribution in space and time of the reactor flux is constant, or if we know the eventual changes exactly. The separation of the sample and the standard in space may cause considerable errors even if they are irradiated simultaneously.

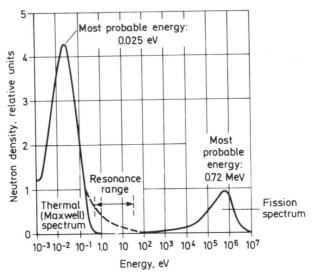

Fig. 4.26. Energy spectrum of reactor neutrons in a nuclear reactor moderated with light water

The irradiations can be carried out in the *channels of the reactor* used also for isotope production, or in *pneumatic or hydraulic systems* built specially for this purpose. The dimensions and cooling of the reactor channels are, as a rule, satisfactory and several samples can be irradiated simultaneously, even for a longer time. The removal of the samples from the channels and their delivery to the laboratory is difficult without a pneumatic system; therefore, this technique is suitable for activation of isotopes with half-lives of about 1 h or longer.

A great advantage of the *pneumatic system* is that it permits one to carry out activation for shorter times and the measurement can be commenced a few seconds after stopping the irradiation. This renders possible the determination of radioactive isotopes with half-lives of a few minutes or even shorter.

Neutron generators (Fig. 4.27) are not too large accelerators producing a deuteron ion current of 1–2 mA intensity by means of a voltage of 100–200 kV. By appropriate selection of the ion source, ion energy and the target, neutrons of different energy can be produced. One of the most frequently used method produces *fast neutrons* of an energy of 14.7 MeV by irradiating a tritium target by deuterium ions, based on the nuclear reaction: $^3_1\text{H}(\text{d}, \text{n})^4_2\text{He}$. If the tritium target (of $5-25 \times 10^{11}$ Bq $= 15-70$ Ci activity) is absorbed by titanium or zirconium, most facilities produce a flux rate of about 10^{12} $1/(\text{m}^2 \text{ s})$.

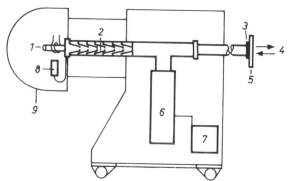

Fig. 4.27. Main units of a neutron generator. *1* — Ion source; *2* — Accelerator tube; *3* — Target holder; *4* — Joint for water cooling of the target; *5* — Sample holder (terminal of the pneumatic system); *6* — Vacuum pump; *7* — Forevacuum pump; *8* — Automated heavy water decomposer (deuterium source); *9* — High voltage source joint

If slow neutrons are to be used, the target is surrounded by paraffin or water. The sample to be irradiated is fed to the neutron generator by a pneumatic system; the same carries the sample after irradiation to the scintillation counter.

High energy and high intensity neutron sources require considerable radiation shielding: Figure 4.28 shows the radiation protection of a 100 kV generator. Heavy concrete walls, concrete bricks and paraffin blocks are used for this purpose. A wall thickness of 1 m decreases the flux of fast neutrons below the permissible value of 10^5 $1/(m^2 \, s)$.

About 60 elements can be determined by irradiation by a neutron generator in the case of macroquantities; the irradiations should not exceed 10 min. About 20 elements can be determined in the range of about 100 μg. Table 4.6 contains the detection limits of elements

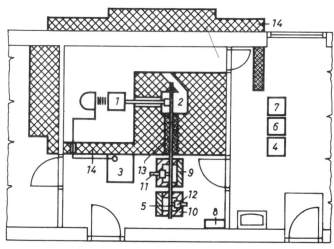

Fig. 4.28. Arrangement of a laboratory for neutron generator activation analysis. *1* — Neutron generator; *2* — Target; *3* — High voltage power source (125 kV); *4* — Generator control unit; *5* — Two-channel sample transporter; *6* — Electronic program control unit; *7* — Multichannel analyzer and data transmitter system; *8* — Pneumatic program control unit; *9* — Low background measuring site for the sample; *10* — Low background measuring site for the monitor; *11* — Sample detector; *12* — Monitor detector; *13, 14* — Shielding

Table 4.6. Detection limit of elements by a neutron generator ($\Phi = 10^{13}$ $1/(m^2$ s), energy: 14 MeV)

Atomic number	Element	Nuclear reaction	Half-life	Sensitivity, μg
3	Lithium	$^7Li(n, d)^6He$	0.83 s	1
4	Beryllium	$^9Be(n, \alpha)^6He$	0.83 s	3
5	Boron	$^{11}B(n, \alpha)^8Li$	0.84 s	1
6	Carbon	$^{12}C(n, p)^{12}B$	20 ms	2
7	Nitrogen	$^{14}N(n, 2n)^{13}N$	10.0 min	8
8	Oxygen	$^{16}O(n, p)^{16}N$	7.2 s	1
9	Fluorine	$^{19}F(n, 2n)^{18}F$	110 min	3
10	Neon	$^{20}Ne(n, p)^{20}F$	11.0 s	0.6
11	Sodium	$^{23}Na(n, \alpha)^{20}F$	11.0 s	0.4
12	Magnesium	$^{24}Mg(n, p)^{24}Na$	15.05 h	11
13	Aluminium	$^{27}Al(n, p)^{27}Mg$	9.5 min	1
14	Silicium	$^{28}Si(n, p)^{28}Al$	2.3 min	0.4
15	Phosphorus	$^{31}P(n, \alpha)^{28}Al$	2.3 min	1
16	Sulphur	$^{34}S(n, \alpha)^{31}Si$	2.6 h	60
17	Chlorine	$^{37}Cl(n, \alpha)^{34}P$	12.5 s	2
18	Argon	$^{40}Ar(n, p)^{40}Cl$	1.4 min	6
19	Potassium	$^{41}K(n, p)^{41}Ar$	1.85 h	2
20	Calcium	$^{44}Ca(n, p)^{44}K$	22 min	140
21	Scandium	$^{45}Sc(n, 2n)^{44}Sc$	4 h	10
22	Titanium	$^{48}Ti(n, p)^{48}Sc$	1.8 d	20
23	Vanadium	$^{51}V(n, p)^{51}Ti$	5.8 min	5
24	Chromium	$^{52}Cr(n, p)^{52}V$	3.77 min	2
25	Manganese	$^{55}Mn(n, \alpha)^{52}V$	3.77 min	5
26	Iron	$^{56}Fe(n, p)^{56}Mn$	2.58 h	6
27	Cobalt	$^{59}Co(n, \gamma)^{60}Co$	10.5 min	1
28	Nickel	$^{58}Ni(n, p)^{58m}Co$	9 h	3
29	Copper	$^{63}Cu(n, 2n)^{62}Cu$	9.9 min	0.5
30	Zinc	$^{64}Zn(n, 2n)^{63}Zn$	38 min	3
31	Gallium	$^{69}Ga(n, 2n)^{68}Ga$	68 min	1
32	Germanium	$^{76}Ge(n, 2n)^{75m}Ge$	49 s	3
33	Arsenic	$^{75}As(n, 2n)^{74}As$	18 d	1
34	Selenium	$^{82}Se(n, 2n)^{81m}Se$	57 min	2
35	Bromine	$^{81}Br(n, 2n)^{80m}Br$	4.5 h	4
36	Krypton	$^{82}Kr(n, \gamma)^{83m}Kr$	1.86 h	70
37	Rubidium	$^{85}Rb(n, 2n)^{84m}Rb$	20 min	0.5
38	Strontium	$^{88}Sr(n, \alpha)^{85m}Kr$	4.4 h	31
39	Yttrium	$^{89}Y(n, 2n)^{88}Y$	108 d	83
40	Zirconium	$^{90}Zr(n, 2n)^{89m}Zr$	4.4 min	7
41	Niobium	$^{93}Nb(n, 2n)^{92}Nb$	10 d	20
42	Molybdenum	$^{100}Mo(n, 2n)^{99}Mo$	66 h	15
44	Ruthenium	$^{96}Ru(n, 2n)^{95}Ru$	99 min	30
45	Rhodium	$^{103}Rh(n, \gamma)^{104m}Rh$	42 s	0.5
46	Palladium	$^{110}Pd(n, 2n)^{109m}Pd$	4.8 min	4
47	Silver	$^{109}Ag(n, 2n)^{108}Ag$	2.4 min	1
48	Cadmium	$^{106}Cd(n, 2n)^{105}Cd$	55 min	25
49	Indium	$^{115}In(n, \gamma)^{116}In$	54 min	0.3
50	Tin	$^{124}Sn(n, 2n)^{123}Sn$	40 min	6
51	Antimony	$^{121}Sb(n, 2n)^{120}Sb$	16 min	1
52	Tellurium	$^{130}Te(n, 2n)^{129}Te$	67 min	4
53	Iodine	$^{127}I(n, \gamma)^{128}I$	25 min	175
54	Xenon	$^{136}Se(n, 2n)^{135m}Xe$	15 min	5
55	Cesium	$^{133}Cs(n, 2n)^{132}Cs$	6.6 d	0.4
56	Barium	$^{138}Ba(n, 2n)^{137m}Ba$	2.6 min	0.5

Atomic number	Element	Nuclear reaction	Half-life	Sensitivity, μg
57	Lanthanum	$^{139}La(n, \gamma)^{140}La$	40.2 h	15
58	Cerium	$^{140}Ce(n, 2n)^{139m}Ce$	55 s	0.4
59	Praseodymium	$^{141}Pr(n, 2n)^{140}Pr$	3.4 min	0.2
60	Neodymium	$^{150}Nd(n, 2n)^{149}Nd$	1.8 h	4
62	Samarium	$^{144}Sm(n, 2n)^{143}Sm$	9 min	15
63	Europium	$^{153}Eu(n, 2n)^{152}Eu$	96 min	9
64	Gadolinium	$^{160}Gd(n, 2n)^{159}Gd$	18 h	35
65	Terbium	$^{159}Tb(n, 2n)^{158m}Tb$	11 s	0.6
66	Dysprosium	$^{164}Dy(n, \gamma)^{165m}Dy$	1.3 min	0.8
67	Holmium	$^{165}Ho(n, 2n)^{164}Ho$	39 min	0.3
68	Erbium	$^{168}Er(n, 2n)^{167m}Er$	2.5 s	4
69	Thulium	$^{169}Tm(n, 2n)^{168}Tm$	86 d	93
70	Ytterbium	$^{174}Yb(n, \gamma)^{175}Yb$	4.2 d	28
71	Lutetium	$^{175}Lu(n, \gamma)^{176}Lu$	3.7 h	2
72	Hafnium	$^{178}Hf(n, \gamma)^{179m}Hf$	19 s	4
73	Tantalum	$^{181}Ta(n, 2n)^{180}Ta$	8.1 h	7
74	Tungsten	$^{182}W(n, \gamma)^{183m}W$	5.5 s	10
75	Rhenium	$^{187}Re(n, 2n)^{186}Re$	90 h	6
76	Osmium	$^{192}Os(n, \gamma)^{193}Os$	31.5 h	400
77	Iridium	$^{191}Ir(n, \gamma)^{192m}Ir$	1.4 min	1
78	Platinum	$^{198}Pt(n, 2n)^{197}Pt$	20 h	7
79	Gold	$^{197}Au(n, 2n)^{196m}Au$	10 h	10
80	Mercury	$^{198}Hg(n, 2n)^{197m}Hg$	25 h	17
81	Thallium	$^{203}Tl(n, 2n)^{202}Tl$	12 d	47
82	Lead	$^{208}Pb(n, 2n)^{207m}Pb$	0.85 s	0.8
83	Bismuth	$^{209}Bi(n, 2n)^{208m}Bi$	2.6 ms	0.9
90	Thorium	$^{232}Th(n, 2n)^{231}Th$	26 h	2
92	Uranium	$^{238}U(n, 2n)^{237}U$	6.75 d	20

Comments:

1. Nuclear reactions with the maximum yields and minimum disturbing effect are shown; they are generally (n, 2n), (n, p) and (n, γ) reactions, except for the (n, d) process for Li and the (n, α) processes for B, Be, Cl, Mn, Na, P, S and Sr.

2. Isotopes with very short half-lives are not shown except for C and Bi.

3. Sensitivity values (detection limits) are calculated data, free of disturbing effects. The assumed detector efficiency is equal to unity.

measurable by means of neutron generators. The sensitivity values are related to a flux of 10^{13} $1/(m^2 \text{ s})$ and to a 75×75 mm NaI(Tl) detector.

The selectivity of the determination can be increased if the activation is carried out by *neutrons of different energies*. By changing the material of the target, the type or the energy of the accelerated positive ions, it is possible to produce nearly monoenergetic neutrons in the energy range between a few keV and several MeV. For example, if fluorine is determined by means of 14 MeV neutrons in the presence of oxygen, the necessary ^{16}N isotope is produced not only in the reaction $^{19}F(n, \alpha)^{16}N$, but also in the process $^{16}O(n, p)^{16}N$. If the bombardment is carried out by 10 MeV neutrons, oxygen does not interfere any more, because a 10 MeV neutron energy is sufficient to effect the $^{19}F(n, \alpha)^{16}N$ reaction, but the activation of oxygen requires neutron energies above 10.4 MeV. The fluorine content of oxygen can thus also be determined in this way.

Isotopic neutron sources. The most frequent types utilize an (α, n) nuclear reaction; their application is, for the time being, limited because most sources available at present have much lower neutron fluxes than neutron generators (Section 1.4.2.4). Isotopic neutron sources are well-suited for direct industrial application, e.g., for the continuous monitoring of bauxite or stainless steel, because the handling of the source is simple.

Nuclear reactions with relatively low threshold energies are suitable neutron sources. Such is the reaction $^9Be(\gamma, n)^8Be$ with a threshold energy as low as 1.67 MeV. This reaction permits one to determine low amounts of beryllium by means of ^{88}Y or ^{124}Sb isotopes.

In neutron sources of *spontaneous fission*, the isotope itself produces neutrons: 1 mg of ^{252}Cf emits about 2.3×10^9 1/s. This is about 100 times higher than the neutron yield of the most active isotopic source known at present. The spectrum of these neutrons is nearly the same as that of the unmoderated fission neutrons of ^{235}U: the mean neutron energy is 1–2 MeV.

For *γ-activation studies, bremsstrahlung* induced by linear electron accelerators, betatrons or microtrons, as well as the γ-radiation of isotopes are used.

4.2.3. MEASUREMENT OF RADIOACTIVE SAMPLES

In most cases of activation analysis, the γ-radiation emitted by the activated sample is measured, because this is the method where the radiation energy can be best determined, and self-absorption in the sample is either negligible or can be taken into correction. In the few special cases when the element to be determined has no suitable γ-radiating isotope and the measurement should, therefore, be carried out on β-emitting systems, the latter are measured, as a rule, as radiochemically pure samples, after an appropriate "cooling" (decay) period and chemical separation. It is sufficient here to detect the β-particles (sometimes together with energy discrimination) by means of, e.g., a thin-window GM-tube or a scintillation counter suitable for β-counting.

4.2.3.1. DETECTORS

The spread of activation analysis is due, first of all, to its high sensitivity; this, however, can be achieved by high efficiency detection only.

The most frequently used detectors in activation analysis are high efficiency *crystal scintillation counters* (e.g., NaI(Tl); see Section 2.1.2). Individual γ-radiating isotopes can be *identified by means of the peak corresponding to the total energy loss.* The probability of photoelectric absorption decreases rapidly with increasing photon energy; thus, in order to increase the intensity of the total energy peak, the probability of secondary absorption should be increased by increasing the dimensions of the crystal. Above a certain limit, this, however, leads to the deterioration of the resolving power, due to technical reasons. A practicable compromise is to apply a 75×75 mm NaI(Tl) scintillation detector as the spectrometer crystal. In order to decrease background radiation, the detector is surrounded by a large lead or iron shielding having thick walls. If the resolving power of the NaI(Tl) crystal is not sufficient, i.e., neighbouring peaks originating from different adioactive isotopes cannot be distinguished, the components have to be *separated by chemical operations* (see Section 4.2.3.3), and the spectrometry has to be carried out with ystems which are, at least partially, radiochemically pure. Hence the fact whether chemical separation, i.e., a destructive activation analysis, is necessary or not, is letermined partly by the resolving power of the detector used.

The application of *semiconductor detectors* (see Section 2.1. . . .) helps to overcome this lifficulty, although the efficiency of such detectors available at present is only 5–10% of hat of the 75 × 75 mm scintillation detectors in the energy range analysed most frequently, .e., around 1 MeV. The use of semiconductor detectors does not eliminate the necessity of chemical separation, but the range where these laborious operations are necessary can be onsiderably decreased. A lithium-drifted germanium detector, e.g., can distinguish etween two γ-quanta with an energy difference as low as 2–4 keV; with scintillation letectors an energy difference higher by an order of magnitude can hardly be detected.

4.2.3.2. RECORDING AND EVALUATION OF γ-SPECTRA

The heights of the signals of a detector are proportional to the energy of the radiation absorbed in the detector. Therefore, in the measurement of particle energy, the detector ignals should be classified according to height. The most suitable instruments for this purpose are multichannel analysers.

Multichannel analysers (see also Section 2.2.3) are suitable for storing the signals letected and amplified in several, as a rule, *256–4000 channels,* according to their quantum nergy, and measure the intensity in each channel. This renders possible the simultaneous

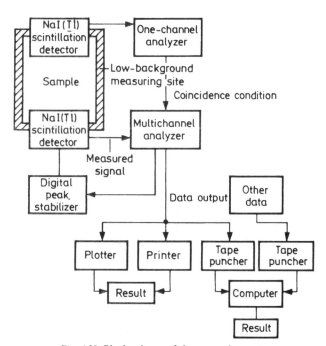

Fig. 4.29. Block scheme of the measuring system

recording of the total energy spectrum of the γ-radiating isotopes produced. Thi
considerably accelerates and facilitates activation analysis studies. For some specia
purposes, the use of a *single-channel analyzer* (see Section 2.2.3) can be the mos
economical.

It is possible to connect *various accessories* to up-to-date analysers. For example, the
automatically *draw, print or record* e.g. *on magnetic tape* the data stored in the memor
after completion of the measuring cycle. Figure 4.29 shows the block scheme of such
system. It is advantageous, especially in the case of totally unknown samples, if the
spectrum can be monitored continuously *on the oscilloscope screen* of the analyzer during
measurement.

If a monoenergetic γ-radiation is measured and the memory content of the
multichannel analyser is drawn, a spectrum like the one shown in Fig. 4.30 is obtained
This simplest γ-spectrum consists of two main parts: the *total energy peak* (so-calle
"*photopeak*") and the Compton-range. A pulse corresponding to the total energy peak i
induced if the total energy of the radiation to be measured is absorbed in the detector; thu
the determination of the position of the peak can be used for measurement of the radiation
energy and, by doing so, often for the identification of the emitting radionuclide.

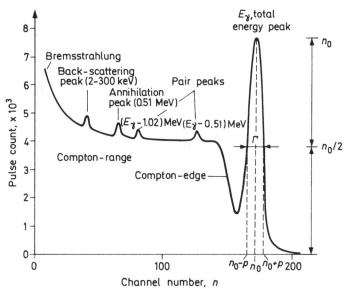

Fig. 4.30. Spectrum of a monoenergetic γ-radiation as measured by a scintillation detector

The Compton-range, beginning from zero energy, can be attributed to the elasti
scattering of the γ-photon on one of the electrons of the detector; the photon transfers it
energy to the electron only partially. The following peaks are in the Compton-range: th
back-scattering peak (energy 2–300 keV):

$$E_v = \frac{E_\gamma}{\dfrac{2E_\gamma}{0.51} + 1} \tag{4.44}$$

the annihilation peak due to the annihilation of positrons (0.51 MeV) and, with $E_\gamma > 1.0$
MeV, the pair formation or escape peaks having discrete energy values: $E_\gamma - 0.51$ MeV and

$E_\gamma - 1.02$ MeV. Although every part of the spectrum is characteristic of a given γ-energy, only the total energy peak is evaluated in practice; the rest of the spectrum is regarded as an unwanted background even if the energy measured is lower in this way.

Owing to the statistical character of radioactive transmutation and the processes taking place in the detector, the spectral line corresponding to the total energy peak widens. The envelope curve of the peak can be approximated by a Gaussian curve:

$$n_n = n_0 \exp\left[-\frac{(n-n_0)^2}{2\sigma_{di}^2} \right] \tag{4.45}$$

where n_n denotes the pulse count in the n-th channel; n_0 the channel corresponding to the peak of the curve; n_0 the pulse count measured in the channel of the peak—i.e., the maximum count; σ_{di} is the deviation owing to the statistical character of the radioactive disintegration; this latter can be characterized by a Poisson-distribution.

The peak area is

$$n_\Sigma = 1.064\, n_0\, \Gamma \tag{4.46}$$

where Γ is the half-width $= 2.35\, \sigma_{di}$ (the difference of channel numbers belonging to pulse counts equal to $n_0/2$, see Fig. 4.30).

In the simple case when the total energy peak does not overlap other peaks in the γ-spectrum of the activated sample and the background below the peak can be regarded as linear, the quantitative determination can be carried out simply by comparing the maximum n_0 counts or the n_Σ areas of the corresponding total energy peaks of the sample and the standard. The determination of n_0 is generally less suitable for intensity measurements, because various effects causing line widening (instability, pulse count changes, partial peak overlapping) result in a decrease of the pulse count in the peak, whereas the n peak area remains the same.

The peak area method is also suitable for manual calculation, with a limited accuracy. Figure 4.31 shows a total energy peak over a background which can be regarded as linear. If p channel number is considered on both sides of the peak maximum, the n_Σ peak area, i.e., the detected pulse number is:

$$n_\Sigma = \sum_{i=n_0-p}^{n_0+p} n_i - \frac{n_{n_0-p}+n_{n_0+p}}{2}(2p+1) \tag{4.47}$$

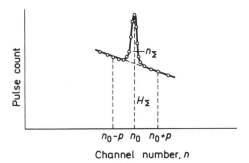

Fig. 4.31. Determination of the total energy peak by the peak area method

Assuming that the pulse counts detected in $(2p+1)$ channels follow a Poisson-distribution, the error of pulse count number under n_{Σ} is as follows:

$$\sigma_{n_{\Sigma}} = \sqrt{n_{\Sigma} + (n - 1/2)(n + 1/2)(n_{n_0 - p} + n_{n_0 + p})} \qquad (4.48)$$

With complex γ-spectra, the peak area calculation method cannot be applied generally. For example, in the γ-spectrum taken by a *scintillation detector* shown in Fig. 4.32, the

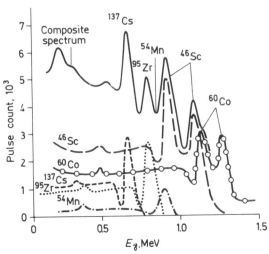

Fig. 4.32. Analysis of the spectrum of composite γ-radiation as measured by a scintillation detector

intensive radiations of ^{46}Sc, ^{60}Co and ^{95}Zr overlap the total energy peak of ^{54}Mn so that the latter does not even appear; at the same time, the peak at about 1.1 MeV is composed of the lines of ^{46}Sc and ^{60}Co. Several processes have been developed for decomposing complex γ-spectra.

The stripping method. The spectrum of each component is deducted in an appropriate x ratio, commencing from the peak of highest energy; the deduction is carried out subsequently, separately for each component and in a stepwise way; finally nearly zero count remains in each channel (Fig. 4.33). The amount of the component investigated can be calculated from the spectra of the sample in comparison with that of the standard spectra, knowing the above-mentioned x ratio. By up-to-date multichannel analyzers this operation can be carried out directly and the result of each deduction can be followed on the screen of the analyzer visually.

The greatest disadvantage of the method is that the statistical error of low energy peaks increases considerably as a consequence of the subsequent deductions, and the result becomes inaccurate with more complex spectra.

The method of least squares can be applied for spectrum decomposing if the number of components in the sample and their spectra are known. The description of spectrum decomposition can be found in several handbooks, therefore only the most important steps will be mentioned here.

Assume that a spectrum of $1 \ldots j \ldots m$ components is measured by a spectrometer, in $1 \ldots i \ldots n$ channels and the n channel spectra of standards of each component with

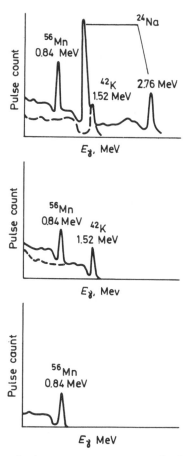

Fig. 4.33. Decomposing of a three-component γ-spectrum by the stripping method

known $A_1 \ldots A_j \ldots A_m$ activity are available. The standards and the sample are measured under identical conditions.

Denote the activity of the standard of component j in the i-th channel by A_{ji}, the total activity in the i-th channel by A_i. If the ratio of the unknown activity of component j to that of the appropriate standard is denoted by x_j, it can be written:

$$A_i = \sum_{j=1}^{m} A_{ji} x_j + \chi_i \tag{4.49}$$

where χ_i denotes the total error of the measurement.

The total activity A is obtained by summing Eq. (4.48) for i. This spectrum artificially composed from the A_j standards approximates the best the spectrum measured originally, if the quadratic sum

$$\chi^2 = \sum_{i=1}^{n} \chi_i^2 \tag{4.50}$$

is at its minimum, i.e.,

$$\chi^2 = \sum_{i=1}^{n} \chi_i^2 = \sum_{i=1}^{n} \left[A_i - \sum_{j=1}^{m} A_{ji} x_j \right]^2 = \min \tag{4.51}$$

In order to solve this problem of extreme value calculation, the expression (4.51) should be differentiated according to parameters $x_1 \ldots x_m$, one after the other; equating the derivatives to zero, a system of equations consisting of m equations and containing m unknowns is obtained. The differentiation gives the following result, for example according to a parameter with index k (x_k):

$$\frac{\partial \chi^2}{\partial x_k} = 2 \sum_{i=1}^{n} \left[A_i - \sum_{j=1}^{m} A_{ji} x_j \right] A_{ki} = 0 \tag{4.52}$$

and consequently:

$$\sum_{j=1}^{m} x_j \sum_{i=1}^{n} A_{ki} A_{ji} = \sum_{i=1}^{n} A_{ki} A_i \tag{4.53}$$

Since the k running index can take every value between 1 and m, a homogeneous linear system of equations is obtained to calculate x_k relative components.

In the simple case when the number of unknown components is three, the following equations will be obtained from (4.53):

$$x_1 \sum_i A_{1i}^2 + x_2 \sum_i A_{1i} A_{2i} + x_3 \sum_i A_{1i} A_{3i} = \sum_i A_{1i} A_i \tag{4.54a}$$

$$x_1 \sum_i A_{1i} A_{2i} + x_2 \sum_i A_{2i}^2 + x_3 \sum_i A_{2i} A_{3i} = \sum_i A_{2i} A_i \tag{4.54b}$$

$$x_1 \sum_i A_{1i} A_{3i} + x_2 \sum_i A_{2i} A_{3i} + x_3 \sum_i A_{3i}^2 = \sum_i A_{3i} A_i \tag{4.54c}$$

The system of Eqs (4.55) can be solved for the wanted components x_1, x_2, x_3. If the number of components is $m \leq 3$, the system of equations can be solved manually, or by a small calculator; with higher m values, however, the lengthy calculation requires a high capacity computer.

Practical spectrum decomposing programs are based on the principles discussed above, but they contain several reasonable modifications and auxiliary steps. It should be considered, for example, that the standard deviation of pulse counts in individual channels is the function of the number of counts in the channel, thus various reliability weight factors (ω) should be selected for each channel to minimize the squares of differences between the measured and composed program. If the statistical error of the pulse counts in the standard spectrum is negligible as compared to that in the composite spectrum, i.e., $\sigma_{A_{ki}} \ll \sigma_{A_i}$, then $\sigma_i^2 = A_i$ and the form of Eq. (4.53) will be, considering also the weight factor, $\omega_i = 1/\sigma_i^2 = 1/A_i$:

$$\sum_{j=1}^{m} x_j \sum_{i=1}^{n} \omega_i A_{ki} A_{ji} = \sum_{j=1}^{m} x_j \sum_{i=1}^{n} \frac{A_{ki} A_{ji}}{A_i} = \sum_{i=1}^{n} A_{ki} \tag{4.55}$$

In order to decrease statistical errors, the standard *spectra are generally smoothed before fitting*. If the smoothing process considers, e.g., 5 points, the pulse count in the i-th channel of the smoothed spectrum is obtained by fitting a parabola to the 5 experimental points corresponding to the channels between $(i-2)$ and $(i+2)$. The fitting is carried out by

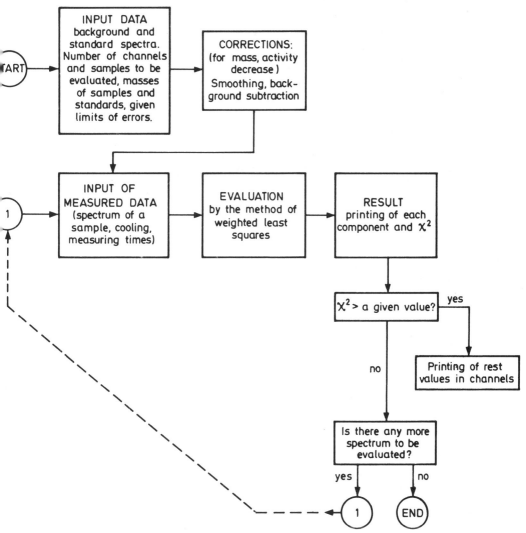

Fig. 4.34. Block scheme for the computer program of spectrum decomposing based on the method of least squares

the method of the least squares; the value of the fitted parabola in the i-th channel is regarded as the actual pulse count in the i-th channel.

The main steps of *spectrum decomposing programs* are shown in Fig. 4.34. The number of channels and samples to be evaluated and the masses of the samples and standards, as well as the spectra of the background and the standards are given as input data. The spectra are smoothed in order to decrease the statistical error, then the background and the standards are normed to unit measuring time and unit mass. Since these spectra are stored during program running in a magnetic memory, by using them, complex spectra can be decomposed one after the other; from the measured data, the machine calculates the amount of each impurity and the error of the determination, using the method of least squares.

19 Földiák

If the fitting is not satisfactory, the equipment prints the difference between the fitted and measured spectra; this facilitates to discover the reason why the fitting was poor; e.g., the spectrum measured contained a component the standard of which had not been supplied. The disadvantage of the method of weighted least squares is that if only one component is missing from among the standards, the fitting can only be carried out with errors or not at all. If there is a large number of components (10–20), the preparation, irradiation and measurement of the standards present a problem.

The increasingly widespread use of *semiconductor detectors* simplifies computer evaluation. As opposed to the poor resolving power of scintillation detectors, the total energy peak measured by semiconductors is limited to such a narrow range where the background can be regarded as linear (or, sometimes, quadratic). Thus computer fitting may be limited to the total energy peaks instead of the full spectrum. This facilitates the solution, because it allows the use of a method analogous to the peak area method discussed for evaluation of monoenergetic spectra.

If Ge(Li) detectors of high resolving power are applied, at least 3000–4000 channels are necessary to obtain a spectrum in the range between 0 and 2 MeV; this large number of data can only be evaluated by a computer. Therefore, the recent trend is that small computers are applied instead of analysers containing a very large number of channels. A part of the memory of the computer (3000–4000 bytes) takes the spectrum directly; the rest of the memory stores the programs necessary for the controlling and evaluation of the measurement. The advantage is that, after having finished the measurement, the machine carries out evaluation and commands to start the next measurement. This method has a bright future, especially in serial analyses.

EXAMPLE 4.13

Let us assume that the following spectrum was determined by a semiconductor detector:

Channel number, n	Pulse count, $\boldsymbol{n}$	
110	75	
111	79	
112	76	
113	95	
114	192	
115	856	
$n_0 = 116$	$n_0 = 1520$	
117	712	$n_\Sigma = 3624$
118	164	
119	85	
120	81	
121	75	
122	74	

The position of the peak is $n_0 = 116$th channel; the range to be evaluated is $p = \pm 3$ channels. The starting data for the calculation are:

$$\text{pulse count in channel } n_0 - p = 113 : \boldsymbol{n}_{n_0 - p} = 95$$

$$\text{pulse count in channel } n_0 + p = 119 : \boldsymbol{n}_{n_0 + p} = 85$$

According to Eq. (4.46):

$$n_\Sigma = 3624 - \frac{95+85}{2} \times 7 = 3624 - 630 = 2994$$

The standard deviation of the measurement according to Eq. (4.48) is

$$\sigma_{n_\Sigma} = \sqrt{2994 + \frac{5}{2} \times \frac{7}{2} \times 180} = 67.6$$

The calculated peak area is

$$n_\Sigma = 2994 \pm 68 \text{ pulses}$$

4.2.3.3. CHEMICAL SEPARATIONS (DESTRUCTIVE METHODS)

Chemical separation is necessary in the following cases:

— the sample contains more than one element producing *radionuclides emitting radiations with similar energies and half-lives;* the total γ-energy peaks of these elements are so close to one another that their separation is impossible in the instrument of a given resolving power;

— the sample contains *such elements whose radiations overlap* the radiation of the element to be investigated, although the spectra of the interfering elements may be different from the spectrum of the element to be analyzed; this may be due to the activation of the disturbing elements to a high degree, owing to their amount, cross-section or half-life; this is the situation, e.g., in the case of sodium or tungsten when trace elements are determined in them;

— the activation *produces β-radiating isotopes only* from the element wanted, and this cannot be separated from the other types of β-radiation of similar energy in the sample;

— *the maximum sensitivity of measurement requires* the separation of the element to be investigated in a radiochemically pure form, because all other radiations should be regarded as a background in the measurement, and this decreases the sensitivity.

Chemical separations are usually carried out after activation, in order to avoid any error owing to the presence of impurities in the reagents. Methods well proved in classical chemical analysis are applied here; but whereas in classical analysis the specific and quantitative character of the separations is very important, these requirements are less important in activation analysis. If the isotope to be measured can be separated from those emitting interfering radiations, this selectivity is sufficient in view of the accuracy of activation analysis.

Of the methods of chemical separation, the most suitable has to be selected; this depends mainly on the chemical composition of the elements present in the sample. The fact should also be often considered that the short half-life of the isotopes to be measured demands very rapid work in chemical operations. The processes used most frequently are the following.

Precipitation separation is one of the most common methods in classical chemical analysis. In activation analysis the direct precipitation of very low amounts is, in general, necessary, therefore a sufficient amount of carrier is added to the sample after dissolving it, or prior to its precipitation. The carrier is usually chemically identical with the element to be determined, but sometimes it is sufficient if their chemical properties are similar. The precipitation method is especially advantageous if an appropriate carrier prevents the precipitation of the disturbing components. Precipitations are generally carried out in acid medium; separation is here more selective.

Solvent extraction is a very rapid and simple separation method (see also Section 4.3.2.1). It is based on the partition of the dissolved substance between two immiscible solvents, whereby the concentrations will be different in the two phases. Rapid separation requires a simple, special device, which is particularly important when processing highly radioactive samples. Up to now, extraction methods have been developed for the separation of more than 75 elements. Various solvents are used for this purpose; in several cases chelates or ion association complexes are formed prior to extraction. Of complexing agents, dithizone, cupferron, diethyldithiocarbamate are particularly suitable for the separation of metals or groups of metals. Figure 4.35 shows the scheme of separation of impurities in gallium arsenide by means of extraction.

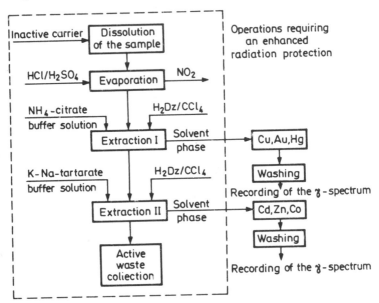

Fig. 4.35. Block scheme of the extraction separation of impurities in gallium arsenide

Very *selective substoichiometric extraction* methods (see Section 4.3.2.2) are efficient means of radiochemical separation.

Ion exchange methods can also be used for selective separation, especially in the case of rare earth metals, actinides and alkali metals; since the method is rather time-consuming, the precondition of its application is that the isotopes should have rather long half-lives (Section 4.3.5.1).

Various *chromatographic* methods (column-, paper- and thin-layer chromatography) can be equally suitable for the separation of radioactive isotopes. The chromatograms are evaluated on the basis of the activity of the individual fractions. The disadvantage of the method is that the separation usually requires long times (see also Sections 4.3.5.3 and 4.3.5.6).

Electrolysis can also be applied to separate disturbing elements in destructive activation analysis. The advantage of the method is that it is relatively rapid and selective in separating the occurring small amounts; it can be applied with success for the separation of, e.g., Ag, Au, Co, Cu, Hg, Pb and Pt and some other metals.

If isotopic exchange is applied, this is based on the fact that different isotopes of the same element are exchanged between various chemical compounds and phases; after the establishment of the equilibrium, the ratio of isotopes will be the same in every phase or component. The main advantage of isotopic exchange is its rapidity (see Section 4.3.6).

Distillation is used if one has to get rid of the bulk of a high activity main component. This is, e.g., a successful method for the separation of the bulk of arsenic and gallium if trace elements are determined in GaAs. It is also possible to utilize the volatility differences of mainly halogen compounds in order to separate noble metal isotopes.

Destructive chemical methods are more sensitive, but they require much time and are laborious. For routine analysis, the application of nondestructive methods without any chemical separation are more advisable but—especially in the case of the determination of several elements in the presence of each other—the accuracy of chemical methods exceeds that of γ-spectrometry. The question whether the analysis be solved by destructive or nondestructive method is decided in many cases by the nature of the problem. *By combining* the methods, the advantages of both processes may be exploited, e.g., by using a chemical separation of the dissolved sample in a way that only a few (3–4) elements remain together in each group. The γ-energies of these elements should be sufficiently far from each other; the resulting, not very complicated γ-spectrum can then be evaluated rather easily.

4.2.4. PRACTICAL APPLICATIONS

Table 4.7 summarizes methods applied in *telecommunication* practice. Semiconductor diodes and transistors of good quality can only be prepared of very high purity materials, because the electric properties are considerably influenced even by trace impurities. Earlier the products were prepared of raw materials whose purity or appropriate composition was checked on the basis of chemical analysis; subsequent experimental runs decided whether the raw material was satisfactory or not. By using activation analysis, as low amounts as $10^{-6}\%$ of the most important impurity, copper, can be detected; this

Table 4.7. Main impurities in telecommunication materials to be determined by activation analysis

Material investigated	Element to be determined
Silicium	As, Au, Cu, Ga, Ge, La, Na, P, Sb
Germanium	As, Au, Cl, Cu, Mo, Sb, W
Gallium arsenide	Au, Cd, Co, Cu, F, Ge, Hg, Mn, Ni, Se, Te, Zn
Zinc sulphide	Al, Cl, Cu, Mn, Na, Sr
Nickel	Al, As, Bi, Cu, Mg, Mn, Sb, Zn
Tungsten	Al, As, C, Cu, Ga, In, K, Na, Ni, Re, Se, Si
Copper	Ag, As, Mn, O, P, Sb, Se, Sn, V
Aluminium	Ag, Au, Ba, Cd, Co, Cu, Fe, Mn, Ni, Pt, Os, Se, U, W, Zn
Molybdenum	Ag, Cd, Co, Cu, Fe, Hg, Mn, Zn
Indium	As, Au, Cu, Sb, Zn
Yttrium oxide	Ca, Ce, Co, Cu, Dy, Eu, Fe, Gd, La, Mn, Nd, Ni, Tb, V
Uranium oxide	Ag, Au, Cd, Co, Cr, Cu, Fe, Mn, Mo, Ni, P
Thorium oxide or nitrate	Cd, Co, Cr, Cu, Mn, Zn
Zirconium nitrate	Cd, Co, Cr, Cu, Mn, Zn

294

Table 4.8. Amount of impurities found in silicon

Average impurities investigated regularly		Average impurities investigated from time to time	
Element	%	Element	%
As	1.3×10^{-5}	Ce	$10^{-5}-10^{-2}$
Au	3×10^{-6}	Ga	$10^{-5}-10^{-3}$
Cu	8×10^{-5}	La	$10^{-5}-10^{-3}$
Na	6×10^{-5}	W	$10^{-5}-10^{-4}$

satisfied technological requirements (Table 4.8). The purity of *distilled water* and various reagents (acids, alkalis, organic solvents) in semiconductor production has to be often determined by the same method and accuracy.

Activation analysis is successfully applied for the determination of trace elements in *pure metals* (Al, Cu, Ni, W) and alloys containing additives in as low amounts as $10^{-6}\%$, in fine metallurgy for telecommunication purposes.

Activation analysis often finds increasing application in industrial practice to determine the main *components of minerals and ores* (e.g., Al, Si), considering the demands of *mining* (e.g., of bauxite) and *geochemical research*. A similar method can be applied for the analysis of *meteorites* and *lunar dust*.

Trace elements in *agricultural products* (foodstuffs), e.g., pesticide residues, such as As, Br, Hg are determined more and more often by measurements following neutron activation.

In biological systems, substances controlling life functions are often present in such low amounts that they can only be detected by activation analysis. Processes have been developed for the determination of about 20 elements so far, e.g., in blood, hair and skin. For example, zinc is very important in living organism being the constitutent of ferments, hormones, vitamins. The absence of zinc is harmful to animal and plant organisms. The nondestructive method for determination of zinc is based on the measurement of ^{65}Zn. The relatively high amount of iron in blood interferes with the measurement, therefore, using the 1.3 MeV total energy peak, the amount of iron is deducted from the spectrum, and the iron content is determined simultaneously.

In forensic analysis, the determination of the amount and distribution of antimony and barium may clarify the question what type of weapon was fired, how many times, and in what body position. On the basis of traces found on the victim, the distance of firing the shot can also be determined. Fractions of Ba and Sb leave the barrel of the gun together with the bullet and are distributed over a funnel-like volume. The higher the diameter of the circle containing these elements around the bullet on the victim, the farther the shot was fired.

Of museum pieces, antiquities, the composition of various coins, metal objects (damask steel), ceramic utensils, enamels can be determined by nondestructive activation analysis. The age of paintings can also be established on the basis of determination of trace elements in white lead, because their ratio was different in different ages.

Activation analysis can *monitor the concentration of substances (not necessarily trace components)* during production processes in cases when conventional methods of chemical analysis are very lengthy. Thus, if a composition different from the prescribed one is detected, the necessary intervention can be done, possibly automatically, before producing higher amounts of refuse.

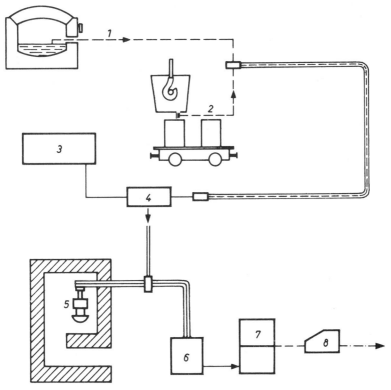

Fig. 4.36. Block scheme for oxygen determination in steel. *1* — Sample from the furnace; *2* — Sample from casting; *3* — Sample from rolling; *4* — Sample preparation; *5* — Neutron generator; *6* — Detector; *7* — Measuring system and automatic control; *8* — Printer

The oxygen content of steels can be determined relatively rapidly by activation analysis in steel works. Under the effect of fast neutrons of a neutron generator, an (n, p) reaction produces ^{16}N from ^{16}O; the former has a half-life of 7 seconds and emits a γ-radiation of about 6 MeV energy. Since all other elements present in steel emit photons with much lower energy, the intensity of the high energy γ-radiation serves as a good basis for determination of the oxygen content. Thus a simple single-channel analyzer is sufficient for this measurement. The neutron generator laboratory and its connection with the steel work is shown in Fig. 4.36.

On-line *slurry analysis* is carried out usually by ^{252}Cf neutron sources ensuring high local neutron flux rates, because they are nearly pinhole sources. The irradiating and measuring cells constructed for this purpose must ensure the homogeneous composition throughout the total volume—i.e., no settling must occur. The most important elements in the system are hence the irradiating and measuring cells. Multistep counterflow mixers are built in the cells; the radiation source is located in the cavity in the axis of the mixer in the irradiating cell, the scintillation counter is placed into the measuring cell.

The method has been successfully used for CaF_2 analysis in a *flotation plant*; the two nuclear reactions used for this purpose are

$$^{48}Ca(n, \gamma)^{49}Ca \qquad (t_{1/2} = 8.8 \text{ min})$$

$$^{19}F(n, \gamma)^{20}F \qquad (t_{1/2} = 10.7 \text{ s})$$

Because of the short residence times, both radioactive isotopes can be measured and, by applying a two- or three-channel measurement, the apparent shift of the Ca: F ratio may provide additional information:

— if the volume flow rate (v) is constant or is measured separately, the change of the total calcium content can be determined;

— if v is not constant or is not measured separately, it can be determined on the basis of the changes of the relative dose rates of the two radionuclides, after calibration.

4.3. ANALYTICAL APPLICATIONS OF TRACER METHODS

Chapter 3 has discussed industrial uses of tracer studies and also applications in some related fields.

The present section will deal mainly with *microanalytical methods* based on the tracer technique: these include isotopic dilution, the application of radioreagents, radiometric titrations, radiochromatography, as well as the analytical applications of isotope exchange reactions.

These methods are *independent analytical procedures*, but separation methods (extraction, ion exchange chromatography, isotopic exchange, etc.) can also be used as *auxiliary methods in activation analysis* (see Destructive activation analysis, Section 4.2.3.3).

4.3.1. ISOTOPIC DILUTION

The method is of great importance in the analysis of *structurally similar*, difficultly separable substances. It can be used successfully for solving analytical problems of this kind in *organic and biochemistry*, especially since compounds labelled with ^3H, ^{14}C, ^{35}S, etc. have become available. The rules of isotopic dilution are valid also for the carrier technique applied in destructive activation analysis. The principle of the method is applied in industrial tracer technique, e.g., for determination of the amount of masses (see Section 3.5.1).

During the application of isotopic dilution methods, a solution of the sample is mixed with a radioactive isotope of known amount and known activity. The isotope of the component to be determined should be added in the form of the same compound or as an ion in the same state of oxidation as characteristic of the unknown component. Subsequently, the component is separated from the system in a state as pure as required for determination by γ- or β- spectrometry. The separation need not be quantitative and any appropriate separation method may be applied (extraction, sublimation, precipitation, electrolysis, chromatography, etc.). The total activity remains unchanged during chemical operations, but the specific activity decreases owing to the dilution with an inactive substance. Several variants of isotopic dilution are known; selection between the methods depends on the sample to be analysed.

4.3.1.1. SIMPLE ISOTOPIC DILUTION

The unknown amount of an inactive substance can be determined by adding the radioactive isotope of the substance to be determined to the system (see Section 3.5.1).

According to those mentioned in the introduction:

$$m_1 : (m_x + m_1) = a_2 : a_1 \tag{4.56}$$

$$m_x = m_1 \left(\frac{a_1}{a_2} - 1 \right) \tag{4.57}$$

where m_x is the amount of the inactive sample to be determined, g or mg; m_1 the known amount of the radioactive isotope added, g or mg; $m_1 = m_{carr} + m^*$ where m_{carr} is the amount of the carrier, m^* that of the active isotope; as a rule, $m^* \ll m_{carr}$; $a_1 = A_1/m_1$, the specific activity of the radioactive isotope added; $a_2 = A_2/m_2$ the specific activity of the component separated in appropriate form, expressed in the same units; m_2 is the amount of the separated material, g or mg.

In the case of a *carrier-free radioactive isotope*, the *pulse counts* can also be used for calculation, if they are measured under identical conditions:

$$m_x = m_2 \frac{n_1}{n_2} \tag{4.58}$$

Simple isotopic dilution analysis is not suitable for the determination of micro-amounts. If a very low amount of m_x inactive substance is added to the relatively high amount of m_1 radioisotope, a_2 and a_1 will be almost equal, and the accuracy of the determination will be very low.

EXAMPLE 4.14

Phosphate ions are determined by simple isotopic dilution. One cm^3 of carrier-free radioactive phosphate solution (^{32}P) is added to the solution to be analyzed and the volume is increased up to 10 l. Then 1.00 cm^3 of stock solution is evaporated to dryness on an aluminium plate, under an infrared heat lamp, and the activity is measured. I_{rel} = 2530 imp/min. Phosphate ions are precipitated from 5.00 cm^3 stock solution as $MgNH_4PO_4 . 6 H_2O$. The precipitation need not be quantitative. The dried precipitate is powdered and its given amounts are measured into aluminium holders. In order to avoid the error due to self-absorption, maximum 120 mg of the precipitate may be measured into a holder of usual dimension (~ 30 mg/cm^2).
Measured data:

$$I_{rel, 1} = 12\,650 \text{ imp/min (for 5 } cm^3 \text{ solution)}$$

Precipitation	m_2, mass of $MgNH_4PO_4 \cdot 6 H_2O$, mg	$I_{rel, 2}$ imp/min	Total $MgNH_4PO_4 \cdot 6 H_2O$, mg	Total PO_4^{3-} content, mg
Sample 1	90.3	3620	315.5	122.1
Sample 2	79.4	3209	313.0	121.4
Sample 3	108.5	4327	317.2	123.1
			Mean value:	122.3

The PO_4^{3-} content of the solution is 244.6 mg/10 cm^3.

4.3.1.2. REVERSE AND DOUBLE ISOTOPIC DILUTION

Reverse isotopic dilution can also be applied as a *micro method*; it is used when the amount of the radioactive isotope is unknown and is determined by diluting it with a known amount of inactive isotope:

$$m_x = \frac{m_1}{\left(\dfrac{a_1}{a_2} - 1\right)} \tag{4.59}$$

where m_x is the unknown amount of radioactive isotope; $m_x = m_{carr} + m^*$, where m_{carr} is the amount of the carrier, m^* that of the active isotope; as a rule, $m^* \ll m_{carr}$; m_1 is the known amount of the inactive diluent; a_1 the specific activity of the component to be determined in the mixture to be analyzed; a_2 the specific activity of the substance separated.

The limit of detection of m_x is decided by the material requirement of the measurement having a_1 specific activity. From the analytical point of view, reverse isotopic dilution is valuable not only as a micromethod. It can be applied for several determinations where the microcomponent is made to react with a compound labelled by an appropriate radioactive isotope and the amount of this new, already radioactive compound obtained is determined by the method of reverse isotopic dilution. This method, called *derived isotopic dilution* in the literature, is used first of all in the analysis of *organic compounds*.

The specific activity a_1 cannot be measured in several cases because the amount of substance is so low that its separation in a pure form is impossible. The measurement of the specific activity can be obviated by the method of double isotopic dilution.

If the method of double isotopic dilution is applied, the mixture to be analyzed is divided into two identical aliquot parts; both contains the same m_x amount of the radioactive substance to be determined. Of the inactive diluent, m_{11} and m_{12} amounts are added, respectively, to the two aliquots; $m_{11} \neq m_{12}$. The total mass of the substance to be determined in the two aliquot parts is:

$$(m_x + m_{11}) \quad \text{and} \quad (m_x + m_{12}) \tag{4.60}$$

The component wanted is separated from both solutions in pure form, and the specific activities are determined (a_{21} and a_{22}). According to Eq. (4.56):

$$m_x = \frac{m_{11}}{\left(\dfrac{a_1}{a_{21}} - 1\right)} \quad \text{and} \quad m_x = \frac{m_{12}}{\left(\dfrac{a_1}{a_{22}} - 1\right)} \tag{4.61}$$

By combining these two equations, the unknown a_1 can be eliminated:

$$m_x = \frac{m_{12}a_{22} - m_{11}a_{21}}{a_{21} - a_{22}} \tag{4.62}$$

EXAMPLE 4.15

Let us determine the phosphate content of a phosphate solution labelled by ^{32}P by double isotopic dilution. Solutions containing 60 and 100 mg inactive phosphate ions, respectively, are added to two 10 cm^3 samples of the radioactive phosphate solution. The phosphate ions are separated as a $MgNH_4PO_4 \cdot 6 H_2O$ precipitate from both solutions (the separation need not be quantitative). The activity of 100 mg precipitate each is measured on aluminium sample holders, after drying and powdering: $I_{rel, 100} = 8950$ imp/min, $I_{rel, 60} = 12\,150$ imp/min. The specific activities are substituted by pulse counts in Eq. (4.62):

$$m_x = \frac{12\,150 \times 60 - 8950 \times 100}{8950 - 12\,150} = 51.9 \text{ mg PO}_4^{3-} \text{ in 10 cm}^3 \text{ solution}$$

4.3.1.3. PRACTICAL APPLICATIONS

Of *inorganic* substances, the analysis of alkali metal salt mixtures may be of interest. By using ^{42}K, potassium can be determined in such mixtures with 1% accuracy; the separation can be achieved in the form of potassium perchlorate. The following method has been developed for the determination of strontium and arsenic in *sea water*: the tracer is ^{89}Sr as strontium chloride; separation is effected as strontium sulphate; its precipitation can be promoted by co-precipitation together with barium sulphate. Traces of arsenic can be determined by means of arsenic acid labelled by ^{76}As: the activity of the reduced arsenic mirror is determined. Several works deal with the measurement of phosphates by means of ^{32}P, e.g., pyro- and trimetaphosphates can be determined by appropriate labelled phosphates.

Isotopic dilution methods are suitable for solving analytical problems in *organic chemistry and biochemistry*. For instance, various reaction products: alcohols, aldehydes, peroxides are produced in hydrocarbon oxidation. The determination of alcohols is important in elucidating the reaction mechnism. ^{14}CH$_3$OH and CH$_3$^{14}CH$_2$OH radioactive preparations are necessary for isotopic dilution analysis, and the reagent is 3,5-dinitrobenzoyl chloride; this gives benzoates with alcohols. Methyl and ethyl dinitrobenzoate can be separated by paper chromatography. The penicillin content of various penicillin products is determined by means of ^{35}S-labelled benzylpenicillin which can be readily synthesized. The penicillin is precipitated partly as the N-ethylpiperidine salt during isotopic dilution. The analysis of amino acid mixtures resulting from protein hydrolysis is a very difficult problem. The compound called pipsyl chloride (p-iodobenzenesulphonyl chloride) is used most frequently as a radioactive reagent; this gives labelled monopipsyl derivatives with amino acids. A known amount of the p-iodophenylsuphonyl derivative of the amino acid to be determined is added as a diluent to the mixture. The separation can be carried out by multistep extraction and repeated recrystallization. Paper chromatographic separation may also be used for low amounts.

Other examples of applications have been described in Sections 3.3.3. and 3.5.

4.3.2. SUBSTOICHIOMETRIC ISOTOPIC DILUTION

Substoichiometric isotopic dilution as a simple isotopic dilution renders possible the high sensitivity determination of trace amounts of *metals*; as a reverse isotopic dilution, it can be applied as a separation method for destructive *activation analysis*. The principle of substoichiometry, i.e., the application of the reagents in amounts lower than stoichio-metric, makes possible the separation of exactly identical amounts of metal ions from solutions of different metal ion concentration.

Simple isotopic dilution: inactive *metal traces are determined* by means of a solution containing known concentrations of radioactive metal ions. The separation is carried out by identical amounts of reagents, less than stoichiometric, from two solutions: one containing a known amount of the active substance, and the other an identical amount of the active substance diluted with the inactive substance to be measured. Owing to the dilution with inactive substance, the activity becomes lower. Because identical amounts were separated from both solutions, the measurement of specific activities is not necessary. The m_x amount wanted can be calculated on the basis of the activity values or the count rates measured under identical conditions:

$$m_x = m_1 \left(\frac{A_1}{A_2} - 1 \right)$$ (4.63)

and

$$m_x = m_1 \left(\frac{I_{rel,1}}{I_{rel,2}} - 1 \right)$$ (4.64)

where m_1 is the known metal content of the solution containing the labelled metal ion; A_1 and $I_{rel,1}$ are the activity and count rates, respectively, of the separated fraction before dilution, A_2 and $I_{rel,2}$ are the same after dilution.

EXAMPLE 4.16

Let us determine zinc by extraction of the zinc dithizonate complex by means of carbon tetrachloride.

The amount of Zn in the zinc stock solution labelled by ^{65}Zn is $m_1 = 3.08$ µg/cm^3.

In a separating funnel, 4.00 cm^3 of dithizone solution in carbon tetrachloride (about 20 µM), 2.00 cm^3 of radioactive zinc stock solution and 3 cm^3 of 10% sodium acetate solution, purified previously by means of dithizone, are mixed and the volume of the aqueous phase is made up to 15 cm^3 by ion-exchange purified distilled water. After shaking for 20 minutes, the carbon tetrachloride phase is transferred into a 20 cm^3 volumetric flask and made up to the mark with carbon tetrachloride. The activity of 10 cm^3 of the carbon tetrachloride phase is measured. $I_{rel,1} = 1279$ imp/100 s (background subtracted).

The extraction is repeated with 4.00 cm^3 of dithizone solution (about 20 µM), 2.00 cm^3 of radioactive zinc stock solution, and 2.00 cm^3 of unknown zinc solution containing inactive zinc. $I_{rel,2} = 994$ imp/100 s (background subtracted).

$$m_x = 3.08 \left(\frac{1279}{994} - 1 \right) = 0.88 \text{ µg/cm}^3 \text{ Zn}$$

The concentration of the dithizone solution need not be adjusted accurately. The stoichiometric dithizone consumption of the solution containing the metal ions can be established by a simple titration with colour indication.

Reverse isotopic dilution. The advantage of the substoichiometric principle as a separation method in *destructive activation analysis* is that the chemical yield need not be determined and the selectivity of the separation is enhanced.

The sample and the standard activated simultaneously are dissolved, the carrier of the radioactive element to be determined is added, and the solutions are allowed to react with the appropriate reagent, applied in a less than stoichiometric amount. The activity of the element to be determined in the sample is A_x.

$$A_x = a_m \frac{m_{carr, m}}{m_m} \tag{4.65}$$

where a_m is the activity of the separated amount, m_m; $m_{carr, m}$ denotes the amount of carrier added to the sample.

The activity of the element to be determined in the standard is A_{st}.

$$A_{st} = a_{st} \frac{m_{carr, st}}{m_{st}} \tag{4.66}$$

where a_{st} is the activity of the separated amount, m_{st}; $m_{carr, st}$ is the amount of the carrier added to the standard.

If $m_{carr, m} = m_{carr, st}$ and $m_m = m_{st}$, then for simultaneous irradiation, the following equation is valid:

$$y_x : y_{st} = a_m : a_{st} \tag{4.67}$$

$$y_x = \frac{a_m}{a_{st}} y_{st} \tag{4.68}$$

where y_x is the unknown amount of the element to be determined in the sample and y_{st} its known amount in the standard.

The condition $m_{carr, m} = m_{carr, st}$ can be met easily because an identical amount of carrier can be added to the sample and the standard without any difficulty. The condition $m_m = m_{st}$ is fulfilled due to the substoichiometric separation.

For *substoichiometric separation*, the following methods can be used: solvent extraction of a metal chelate, extraction of ion association complexes, formation of chelates soluble in water (ion exchange resins or solvent extraction of the metal chelates can be used for separation from the free metal ions), displacement substoichiometry, etc.

The typical concentration of solutions in destructive activation analysis is 1–10 mM, if, however, inactive metal traces are determined, the concentration is, as a rule, lower than $1 \mu g/10 \, cm^3$. The chemical conditions of separation are much more severe in the latter case.

4.3.2.1. SOLVENT EXTRACTION OF METAL CHELATES

The extraction of chelate complexes is one of the best methods for substoichiometric separation of *metal ions*. Let us investigate the conditions of extractability, with particular emphasis on the *pH* range of extraction and on the effect of auxiliary complexing agents, and compare the selectivity of substoichiometric extraction with that carried out by an excess of reagents.

The pH range of extraction. The basic equation of extraction is as follows:

$$M + n(HA)_{org} \rightleftharpoons (MA_n)_{org} + nH$$

where M is a metal ion, HA an organic chelating agent, e.g., dithizone; MA_n stands for the metal chelate (charges have not been denoted, for clarity).

The equilibrium constant of the process is the extraction constant:

$$K_M = \frac{[MA_n]_{org} [H]^n}{[M] [HA]_{org}^n} \tag{4.69}$$

If the value of the extraction constant (Table 4.9), the equilibrium concentrations of the metal complex ($[MA_n]_{org}$) and the organic chelating ($[HA]_{org}$) in the organic phase and the equilibrium concentration of the metal ion ($[M]$) in the aqueous phase are substituted into Eq. (4.66), the lowest pH at which the extraction can be carried out can be calculated.

Table 4.9. Logarithms of extraction constants of metal chelates (log K)

Ion	Diethyl-dithiocarbamic acid, CCl_4	Dithizone, CCl_4	Dithizone, $CHCl_3$	Cupferron, $CHCl_3$	α-Nitroso-β-naphthol, $CHCl_3$	Oxine, $CHCl_3$	Thenoyltrifluoroacetone, C_6H_6
Ag^+	11.9	7.18	6.0			−4.51	−5.23
Al^{3+}				−3.50		−5.22	
Ba^{2+}						−20.9	−14.4
Be^{2+}				−1.54		−9.62	−3.2
Bi^{3+}	16.79	9.98	5.4	5.08		−1.2	
Ca^{2+}						−17.89	−12.0
Cd^{2+}	5.41	2.14	0.5			−5.29	
Co^{2+}	2.33	1.53	−0.5	−3.65		−2.16	−6.7
Cu^{2+}	13.70	10.5	6.5	2.66		1.77	−1.32
Dy^{3+}							−7.03
Er^{3+}							−7.2
Eu^{3+}							−7.66
Fe^{2+}	1.20						
Fe^{3+}				9.8		4.11	3.3
Ga^{3+}		−1.3	−1.3	4.92		3.72	−7.57
Hg^{2+}	31.94	26.8		0.91		−3	
Hf^{4+}							7.80
Ho^{3+}							7.25
In^{3+}	10.34	4.84	1	2.42		0.89	−4.34
La^{3+}				−6.22		−16.37	−10.31
Lu^{3+}							−6.77
Mg^{2+}						−15.13	
Mn^{2+}	−4.42					−9.32	−1
Nd^{3+}							−8.58
Ni^{2+}	11.58	1.18	−2.5			−2.18	
Pb^{2+}	7.77	0.44		−1.53		−8.04	−5.2
Pd^{2+}	>32	>27				15	
Pm^{3+}							−8.05
Pr^{3+}							−8.85
Sc^{3+}				3.34		−6.64	−0.77
Sm^{3+}						−13.41	−7.68
Sn^{2+}		2					
Sr^{2+}						−19.71	−14.1
Tb^{3+}							−7.51
Th^{4+}				4.44	−1.64	−7.18	0.8
Tl^+	−0.53	−3.3	−3.8				−5.2
Tl^{3+}				3		5.0	
Tm^{3+}							−6.96
U^{4+}							5.3
Y^{3+}				−4.74			−7.39
Yb^{3+}							−6.72
Zn^{2+}	2.96	2.3	0.8			−2.41	
Zr^{4+}							9.15

The $[MA_n]_{org}$ equilibrium concentration is calculated on the basis of the assumption that more than 99.9% of the organic chelating agent added substoichiometrically is consumed in complexing the metal:

$$[MA_n]_{org} V_{org} \geqq 0.999 \frac{c_{HA}}{n} V_{org} \tag{4.70}$$

where V_{org} is the volume of the organic phase, and c_{HA} the initial concentration of the chelating reagent in the organic phase. This condition ensures that the same amount of metal ion is separated with an accuracy higher than 0.1%.

The equilibrium concentration of the metal ion, $[M]$, is given by the following equation:

$$[M] V = c_M V = \frac{c_{HA}}{n} V_{org} \tag{4.71}$$

where V is the volume of the aqueous phase and c_M denotes the initial concentration of the metal ion in the aqueous phase.

The following relation holds for the equilibrium concentration of the organic chelating agent, $[HA]_{org}$:

$$[HA]_{org} V_{org} \leqq 0.001 \, c_{HA} V_{org} \tag{4.72}$$

Substituting expressions (4.70), (4.71) and (4.72) into Eq. (4.69) and expressing the hydrogen ion concentration from the latter, applying negative logarithms, the following equation is obtained for the lower pH limit of the extraction:

$$pH \geqq \frac{1}{n} \log \left(\frac{c_{HA}}{n} \right) - \frac{1}{n} \log \left(c_M - \frac{c_{HA}}{n} \times \frac{V_{org}}{V} \right) -$$

$$- \frac{1}{n} \log K_M - \log (0.001 \, c_{HA}) \tag{4.73}$$

An analysis of Eq. (4.70) shows that the decisive factor in determining the lower pH limit is represented by the two latter terms of the inequality.

In destructive activation analysis, generally 1–10 mg of carrier is added to the sample; the solution will then have a concentration of about 1–10 mM. The concentration of the organic chelating agent should be in the same order of magnitude, thus the simplified form of Eq. (4.73) will be

$$pH \geqq 6 - \frac{1}{n} \log K_M \tag{4.74}$$

Correlation (4.73) is valid strictly only if the organic reagent is not dissolved in the aqueous phase, owing to its dissociation, i.e., if:

$$V[A] < [HA]_{org} V_{org} \tag{4.75}$$

This condition is fulfilled if the pH of the aqueous phase is:

$$pH \leqq pK_{HA} + \log q_{HA} + \log \frac{V_{org}}{V} \tag{4.76}$$

where pK_{HA} is the dissociation exponent of the organic reagent as an acid, and q_{HA} is the partition coefficient (Table 4.10). Eq. (4.73) gives the upper pH limit of extraction with

Table 4.10. The pK_{HA} and $\log q_{HA}$ values for some chelating reagents

Reagent	Solvent	pK_{HA}	$\log q_{HA}$
Diethyldithiocarbamic acid	CCl_4	~3.8	2.38
Dithizone	CCl_4	4.5	4.0
	$CHCl_3$	4.5	5.7
Cupferron	$CHCl_3$	4.16	2.18
α-Nitroso-β-naphthol	$CHCl_3$	7.63	2.97
Oxine	$CHCl_3$.	9.9	2.6
		(pK_1)	
		5.0	
		(pK_2)	
Thenoyltrifluoroacetone	C_6H_6	6.23	1.6

respect to the organic reagent. For example, extraction with dithizone dissolved in carbon tetrachloride can only be carried out if

$$pH \leq 8.5 + \log \frac{V_{org}}{V}$$

otherwise dithizone undergoes dissociation and is dissolved by the aqueous phase.

EXAMPLE 4.17

What is the lower *pH* limit of the extraction if bismuth ions are extracted by means of dithizone solution in carbon tetrachloride? $K_{Bi} = 10^{9.98}$; $BiDz_3$, i.e., $n = 3$, $c_{HA} = 1$ mM.

$$pH > -\frac{1}{n} \log K_{Bi} - \log 10^{-6} > 2.7$$

The effect of auxiliary complexing agents. The upper *pH* limit of extraction, apart from Eq. (4.73), also depends on the fact whether *insoluble metal hydroxides* are formed or not when the aqueous solution is made alkaline. This disturbance can often be prevented by using auxiliary complexing agents, such as cyanide ions, oxalic acid, tartaric acid, nitrilotriacetic acid, ethylenediamine tetraacetic acid (EDTA), 1,2-diaminocyclohexane tetraacetic acid, etc.

If the auxiliary complexing agent gives a water soluble complex with the metal ion, its effect can be considered by means of the α_M functions:

$$\alpha_M = \frac{[M']}{[M]} = \frac{[M] + [MB] + [MB_2] + \ldots + [MB_n]}{[M]} \qquad (4.77)$$

where $[M']$ is the total concentration of the metal and $[M]$ is the free metal ion concentration in the aqueous phase. Rearranging Eq. (4.77) considering the (4.75) formulae

$$\beta_1 = \frac{[MB]}{[M][B]} = K_1; \qquad \beta_2 = \frac{[MB_2]}{[M][B]^2} = K_1 K_2, \ldots$$

$$\ldots \beta_n = \frac{[MB_n]}{[M][B]^n} = K_1 K_2 \ldots K_n \qquad (4.78)$$

we obtain:

$$\alpha_{M(B)} = 1 + \beta_1[B] + \beta_2[B]^2 + \ldots + \beta_n[B]^n \tag{4.79}$$

where $\beta_1, \beta_2, \ldots \beta_n$ are the product of the stepwise stability constants $(K_1, K_2, \ldots, K_n)$ of those metal complexes which are formed by the metal ion and the H_nB auxiliary complexing agent; $[B]$ is the equilibrium concentration of the auxiliary complexing anion at the given pH value. Table 4.11 shows $\log \beta$ values for some metal ions and some auxiliary complexing agents.

The c_B total concentration of the auxiliary complexing agent can be included into the calculations if its protonation is considered:

$$\alpha_{B(H)} = 1 + K_1[H^+] + K_1 K_2[H^+]^2 + \ldots \tag{4.80}$$

where $K_1, K_2 \ldots$ are the corresponding protonation constants given in Table 4.12.

If the protonation of the auxiliary complexing agent is considered, Eq. (4.76) can be written as follows:

$$\alpha_{M(B)} = 1 + \beta_1\left(\frac{c_B}{\alpha_{B(H)}}\right) + \beta_2\left(\frac{c_B}{\alpha_{B(H)}}\right)^2 + \ldots + \beta_n\left(\frac{c_B}{\alpha_{B(H)}}\right)^n \tag{4.81}$$

In the presence of a H_nB auxiliary complexing agent, the apparent extraction constant of the metal chelate should be used in Eqs (4.73) and (4.74):

$$K'_M = \frac{K_M}{\alpha_{M(B)}} \tag{4.82}$$

The pH range of the extraction is usually shifted towards higher pH values in the presence of auxiliary complexing agents.

EXAMPLE 4.18

Extract gallium(III) ions by means of an oxine solution in chloroform, using the substoichiometric method. The composition of gallium oxinate is $Ga(Ox)_3$, its extraction constant $K_M = 10^{3.72}$; if we use a reagent solution of 1 mM concentration, the lower pH limit of extraction is, according to Eq. (4.71):

$$pH \geq 6 - \frac{1}{n} \log K_M \geq 6 - \frac{1}{3} \log 10^{3.72} \geq 4.8$$

Gallium suffers, however, hydrolysis commencing from $pH \sim 4$. The extraction is therefore carried out from a tartarate medium. Let us calculate the values of the $\alpha_{B(H)}$ and $\alpha_{M(B)}$ functions by means of expressions (4.77) and (4.78) for $pH = 5$, and the lower pH limit of extraction with the apparent extraction constant; $c_B = 5$ mM. The protonation constants of tartaric acid are: $K_1 = 10^{4.54}$; $K_1 K_2 = 10^{7.56}$, the complex product of the tartarate complex of gallium(III) ions is $\beta_2 = 10^{9.8}$.

$$\alpha_{B(H)} = 1 + 10^{4.54} \times 10^{-5} + 10^{7.56} \times 10^{-10}; \quad pH = 5,$$

$$\alpha_{B(H)} = 1.350$$

$$\alpha_{M(B)} = 1 + 10^{9.8}\left(\frac{5 \times 10^{-3}}{1.350}\right)^2$$

$$\alpha_{M(B)} = 10^{4.94}$$

Table 4.11. The log β values of complexes of some cations with auxiliary complexing agents H_3X: nitrilotriacetic acid; H_4Y: ethylenediamine tetraacetic acid; H_4Z: 1,2-diaminocyclohexane tetraacetic acid

Ion	Oxalic acid			Tartaric acid		HCN			H_3X		H_4Y	H_4Z
	$\log \beta_1$	$\log \beta_2$	$\log \beta_3$	$\log \beta_1$	$\log \beta_2$	$\log \beta_2$	$\log \beta_4$	$\log \beta_6$	$\log \beta_1$	$\log \beta_2$	$\log \beta_1$	$\log \beta_1$
Ag^+	3.7					21			5.2		7.1	8.2
Bi^{3+}	4.7				11.3						26	
Cd^{2+}		5.7		2.7		10.6	18		9.5	15.5	16	18.9
Co^{2+}		6.8		2.8				19	10.8	14.3	16.6	
Cu^{2+}		10.5		3.0	5.1		25		13.0		18.9	21.6
Fe^{2+}	4.7	6.9			4.8				8.8		14.3	
Fe^{3+}	9.4	16.2	20.5		11.9			31	15.9	24.6	25.1	
Ga^{3+}			18.0		9.8						20.3	22.9
Hg^{2+}							41.4				21.8	24.3
Ho^{3+}											18.7	
In^{3+}	3.8		14.7	4.5						24.4	25.0	
Mn^{2+}		5.2		2.9					7.4		12.9	14.7
Ni^{2+}		7.9			5.4		22		11.5		18.4	19.7
Pb^{2+}		6.6		2.9			10		11.5		18.3	19.5
Sc^{3+}	8.0	13.4	16.3		12.5					24.1	23.0	25.4
Tl^+	<2										6.6	
Tl^{3+}			16.9				35				23.2	
Y^{3+}	6.5	9.3							11.4		18.0	19.2
Yb^{3+}	7.3	10.1	13.9						12.1	21.3	19.5	21.1
Zn^{2+}	4.9	7.6		2.7			19		10.4		16.3	18.7

$$B^{2-} + H^+ \rightleftharpoons HB^-; \quad K_1 = \frac{[HB^-]}{[B^{2-}][H^+]} = \frac{1}{k_2}$$

$$HB^- + H^+ \rightleftharpoons H_2B; \quad K_2 = \frac{[H_2B]}{[HB^-][H^+]} = \frac{1}{k_1}$$

$k_1, k_2, \ldots$ are the acid dissociation constants

Table 4.12. Protonation constants (K) of H_nB auxiliary complexing agents

Auxiliary complexing agent	$\log K_1$	$\log K_2$	$\log K_3$	$\log K_4$
HCN	9.2			
Oxalic acid	4.21	1.19		
Tartaric acid	4.54	3.02		
Nitrilotriacetic acid	9.73	2.49	1.89	
Ethylenediamine tetraacetic acid	10.3	6.2	2.0	2.0
1,2-Diaminocyclohexane tetraacetic acid	11.78	6.20	3.60	2.51

First approximation:

$$pH \geq 6 - \frac{1}{3} \log \frac{K_M}{\alpha_{M(B)}} = 6 - \frac{1}{3} \log \frac{10^{3.72}}{10^{4.94}}$$

$$pH \geq 6.4$$

Second approximation: the values of $\alpha_{B(H)}$ and $\alpha_{M(B)}$ should be calculated for $pH = 6.5$:

$$\alpha_{B(H)} = 1.011, \qquad \alpha_{M(B)} = 10^{5.19}$$

$$pH \geq 6 - \frac{1}{3} \log \frac{10^{3.72}}{10^{5.19}}$$

$$pH \geq 6.4$$

Selectivity. A metal of higher extraction constant can be separated from another metal having a lower extraction constant. Substoichiometric separation has a higher selectivity than that carried out with excess reagent.

Let two metal ions, charged identically, be present in the solution. It follows from the extraction equilibria that

$$\frac{[M'A_n]_{org}}{[M''A_n]_{org}} = \frac{K'_M[M']}{K''_M[M'']} \qquad K'_M > K''_M \tag{4.83}$$

The requirements of the quantitative separation of M' from M'' are:

$$\frac{[M'A_n]_{org}}{[M''A_n]_{org}} \geq 100 \quad \text{and} \quad \frac{[M']}{[M'']} \leq 0.01 \tag{4.84}$$

Substituting preconditions (4.84) into Eq. (4.83), it can be seen that the precondition of a quantitative separation by means of excess reagent is that:

$$K'_M/K''_M \geq 10^4 \tag{4.85}$$

20*

Applying substoichiometric separation, if only 50% of the metal with higher extraction constant is extracted:

$$\frac{[M']}{[M'']} = 0.5; \qquad \frac{[M'A_n]_{org}}{[M''A_n]_{org}} \geqq 100 \qquad (4.86)$$

the ratio of the extraction constants according to Eq. (4.80) will be

$$K'_M/K''_M \geqq 200 \qquad (4.87)$$

If metal ions with different charges are separated, the selectivity is a function of also the pH value of the aqueous phase and the starting concentration of the organic chelating agent.

The selectivity can be increased by adding *masking reagents*. Copper determination by means of dithizone is disturbed by the presence of mercury(II) ions. If the extraction is carried out from a solution of 0.1M potassium iodide, dithizone extracts copper only, because it cannot decompose the iodomercurate complex (HgI_4^{2-}; $K = 10^{30.5}$).

If inactive metal traces (10^{-8}–10^{-5} g) are determined by the single isotopic dilution method, the lower pH limit of the extraction is shifted toward higher pH values according to the approximation

$$pH \geqq -\frac{1}{n} \log K_M - \log (0.001\ c_{HA}) \qquad (4.73)$$

since the values of c_{HA} should also be between 10^{-8}–10^{-5} M. The upper pH limit remains unchanged, thus the pH range of extraction is narrowed. Only such organic chelating agents can be applied for which the extraction constants of the chelate complexes are high enough so that the pH of the aqueous phase need not be adjusted to a too alkaline value, which would lead to hydrolysis or to the adsorption of the trace metal on the wall of the vessel. When selecting the reagent it should also be considered that it must be stable against light, oxidizing agents, etc., even in the very large dilution required, although such effects may be eliminated by simultaneous extraction of the sample and the standard. Well proved organic chelating agents are: dithizone ($Pb^{2+}, Cu^{2+}, Hg^{2+}, Ag^+, Zn^{2+}$), α-nitroso-β-naphthol (Co^{2+}), cupferron (Fe^{3+}), etc.

4.3.2.2. OTHER SUBSTOICHIOMETRIC SEPARATIONS

Solvent extraction of ion association complexes formed with heavy organic cations such as tetraphenylphosphonium, -arsonium and -sulphonium can be successfully used for the substoichiometric separation of anions. The simplified precondition of reproducible separation is that the extraction constant should correspond to Eq. (4.85):

$$K_M = 5 \times 10^2/c_T \qquad (4.88)$$

where c_T is the starting concentration of the heavy inorganic cation. In the case of activation analysis, $c_T = 1$–10 mM, thus

$$K_M \geqq 5 \times 10^4 - 5 \times 10^5 \qquad (4.89)$$

On the basis of the extraction constants, manganese can be extracted as permanganate, rhenium as the perrhenate ion, and also other metals which form stable chlorocomplexes:

$AuCl_4^-$, $HgCl_4^{2-}$, $PdCl_4^{2-}$, etc. In the case of simple isotopic dilution

$$c_T = 10^{-9} - 10^{-6} M \text{ and thus}$$

$$K_M = 5 \times 10^8 - 5 \times 10^{11} \tag{4.90}$$

Water soluble chelate complexes, if they have a negative charge, can be separated from free metal ions which did not react in the complexing process by means of cation exchange resins, e.g., Dowex 50. The following requirement should be fulfilled by the stability constant of the chelate complex:

$$\beta_{MY} \geq \frac{1}{10^{-3} c_{H_nY}} \tag{4.91}$$

where c_{H_nY} is the starting concentration of the chelating agent. In *activation analysis*, $c_{H_nY} = 1$–10 mM, thus

$$\beta_{MY} \geq 10^6 \tag{4.92}$$

In the case of *simple isotopic dilution*, $c_{H_nY} = 10^{-10}$–10^{-5} M, thus

$$\beta_{MY} = 10^8 - 10^{13} \tag{4.93}$$

Several metal complexes of EDTA and 1,2-diaminocyclohexane tetraacetic acid fulfil the requirements concerning the stability constant. Processes have been developed mainly for the determination of iron(III) and indium(III) ions.

For selectivity, those described in Section 4.3.2.1 apply to the formation of both ion-association and chelate complexes.

The principle of *displacement substoichiometry* can be illustrated by the following example:

$$n(MA_m)_{org} + mN \rightleftharpoons m(NA_n)_{org} + nM$$

where M and N are cations of m^+ and n^+ charge, respectively; A is the one-valence anion of the chelating agent. For example, the determination of gold traces in high purity lead, biological substances, in silicon of semiconductor grade purity is possible by destructive activation analysis ($^{197}Au(n, \gamma)^{198}Au$) by reacting the gold(III) ions in a displacement reaction with a substoichiometric amount of copper diethyldithiocarbamate. The amount of gold can be obtained on the basis of the measurement of the activity in the organic phase.

The substoichiometric principle can also be applied for *precipitate separation* in activation analysis; by doing so, the selectivity of precipitation methods can be increased.

4.3.3. RADIOREAGENTS

One possible application of radioreagents can be characterized by the following schematic reaction:

$$A + B = C$$

The labelled reagent B (of a known total activity) is added to substance A to be determined and, after separation of the reaction product C, either its activity is measured, or the decrease of the activity of B is used to estimate the amount of substance A. The evaluation method is the construction of *calibration curves* with solutions of A of known concentrations.

Most determinations of this type are based on *precipitation*. The precipitate is separated from the excess reagent by filtration or centrifuging.

In the first experiments natural radioactive elements were used as tracers. For example, chromate ions were precipitated by an excess of a solution of lead ions labelled with ^{212}Pb (ThB):

$$CrO_4^{2-} + Pb^{2+} = PbCrO_{4(s)}$$

and the activity of the solution separated from the precipitate by centrifuging was measured. This method was called originally *radiogravimetry*, although it lacked just the weighing process characteristic of gravimetry.

The abundant supply of artificial radioisotopes permits one to carry out several such determinations. The determination of thallium is important from the *toxicological* point of view. Thallium(I) ions can be precipitated as thallium(I) iodide:

$$Tl^+ + {}^{131}I^- = Tl^{131}I_{(s)}$$

It is advantageous to apply dismountable polyethylene centrifuge tubes of about 10 cm length for centrifuging the low amounts of precipitates; they have a small metal plate at their bottom permitting one to measure the principitate directly. If [hexaminecobalt(III)] chloride is used, a precipitate with the composition of $[Co(NH_3)_6]TlCl_6$ can be separated. The reagent can be labelled with ^{60}Co and the precipitate dissolved in the mixture of hydrogen peroxide and acetic acid; the activity of the evaporation residue is measured.

The determination of potassium (sometimes that of low amounts of barium) is important in liquids or samples of *biological* origin. Sodium [hexanitrocobaltate(III)] labelled with ^{60}Co is a sensitive reagent for potassium; barium ions can be determined by means of $^{35}SO_4^{2-}$ reagent.

The determination of low amounts of fluoride ions is still a problem for the analyst. By means of $^{45}CaCl_2$ radioreagent the determinations can be carried out rapidly and with 1% accuracy. The solutions precipitated are centrifuged for 5 min. If the activity of the pure reagent solution and that of the solution over the precipitate is measured, the fluoride content can be read from preconstructed calibration straight lines (Fig. 4.37).

If the component to be determined is gaseous, it is adsorbed by a solution of radioreagent forming a precipitate.

The sulphur content of steel and cast iron can be determined by the radioactive variation of the well-known *Schulte* method as follows: hydrogen sulphide liberated from

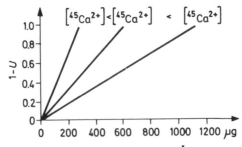

Fig. 4.37. Calibration curves for the determination of fluoride. $U = \dfrac{I_{rel,r}}{I_{rel,e}}$; $I_{rel,e}$ — the original count rate of the added reagent; $I_{rel,r}$ the count rate of excess reagent

the sample by concentrated hydrochloric acid is led into cadmium acetate solution labelled with ^{115}Cd. The ^{115}CdS is filtered off or centrifuged, and the activity of the precipitate is measured. The method is simple and rapid.

The radioreagent may also be gaseous. Phase separation means no difficulty in this case, because the excess gas can be removed easily from the system. Of course, it is the activity of the reaction products only which can be measured in this case. The determination of metal traces by ^{35}S-labelled hydrogen sulphide can be mentioned as an example.

Radio exchange reactions represent another possible field of application of radioreagents. An excess of the radioreagent in another phase is added to the solution containing the component to be determined. As a result of a *substitution* or *redox reaction* occurring between the radioreagent and the component in question, an equivalent amount of the radioactive component is liberated and transferred to the other phase. Calibration curves are necessary here, too, for concentration measurements.

As an example for the exchange between a *solid and liquid*, the determination of oxygen dissolved in water by means of ^{204}Tl-labelled thallium metal can be mentioned. Under the oxidizing effect of oxygen, an equivalent amount of labelled thallium(I) ions are dissolved:

$$4\,^{204}Tl_{(s)} + O_{2(sol)} + 2\,H_2O = 4\,^{204}Tl^+_{(aq)} + 4\,OH^-$$

The sensitivity of the determination is $2 \times 10^{-5}\%$ of oxygen.

The following exchange reaction is suitable for determining microamounts of chloride, bromide and cyanide ions:

$$^{203}Hg(IO_3)_{2(s)} + 2\,X^- = ^{203}HgX_{2(aq)} + 2\,IO^-_{3(aq)}$$

where $X = Cl^-$, Br^- or CN^-. The solubility of ^{203}Hg-labelled mercury iodate precipitate can be suppressed by adding inactive iodate solution.

The following process was developed for determination of the thiosulphate and polythionate contents of sulphite liquors in paper industry:

$$m\,^{110m}AgSCN_{(s)} + n\,S_2O^{2-}_{3(aq)} = \,^{110m}Ag_m(S_2O_3)^{(2n-m)}_{n(aq)} + m\,SCN^-_{(aq)}$$

The exchange of radioprecipitates can be combined with other analytical processes, e.g., the reaction

$$^{109}CdS_{(s)} + Cu^{2+}_{(aq)} = CuS_{(s)} + ^{109}Cd^{2+}_{(aq)}$$

can be carried out by *paper chromatography in a circular oven*; in this case, traces of metals—silver, copper, etc.—forming a more insoluble sulphide than cadmium, can be determined by measuring the activity of the zone containing cadmium ions.

By exchange between *solid and gaseous components*, the determination of low amounts of gases is possible. The gas mixture to be analyzed is passed through a column. The column packing gives off a radioactive gas when it reacts with the gas to be determined. The measurement of the activity of the secondary gas renders possible the determination of the original gas concentration by means of a calibration curve. The sensitivity of the method is determined by the conversion factor of the exchange reaction and by the specific activity of the radioactive isotope.

The measurement of ozone may be necessary *in monitoring air pollution.* This can be determined by the solid ^{85}Kr hydroquinone *clathrate:*

$$[C_6H_4(OH)_2]_3 \cdot \,^{85}Kr_{(s)} + O_{3(g)} = 3\,C_6H_4O_{2(s)} + 3\,H_2O_{(l)} + \,^{85}Kr_{(g)}$$

The method is as sensitive as $10^{-8}\%$ ozone.

Sulphur dioxide which is an important component in analyses for *environmental protection*, can be determined indirectly in as low concentrations as $10^{-7}\%$:

$$SO_2 + ClO_2^- \rightarrow ClO_2$$

where chlorine dioxide oxidizes then a radioactive clathrate. Traces of oxygen can be determined by kryptonated copper, etc.

4.3.4. RADIOMETRIC TITRATIONS

Radiometric *titration curves* show the changes of activity of a solution to be titrated as a function of the volume of the titrating solution (percent neutralization, percent oxidation or reduction, degree of titration, etc.). Equivalent consumption can be determined from the break point on the curve. Since the total activity of the solution remains unchanged during titration, end-point indication is possible only if phase exchange can be carried out by means of precipitation, solvent extraction, etc.

In precipitation titrations (e.g., radioargentometry), phase exchange follows from the nature of the chemical reaction. In all other types of titration (*complexing, redox and acid-base* titrations) the titration itself does not involve a phase exchange in the vast majority of cases; this must be ensured by an appropriate auxiliary process, e.g., *solvent extraction, ion exchange*, or the application of *labelled solid indicators*.

It is worth carrying out radiometric titrations in coloured, turbid solutions, in nonaqueous media, where radioactive indication has objective advantages. One of the advantages is that *two or three components can be determined simultaneously*. In trace determinations, extractive radiometric titrations based on the formation of metal chelates are important.

4.3.4.1. RADIOACTIVE ISOTOPES SUITABLE AS INDICATORS

Suitable radioactive indicators are, as a rule, those which have a half-life between one week and a few months. Those with longer half-life have insufficient specific activity; if the half-life is too short, activity values measured during titration should be corrected for radioactive decay. The use of harder β- ($E_{max} > 0.3$ MeV) and γ-emitters is convenient in view of the measuring technique.

The indicator should be radiochemically pure and, if possible, should be in an identical chemical form as the element to be labelled. This is important especially with elements of varying oxidation number. The specific activity of the component to be labelled should be adjusted experimentally, if possible, by adding a carrier-free preparation. The high sensitivity of radiometric titrations can be attributed to the tracer technique; its full utilization can be ensured by the appropriate selection of the conditions of the chemical reaction.

4.3.4.2. EXTRACTION TITRATIONS

Extractive radiometric titrations are carried out utilizing such *reactions of complex formation* the product of which can be extracted by an *organic solvent* immiscible with water. The activities of both phases change as a function of the volume of the titrant. Two

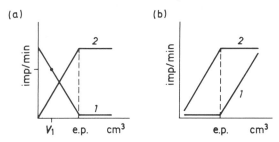

Fig. 4.38. Extractive radiometric titration curves. (a) — The titrated solution is radioactive, the titrating solution is inactive; (b) — The titrated solution is inactive, the titrating solution is radioactive; *1* — Aqueous phase; *2* — Organic phase

types of titration curves can be obtained if one component is titrated; their idealized shape is shown in Fig. 4.38.

The component or ion to be determined is labelled with a radioisotope and the solution is titrated by means of an *inactive standard solution* dissolved in an organic solvent. Another possibility is the use of an aqueous titrating solution in the presence of an organic solvent. The reaction product can be extracted by the *organic solvent which is immiscible with water.* The activity of the aqueous phase decreases, that of the organic phase increases during titration. After the point of equivalence has been reached, the activity of the organic phase becomes constant; that of the aqueous phase will be equal to the background. If the volumes are equal, the activity of the organic phase will be nearly as high as the initial activity of the aqueous phase. If a preparation with a carrier is used for labelling, the volume of titrant consumed by the indicator should be taken into account as a correction.

If an inactive solution is titrated by a radioactive titrating solution and if the product is soluble in the organic solvent, the activity of the aqueous phase is practically zero (or rather, equal to the background) until the equivalence point has been reached, afterwards it increases sharply. The activity of the organic phase increases gradually up to the equivalence point, it remains then constant. The ascendent sections of the two titration curves are parallel to each other.

Titrations are generally carried out step by step, in *separating funnels.* Continuous titration and its automatization are difficult because of the extraction step.

The end-point can be determined by graphical evaluation of the titration curves, but an extrapolation method of calculation based on the determination of two points can also be used. If the activity of the aqueous phase is measured, the end-point determined by extrapolation for curve *a* in Fig. 4.38 (V_{ex}) is:

$$V_{ex} = V_1 \frac{I_{rel, 0}}{I_{rel, 0} - I_{rel, 1}} \qquad (4.94)$$

where $I_{rel, 0}$ is the count rate in the aqueous solution before titration, imp/min; $I_{rel, 1}$ denotes the count rate after adding V_1 cm^3 of the titrant ($V_1 < V_{ex}$).

Dithizone (diphenylthiocarbazone, abbreviated as HDz) can be used to advantage for the titration of several kinds of labelled metal ion, in carbon tetrachloride or chloroform solution (see Fig. 4.38, curve *a*). In the appropriate *pH* range, dithizone forms coloured internal complexes with more than thirty different cations; the formation of "full" dithizonates—with composition MDz$_n$, where *n* is the valence of the metal—can be used

Table 4.13. Metal dithizonates CCl_4

Metal dithizonate	Tracer	Colour	pH range of extraction	Extraction constant, log K
AgDz	$^{110m}Ag^+$	Orange	−1–8	7.18
$CdDz_2$	$^{115}Cd^{2+}$	Pink	6–14	2.14
$CoDz_2$	$^{60}Co^{2+}$	Red-violet	6–8	1.53
$CuDz_2$		Red	1.5–5	10.5
$HgDz_2$	$^{203}Hg^{2+}$	Yellow	−1–4	26.8
$NiDz_2$		Brownish-violet	6–8	1.18
$PbDz_2$		Red	7–10	0.44
$ZnDz_2$	$^{65}Zn^{2+}$	Red	6–9.5	2.3

for titration. Table 4.13 contains characteristic properties of a few metal dithizonates and the appropriate tracer. The general equation of the titrations is:

$$*M^{n+}_{(aq)} + n\,HDz_{(org)} = *MDz_{n\,(org)} + nH^+_{(aq)}$$

where (aq) denotes the aqueous, (org) the organic phase.

Processes have been developed for the titration of mercury and silver in acidic medium, of cobalt in acetate buffer, of zinc in nearly neutral solution, etc. Cobalt labelled with ^{60}Co can be titrated in pyridine with a titrating solution containing ammonium thiocyanate. The [cobalt–pyridine–thiocyanate] complex can be extracted by chloroform.

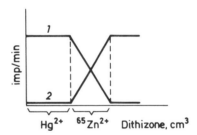

Fig. 4.39. Simultaneous titration of two metal ions *1* — Aqueous phase; *2* — Organic phase

Two metal ions can be determined simultaneously if the values of the extraction constants are sufficiently different. The metal ion with the lower extraction constant should be labelled. Figure 4.39 shows the titration curves of mercury (II) and zinc ions with dithizone, the tracer is ^{65}Zn. Simultaneous titration with dithizone is possible also for the pairs silver and zinc (^{65}Zn), copper and zinc (^{65}Zn), zinc and cobalt (^{60}Co), lead and cobalt (^{60}Co), nickel and cobalt (^{60}Co), etc.

The titration of *three components* can also be carried out simultaneously; in a favourable case, also the accuracy is sufficient. The metal ions with the lowest and highest extraction constants should be labelled here. Figure 4.40 shows the titration curves of mercury (^{203}Hg), silver and zinc (^{65}Zn) in aqueous solution, as an example.

An example for curve *b* in Fig. 4.38 is represented by the titration of thiocyanate ions by a ^{60}Co-labelled cobalt sulphate titrant. The complex produced can be extracted by iso-amyl alcohol.

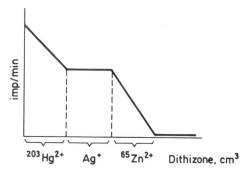

Fig. 4.40. Simultaneous titration of three metal ions; aqueous phase

The sensitivity and accuracy of extractive radiometric titrations depend on all factors to be considered in complex equilibria and extraction technique. Under proper conditions, amounts as low as a few µg can be titrated. The error is generally ± 1 rel.%.

4.3.4.3. EDTA TITRATIONS

A ^{110m}Ag-labelled silver iodate precipitate can be used as an end-point indicator of titrations carried out with EDTA (ethylenediamine tetraacetic acid, Na_2H_2Y). EDTA reacts first with the "free" metal ions quantitatively, giving a water soluble complex, e.g.,

$$Ca^{2+} + Y^{4-} = CaY^{2-}$$

It reacts with the labelled silver iodate precipitate only after the end-point has been reached:

$$^{110m}AgIO_{3(s)} + Y^{4-} = {}^{110m}AgY^{3-} + IO_3^-$$

Before the end-point, the activity of the solution over the precipitate is minimum, afterwards it increases gradually, owing to the dissolution of silver iodate.

4.3.4.4. PRECIPITATION TITRATIONS

In precipitation titrations, the activity of the liquid phase over the precipitate is measured as a function of the volume of the titrating solution added. Three types of titration curves may be obtained, depending on the way of labelling (Fig. 4.41).

If several points of the titration curve have to be determined experimentally, one titration may last as long as 40–50 minutes. For serial analysis, the well-working extrapolation method is rather applied.

In the case of an L-curve:

$$V_{ex} = \frac{V_1(I_{rel,0} - I_{rel,F})}{I_{rel,0} - I_{rel,1}}; \qquad V_1 < V_{ex} \qquad (4.95)$$

where V_{ex} is the equivalent consumption, cm^3; $I_{rel,0}$ the initial count rate of the solution to be titrated, imp/min; $I_{rel,1}$ the count rate after adding V_1 cm^3 of the titrant.

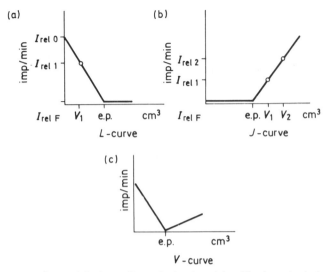

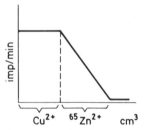

Fig. 4.41. Titration curves for precipitation radiometric titrations. (a) — The titrated solution is radioactive, the titrant is inactive; (b) — The titrated solution is inactive, the titrant is radioactive; (c) — Both the titrated and the titrating solutions are radioactive

In the case of a J-curve:

$$V_{ex} = \frac{V_1(I_{rel, 2} - I_{rel, F}) - V_2(I_{rel, 1} - I_{rel, F})}{I_{rel, 2} - I_{rel, 1}}$$ (4.96)

$$V_{ex} < V_1 < V_2$$

$I_{rel, 1}$ is the count rate in the aqueous phase after adding V_1 cm^3 and $I_{rel, 2}$ the same after adding V_2 cm^3 of the titrating solution.

$I_{rel, F}$ is the count rate in the pure solution saturated with the precipitate; its value can be read from a basic titration curve predetermined by means of several experimental points. This value can be used well in routine analysis when amounts not very different from each other are to be determined.

For simultaneous determination it is necessary that the solubilities of the precipitates should differ considerably; this can be achieved by the proper adjustment of the *pH* or by using complexing agents.

Figure 4.42 shows the titration curve of a solution containing copper(II) and zinc ions; the titrating solution is [hexacyanoferrate(II)]. First, the more insoluble copper

Fig. 4.42. Simultaneous precipitation titration of two ions

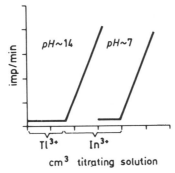

Fig. 4.43. Simultaneous precipitation titration of two ions by a radioactive titrating solution

precipitate is separated during titration; separation of the less insoluble zinc precipitate starts only at the first end-point. The less insoluble precipitate should be labelled, therefore [65]Zn is the tracer in our case.

Thallium(III) and indium(III) ions can be titrated simultaneously by ^{35}S-labelled 1-dithiocarboxy-3-methyl-3-phenylpyrazoline sodium salt, as the titrating solution. Under its effect, first thallium ions precipitate from a solution of about *pH* 14, then the activity of the solution over the precipitate begins to increase after the "Tl end-point". Then the *pH* of the solution is adjusted to about 7; this is accompanied by the precipitation of the white indium precipitate, whereupon the activity of the solution decreases to a minimum value corresponding to the solubility product; it begins to increase again only after the total precipitation of indium (Fig. 4.43).

Sulphide and iodide ions can be titrated simultaneously, after labelling with ^{131}I, by an inactive silver nitrate solution. First, the more insoluble silver sulphide precipitates; the activity of the solution is constant until the "sulphide end-point" is reached, then the precipitation of silver iodide begins and the activity of the clear solution decreases gradually until the "iodide end-point" has been reached; at the end-point and after it, a practically constant, minimum value is measured which corresponds to the solubility of silver iodide.

In developing simultaneous determinations, *chemical and radiochemical possibilities* can be *combined* successfully.

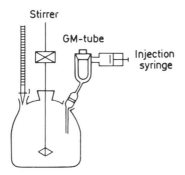

Fig. 4.44. Titrating device for precipitation titrations

During *titrations*, the count rate of the solutions can be measured, e.g., by a scintillation detector, after filtration or centrifuging of the active precipitate. Radiometric titration devices should be constructed in the laboratory. Figure 4.44 shows such an apparatus.

Precipitation titrations can be used in the mg range, their error is $\pm 3\text{--}4$ rel.%.

4.3.4.5. REDOX AND ACID-BASE TITRATIONS

The end-point of *redox titrations* can be detected by a *labelled solid indicator, by solvent extraction or labelled amalgam*. Each principle will be illustrated by an example. The indicator in the titration of iodine solution by thiosulphate can be ^{110m}Ag-labelled silver thiocyanate precipitate; this is gradually dissolved under the effect of the titrant as complex; the activity of the solution sharply increases after having reached the end-point. The determination can be carried out with 10^{-4}M solutions.

The principle of *solvent extraction* can be utilized in radiometry: if a slightly acidic labelled iodine solution is shaken with an equal volume of carbon tetrachloride, 99.5% of the iodine is transferred to the organic phase. If, however, the aqueous phase contains labelled iodide ions only, a minimum activity is dissolved in the organic phase. Thus, e.g., thiosulphate is titrated by labelled iodine solution, the activity of the organic phase begins to increase only after the reaction

$$2\,Na_2S_2O_3 + ^{131}I_2 = Na_2S_4O_6 + 2\,Na^{131}I$$

has taken place, and excess iodine is present.

Labelled cadmium (^{115}Cd) or zinc (^{65}Zn) *amalgam* can be used as an indicator for hydroquinone determination by means of iron(III) chloride titrant. After having reached the point of equivalence, cadmium and zinc ions are dissolved from the amalgam under the effect of the excess oxidizing titrating solution, thus the activity of the solution begins to increase.

For example, *radioactive kryptonates* such as [Mg^{85}Kr] and [Zn^{85}Kr] can be used as end-point indicators of *acid-base titrations*: the strong acid titrating solution liberates ^{85}Kr, whose radioactivity is measured by an appropriate detector. The activity is very low until the end-point has been reached, afterwards it increases. The end-point can be indicated also by labelled metal salts, first of all a zinc salt (^{65}Zn label); ^{65}Zn precipitates as a hydroxide at $pH \approx 7$; if a strong acid is titrated with a strong base, the activity of the solution begins to decrease at the end-point. If the concentration is lower than 0.05 M, ^{65}Zn-labelled zinc dithizonate can be used as the indicator. The zinc dithizonate can be extracted by carbon tetrachloride between $pH = 5\text{--}8$, i.e., in the range of equivalence point of acid-base titrations.

4.3.5. RADIOCHROMATOGRAPHIC METHODS

Chromatography is an efficient tool for the separation of small amounts of substances. The following sections give information about the main application possibilities of ion exchange radiochromatography, radio paper chromatography, radio paper electrophoresis, focussing paper electrophoresis, radio thin-layer chromatography and radio gas chromatography. Knowledge of the corresponding inactive methods is supposed.

4.3.5.1. ION EXCHANGE RADIOCHROMATOGRAPHY

Ion exchange chromatography is widely applied as a radioanalytical method. Ion exchange techniques permit one to separate, concentrate and purify various radioactive species on a scale from trace amounts up to a few moles.

Of polymerization type *ion exchange resins*, the polystyrene–divinylbenzene copolymer-based cation and anion exchange resins are most often used. High selectivity and high radiation resistance are characteristic of *inorganic ion exchangers* such as zirconium phosphate, chromium(III) phosphate, etc., but their chemical resistance is low.

The elution method of *ion exchange column chromatography* is most suitable for the separation of ions of similar charge and similar properties. The ions to be separated are eluted in a stepwise or a gradual manner either by a solution of an ion bonded less strongly, or a solution of a complexing agent of appropriate concentration or *pH. Metal* ions can be chromatographed on an anion exchange column, too, if appropriate complexing agents are used. As the stability of neutral or negatively charged complexes is higher, the partition coefficients of the metal ions on an anion exchanger will also increase. The order of elution of the ions to be separated will be opposite on a cation and an anion exchange column. In practice, cation and anion exchangers complete each other well. Of anion exchange resins, those with strong basicity can be applied.

Chloride ion, which forms chlorocomplexes of various composition and stability with several metal ions, is often used as a complexing reagent. The metal ions of the chlorocomplexes can be bound on anion exchange resins with high selectivity, even from solutions of very low concentrations. The extent of binding can be regulated by changing the concentration of the carrier electrolyte containing the halides.

The binding of an ion on an ion exchange resin can be characterized by the *partition coefficient (D)*:

$$D = \frac{[M]_r}{[M]_{aq}} \tag{4.97}$$

where $[M]_r$ and $[M]_{aq}$ denote the metal ion concentration in the resin phase and in the aqueous phase, respectively. Partition coefficients are well utilized by the analysis, because they permit them the planning of separations by ion exchange resins, i.e., the selection of experimental parameters theoretically. Equation (4.98) describes the adjustment of partition coefficients by means of experimental parameters such as complexing, *pH*, etc.:

$$\log D = \log K - \log \alpha_{M(B)} + n \log Q - n \log [A] \tag{4.98}$$

where K is the equilibrium constant of the

$$n\,RA + M^{n+} \rightleftharpoons R_nM + nA^+$$

ion exchange process, n the valence of the metal ion, Q the capacity of the ion exchange resin, mval/cm^3, [A] the concentration of the eluting ion—e.g., NH_4^+, Na^+, H^+, etc.—in the eluent; for functions $\alpha_{M(B)}$ and $\alpha_{B(H)}$ (the latter should be considered if the complexing reagent is protonated), see Section 4.3.2.1.

The technique of measuring. The amount of radioactive substances is determined in the *eluate* leaving the ion exchange column by measuring the count rate. The simplest equipment for this consists of a test tube and a well-type scintillation crystal. The continuous measurement of the activity can also be readily effected by passing the eluate through a glass spiral placed in a well-type scintillation crystal. A ratemeter can be used for

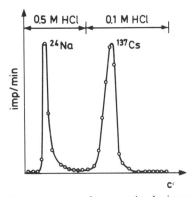

Fig. 4.45. Elution curve of a separation by ion exchange

continuous activity monitoring; a recorder attached to it can plot the elution curve afte.
appropriate amplification.

Practical applications. For analytical purposes, ion exchange chromatography i
advantageous for separating components of complex systems. Figure 4.45 shows the
elution curve of the separation of ^{24}Na and ^{137}Cs from each other; the ion exchanger is a
superficially sulphonated resin in hydrogen form, the eluent is hydrochloric acid. The
separation of alkali earth metal ions $(Mg^{2+}, Ca^{2+}, Sr^{2+}, Ba^{2+})$ can be achieved by mean
of a cation exchange column in the NH_4-form, by elution with ammonium acetate–aceti
or ammonium lactate–lactic acid buffers. Rare earth metals of similar behaviour can be
separated by eluents of appropriate *pH,* such as lactate, citrate, α-hydroxyisobutyrate, etc
Transuranium elements can be separated similarly to the rare earth metal ions
Complexing agents allow the separation also of *groups of elements;* for example, if si:
eluents of different *pH* and containing different complexing agents are used, 36 fission
products can be separated into six groups during about 3.5 h.

Ion exchange chromatography is an advantageous method of separation in *destructive*
activation analysis. The complex system produced during irradiation is separate
generally to groups of elements, which can be easily examined by γ-spectroscopy. No o
minimum amounts of carriers should be used.

Ion exchange chromatography can be favourably applied for the separation of *carrier*
free daughter elements from their mother element, e.g., for the separation of ^{140}Ba from
^{140}La or ^{90}Sr from ^{90}Y.

Ion exchange resins are well-suited for the *preparation of radioactive isotopes* with lon
half-lives. For example, radium should be separated mainly from barium, present in th
original raw material or added as a carrier; strongly acidic Dowex 50 cation exchang
resin can be applied for this purpose. Every rare earth metal can be separated from th
fission products of uranium, in a measurable quantity and high purity. The method i
useful in the preparation of transuranium elements, such as curium, berkeliun
einsteinium, etc.

The application of *labelled ions* greatly facilitates the development of new ion exchang
techniques and the determination of partition coefficients.

4.3.5.2. EXTRACTION RADIOCHROMATOGRAPHY

Extraction chromatography is a *partition chromatography with reversed phases.* The stationary phase is a liquid extracting agent, chelating reagent or liquid ion exchange resin on a hydrophobic support, such as polytrifluorochloroethylene, polytetrafluoroethylene, polyethylene, polyvinyl chloride or even hydrophilic cellulose powder; the moving phase is an aqueous solution which may also contain a complexing agent to improve selectivity. A column of appropriate dimensions may comprise as many as 500 theoretical plates. The separation is thus much more efficient than attainable by repeated batch extraction. The method is especially advantageous in hot chambers, where remote manipulation can be used.

Practical applications. Extraction chromatography can be applied advantageously in *analyses* of *fission products, fuel reprocessing, burn-up determinations*; in *radiotoxicology,* for the determination of incorporated radioisotopes (such as uranium, thorium, enriched uranium, plutonium) in daily urine; for the *separation of radioactive elements* of similar properties (e.g., ^{45}Zr and ^{95}Nb, ^{233}Pa, ^{95}Nb and ^{182}Ta), etc.

4.3.5.3. RADIO PAPER CHROMATOGRAPHY

Paper chromatography is a method of high efficiency suitable for the *separation of very small amounts* of material. Radio paper chromatography and common paper chromatography differ from each other only as far as the method of evaluation is concerned. Ascending or descending, as well as two-dimensional paper chromatography can equally be applied. Evaluation may involve autoradiography or radiometry.

Autoradiographic evaluation. For contact autoradiography, in general, an X-ray film is used. This method is discussed in detail in Section 7.6. The duration of the exposure depends, besides the nature and energy of the radioactive radiation, on the radioactivity of the chromatographic spot, the properties of the emulsion and its distance from the chromatogram, as well as on the thickness of the chromatographic paper. Radioactive spots appear on the X-ray film as dark spots. Quantitative evaluation is based on densitometry (Section 7.6.3).

Radiometric evaluation is a more sensitive, more rapid and more accurate method than autoradiography. If the radioactive spots have been previously located by autoradiography or chemically, the substance to be determined can be eluted after cutting out the spot, or the spot can be mineralized. This is followed by activity measurement.

A rather frequently used evaluation method consists in moving the paper strip at a uniform rate automatically, beginning from the starting line and proceeding to the front, *under a detector with a transversal slit.* The activity is continuously monitored by a ratemeter, and a recorder connected to it takes the radio paper chromatogram (Fig. 4.46).

Besides continuous evaluation, a stepwise forwarding technique can also be applied; e.g., counts are measured in every section of the paper for a preset time, or the strip is stopped from time to time until a preset number of counts have been reached. The time values belonging to the preset counts reflect the activity values. Several commercial devices are available for evaluating radio paper chromatograms.

The value of the R_f retention factor on the *radio paper chromatogram* is obtained by dividing the distance of the maximum activity from the starting line by the distance between the solvent front and the starting line. The activities of the individual spots, i.e.,

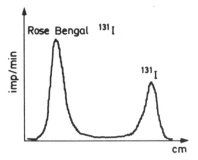

Fig. 4.46. Radio paper chromatogram of Rose Bengal contaminated with iodide ion

the amounts of substances chromatographed are proportional to the peak areas. The base line should be drawn at the level of the average of the background. For quantitative determinations, calibrating chromatograms should be made with solutions of known concentrations.

Practical applications. It is possible to chromatograph a mixture of radioactive substances, i.e., to apply paper chromatography as a method of *separation in destructive activation analysis*; one can add a radiotracer to the sample to be investigated, combining thus isotopic dilution (Section 4.3.1) with paper chromatography; it is possible to prepare radioactive derivatives of the substances to be investigated by appropriate radioactive reagents, and separate them afterwards by paper chromatography.

It is also possible to run an inactive mixture; the developed chromatogram is then treated with a radioactive reagent, or subjected to neutron activation.

If paper chromatography is applied in *destructive activation analysis* as a separation method, it is advantageous to carry out chromatography *after irradiation*. Separation can be carried out together with small amounts of a carrier or without it if the mixture of solvents contains a strong acid. Selection of the proper developing solvent mixture is very important: e.g., a mixture of 420 cm^3 of ethanol, 30 cm^3 of conc. HCl and 150 cm^3 of H$_2$O is suitable for the separation of 18 metal ions in one single run. Paper chromatography can be well combined with γ-spectrometry. After a preliminary approximative evaluation, the ready radiochromatogram is cut into several parts which can be evaluated accurately by γ-spectrometry.

Neutron activation after development is suitable, e.g., for the determination of *impurities in uranium compounds*. The bulk of uranium is removed prior to irradiation by paper chromatography using such experimental conditions that impurities should remain at the starting line. After incinerating the paper strip it is activated by neutrons; the active impurities are separated by repeated paper chromatography. Neutron activation by irradiation of the paper is generally less advantageous because radiation can easily damage the paper. Trace impurities of the paper also disturb.

Radio paper chromatography can be applied for testing the radiochemical purity of organic and inorganic radioactive preparations used in *medicine* for *diagnostics* and *therapy*. The method is suitable for the separation of amino acids; even amino acid isomers can be separated. It may also be used for the determination of nucleic acids and their components, vitamins, hormones, etc. Monosaccharides and polyols can be determined by means of *Tollens* reagent (silver nitrate containing ammonia). The developed paper chromatogram is sprayed with *Tollens* reagent; dispersed elementary silver is formed on the reducing spots which reacts rapidly with ^{131}I.

4.3.5.4. RADIO PAPER ELECTROPHORESIS

In paper electrophoresis the migration of charged particles under the effect of an electric field is utilized to achieve separation. Generally a high field strength (50 V/cm) is used; thus the duration of the run is much shorter and a sharper separation is ensured; the higher Joule heat should be removed.

Radio-electrophorograms can be evaluated by autoradiography (Section 7.6) or radiometrically (Section 4.3.5.3).

Practical applications. The method is used both in inorganic and organic analyses. A few interesting applications are as follows: very rapid *separation* of short half-life radioactive isotopes; the separation of a minimal amount of a daughter element from the irradiated target; separation of transuranium elements; separation of fission products from each other; investigation of radiochemical impurities in radioactive preparations used in *medicine*; the study of incorporation of ^{32}P in individual phosphate components of haemolysates of red blood corpuscles; the investigation of bonding of ^{131}I and ^{131}I-labelled thyroxine and triiodothyronine to proteins. The electrophoretic technique is especially advantageous for the separation of proteins.

^{137}Cs and ^{137m}Ba can be separated from each other within 30 s by *high voltage paper electrophoresis*. The time of separation is shorter and its efficiency higher if the two components of a given mixture travel in opposite directions. This can be ensured by appropriate complexing electrolytes, e.g., by EDTA. The end-products of the following nuclear reactions have been separated with high efficiency, in a carrier-free state:

$$^{160}Gd(n, \gamma)^{161}Gd \xrightarrow{\beta} {}^{161}Tb$$

$$^{170}Er(n, \gamma)^{171}Er \xrightarrow{\beta} {}^{171}Tm, \text{ etc.}$$

The ratio of irradiated and isolated amounts of substances is $10^8:1$.

I_2 and IO_3^- may form from carrier-free $Na^{131}I$ during storage. Iodine in various states of oxidation can be readily detected by paper electrophoresis. The method is suitable for the detection of the Hg^{2+} ion content of organic mercury preparations labelled with ^{197}Hg or ^{203}Hg, of Cr^{3+} impurities in $Na_2{}^{51}CrO_4$, the iodide ion content of ^{131}I-thyroxine, etc.

4.3.5.5. FOCUSSING PAPER ELECTROPHORESIS

Focussing paper electrophoresis is suitable for separation of amounts as little as a few micrograms. The scheme of an apparatus is shown in Fig. 4.47.

The *cathode vessel* contains a B^{m-} complexing anionic solution, e.g., NTA, EDTA, giving a negatively charged complex with the metal ion; the *anode vessel* contains an acid solution decomposing the complex, e.g., HCl, HNO_3. The two vessels are connected by a paper strip cooled in a stirred carbon tetrachloride bath. Before inserting the paper strip into the vessels, one half of it is wetted by the anodic solution, the other by the cathodic solution, so that the whole strip be wetted by one or the other of the electrolytes. Switching about 500 V direct voltage on the electrodes, H_3O^+ and B^{m-} ions travel toward each other on the paper strip and react with each other. As a result, a p_H-p_B gradient is formed on the paper strip, similar to the titration curve of the complexing agent by a strong acid.

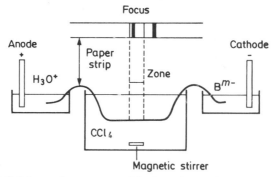

Fig. 4.47. Scheme of apparatus used for focussing paper electrophoresis

The metal ion is present on the anodic section of the *paper strip* in the form of M^{n+} cations, on the cathodic section, in turn, as a complex anion. The electric field causes these ions to travel toward each other; at a certain p_H–p_B value, the resulting movement is zero. Here is the so-called *focus*. Its location on the paper strip is determined mainly by the stability constant of the metal complex. If more than one metal ion is present in the solution, each of them forming complexes of various stability with the ligand, each metal will appear in a well-defined focus. The more stable the complex, the nearer the focus to the anode.

Practical applications. The method can be used for *separating* trace impurities in destructive activation analysis if the bulk of the matrix could previously be separated from the impurities. It is possible to separate fission products, natural radioactive substances, ^{241}Am and ^{242}Cm from the rare earth metal carrier, etc.

4.3.5.6. RADIO THIN-LAYER CHROMATOGRAPHY

The advantage of thin-layer chromatography (TLC) as compared to paper chromatography is that the similarly small amounts of substances can be separated within much shorter periods (10–30 min); a further advantage is that the spots or lines are separated from one another much more sharply. Thin-layer chromatography can be used both for analytical and micropreparative purposes.

Radio thin-layer chromatography is suitable for measuring low activities, as well as for operations with soft β-emitters. The most generally used isotopes are as follows: ^3H, ^{14}C, ^{32}P, ^{35}S, ^{125}I, ^{131}I. The conditions of separation are the same with radioactive substances as with inactive ones; evaluation, in turn, can be carried out by autoradiography or radiometry (see Sections 7.6 and 4.3.5.3, respectively).

Autoradiographic evaluation. A "No-Screen Medical X-ray Safety Film" grade is used usually for autoradiography (see Section 7.6); this film has a very sensitive emulsion. A good contact should be ensured between the film and the thin-layer chromatogram. Traces of solvents have to be removed carefully and reducing spots must be absent on the chromatogram. It is advantageous to coat the plate by silica gel mixed with anthracene if the chromatogram is run with ^3H-labelled compounds. These "*fluorograms*" are very sensitive, especially at lower temperatures.

Radiometric evaluation. If the position of the spot is known, the determination can be carried out by removing it, followed by the elution of the radioactive substance with an appropriate solvent and measurement of the activity of this solution. It is advantageous to apply the *liquid scintillation* technique for measuring soft β-emitters. Several scintillators can be used as eluting agents, too. If the material of the removed spot is mixed to a liquid scintillator, the radioactivity of this suspension can be measured. Alternatively, the developed chromatogram is sprayed with aqueous polyvinyl propionate. After drying, the thin layer adheres to the thin plastic film and can be removed from the glass plate. Then the other side of the thin layer can also be sprayed and the radiometric evaluation carried out. This process is suitable for isotopes emitting harder radiation. Devices suitable for the direct quantitative evaluation of radio thin-layer chromatograms are available commercially. Some devices designed for the evaluation of radio paper chromatograms have been adapted for evaluating thin-layer chromatograms, too.

The sensitivity of the method with a given radioactive isotope depends on the *detector*; open methane flow-through proportional counters have the highest efficiency and are suitable also for the measurement of ^3H. The accuracy of the measurement and the resolving power of the apparatus depend on the rate of movement of the plate and on the width of the slit of the detector. On well-designed devices these two parameters are adjustable.

Practical applications. It is possible to apply onto the thin layer a radioactive mixture, i.e., labelled substances, to use the isotopic dilution method, or to prepare a radioactive derivative by means of an appropriate reagent. An inactive mixture may also be run and the ready chromatogram is then treated with a radioactive reagent.

Thin-layer chromatography has been applied in inorganic analysis for the *separation* of several radioactive cations. It is widely used in organic analysis to separate labelled compounds of the most various structures especially for the elucidation of *biochemical, pharmacological and toxicological* problems; it is also used to check the *purity of pharmaceutical and diagnostic preparations.*

4.3.5.7. RADIO GAS CHROMATOGRAPHY

Radio gas chromatography is applied first of all for the separation and determination of ^3H- and ^{14}C-labelled compounds. The technique of the method differs from inactive gas chromatography only in that the radioactivity of gases or vapours leaving the column is also monitored by an appropriate radioactivity detector. For separation, the elution method is used almost exclusively.

Batch type analysis is carried out, in general, with low specific activities. In this method, individual fractions leaving the column are trapped separately, e.g., in traps cooled by dry ice or liquid nitrogen, or by adsorption on activated carbon or molecular sieves of the proper pore size.

Radioactive isotopes emitting hard β- or γ-radiation are detected by a GM-tube or by a NaI(Tl) scintillator; for monitoring soft β-emitters such as ^3H and ^{14}C, the vapours leaving the column are condensed, e.g., over anthracene crystals coated with silicon oil. The separated fraction is dissolved in the silicon oil, its radioactivity may be measured by a scintillation spectrometer. A gas fraction collecting adapter is available, for special liquid scintillation spectrometer.

In the case of *continuous monitoring*, the radioactivity detector is connected after one of the usual gas chromatographic detectors. The latter measures all components, the former labelled compounds only. The two measurements occur in a parallel way; signals are recorded by two synchronized recorders.

The properties of the radioactive radiation and the experimental conditions of gas chromatography must be considered together in determining the type of radioactivity *detector* to be used. Hard β- or γ-radiations can be monitored in the effluent by a GM-tube or by a well-type NaI(Tl) scintillation detector. Ionization chambers, gas flow-through proportional counters, liquid or plastic scintillators can be used for measuring the activity of soft β-emitters (^3H, ^{14}C).

Gas flow-through proportional counters have an efficiency of 100% for ^{14}C and 60% for ^3H with a sensitivity of 4–40 Bq (0.1–1 nCi). Methane gas ensures an excellent counting efficiency, but it is disadvantageous in view of the sensitivity of the gas chromatographic detector and may deteriorate, in general, the whole chromatographic system. The usual carrier gases such as H_2, He and Ar are disadvantageous for the counting efficiency. An appropriate mixture of the counting and carrier gases is used, as a rule. It is advisable that the radioactive component should have as long a time of residence in the radioactivity detector as possible. This time is determined by the volume of the detector and the flow rate of the carrier gas. Elution should be effected with low gas volumes in order to get a significant signal as compared with the background noise.

The specific activity can be determined from the ratio of the peaks on the mass chromatogram and the radiochromatogram, comparing this ratio with a standard. The position and the shape of the two peaks help to identify the radioactive substances and their purity can be checked. If traces of radioactive impurities are present, they may distort the peak on the radiochromatogram as compared with the mass chromatogram.

The method of burning has some advantages over the direct activity measurement of the effluent. Organic substances leaving the chromatographic column are burned to give carbon dioxide and water; the latter is reduced to hydrogen gas and the activities of $^{14}CO_2$ and ^3H are then measured separately or simultaneously. For this the apparatus need be calibrated for two gases only, the most suitable radioactivity detector can be applied, etc.

Practical applications. Radio gas chromatography is suitable for *gas analysis*, e.g., it is possible to distinguish hydrogen gases of different isotopic composition. Other applications are the separation of artificial radionuclides obtained by activation with neutrons or charged particles, the analysis of *hydrocarbons* labelled with ^3H and/or ^{14}C

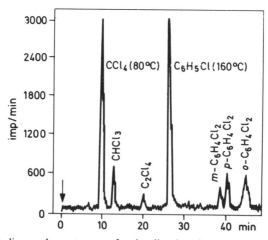

Fig. 4.48. Radio gas chromatogram of an irradiated equimolar CCl_4–C_6H_5Cl system

(Figure 4.48), purity control of labelled compounds, the preparation of labelled, mainly organic compounds. It can be used advantageously in *hot atom chemistry* as a method of separation, especially for the isolation of short-lived radioactive isotopes.

4.3.6. ISOTOPIC EXCHANGE REACTIONS

Isotopic exchange reactions (see also Sections 3.6.2.1 and 3.8.2) are *exclusively radiochemical methods*; they have no inactive analogies. They can be used first of all as rapid separation processes in *destructive activation analysis*, but *independent analytical procedures* have also been developed for trace determinations on the basis of isotopic exchange.

Every reaction is called isotopic exchange reaction where different isotopes of a given element are exchanged between two or more states of the element, being in equilibrium apart from isotopic distribution, e.g.,:

$$Hg_{(vapour)} + {}^{203}Hg^{2+}_{(aq)} \rightleftharpoons {}^{203}Hg_{(vapour)} + Hg^{2+}_{(aq)}$$

Different states may mean different physico-chemical states, different molecules of different types of bonds.

Heterogeneous isotopic exchange reactions are suitable for separation: amalgam exchange, precipitate exchange and isotopic exchange on ion exchange resins.

The general equation for *amalgam exchange* is as follows:

$$M^0(Hg) + {}^*M^{n+}_{(aq)} \rightleftharpoons {}^*M^0(Hg) + M^{n+}_{(aq)}$$

where $M^0(Hg)$ is the appropriate inactive liquid amalgam and ${}^*M^{n+}_{(aq)}$ denotes the radioactive isotope in the aqueous phase. When shaken, the radioactive isotope goes from the solution to the amalgam, e.g., ${}^{65}Zn^{2+}$ ions into the zinc amalgam. If a carrier-free radioactive isotope is to be separated, the ratio of inactive to active atoms can be as high as $10^6:1$; this makes possible a very high efficiency. Amalgam exchange takes place within a few minutes, it is rather selective and—even though it is not always quantitative—it is useful because of its simplicity and rapidity, especially in the separation of short-lived radioactive isotopes. If the task is more complicated than to separate disturbing metals and also includes the determination of the metal incorporated into the amalgam, then, after the exchange, the radioactive isotope is displaced from the isolated amalgam by an inactive solution containing the same of foreign ions. Amalgam exchange processes have been developed first of all for the radiochemical separation of Bi, Cd, Ga, Tl and Zn. Active Hg(II) ions can also be bound by metallic mercury even if the mercury is present as a complex in the solution.

Ionic precipitations can be used to advantage in several cases to bind radioactive trace elements from their ionic solutions. Binding is the result of various processes such as isotopic exchange, adsorption, redox reaction, ion exchange or selective sorption. Some examples are the following: the reaction between silver chloride precipitate and silver ions can be used to separate silver; the exchange between silver iodide precipitate and iodide ions is suitable for the isolation of iodides; strontium can be separated by means of strontium sulphate or oxalate, antimony by antimony trioxide, tin by tin dioxide hydrate. By using metal sulphide precipitates corresponding to the order of the solubility products, it is possible to separate the corresponding radioactive isotopes from the solution. Among others, the separation of ${}^{203}Hg^{2+}$, ${}^{65}Zn^{2+}$ and ${}^{24}Na^+$ is possible as follows: the strongly acidic solution is passed first through a HgS precipitate which binds radioactive mercury; the acidity is diminished and the radioactive zinc is isolated by ZnS; disturbing radioactive sodium remains in the solution. Radioactive ions present in trace amounts are practically totally bound by the relatively high amounts of inactive precipitates.

The flow-through technique is advantageous for carrying out these reactions rapidly; a small filtration crucible required for this purpose is depicted in Fig. 4.49. By measuring the activity of the separated *precipitates*, it is possible to determine the amount of impurity in question when simultaneously irradiated standards are also used. If *biological substances* are activated by thermal neutrons, the activity originating from ^{24}Na is, as a rule, very disturbing. This can be bound by hydrated antimony pentoxide (HAP) used as a chromatographic column through which a strongly acidic solution is passed.

Ion exchange resins can also be used as a solid phase for separating a radioactive matrix element from radioactive trace elements. If the resin binds the matrix element selectively, then the solution is passed after destruction through a column containing the ion exchanger in the form of the matrix element; the breakthrough of the radioactive matrix element is delayed owing to isotopic exchange, whereas radioactive trace elements are readily eluted from the column.

Heterogeneous isotopic exchange *between a solution and a gas*, between metal components in an *aqueous and an organic solution* can also be applied for analytical purposes. The exchange reaction between mercury(II) ions in acidic aqueous phase and mercury di(n-butylthiophosphate) in carbon tetrachloride can be utilized for the determination of small amounts of mercury (10^{-7}–10^{-2} g); the tracer is ^{203}Hg. Similarly, mercury can be determined in biological samples (urine, tissues) by adding a ^{203}Hg^{2+} solution to the sample, binding the mercury by cysteine and leading mercury vapour through the solution. Isotopic exchange leads to the appearance of radioactive mercury in the vapour phase, and this can be monitored continuously. The half-life of the exchange reaction ($t_{1/2}$) is proportional to the mercury concentration of the aqueous phase:

$$t_{1/2} = \frac{0.693 \, V c_{Hg(aq)}}{v_m} \qquad (4.98)$$

where V is the volume of the solution, cm^3; v_m the rate of flow of the mercury vapour, µg/min; $c_{Hg(aq)}$ the mercury concentration of the aqueous phase, µg/cm^3.

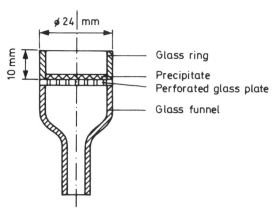

Fig. 4.49. Filtration crucible for isotopic exchange

Isotopic exchange in the *homogeneous phase* can be used for analytical purposes only if one of the components can be separated after the exchange equilibrium has been established. According to one of the processes developed for the determination of micro-quantities of iodine, an iodide solution labelled with carrier-free ^{131}I is allowed to react with methyl iodide, then, after an hour, methyl iodide is removed by heating. The activity of the remainder dissolved in acetone is measured:

$$c_x = c_1 A_\infty/(A_0 - A_\infty) \qquad (4.99)$$

where c_x is the unknown amount of iodide to be determined, mol; c_1 means the molar concentration of the methyl iodide solution; A_0 and A_∞ are the activity of the sample to be investigated and that of the remainder, respectively.

The rapid exchange reaction between the chelate complex and the simple coordination complex of a metal in organic phase can be utilized for the determination of traces of metal. The two types of complex can be readily separated: simple coordination complexes can be transferred into the aqueous phase by a slightly acidic washing. The specific activity of the metal in the two complexes is the same after the isotopic equilibrium has been attained:

$$\frac{A_1}{m_1} = \frac{A_2}{m_2} \qquad (4.100)$$

where A_1 and A_2 are the activities of the phases containing the metal chelate and the coordination complex, respectively; m_1 and m_2 are the corresponding amounts of metal in the two types of complex.

The principle can be used for determination of traces of bismuth. Bismuth is extracted from the sample to be investigated into diethyl dithiocarbamate in carbon tetrachloride, the excess reagent is removed by alkaline washing, then the solution of the chelate complex in carbon tetrachloride is shaken with a known amount of ^{210}Bi (RaE)-labelled $[HBiI_4]$ complex dissolved in amyl acetate. The two solvents are miscible. After the exchange equilibrium has been established, the $[HBiI_4]$ complex is transferred into the aqueous phase by an acetate–acetic acid buffer of *pH* 4 and the radioactivity is measured in the organic and the aqueous phase.

Traces of other metals can be determined according to the same principle.

REFERENCES

[4.1] *Advances in X-Ray Analysis.* Vols 15, 16. Plenum Press, New York, 1972, 1973.

[4.2] Azaroff, L. V. (Ed.), *X-Ray Spectroscopy.* McGraw-Hill, New York, 1974.

[4.3] Brunner, G., *Isotopenpraxis,* **5,** 328 (1969).

[4.4] Brunner, G., Dahn, E. and Geisler, M., *Analyse von Mineral- und Syntheseölen mit radiometrischen Methoden.* Akademie-Verlag, Berlin, 1968.

[4.5] Burger, K., (Application of Mössbauer Spectroscopy in Complex Chemistry) (In Hungarian). In *A kémia újabb eredményei,* Vol. 9, Akadémiai Kiadó, Budapest, 1972.

[4.6] Criss, J. W., *Anal. Chem.* **48,** 179 (1976).

[4.7] *Handbook of Spectroscopy.* CRC Press, Cleveland, Ohio, 1974.

[4.8] Hartmann, W., *Meßverfahren unter Anwendung ionisierender Strahlung.* Akademische Verlagsgesellschaft Geest & Portig K.-G., Leipzig, 1969.

[4.9] Jenkins, R., *An Introduction to X-Ray Spectrometry.* Heyden and Sons Ltd., London, 1975.

[4.10] Jentzsch, D. and Otte, E., Detektoren in der Gas-Chromatographie. In *Methoden der Analyse in der Chemie.* Vol. 14. Akademische Verlagsgesellschaft, Frankfurt am Main, 1970.

[4.11] May, L., *An Introduction to Mössbauer Spectroscopy.* Adam Hilger, London.

[4.12] Mott, W. E., *Isotopic Techniques in the Study and Control of Environmental Pollution.* IAEA-SM-142a/1.

[4.13] Müller, R. O., Spectrochemical Analysis by X-Ray Fluorescence. Adam Hilger Ltd., London, 1972.
[4.14] Nagy, L. Gy.,(Radiochemistry and Isotopic Technique)(In Hungarian). Tankönyvkiadó, Budapest, 1983. 1983.
[4.15] Radioisotope X-Ray Fluorescence Spectrometry. IAEA-STI/DOC/10/115. Vienna, 1970.
[4.16] Rasberry, S. D. and Heinrich, K. F. J., Anal. Chem. 46, 81 (1974).
[4.17] Tölgyessy, J.,(Nuclear Radiation in Chemical Analysis)(In Hungarian). Műszaki Könyvkiadó, Budapest, 1965.
[4.18] Uchida, K., Tominaga, H. and Imamura, H., IAEA-SM-68/6.
[4.19] Victoreen, J. A., J. Appl. Phys. 20, 1141 (1949).
[4.20] Watt, J. S., Radioisotope On-stream Analysis. in Atomic Energy in Australia, 16(4), October, 1973.
[4.21] Woldseth, R., X-Ray Energy Spectrometry. Kevex Corp., Burlinghame, Calif., 1973.
[4.22] Birks, L. S. and Gilfrich, J. V., Anal. Chem. 44, 557R (1972); 46, 360R (1974).
[4.23] Birks, L. S. and Gilfrich, J. V., Anal. Chem. 48, 273R (1976).
[4.24] Tölgyessy, J. and Varga, S., Nuclear Analytical Chemistry. VEDA, Bratislava, 1975.
[4.25] Henley, E. J. and Lewis, J. (Eds), Advances in Nuclear Science and Technology. Vols 8, 9. Academic Press, New York, 1975.
[4.26] Coomber, D. I. (Ed.), Radiochemical Methods in Analysis. Plenum Press, New York, 1975.
[4.27] ATOMKI Közlemények 17, No. 3 (1975)
[4.28] Berei, K. and Vasáros, L., Isotopenpraxis 4, 19 (1968).
[4.29] Bereznay, T., Bodis, D. and Keömley, G., J. Radioanal. Chem. 36, 59 (1977).
[4.30] Bowen, H. I. M. and Gibbons, D., Radioactivation Analysis. Clarendon Press. Oxford, 1963.
[4.31] Braun, T., Die Analysen mit radioaktiven Reagenzien und die radiometrischen Titrationen. Radiochemische Methoden in der analytischen Chemie SVA. Separate from Chimia, 21, Nr. 1, 3, 4 (1967).
[4.32] Braun, T. and Tölgyessy, J., Radiometric Titrations. Akadémiai Kiadó, Budapest, 1967.
[4.33] Braun, T. and Ghersini, G., Extraction Chromatography. Akadémiai Kiadó, Budapest, 1975.
[4.34] Broda, E. and Schönfeld, T., Radiochemische Methoden der Mikrochemie. Springer Verlag, Wien, 1955.
[4.35] Bujdosó, E., Fizikai Szemle 25(3), 85 (1975)
[4.36] Burger, K., Organic Reagents in Metal Analysis. Akadémiai Kiadó, Budapest, 1973.
[4.37] Cali, I. P., Trace Analysis of Semiconductor Materials. Pergamon Press, Oxford, 1964.
[4.38] Crouthamel, C. E., Applied Gamma-Ray Spectrometry. Pergamon Press, Oxford, 1960.
[4.39] De Voe, J. R. (Ed.), Modern Trends in Activation Analysis. NBS Vols I–II, Gaithersburgh, 1969.
[4.40] Erdey, L., Gravimetric Analysis. Vols I–III. Akadémiai Kiadó, Budapest, 1960.
[4.41] Girardi, F., Pietra, R. and Sabbioni, E., J. Radioanal. Chem. 5, 141 (1970).
[4.42] Hais, I. M. and Maček, K., Handbuch der Papierchromatographie. VEB Gustav Fischer Verlag, Jena, 1958.
[4.43] Inczédy, J., (Applications of Ion Exchangers in Chemical Analysis), Műszaki Könyvkiadó, Budapest, 1962.
[4.44] Inczédy, J., (Analytical Applications of Complex Equilibria) (In Hungarian). Műszaki Könyvkiadó, Budapest, 1970.
[4.45] Inczédy, J., Analytische Anwendungen von Ionenaustauschern. Akadémiai Kiadó, Budapest, 1964.
[4.46] Iwantscheff, G., Das Dithizon und seine Anwendung in der Mikro- und Spurenanalyse. Verlag Chemie, Nürnberg, 1972.
[4.47] Koch, R. C., Activation Analysis. Academic Press, New York–London 1960.
[4.48] Kőrös. E., MTA Kém. Tud. Oszt. Közl., 33, 223 (1970).
[4.49] Lengyel, T. and Jász, Á., (A Pocket Book for Isotope Laboratories) (In Hungarian. Műszaki Könyvkiadó, Budapest, 1966.
[4.50] Lyon, W. S., Guide to Activation Analysis. Van Nostrand, Princeton, N. Y., 1964.
[4.51] Matus, L. and Szabó, E., Kémiai Közl. 31, 305 (1969).
[4.52] Minczewski, J., Preconcentration in Trace Analysis (Eds Meinke, W. W., Scribner, B. P.), Trace Characterisation, Chemical and Phys. US. N.B.S. Washington, 385, 1967.
[4.53] Morrison, G. H. (Ed.), Trace Analysis, Physical Methods. Interscience Publ., New York–London–Sydney, 1965.
[4.54] Nagy, L. Gy. and Szokolyi, L. (Neutron Activation Studies) (In Hungarian). Műszaki Könyvkiadó, Budapest, 1966.
[4.55] Pierce, D. B., Radioanal. Chem. 12, 23 (1972); 17, 55 (1973).
[4.56] Proceedings GDCh 1976 International Conference. Modern Trends in Activation Analysis, Vols I–II, München, FRG, 1976.
[4.57] Ruzička, J. and Stary, J., Substoichiometry in Radiochemical Analysis. Pergamon Press, Oxford, 1968.
[4.58] Salamon, A. and Szabó, E., Kémiai Közl., 33, 179 (1970).
[4.59] De Soete, D., Gijbels, R. and Hoste, J., Neutron Activation Analysis. John Wiley, London, 1972.

[4.60] Stary, J., *The Solvent Extraction of Metal Chelates*. Pergamon Press, Oxford, 1964.

[4.61] Stary, J., Kratzer, K. and Zeman, A., *J. Radioanal. Chem.*, **5**, 71 (1970).

[4.62] Szabó, E., *Kémiai Közl.*, **39**, 306 (1968).

[4.63) Szabó, E. and Simonits, A., *(Activation Analysis)* (In Hungarian). Műszaki Könyvkiadó, Budapest, 1973.

[4.64] Tölgyessy, J., *(Nuclear Radiation in Chemical Analysis)* (In Hungarian). Műszaki Könyvkiadó, Budapest, 1965.

[4.65] Tölgyessy, J., Braun, T. and Kyrš, M., *Isotope Dilution Analysis*. Akadémiai Kiadó, Budapest, 1972.

[4.66] Tölgyessy, J. and Varga, S., *Nuclear Analytical Chemistry*, Vols I–II. University Park Press, New York–Publ. House of the Slovak Acad. Sci., 1972.

[4.67] Török, G., *(Programmed Destructive Neutron Activation Analysis. Method, Elaboration and Application of Apparatus)* (In Hungarian). Ph. D. Thesis, Budapest, 1977.

[4.68] Vámos, E., *(Chromatography)* (In Hungarian). Műszaki Könyvkiadó, Budapest, 1959.

[4.69] Vámosné, Vigyázó L., *(Paper Electrophoresis)* (In Hungarian). Műszaki Könyvkiadó, Budapest, 1967.

[4.70] Veres, Á. and Pavlicsek, I., *Kémiai Közl.*, **33**, 243 (1970).

[4.71] *VI. Internationales Symposium für Mikrochemie. Graz, Österreich*. Vol. E. 9, Verlag der Wiener Medizinischen Akademie, Vienna, 1970.

[4.72] Zolotov, Yu. A., *(Extraction of Internal Complex Compounds)* (In Russian). "Nauka" Publ. House, Moscow, 1968.

[4.73] *(Scientific-Technical Conference on the Apparatus for Activation Analysis)* (In Russian). Collected lectures, Moscow, 1969.

5. NUCLEAR BOREHOLE GEOPHYSICS

Well logging (borehole geophysics) includes various physical and physico-chemical measurements and techniques applied *in drilled holes* for the determination of different properties of the penetrated rocks, the conditions of the well, etc. Well logging is carried out with a *sonde* containing transmitter–receiver pairs; the sonde is lowered into the borehole using a *logging cable* of appropriate breaking strength and electrical insulation

Nuclear logging is a branch of well logging that is of great importance in mining for *oil coal* and *ores* as well as in the exploration of *water.* The main *advantages* of nuclear methods can be summarized as follows:

— they give information even on the chemical composition of the rocks penetrated by the hole, whereas this property is inaccessible to most other logging methods;

— they can be applied in cased holes and in wells drilled by using non-conductive drilling fluids.

Their *disadvantages* are:

— their limited radius of investigation;

— their application tends to be labour-consuming in comparison to other logging methods;

— their response is strongly influenced by statistical fluctuations;

— a quantitative interpretation of the measurement results is complicated and not easily understandable;

— extreme care needs to be exercised in view of the possibility of radiation hazard

The principal design of the instruments applied to nuclear logging is generally the same as that used in other fields of isotope techniques (Section 2.1), though obviously the way they are constructed differs.

The most important peculiarities of *nuclear logging instruments* are as follows:

— the detectors are operated at a considerable distance (up to several kilometres) from the recording equipment; thus, the transmission of the signal is by no means perfect;

— only a limited number (1–7) of conductors are available for power supply, control of the sonde, and transmission of the detected signal;

— the sondes need to be able to withstand severe operating conditions, viz. a temperature of 250–300 °C pressures of 120–160 MPa (1200–1600 bar);

— the dimensions of borehole instruments are limited; standard logging tools have diameters of 65–90 mm, but considerable benefit would be gained by their reduction to a range of, say, 24–40 mm.

During logging, the sondes, which contain detectors, an electronic cartridge (high voltage power supply, amplifier, etc.) and a suitable radiation source (Fig. 5.1.), are lowered into the borehole using a cable of 5–80 kN breaking strength which is spooled off and on by a truck-mounted winch.

The detected signals are transmitted through the cable to a suitable surface panel where, after having been amplified, selected, or unified and integrated, they are finally recorded as a function of the depth.

Earlier, GM-counters were exclusively used in well logging for detecting γ-photons but

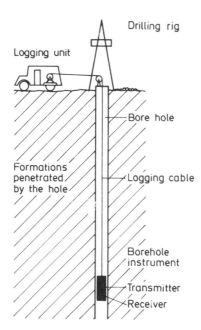

Fig. 5.1. Principal scheme of well logging methods

now, in boreholes with a bottom temperature not higher than 200 °C, scintillation detectors are more commonly used. Scintillation detectors have been found to be the most suitable for detecting neutrons but helium proportional counters are also gaining in importance. Semi-conductor detectors have been applied (in through-turbing instruments) but mainly for studying slim holes.

In well-logging *practice*, the nuclear methods can be divided into three main groups:
— measurement of *natural radioactivity*;
— detection of radiation *induced by outer*;
— *isotope tracer techniques* used in boreholes. These groups could be subdivided even further in accordance with the emitted and detected radiation (Table 5.1).

According to the purpose and conditions of application these measurements can be grouped as *exploration logging, production logging, or open hole and cased hole logging.*

The methods will be discussed here in accordance with the system presented in Table 5.1; reference will be made to the field and the way the methods are applied.

5.1. NATURAL γ-RAY LOGGING

The measurement of natural γ-radiation along the borehole, is called natural γ-ray logging. Two types of this method can be distinguished, namely
— *total intensity measurement* (independent of energy) is called natural γ-ray logging or, in brief, γ-logging (symbol: GR);
— *energy selective measurement* which is also known as *γ-ray spectral logging* (more exactly, *natural γ-ray spectral logging*).

The radioactivity of rocks results from radioactive elements accumulated in them; such elements are uranium, thorium as well as the products of their disintegration, and

Table 5.1. Nuclear methods in borehole geophysics

Detected radiation	External source					
	Unnecessary	Sealed natural source	Artificial source	Sealed natural neutron source	Accelerator neutron source	Radioactive chemicals
γ	Natural γ-method	γ-γ method	Photo-activation method	Neutron-γ method	Inelastic scattering method	
		Selective γ-γ method	γ-activation analysis	Neutron-γ spectroscopy	Fast neutron activation method	Tracer techniques
	Natural γ-spectroscopy	Other "weak" methods	Other "hard" methods	Neutron activation method		
Neutron	—		γ-neutron method	Neutron-thermal neutron method	Neutron life-time logging	
				Neutron-epithermal neutron method		

potassium. When radioactive isotopes disintegrate, α-, β- and γ-radiation is emitted, but only γ-radiation has the ability to penetrate the media so that it can be detected under borehole conditions.

In the decay of some isotopes neutron emission can also occur; its detection, however, has not become widely used because of the accompanying γ-radiation.

5.1.1. GEOCHEMICAL CHARACTERISTICS OF ROCKS

Uranium (^{238}U) is present in nature in both tetra-valent and hexa-valent forms. It has an amphoteric character, i.e., it forms uranates and diuranates in alkaline media; these products hardly dissolve in water.

Radium (^{226}Ra), when reacted with acids, produces salts such as carbonates and sulphates which are almost insoluble, as well as halides and nitrates which easily dissolve in water. It does not form minerals itself, but occurs generally in a dispersed form.

Table 5.2. Radioactive material content

(a) Igneous rocks

Rock	Radioactive elements, kg/kg			
	Ra, 10^{-12}	U, 10^{-6}	Th, 10^{-6}	K, 10^{-2}
Acidic	1.40	4.0	13.0	2.6
Neutral	0.51	1.4	4.4	2.0
Basic	0.38	1.1	4.0	1.4
Ultrabasic	0.20	0.6	2.0	0.4

(b) Sedimentary rocks

Rock	Radioactive elements, kg/kg		
	U, 10^{-6}	Th, 10^{-6}	K, 10^{-2}
Shale	3.00	11.40	2.40
Limestone	1.38	1.16	0.20
Dolomite	1.20	0.47	0.13
Sandstone	1.20	5.00	1.20

Thorium (^{232}Th) is always a tetra-valent element and it can commonly be found in nature in the form of insoluble oxides, silicates (e.g., thorite) and of complex salts (e.g., monazite).

Potassium (^{40}K) is present in the majority of minerals (feldspars, micas, glauconite, etc.) although generally in small quantities. In sedimentary rocks, mostly in shales, besides the chemically bonded potassium, the amount of adsorbed potassium is also significant.

The specific *natural γ-radioactivity* of a rock, which is generally measured in Bq/kg or 10^{-12} g Ra-equ/g, is determined by the amount of the radioactive isotopes accumulated in them. Since the main radioactive elements occur in various forms of minerals, the radioactivity may change even within the same minerals.

Table 5.2 shows the radioactive material content of some igneous (a) and sedimentary (b) rocks. Figure 5.2 gives an overview of the radioactivity of different minerals.

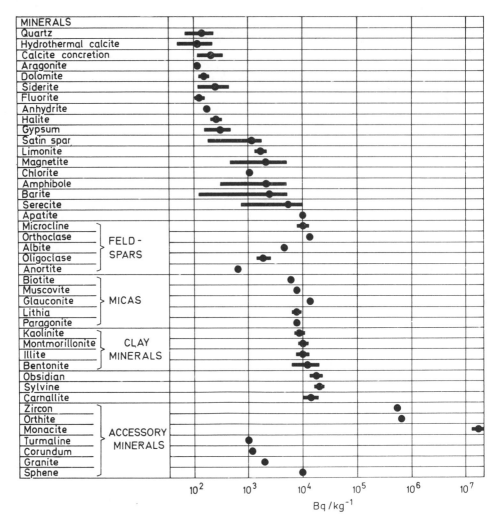

Fig. 5.2. Radioactivity of some rock-forming minerals

5.1.2. THEORY OF THE METHOD

The data obtained in well logging are influenced by many factors; because of this, an exact mathematical–analytical solution of the problem has not been found for the general case. In a system of *cylindrical symmetry*, commonly used in borehole geophysics, the problem can be described by an exponential integral equation whose numerical solution leads to the conclusions necessary for us (Sections 5.3 and 5.4).

Figure 5.3 shows relative γ-ray curves, obtained with theoretical calculations, based on a system of cylindrical symmetry, for beds of thickness h, penetrated by a borehole of diameter d_0. The following conclusions can be drawn from the figure:

— the value of the anomalies observed in certain beds is dependent on the thickness of each bed;

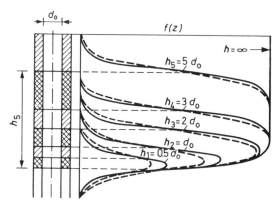

Fig. 5.3. Configuration of $I_{rel\gamma}$ curves against layers of different thickness. ———Point detector; – – – – Detector of 40 cm length

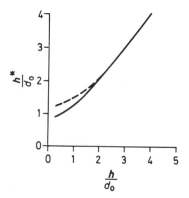

Fig. 5.4. Relationship between apparent thickness h^* and true thickness h. ——— Point detector; – – – – Detector of 40 cm length

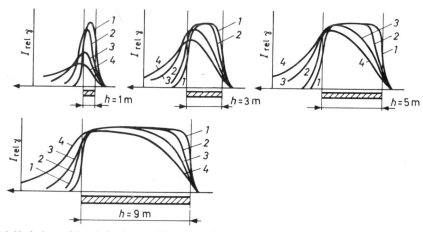

Fig. 5.5. Variations of $I_{rel\gamma}$ deflection at different running speeds (time constant $t_i = 6$ s); (direction of running marked with an arrow); speed $1 = 0$ m/h; speed $2 = 300$ m/h; speed $3 = 600$ m/h; speed $4 = 900$ m/h

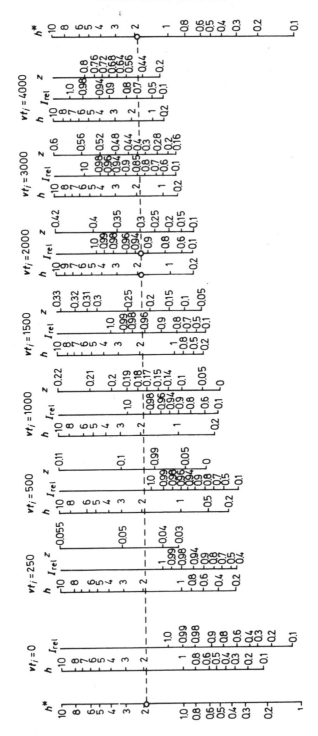

Fig. 5.6. Nomogram for determining true thickness *h* from apparent thickness *h** at different time constants (t_i) and running speed (v)

— the boundaries of thick beds should be located at the half-values of the intensity maxima; those of thinner beds shift towards the intensity-maximum values. The distortion is shown in Fig. 5.4.

Although the above equations were derived assuming that the measurements are performed with a *stationary* instrument, in practical use, the sonde is *run at a speed of 50–900 m/h up-hole.* The movement of the sonde causes *distortions on the curves,* in accordance with the integration time constant of the recorder (Section 2.1.4). The measure of distortion can be calculated on the basis of *Kirchhoff*'s rules; curves derived in this way are shown in Fig. 5.5. As seen in the Figure, with increasing running speed the indication becomes deformed and is shifted in the direction of movement of the sonde. Figure 5.6 presents correction charts necessary for routine work.

The area-method is generally used in ore mining where the thickness of beds is small; the reason for this is that the correction of the apparent values using the diagram in Fig. 5.5 would lead to inaccurate results. If it is assumed that the density of the bed studied and that of the adjacent formations are the same, the following equation can be written for the area below the recorded anomaly:

$$S = 2 \int_0^{h/2} I_{rel_1}(z)dz + 2 \int_{h/2}^{\infty} I_{rel_2}(z)dz \qquad (5.1)$$

where S is the area of the recorded anomaly (Fig. 5.7); h is the thickness of the bed; $I_{rel_1}(z)$ and $I_{rel_2}(z)$ are relative dose rates observed against the actual formation and at a point (z) outside the formation. Making the proper substitution and finding the integral function we get
$$S = I_{rel\gamma} h \qquad (5.2)$$

Whenever S and h are known, the true activity of the bed can be determined.

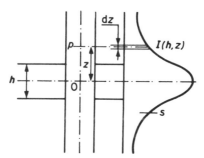

Fig. 5.7. An aid to bed-thickness correction

5.1.3. INTERPRETATION OF NATURAL γ-RAY LOGS

The purpose of interpretation is to locate the formation boundaries, to divide the formation into homogeneous zones and to determine the natural γ-radioactivity and the content of radioactive matter of each bed. In geophysical practice, the latter quantity is generally given in units of N Ra equ/MN (for example, pGy/s, Bq/kg, or μR/h).

In the United States the logging service companies use API (American Petroleum Institute) standard units so that the logs can easily be compared. The "API Gamma Ray Unit" was determined in test-pit at the University of Houston and this represents a two-

Table 5.3. Units of transformation for natural γ-ray logging

Service Company	Company standards	API units
Schlumberger	1 g Ra-equ/t	16.5
	1 g Ra-equ/t	11.7
Lane Wells	1 arbitrary unit	2.16
PGAC (Pan Geo Atlas Co.)	70 nA/kg (1 R/h)	15.0
Mc'Collough	70 nA/kg (1 R/h)	16.4

hundredth part of the difference between the values measured at two different sections of the model. Re-calibration of the logs can be made on the basis of Table 5.3.

Proper quantitative interpretation requires well calibrated instruments. Calibration is carried out with the aid of a point-source located at different distances from the instrument or in a *calibration pit*. The logging speed and the time-constant of the instrument—as far as the statistical fluctuation effect is concerned—should be adequately controlled.

If ε is the permissible error and $\bar{I}_{rel}$ is an average count rate, the integration time constant can be given by the equation:

$$t_i = 1.36 \times 10^5 \frac{1}{\varepsilon^2 \bar{I}_{rel}} \tag{5.3}$$

If the time-constant and the minimum bed thickness are known, the maximum permissible value of the running speed can be given by

$$v = \frac{300 (h_{min} - a)}{b t_i} \tag{5.4}$$

where v is the logging speed, m/h; t is the time-constant of integration, s; h_{min} is the thickness of the thinnest bed, m; a is the length of the γ-detector, m; b is a constant whose value is about 2–3.

The first step of interpretation is the determination of the measure of *statistical fluctuations*. In the classical interpretation, this means the estimation of the probability level of fluctuations. With the use of a computer, however, even some type of *filtering* may be carried out.

The next step is the correction of the logs according to the following parameters:
— thickness of the bed;
— logging speed and time constant;
— borehole diameter and tool excentricity;
— mud density and radioactivity of the borehole fluid;
— features of casing and cement-sheath.

The boundary of the bed has to be marked out at the half-value of the deflection for a formation having a thickness of higher than 1 metre and at four-fifths of the deflection in the case of thinner beds. If necessary (if $vt_i > 1500$), the correction for bed-thickness should be made using Fig. 5.6. The nomogram in this Figure allows a correction for the effects of *logging speed* and time-constant.

The correction related to the diameter of the hole depends on the type of borehole tool. This kind of correction chart is shown in Fig. 5.8, for tools of large diameter (a) and small diameter (b). The position of the sonde in the borehole is not an unimportant point either. The radioactivity measured by a detector located at the hole-axis decreases to the activity level of the mud on increasing the hole diameter; on the other hand, with the detector moving along the hole-wall the arithmetic mean value of radioactivity of the rock and the mud will be approached (Fig. 5.9).

The influence of sonde excentricity is a function of hole diameter too; for this reason, when quantitative evaluation is made, the correction on these two parameters should be performed simultaneously, as seen in Fig. 5.10, where

$$\eta_\gamma = \frac{I_{rel\,\gamma}\,(d=0)}{I_{rel\,\gamma}\,(d=x)}\,; \qquad \Delta d = d - d_m \text{ cm}$$

d_m is the diameter of the tool, cm.

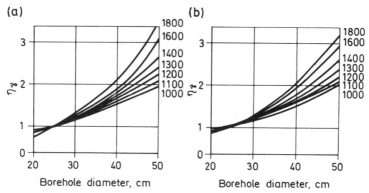

Fig. 5.8. Diagram to correct borehole diameter (η_γ) as a function of mud density (parameter of curves). (a) — 90 mm; (b) — 60 mm

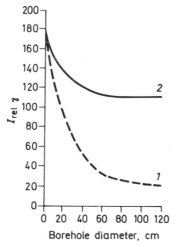

Fig. 5.9. $I_{rel\,\gamma}$ values as a function of detector position. *1* — With a centralized tool in the hole-axis; *2* — With a tool forced against the wall of the borehole

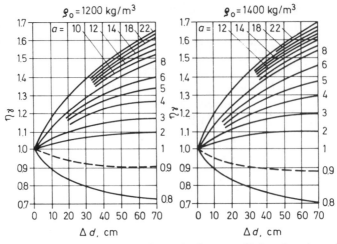

Fig. 5.10. Diagram to correct the effect of Δd, the difference in diameters of hole and sonde, mud weight (ϱ_0) and activity of mud (a)

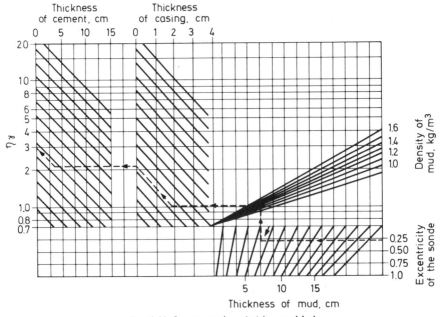

Fig. 5.11. $I_{\text{rel}\gamma}$ corrections (η_γ) in cased hole

When using Fig. 5.10 the *radioactivity of the mud* must be known which has, however, quite often been neglected even by experts in the field; this is determined from the expression:

$$a = a_{\text{clay}} \frac{c_{\text{clay}}}{1 - \Phi_{\text{clay}}} \qquad (5.5)$$

where a is the specific radioactivity of the mud, a_{clay} is the specific radioactivity of the clay used to make the mud, Φ_{clay} is the porosity of the clay, c_{clay} is the volumetric clay content of the mud, kg.

When γ-ray log is run in a *cemented casing*, the absorbing effect of the casing and the cement must also be taken into account. The nomogram in Fig. 5.11 is provided for this purpose.

5.1.4. APPLICATION OF THE γ-RAY METHOD

Among the applications *of general interest*, the most important ones are:

— Identification of geological sections, *locating the boundaries of formations, selecting marker-layers.*

— *Finding correlations between wells.* An advantage of the γ-ray method as compared to other procedures is that it can be used even in cased holes and its indication is not influenced by the fluid content of formations (water, oil, gas), the salinity of formation water and borehole fluid, and, moreover, it can be used for logging in non-conductive mud and empty holes. Its drawback lies in the fact that the statistical fluctuations can obscure fine details.

— *Studies on lithology; preparation of radioactivity maps.* In oil and gas mining, this method makes it possible to reconstruct paleogeographical and sedimentary conditions.

— *Base or reference logging.* In tracer techniques, base (reference) logs are of great importance. Their usage allows a great reduction in the quantity of isotopes needed and thus even minor deviations can be detected.

In oil and gas mining, the γ-ray log is used mainly for the following purposes:

— *Shale content determination* in the formations penetrated by the hole. The principle of the method is based on the fact that the radiation intensity of sandstones is much lower than that of shales (see Table 5.2).

It is not an absolute but a relative method which uses the relationship

$$\Delta I_{rel\gamma} = \frac{I_{rel\gamma} - I_{rel\gamma}^{a=0}}{I_{rel\gamma}^{a=100}\ I_{rel\gamma}^{a=0}} \tag{5.6}$$

where $I_{rel\gamma}^{a=0}$ and $I_{rel\gamma}^{a=100}$ are the relative intensities of clean sand and clay, respectively.

The actual relationship between the above $\Delta I_{rel\gamma}$ and the shaliness of formations must be determined separately in each area.

— *Permeability and productivity estimations.* The method to be followed is nearly the same as that used in the procedure of shaliness determination.

— *In coal-mining* γ-ray measurements have an important role in locating coal-layers. The effectiveness of the method strongly depends on the radioactivity of adjacent formations. This is illustrated in Fig. 5.12 where curves from two different coal basins are shown. As seen in the Figure in one of the curves (case a) the coal layer is easy to locate using a γ-ray log, whereas it is inapplicable in the other case (b).

— *In the exploration of radioactive ores,* γ-ray logging is the most important method. In an infinitely thick bed, the γ-log response is directly proportional to the uranium (radium) content of the layer (c):

$$I_{rel\gamma}^{\infty} = ck100 \tag{5.7}$$

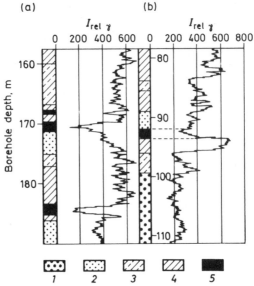

Fig. 5.12. Detection of coal beds using γ-ray log. *1* — Conglomerate; *2* — Sandstone; *3* — Silt; *4* — Shale; *5* — Coal

where *k* is a constant to be determined for each instrument. Substituting Eq. (5.2) into (5.7), we get

$$c = \frac{S}{kh} \times 0.01\% \text{ uranium} \tag{5.8}$$

Equation (5.8) serves only as an illustration of the possibility how, in principle, the uranium content could be determined. In practice, a number of different factors must be taken into consideration (e.g., thorium content, radioactive disequilibrium).

— *Potassium deposits* can be distinguished very well from salt and anhydrite on the basis of γ-ray logs because the latter's radioactivity is the smallest among those of the rocks. To determine the potassium content, a relationship analogous to Eq. (5.8) is used. As a disturbing factor, borehole enlargement must be mentioned.

5.1.5. NATURAL γ-SPECTROSCOPY

γ-Spectroscopy is a more advanced version of natural γ-logging. In a number of applications γ-spectroscopy may give extra information on the structure and mineral composition of formations on the basis of the *characteristic spectra* of their radioactive components. Two groups of borehole spectrometers have been developed:

Total spectrum borehole spectrometers, single- or multichannel type instruments, designed for point-by-point measurements. Their application is rather labour-consuming and is, therefore, limited to special cases only.

Energy-selective borehole spectrometers (equipped generally with 2–4 channels) produce continuous curves. Under borehole conditions, NaI(Tl) scintillation detectors give the best results at the energy levels as follows:

$$^{40}\text{K} \; : \; 1.46 \text{ MeV}$$

$$^{226}\text{Ra}: \; 0.68 \text{ MeV}$$

$$^{232}\text{Th}: \; 0.98 \text{ MeV}$$

The optimum gate-width is 0.2 MeV. In each case it is recommended that a background (b) curve also be recorded up to a maximum energy level of 3 MeV. The following relationships apply to each channel:

$$I_{\text{rel,Ra}} = \alpha_{\text{Ra}} \cdot c_{\text{Ra}} + \beta_{\text{Ra}} \cdot c_{\text{Th}} + \gamma_{\text{Ra}} \cdot c_{\text{K}} + \varDelta_{\text{Ra}} \qquad (5.9)$$

$$I_{\text{rel,Th}} = \alpha_{\text{Th}} \cdot c_{\text{Ra}} + \beta_{\text{Th}} \cdot c_{\text{Th}} + \gamma_{\text{Th}} \cdot c_{\text{K}} + \varDelta_{\text{Th}} \qquad (5.10)$$

$$I_{\text{rel,K}} = \alpha_{\text{K}} \cdot c_{\text{Ra}} + \beta_{\text{K}} \cdot c_{\text{Th}} + \gamma_{\text{K}} \cdot c_{\text{K}} + \varDelta_{\text{K}} \qquad (5.11)$$

$$I_{\text{rel,b}} = \alpha_{\text{Ra}} \cdot c_{\text{Ra}} + \beta_{\text{Th}} \cdot c_{\text{Th}} + \gamma_{\text{K}} \cdot c_{\text{K}} + \varDelta_{3\,\text{MeV}} \qquad (5.12)$$

Coefficients α, β and γ are determined by calibration using appropriately selected and well-defined materials; $\varDelta$ values are estimated by extrapolation from an exponential noise measured up to an energy level of 3 MeV and using energy levels corresponding to Ra, Th and K gates.

The main applications of γ-spectroscopy are as follows:

— *exploration of radioactive materials;*

— solution of *general geological exploration problems* (study of a sedimentation, tectonic regions, correlation between "dumb zones", etc.).

— determination of my clay minerals.

5.2. METHODS USING γ-RADIATION SOURCES

Several methods, although different in their physical principles, belong to this group, their common features being the application of a sealed γ-source (Section 1.4.2) and the fact that their operation is related to the interaction of γ-radiation with matter (see Section 4.1).

According to the *detected radiation* two different methods can be distinguished: the γ–γ method and the γ-neutron method. The former can be subdivided further in accordance with the energy of the source and that of the detected radiation.

5.2.1. γ–γ DENSITY LOGGING

Whenever γ-rays irradiate the formation rocks penetrated by a hole, scattered γ-radiation is generated due to the Compton effect: its recording along the hole-axis is called γ–γ density logging, which is the most important γ-method. In practice, it is known simply as γ–γ or density logging.

The theory behind the method was developed using the diffusion equation used in neutron physics (Section 5.3). On the basis of this theory the following approximation can be derived for a homogeneous medium:

$$I_{\text{rel}\gamma\gamma} = \frac{3\,A(\lambda_2 - \lambda_1)}{4\pi L_0} \frac{\rho}{l} \exp\left(-q \times \rho \times l\right) \qquad (5.13)$$

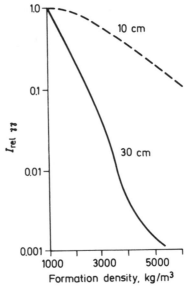

Fig. 5.13. γ–γ Deflections as a function of formation densities; the parameter of the curve is the spacing

where A is activity of the source, ρ the bulk density of the material, l the distance between source and detector, λ_1 and λ_2 are the wavelengths of emitted and absorbed quanta, respectively, L_0 is the total range of a quantum, referring to the average density, and q is a parameter depending on source and medium.

Figure 5.13 shows a plot of $I_{\text{rel}\gamma\gamma} = f(\rho)$ calculated by Eq. (5.13). Note that the relative intensity decreases exponentially with the increase of formation bulk density; the rate of decrease is controlled by the sonde spacing.

5.2.1.1. PROCEDURE AND INTERPRETATION

It was concluded from theoretical and model studies that the *radius of investigation* of the γ–γ procedure is rather small. The basic formula is

$$R = \frac{16.6}{\rho} + 0.66\,l \qquad (5.14)$$

where R is the depth of investigation, i.e., that of radial penetration, l is the distance between source and detector, which is also called spacing.

An increase in the *spacing* results in the improvement of the sensitivity of the method to density, and the range increases, as well (Fig. 5.13). However, the range will be limited by the intensity of the radiation source (50–75 GBq, 1.5–2 Ci). For this reason, in practice, a spacing is chosen not greater than 30–50 cm, the actual value depending on the hole-diameter and the density of formations, and so the investigation radius commonly achieved is only about 15 cm, when ρ is 2600 kg/m³.

It should be borne in mind that the γ–γ method is strongly influenced by the *intermediate zone* (materials located between the detector and the rocks), by the *roughness of the side-wall*, as well as by the presence of *mud cake*. The latter is of particularly great importance in oil mining, since mud cakes readily form on the sidewall against permeable beds. Care should be taken not to forget about the correction for the effect of mud cake; such correction factors are given in Fig. 5.14.

The influence of the intermediate zone can be reduced by forcing the tool with springs against the wall of the hole. In this way the results may improve, however, in some cases deviations which cannot be corrected will persist because of *micro-caverns*. In order to eliminate these errors, a number of procedures are available:

— *The sonde* pressed against the sidewall is equipped with a microcaliber which measures the clearance between the sonde and wall of the hole. This solution is simple and produces good results, but cannot be applied in the presence of mud cake, because of this it is used only in ore-mining.

— Logging with *dual-spacing* sonde. This procedure is complicated, but it has been proved to be suitable even in the case of holes drilled for the exploration of hydrocarbons. This method makes use of the different responses of the sondes with different spacings to changes in density (see Fig. 5.14). For the sake of better differentiation, the short-spaced (5–8 cm) sonde is designed in such a manner that it is affected predominantly by mud cake.

Figures 5.15 (a) and (b), respectively, represent the conditions with both absence and in the presence of mud cake of 1500 kg/m³ density, in a formation density of 2500 kg/m³ density; the general trend for different rocks is shown in Fig. 5.15 (c). If the density of the mud cake is known, the bulk density of the formation can unambiguously be determined even if the thickness of the intermediate zone is not known. This procedure, known as FDC (Formation Density Compensated), utilizes a small analogue computer mounted on the surface panel to calculate the density log from the two recorded curves.

The dual radius method is based on another principle, however, using two detectors in a similar fashion. Here, the single-scattered quanta are detected since it has been verified that the ratio of the detector responses is not affected by the intermediate zone:

$$I_{rel, 1}/I_{rel, 2} = \exp\left(\mu_1 \times u + \mu_2 \times v\right) \tag{5.15}$$

where μ_1 and μ_2 are values of the linear absorption coefficient before and after collision, respectively, and u and v are the excess distances the γ-quanta have to travel before they reach the detector located further, before and after collision, respectively.

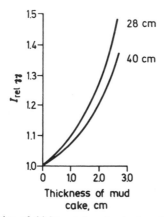

Fig. 5.14. γ–γ Indications as a function of thickness of mud cake in a limestone of 2400 kg/m³ density; the parameter of the curve is the spacing

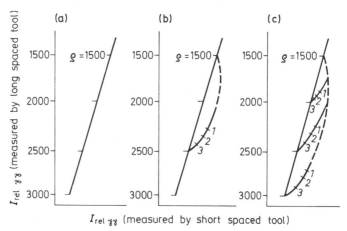

Fig. 5.15. Interpretation of compensated γ–γ log for a mud density of 1500 kg/m³. Thickness of mud cake: *1* — 1.9 cm; *2* — 1.27 cm; *3* — 0.64 cm. Formation density: ϱ

This method has a rather poor radius of investigation when used with sealed sources and it is, therefore, used only in ore-exploration.

In the USSR, the sensitivity of this method to density has been successfully doubled by applying a high-output (10^{12} quanta/s) neutron-generator of 0.5 MeV energy.

5.2.1.2. APPLICATION OF THE METHOD

The γ–γ method has two main fields of application:
— determination of rock density;
— solution of certain technical problems.

γ–γ Logging is extensively used in oil- and gas-drilling and *coal-mining* for the *determination of formation densities*, but it has been also used in exploration for *water* and *ores* (in the latter case it is combined with a selective γ–γ method, as described in Section 5.2.2).

In the *oil and gas* industry, this method is of great importance, since it makes it possible to determine the *porosity* of formations. Porosity can be derived from the relationship:

$$\rho = (1 - \Phi) \times \rho_M + \Phi \times \rho_f \qquad (5.16)$$

where Φ is the formation porosity, ρ is the average bulk density of the rock matrix, and ρ_f is the density of the pore fluid and ρ_M the density of rock-matrix. If the equation is rearranged, the following expression is obtained for the porosity:

$$\Phi = \frac{\rho_M - \rho}{\rho_M - \rho_f} \qquad (5.17)$$

The rock-matrix and fluid density values are found in Table 5.4. The determination of porosity from density logging has the advantage over other methods in that it is not affected so strongly by shaliness.

Table 5.4. Rock-matrix and fluid densities

Rock	Density, kg/m^3
Salt	2160
Gypsum	2320
Quartz	2650
Calcite	2710
Dolomite	2870
Anhydrite	2960
Shales	2500–2950
Feldspars	2610–2640
Petroleum	850–950
Water	1000–1200

In gas-bearing formations the residual *gas content* of the invaded zone makes the porosity calculation inaccurate, especially if the invasion is shallow; the density of gas ranges from 60 to 750 kg/m^3. In combination with sonic or other logs the γ–γ method renders it possible to determine both the porosity and the gas content, simultaneously.

In both *hydrocarbon and coal mining*, density logging is used for *lithology determination*. As seen from Table 5.4, in a carbonate formation limestone can effectively be distinguished from dolomite, and anhydrite from gypsum. An important point is that coal layers can be separated from the barren ones because the densities (Table 5.5) of various types of coal differ considerably from those of the barren rocks. Therefore, γ–γ logging is an effective tool in coal mining.

Table 5.5. Density of organic materials contained in various types of coal

Grade of coal	Density of organic materials, kg/m^3
Rich coal	1200–1250
Coal-coke	1250–1320
Semi-anthracite	1320–1400
Anthracite	1400–1600

γ–γ Logging gives the possibility not only to locate the coal layers (Section 4.1.3), but also to determine the *ash content*. For an estimation of the ash content, among others, the following empirical formula is used

$$\rho = \rho_0 + 0.01 \times a + 0.012 \times S \qquad (5.18)$$

where ρ is the mean value of bulk density, ρ_0 is the density of the organic content, a is the percentage ash-content in coal, and S is the percentage sulphur content in coal.

The nomogram shown in Fig. 5.16 was derived using a similar equation. In order to use this equation, one has to know the actual values of the ash density and the organic content density, from laboratory analysis.

Since the densities of the ash components, SiO_2, Al_2O_3, CaO and MgO are nearly the same, i.e., about 3160 kg/m^3, the analysis of coals having a low ash content is not needed. In the case of highly sulphuric coals, the pyrite of 4900 kg/m^3 density must also be taken into consideration.

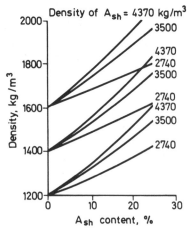

Fig. 5.16. Relationship between coal density and ash content; the parameter of the curves is the density of barre materials

In recent years, the range of technical problems has considerably widened. In oil and gas producing wells, the measurement of the *fluid-density* as a *function* of *depth* has become of great significance. This procedure has proved very useful—among others—for locating the *depth of water, oil or gas inflow* (for two or three phase flows), for the determination of the bubble-point of the gas, etc. These measurements are carried out with strongly collimated beams of a γ-source (generally ^{170}Tm) and a collimated detector. Using a sonde of the appropriate design a closely linear relationship is found between the recorded curves and the bulk density.

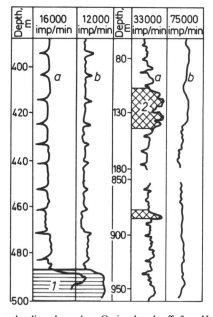

Fig. 5.17. Casing thickness and caliper logs. *1* — Casing break-off; *2* — Heavy corrosion on the casing

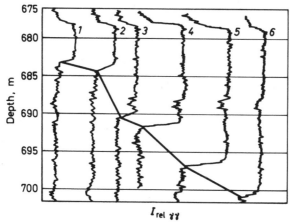

Fig. 5.18. Control of contact-level formed between oil products and salt-water at different times (1–6)

The location of cement top and quality control of cement sheath beyond a casing is based on the shallow penetration range of the radiation. The γ-beams which have passed through the casing reflect only the presence of cement surrounding the pipe (the density of cement is about 1800 kg/m^3; that of mud is much lower), however, owing to the small penetration range of radiation, the beams are not affected by the rocks. For the purpose of this measurement, various types of 3- and 4-channel rotating-detector systems have been developed. With the introduction of sonic procedures, the importance of these methods has been very much reduced.

Recently, for detecting the *corrosion of the casing column* and locating the perforation intervals (in old wells), a special type of γ–γ tool has been developed which uses very short spacing (6–8 cm) and a ^{75}Se source with energy of the order of 0.1 MeV. A typical log is shown in Fig. 5.17.

The determination of the fluid level relating to *artificial underground oil storage in salt domes,* is an important task. In caves formed in salt blocks great amounts of oil products can be stored, which float on the saturated salty solution. This storage process is illustrated in Fig. 5.18 (see also Section 2.3.3). As seen from the data the indication changes markedly at the contact level formed between the petrol and the salt solution.

5.2.2. SELECTIVE γ–γ METHOD

The registration of the absorption of *photoelectrons* brought about by γ-beams in hole-penetrated formations, along the hole-axis, is known as selective γ–γ logging. This method strongly resembles the density logging used in coal and *ore-mining.* However, in contrast to density logging, where use is made of the Compton effect and, therefore, a relatively high energy γ-source (> 1 MeV) is necessary, the present procedure works in the *range of lower energies* (^{75}Se, ^{113}Sn, ^{137}Cs and ^{203}Hg). It is used to obtain information on *density* as well as on the *chemical composition* of the matter studied—first of all on the presence of heavy elements (i.e., elements with high mass number).

The principle of selective γ–γ logging is practically the same as that of density logging (see Section 5.2.1). The results obtained using Eq. 5.13, which has been justified

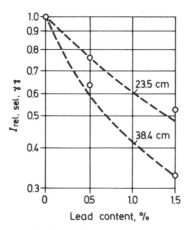

Fig. 5.19. Relationship between spectral $\gamma-\gamma$ log deflection and lead content of a rock; the parameter of the curve is the spacing

experimentally, are shown in Fig. 5.19; note that the lead content of the rocks can also be derived from the selective $\gamma-\gamma$ log. The method has proven to be particularly effective when the materials to be explored differ from the surroundings in atomic numbers rather than in densities. The effective mass number of the rocks ($\mathscr{A}$) can be obtained from the relationship

$$\mathscr{A} = \sqrt[3]{\sum_{i=1}^{n} c_i \cdot \mathscr{A}_i^3} \qquad (5.19)$$

where c is the mass percentage of i-th component, $\mathscr{A}_i$ is the mass-number of the i-th component, and n is the number of components.

The main field of application of selective $\gamma-\gamma$ logs is in *coal- and ore-mining.* In mining for coal it is used to determine the *ash content of coal.* The effective mass-number of the organic content in coals is about 6, independently of the quality of the coal, that of water is about 7.69 and that of the ash content ranges from 12 to 22. Once the ash material is known from laboratory analysis, a relationship can be set up between selective $\gamma-\gamma$ intensity and ash content, so that the ash content can be determined (see Section 4.1.3).

The following relationship is generally used:

$$f(a) = \frac{I_{rel\gamma\gamma,s}^{a}}{I_{rel\gamma\gamma,s}^{0}} \qquad (5.20)$$

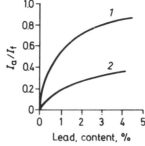

Fig. 5.20. Quantitative determination of lead content using γ-ray log. *1* — Selective $\gamma-\gamma$ log; *2* — $\gamma-\gamma$ log

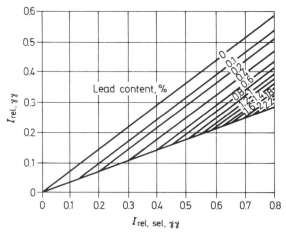

Fig. 5.21. Determination of lead content in formations of different densities; the parameter of the curves is the percentage of lead content

where $I^a_{rel\gamma\gamma,s}$ and $I^0_{rel\gamma\gamma,s}$ are the respective selective γ–γ intensities measured in coal with known ash content and in that with no ash content (graphite); a is the ash content, %.

When *exploring* for ores the percentage of lead or other ores can often directly be established using selective γ–γ logging provided that the density of the barren rocks is constant. It can be seen from Fig. 5.20 that the selective method shows a considerably higher sensitivity than that of the density log.

When the barren density is not constant, it is a straightforward matter to use a combination of the two versions of the γ–γ procedure; this makes it possible to determine the ore grade accurately (Fig. 5.21).

5.2.3. OTHER γ–γ METHODS

Numerous experimental techniques are known that represent further developed versions of the selective γ–γ methods or utilize physical phenomena related to γ-radiations of energies not higher than 1 MeV (these radiations fall into the range of X-rays and the probability of the occurrence of Compton scattering can thus be neglected) or higher than 2 MeV.

The critical energy method using γ-radiation with *lower energy than 1 MeV*, detects γ-quantum energies related to the intensity maximum of the scattered γ-radiation. For elements having a mass number less than 50, the critical energy is in a well-defined relationship with the equivalent mass number of the rock. This is of advantage in coal-mining as well as in lithology determinations.

The spectrum-ratio method consists of the recording of the scattered γ–γ logs corresponding to two different energy intervals (40–90 keV and >120 keV), the latter depending only on density and the former also on mass number.

Spectral logging. The spectra of scattered γ-rays render it possible to detect materials in which the binding energy of the K electrons of the constituent atoms is much higher than the average value such as Ba (37.4 keV), W (69.3 keV), Pb (87.7 keV) and Hg (89.8 keV).

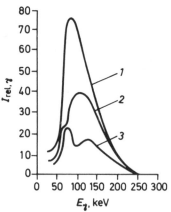

Fig. 5.22. γ-spectrum of scattered γ-radiation. 1 — In sand; 2 — In sand with W content as high as 0.8%; 3 — In sand with Pb content as high as 2%

Some spectra are shown in Fig. 5.22; at energies lower than the binding energy, a rapid decrease of intensity is observed.

The critical energy, the spectrum-ratio and the spectral logging methods use scintillation detectors and 15–40 GBq (0.5–1.0 Ci) ^{137}Cs sources, i.e., the same as those used in the selective γ–γ method.

γ–X-ray logging is a new method in the exploration for ores and is based on the recording of characteristic X-rays generated by γ-radiation (Section 4.1.4). The advantage of the method lies in the higher sensitivity compared with γ–γ spectrum logging, and can be applied to detecting elements with mass numbers higher than 40. Its drawback is that it has a low range of investigation depth (max. 1.5 cm) and requires rather complicated instrumentation.

The Mössbauer effect is also based the X-ray fluorescence (Section 1.4.1.6), and represents a special version of γ–γ logging. Its principle lies in the resonance absorption of radiation. Due to the very weak radiations involved, the investigation depth of the method is as low as 2–3 mm; it can therefore be used in empty holes only and even so a number of technical problems remain to be solved. The resolution of the method is fairly good; for instance, a tin content as low as 0.01% can still be detected.

The X-ray diffusion procedure is characterized by the fact that it does not use a radioisotope for generating low energy (50 keV) primary radiation and that the secondary radiation detected is of very low energy (0.1–50 keV). The latter feature of the method gives rise to great technical difficulties: the mud cake must be removed completely from the wall of the hole and, at the same time, the detectors equipped with minimum shielding, must be forced against the wall of the hole.

Procedures using very high (higher than 3 MeV) energy γ-radiation apply borehole accelerator sources (such as a betatron), because even the highest energy γ-radiation provided by sealed ^{24}Na sources is of an energy not higher than 2.76 MeV (see Section 7.1.1).

In the USA, a linear accelerator has been constructed that enables photons of 20–25 MeV to be produced under borehole conditions. This achievement may give rise to rapid progress in high-energy γ–γ and γ-neutron procedures.

Table 5.6. Minimum quantity of materials which can be detected using photo-activation

Element	Minimum quantity, mg	Element	Minimum quantity, mg
Barium	100.0	Mercury	34.0
Cadmium	1.0	Platinium	3.2
Gold	2.3	Selenium	3.3
Hafnium	0.25	Silver	1.4
Indium	0.11	Strontium	3.4
Iridium	1.3	Yttrium	9.8

Table 5.7. Activation properties of some rock-forming minerals

Isotope to be activated	Percentage found in natural mixtures	Energy of reaction, MeV		Maximum cross-section of reaction	Product of reaction	Half-life	Energy of γ-quantum, MeV
		at threshold	at maximum yield				
^{24}Mg	78.6	16.5	19.4	9.8	^{23}Mg	12 s	0.44
^{26}Mg	11.29	14.2	22.0	25.0	^{25}Na	60 s	0.4–1.61
^{28}Si	92.27	17.14	20.9	21.0	^{27}Si	4.2 s	0.84
^{29}Si	4.68	12.8	19.6	31.0	^{28}Al	2.3 s	1.78
^{30}Si	3.05	12.9	21.0	32.0	^{29}Al	6.6 min	1.28–4.43
^{32}S	95.1	14.7	20.1	15.0	^{31}S	2.6 s	1.27
^{35}Cl	75.4	12.8	19.0	18.3	^{34}Cl^m	32.4 min	0.14–3.3
^{39}K	93.08	13.2	19.3	11.2	^{38}K	7.7 min	2.1
^{40}Ca	96.97	15.9	19.6	15.0	^{39}Ca	0.9 s	2.5
^{54}Fe	5.84	13.8	17.7	67.0	^{53}Fe	9.0 min	0.37
^{63}Cu	69.1	10.8	17.5	11.0	^{62}Cu	9.8 min	0.66–2.24
^{64}Zn	48.89	11.6	18.7	120.0	^{63}Zn	38 min	0.67–2.9
^{90}Zn	57.46	12.37	18.0	150.0	^{80}Zn^m	4.4 min	0.59–1.5
^{107}Ag	51.35	9.5	16.5	320.0	^{106}Ag	24 min	0.51
^{115}In	95.77	9.1	15.0	420.0	^{114}In	72 s	0.13–0.20
^{204}Pb	1.48	8.7	22.0	280.0	^{203}Pb^m	6 s	0.83
^{208}Pb	52.3	10.8	14.0	300.0	^{207}Pb^m	0.8 s	0.57–1.06

The photo-activation method is based on the measurement of radioactivity of excited nuclear isomer nuclei generated by the inelastic (γ–γ) scattering of γ-radiation (Section 4.2.1.2). The method uses sources of high activity since the probability of interaction is low and will be effective only if relatively high energy (2–3 MeV) γ-radiations are applied. It is of primary importance in the detection of *heavy elements*. The minimum masses that can be detected are shown in Table 5.6.

γ-Resonance logging is based on the elastic scattering of γ-radiation. It needs much higher energy radiation sources than the previous methods (e.g., γ–γ logging). The method is used for detecting *light elements*.

Even though it is not yet commonly used in practice, *γ-activation analysis* (Section 4.2.1.2) seems to be fairly promising. It is based on the detection of isotopes of short half-life generated by high-energy (max. 30 MeV) γ-radiation in (γ, n) and (γ, p) nuclear reactions. A significant feature of γ-activation analysis is that both threshold- and resonance energy photoneutron- and photoproton-interactions are characteristic of the elements to be detected. The activation properties of the most easily detectable elements are summarized in Table 5.7.

5.2.4. γ–NEUTRON METHOD

The method known as γ–neutron or *photo-neutron* logging implies the recording of the neutron flux formed along the borehole as a result of (γ, n) reactions brought about by the interaction of high energy γ-radiation with the material of the formations penetrated by the borehole. This reaction (see γ-activation analysis in Section 5.2.3) takes place only when the radiation energy *exceeds a threshold value*. Since in nature only the threshold energies of beryllium and deuterium (1.63 MeV and 2.23 MeV, respectively) fall within the energy range of isotope sources, a logging method has been developed for these two elements only. Other elements have threshold energies as high as 8–16 MeV and, therefore, the use of an accelerator (e.g., a betatron) is needed for their detection.

For the determination of *beryllium* a ^{124}Sb source can be used. It comes from the theory of the method that the density of photoneutrons is proportional to the beryllium content of the formations, but in addition, it strongly depends on their slowing-down and absorption characteristics, as well. This feature must be taken into consideration in the exploration for beryllium. The microscopic cross-section of beryllium for the (γ, n) reaction is low $(\sigma_{\gamma, \text{n}} \approx 9 \times 10^{-32} \text{ m}^2)$ and therefore it is practical to run natural neutron beam and neutron–neutron logs simultaneously with the γ–neutron logging, when the former ones serve as reference or base logs. This procedure is justified by the fact that beryllium-ores are found usually together with lithium and rare earth metals: since the latters' neutron-capture cross-sections are anomalously high, they might falsify the results. The neutron–neutron logging is carried out with a Sb–Be source. In this way, by comparing the γ–neutron log with the neutron–neutron log we can obtain detailed information on the presence of absorbing materials.

Based on the determination of *deuterium* with the γ–neutron method, oil can be distinguished from water. As a source first of all ^{24}Na is used.

5.3. METHODS USING NEUTRON SOURCES

5.3.1. METHODS BASED ON ISOTOPE NEUTRON SOURCES

Rocks are activated by neutrons generated by an isotopic neutron source and a secondary neutron or γ-radiation is detected. Based on the time delay between irradiation and measurement, methods can be grouped into two classes:
— conventional neutron methods;
— neutron activation methods.

5.3.1.1. CONVENTIONAL NEUTRON METHODS

On the basis of *recorded radiation* three conventional neutron methods can be distinguished:
— determination of the intensity of epithermal neutrons slowed down by scattering;
— measurement of the spatial distribution of thermal neutrons evolved in the slow-down of fast neutrons;
— recording of γ-radiation induced by the capture of thermal neutrons.

As a *radiation source*, a Po–Be or Pu–Be systems of 4–5 MeV or Ra—Be systems of 0.5 MeV energy the same energy and of 100–300 GBq activity—are used.

Among the theories the best result is given by the "*transport-theory*" for the spatial distribution of the neutrons (5.10). However, this theory is usually not at all satisfactory for practical purposes, and therefore it can be used only for drawing approximated conclusions of limited validity; the correction curves necessary for more exact geophysical calculations must be determined by model studies. Values obtained for the neutron distribution using the "*neutron–gas diffusion theory*" are close to the results given by the transport theory, especially at larger distances from the radiation source. The diffusion theory is based on the equation

$$\frac{dn}{dt} = q - p - f \tag{5.21}$$

where $\frac{dn}{dt}$ is the change in the number of neutrons in a volume dV during unit time interval; q is the number of particles formed in or entering the volume element dV; f is the number of neutrons leaving the dV element, and p is the number of particles absorbed in unit time in unit volume.

After substitutions, we get the following differential equation for stationary conditions:

$$D\Delta\Phi_n - \Sigma_a\Phi_S + S(l) = 0 \tag{5.22}$$

where D is the diffusion coefficient; Φ_S the neutron flux; Σ_a the macroscopic capture cross-section; $S(l)$ the so-called source density, and Δ is the *Laplace* operator.

The spatial distribution of neutrons for given geometrical conditions can be determined by solving Eq. (5.22). Assuming a homogeneous space of infinite dimensions and a point thermal neutron source with Q neutron/s intensity located in the origin we get the equation

$$\Phi(l) = \frac{Q}{4\pi lD} \exp\left(-\frac{l}{L_D}\right) \tag{5.23}$$

where l is the distance from the origin (source) and L_D is the length of diffusion $\left(L_D^2 = \frac{D}{\Sigma_a}\right)$.

The physical possibility of applications is offered by the different behaviour of the individual rock forming minerals towards neutrons, the most important parameters being the elastic scattering (σ_s) and capture (σ_c) cross-sections. In the majority of studies the macroscopic cross-section (Σ) or other equivalent expressions are used for defining the neutron response of rocks, instead of the microscopic cross-sections.

The macroscopic scattering cross-section of rocks is given by

$$\Sigma_s = \frac{\rho_k A}{100} \sum_{i=1}^{n} \frac{\sigma_{si} c_i}{\mathscr{A}_i} \tag{5.24}$$

the *capture cross-section* can be expressed as

$$\Sigma_c = \frac{\rho_k A}{100} \sum_{i=1}^{n} \frac{\sigma_{ci} c_i}{\mathscr{A}_i} \tag{5.25}$$

where ρ_k is the density of the rocks, A is Avogadro's constant, 6.02×10^{23} molecules/mol; c_i is the percentage of the i-th component; $\mathscr{A}_i$ the mass number of the i-th component, n is the number of components.

Table 5.8. Neutron properties of some rocks and minerals

Rocks and minerals	Composition	Density, kg/m^3	Σ_{Cl}, 1/m	Σ_S, 1/m	L_{sl}, cm	L_d, cm
Clay, wet		2200				4.0
Clay, dry		2880				14.7
Anhydrite	CaSO$_4$	2920	0.0119	0.348	27	8.2
Anthracite		1500				4.9
Brown coal		1300				2.6
Dolomite	CaMg(CO$_3$)$_2$	2850	0.0046	0.441		9.2
Gypsum	CaSO$_4 \cdot$ 2 H$_2$O	2300	0.018	1.51	11	3.2
Hematite	Fe$_2$O$_3$	5100	0.093	0.666	34	2.3
Sand, wet		2050				20
Sand, dry		2300			14	6.0
Calcite	CaCO$_3$	2720	0.0071	0.432	35	9.5
Corundum	Al$_2$O$_3$	4000	0.0102	0.364		9.3
Halite	NaCl	2150	0.711	0.310		1.2
Coal		1350				2.9
Quartz	SiO$_2$	2650	0.0034	0.268	37	17
Magnetite	Fe$_3$O$_4$	5000	0.0948	0.648	19.6	2.3
Oil		875			9.3	2.0
Water	H$_2$O	1000	0.022	2.68	7.7	2.3

The knowledge of *slow-down* (L_{s1}) and *diffusion* (L_d) *lengths* is necessary for the determination of the radius of investigation (i.e., the depth of penetration) of the different methods.

These parameters for some rocks and rock forming minerals are given in Table 5.8.

When using the *neutron-epithermal neutron method*, the neutron intensity measured from a given neutron source depends exclusively on the neutron slowing down properties of rocks. With *elastic scattering*, the energy-loss of neutrons can be calculated from the following equation if the mass number of the material is known

$$\xi = 1 + \frac{(\mathscr{A}-1)^2}{2\mathscr{A}} \frac{\mathscr{A}-1}{\mathscr{A}+1} \tag{5.26}$$

where ξ is the effectivity of slow-down (mean logarithmic energy-loss).

The smaller the mass number $\mathscr{A}$, the higher the effectivity of slow-down ($\xi_H = 1$, $\xi_D = 0.725$, $\xi_{Be} = 0.209$, $\xi_C = 0.158$).

This method is unambiguously suitable for the *detection of hydrogen* as its ξ value is at least five times greater than that of any other rock forming element. With *sedimentary rocks*, the majority of hydrogen is in the water filling up the pore volume of the rock and therefore the hydrogen content of these rocks is directly related to their porosity. The effect of the rock-matrix can be observed only in the low porosity range ($\Phi < 5\%$). This is shown in Fig. 5.23.

In shaly-rocks there is water in the rock-matrix, too. Because of this, the term "hydrogen-porosity" is more appropriate for the porosity determined by neutron methods as this value is influenced both by the effective porosity and the water of the clay minerals.

The *neutron-thermal neutron method* is based on the measurement of the intensity of neutrons slowed down to the thermal energy level. The intensity depends mainly on the *capture cross-section* of the rocks besides their slowing down properties. In the absence of strongly absorbing elements (B, Cl, Li, Mg, etc.), the response of the thermal neutron method is practically equal to that of the epithermal method. The advantage of the

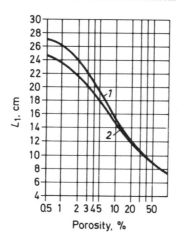

Fig. 5.23. Slow-down length L_1 vs. formation porosity. *1* — Sandstone; *2* — Limestone

procedure is its larger investigation depth range compared with the epithermal method and, therefore, it is less sensitive to the sidewall rugosity, to mud-caking and to hole-enlargements.

The neutron-resonance method was developed primarily to detect minerals containing B, Gd, Sm. Detection is performed using cadmium and hafnium ($E_r = 1.1$ and 7.8 eV, respectively), or indium (1.45 eV) foils equipped with a 2–3 cm thick lead shielding. '

The neutron–γ method is based on the detection of quanta generated by captured thermal neutrons. This is, perhaps, the most complicated procedure, because the intensity of the induced γ-radiation is affected not only by the neutron scattering and capturing properties of the rocks, but also by the energy and intensity of the generated γ-quanta as well as by the interactions between γ-radiation and the rocks. This method is, however, commonly used neutron procedure in the practice of geophysics for the following reasons:

— the radius of investigation is considerably larger than that of any other neutron–neutron method;

— using a proper tool spacing the intensity of γ-radiation is determined primarily by neutron slowing down since the capture cross-section of the rock-matrix is, in general, smaller than that of hydrogen;

— the detection of γ-radiation can be regarded as a much simpler task than that of neutrons.

The main drawback of the neutron–γ method, compared to the neutron–neutron methods, is that the *sensitivity to porosity* of the neutron–neutron method is about ten times higher than that of the neutron–γ method.

In order to present a *correct interpretation* it is important to select an appropriate sonde length, i.e., to choose a proper spacing between source and detector. The thermal neutron density of some materials with different *hydrogen porosity* determined with different values of spacing is shown in Fig. 5.24. The "inversion" of the curves occurs in the 15–30 cm range, the value is depending on porosity. Below this range, I_{rel} is directly, whereas above this range, it is inversely proportional to Φ. For numerous reasons, long spacing tools have become commonly used in practice. Table 5.9 shows spacings used in different methods.

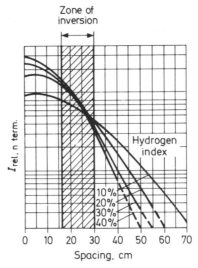

Fig. 5.24. Relation between thermal neutron intensity and spacing; the parameter of the curve is the hydrogen index

Table 5.9. Spacing of tools used in neutron methods

Method	Condition of the borehole	Sonde spacing, cm
Neutron–epithermal neutron	Filled with water	30–40
Neutron–thermal neutron	Filled with water	40–50
	Dry	50–60
Neutron–γ	Filled with water	45–65
	Dry	60–85

Proper calibration of the measuring system is a necessary precondition of quantitative interpretation. During the logging operation arbitrary units or (for an uncalibrated tool) relative dose-rates measured in counts are used. The most simple arbitrary or conditional unit is counts/min, measured in a homogeneous water medium (this is used, for example, in the USSR).

Logging companies in the USA and some other countries calibrate sondes in a test pit at the University of Houston in accordance with API standards. This *artificial well* was built from three rock blocks, each of them with a diameter of 1.8 m and different porosities (Carthago marble: 1.9%; Indiana limestone: 15%; Austin limestone: 26%); thus, the characteristic curve of neutron sondes can be determined. Equation (5.4) can be used to determine the appropriate recording speed of logging by taking the sum of the tool's and the detector's half-length.

The process of interpretation is about the same as that described at the natural γ-method (see Section 5.1.3), the only exception being the neutron–γ method where it is necessary to take into account the background radiation. With thermal neutron and neutron–γ logs, correction should be made regarding the chlorine content of the mud and the water contained in the formation.

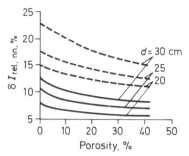

Fig. 5.25. Effect of thickness of mud cake on neutron–neutron log in boreholes of different diameters, $I = 40$ cm
$- - -$ 2 cm; ———— 0.5 cm

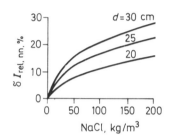

Fig. 5.26. Effect of chlorine content of drilling mud on neutron–neutron log for different hole-diameters, $I = 40$ cm

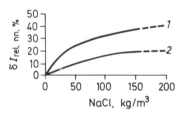

Fig. 5.27. Effect of the salinity of formation water on the neutron–neutron log, $I = 40$ cm, hole-diameter $= 25$ cm
1 — Borehole filled with salty water; *2* — Borehole filled with fresh water

The most important δI_{rel} correction curves are shown in Figs 5.25 (*mud cake*), 5.26 (*chlorine content* of mud), 5.27 (chlorine content of formation water) and in Table 5.10 (borehole diameters).

In recent years several new versions of the neutron–neutron method have been developed and have made interpretation easier and more accurate.

The *sidewall neutron porosity* (SNP) *method* was developed in analogy with the γ–γ apparatus equipped with two detectors (see Section 5.2.1.1). The SNP procedure utilizes a source combined with two detectors pressed against the wall of the borehole.

The compensated neutron log method also utilizes two detectors pressed against the wall (Fig. 5.28) but the spacing is larger. The ratio of counting rates obtained using short-spaced and long-spaced detectors is directly proportional to the porosity. Figure 5.20 suggests that in order to perform a correct porosity determination the constituents of the rock must be known very accurately.

Table 5.10. Porosity corrections for various borehole diameters

Porosity, %	Hole diameter cm	Porosity correction, %					
		NGM-60		NNT-40		NNT-50	
		thickness of caked mud, cm					
		1	2	1	2	1	2
1	15	0.5	0.9	0.4	0.8	0.8	0.7
	20	0.7	1.5	0.6	1.3	1.3	0.6
	25	0.7	1.1	0.5	0.8	0.2	0.4
	30	0.4	0.6	0.2	0.4	0.2	0.4
10	15	0.7	1.6	1.3	2.5	1.6	2.8
	20	1.2	1.8	0.9	1.6	1.6	1.7
	25	0.8	1.6	0.6	1.3	0.9	1.4
	30	0.6	1.2	0.5	1.1	0.8	1.3
15	15	0.7	1.8	1.5	2.2	1.8	2.7
	20	1.2	2.0	1.0	1.6	0.9	1.8
	25	0.9	1.7	0.7	1.2	0.7	1.5
	30	0.8	1.7	0.6	1.0	0.7	1.4
25	15	0.6	1.2	1.2	2.1	1.6	2.6
	20	0.7	1.3	1.1	1.5	1.0	1.6
	25	0.5	1.0	0.7	1.5	0.7	1.3
	30	0.4	0.8	0.5	1.2	0.2	0.5

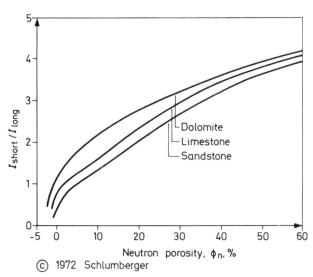

Fig. 5.28. Determination of neutron porosity in different rocks by means of compensated neutron logging

As a consequence of the fact that during measurements the sonde is pressed against the wall of the borehole, the so-called *excavation effect* arises opposite to gas-bearing layers resulting in a neutron porosity value smaller than the real one. The nomogram shown in Fig. 5.29 can be used to make the correction for this effect.

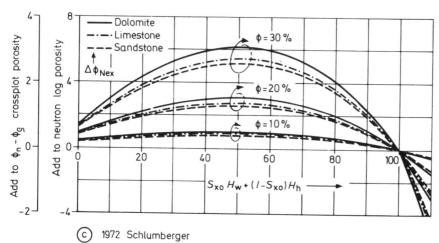

Fig. 5.29. Correction curves for the excavation effect as a function of S_{xo}; $H_h = 0$; for three values of porosity

In the *oil industry*, neutron methods have become widely used:
— to identify the lithological units in the section;
— to measure the porosity of rocks;
— to determine the gas-content of formations;
— to determine the oil–water contact;
— to estimate the clay content of the rocks, in combination with the γ–γ method;
— to control production of gas-wells;
–– to examine technical conditions of wells, etc.

Lithological identification is the same as described at the γ-ray method (see Section 5.1.4).

Porosity determination can be regarded as the most important field of application: a number of procedures are available for this purpose. Nomograms, such as those shown in Figs 5.30 and 5.31, are used together with calibrated sondes to determine hydrogen

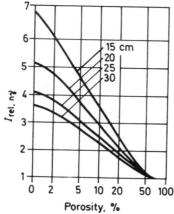

Fig. 5.30. Neutron-γ response as a function of porosity for different borehole diameters, $I = 60$ cm

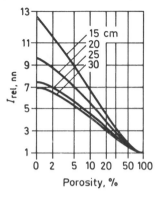

Fig. 5.31. Neutron–neutron response vs. porosity for different borehole diameters, $l = 50$ cm

porosity, after making the necessary correction for hole diameter, mud cake, etc. These types of nomograms must of course be determined individually for each kind of sonde. This is the so-called "*absolute method*" which is, however, less accurate than the relative methods, even if calibrated instruments are used (if the tool is not calibrated, only the latter method can be applied).

Among the *relative procedures* the method of *two basic zones* is the most accurate. The q factor serves as the basis of interpretation:

$$q = \frac{I_{rel,x} - I_{rel,b_1}}{I_{rel,b_2} - I_{rel,b_1}} \tag{5.27}$$

where $I_{rel,x}$ is the measured intensity of the investigated zone; I_{rel,b_1} and I_{rel,b_2} are measured intensities of the two basic zones.

A layer of great hydrogen porosity (clay, gypsum) and a layer of small porosity (compacted limestone: $\Phi = 1$–2%, anhydrite: $\Phi = 0.5$–1%) must be selected for the first (b_1) and second (b_2) base zones, respectively. Knowing the q value, the porosity of each zone can be determined, e.g., using the curves in Fig. 5.32.

The one basic zone method is less accurate: the determination is performed by making a comparison with a zone of known porosity and using the nomograms shown in Figs 5.30 and 5.31.

From all the logging procedures which are presently available nowadays the one which utilize neutron methods are unique for determination of gas content in the formations in that the *gas content of rocks* can be determined by using them. The basis of the method lies in the fact that the hydrogen index of water is taken as 100 and that of oil as 97; the value of the index for a gas can range from 3 to 50 depending on the pressure in the formation and the chemical composition of the gas. Accordingly, gas-bearing zone display extremely high intensities. This effect is particularly strong in cased holes without no mud.

Gas saturation can also be determined in cased holes, if the infiltration has already been drawn back. The nomograms necessary for the interpretation are shown in Fig. 5.3?

To detect gas-bearing zones, the so-called "*dual spacing method*" was developed. In addition to a long-spaced detector which is sensitive primarily to the gas content, a short spaced detector close to the inversion range is also used, this latter having a shallow range

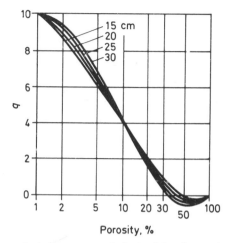

Fig. 5.32. Determination of porosity in limestone on the basis of the q factor; the parameter of the curves is the borehole diameter

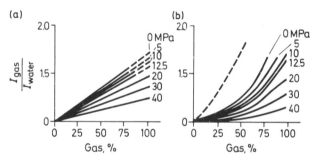

Fig. 5.33. Neutron-γ-response vs. gas content; the parameter of the curves is the formation pressure in MPa, $l = 60$ cm (a) — Cased well: dry; (b) — Cased well: filled with water

of investigation and, consequently, reflecting mainly the parameters of the invaded zone. Gas-bearing zones are reflected by the separation between the two curves. The compensated neutron log method described before in connection with Fig. 5.28, was developed from this technique.

Detection of *oil–water contact* in cased holes is possible only with high formation water salinity (higher than 80 kg/m³ NaCl). The method is based on the great neutron capture cross-section of chlorine ($\sigma_{aCl} = 3.2 \times 10^{-27}$ m² = 32 barn).

The interpretation procedures based on "*crossplot techniques*" have great importance since the last decade. In general, various physical parameters (porosity, clay content, etc.) and the components of the rocks are determined by comparing the results obtained using two different methods: GR-CNL, GR-SNP, GR-SHC, CNL-SNP (see also Section 5.2.1).

Quality control and correction of the logs is performed in the same fashion. The simultaneous determination of porosity and lithology is rendered possible by using the following equations and their combinations:

$$I_{rel,a} = I_{aw}\Phi_w + I_{a1}\Phi_1 + I_{a2}\Phi_2 \tag{5.28}$$

$$I_{rel, b} = I_{bw}\Phi_w + I_{b1}\Phi_1 + I_{b2}\Phi_2 \tag{5.29}$$

$$\Phi_w + \Phi_1 + \Phi_2 = 1 \tag{5.30}$$

where $I_{rel,a}$ and $I_{rel,b}$ are the data measured by methods a and b, respectively; $I_{rel,aw}$, $I_{rel,bw}$, $I_{rel,a1}$, $I_{rel,a2}$, $I_{rel,b1}$, $I_{rel,b2}$ are the data obtained using methods a and b in water, and in the first and second rock-forming mineral, respectively; Φ_w is the porosity, Φ_1, Φ_2 are the relative porosities of the first and second rock-forming mineral, respectively.

Neutron methods have grown in importance also in the field of *ore-mining*. Their application is advantageous mainly in cases when particularly high or low neutron capture cross-sections are involved.

A classical field of application is *boron-prospecting* ($\sigma_B = 7.52 \times 10^{-26}$ m^2). Rocks with 0.05% boron content can also be reliably detected by the thermal neutron method. The resonance method, however, has recently superseded the thermal neutron method because of its better linearity and higher accuracy. Using the same principle we can estimate the possibility of prospecting some other materials as follows

$$c_x = \frac{c_B \sigma_B \cdot \mathscr{A}_x}{\sigma_x \cdot \mathscr{A}_B} \tag{5.31}$$

where σ_B and σ_x are neutron capture cross-sections of boron and the material to be prospected, $\mathscr{A}_B$ and $\mathscr{A}_x$ are the mass numbers of boron and the material to be prospected; c_B is 0.05% and c_x is the minimum amount of the material studied detectable using the method.

The detection limits computed by Eq. (531) are shown in Table 5.11.

The following *conclusions* can be drawn on the basis of the Table:

— *lithium* can be detected only in materials in which it has accumulated to a great extent;

— the possibility of detecting *manganese* is satisfactory for industrial needs;

— sensitivity to *cobalt, silver* and *cadmium* is one or two orders of magnitude lower than that required by industry;

— *rare earths* can be detected in any concentration important for industry. Thus, elements such as *titanium, niobium* and others having no uncommon neutron capture cross-sections but occurring together with rare earths can also be detected.

No reference has been made to *iron* in the Table, although it can be measured ($\mu_{Fe} = 2.53 \times 10^{-28}$ m^2 = 2.53 barn), in principle. Determination of iron is difficult because of the variable and locally high porosity of iron ores ($\Phi = 3$–50%) causing serious alterations in the slowing down of neutrons.

Detection of *magnesite* is rendered possible by the extremely small neutron capture cross-section of magnesium ($\mu_{aMg} = 6.23 \times 10^{-30}$ m^2 = 62.3 mbarn). It follows that the effective range of neutrons is four times longer in magnesite than in limestone.

Table 5.11. Detection limits of elements based on their neutron capturing properties, %

Element	B	Li	Cl	Mn	Co	Ag	Cd	Rare earths
c_x	0.05	0.34	3.9	15.2	5.82	5.85	0.16	0.01

5.3.1.2. NEUTRON–γ SPECTROSCOPY

With neutron–γ-spectrum logging, the γ-spectrum generated by capture of thermal neutrons is recorded along the borehole. In the *oil-industry* only one version of the method, the so-called *"chlorine logging"*, is used to determine oil–water contact. Logging is carried out in the 5–6 MeV energy range, where ^{38}Cl displays several peaks. Although the majority of γ-peaks emitted by chlorine falls into the energy range below 3 MeV, this fraction of the spectrum cannot be utilized because of the presence of hydrogen and rock minerals. Figure 5.34 shows a log and a spectrum run in an oil-prospecting borehole.

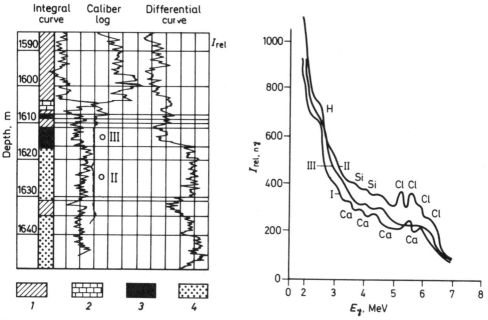

Fig. 5.34. Neutron-γ longs and spectra. *1* — Shale; *2* — Limestone; *3* — Oil-bearing sandstone; *4* — Water-bearing sandstone. I — Comparative cement spectrum; II — Spectrum of water-bearing sandstone; III — Spectrum of oil-bearing sandstone

In *ore-mining*, neutron–γ spectroscopy is used to detect Fe, Hg and Ni. In these cases, logging is carried out in the energy ranges: 5.0–6.5, 6.3–6.9 and 8.2–8.8 MeV. Sensitivity of the method to Hg is 0.1%.

A modern digital data recording equipment can record the whole spectrum along the borehole and then the spectra belonging to different depth points can be evaluated by the computer.

5.3.1.3. NEUTRON ACTIVATION METHODS

Neutron activation logging can be regarded as a variant of neutron activation methods, adapted specifically to borehole measurements—the original technique being developed for laboratory analysis. The principles of this method are discussed in Section 4.2. With respect to well-logging the following limitations of the method should be considered:

— under borehole conditions, high activity radiation sources cannot be used; in general, Po–Be and Po–B sources of 70–400 GBq (2–10 Ci) activity have proved good;

— detection is limited to elements with their corresponding isotopes having a half-life shorter than 1–2 h;

— when selecting the appropriate method attention should be paid to the fact that, besides the element to be studied, some (or some tens of) other elements are also present, in unknown concentrations.

From a technological point of view, two methods can be distinguished:

— measurements carried out at discrete points (stations), and

— continuous logging.

The *station measuring* technique is used only for the solution of special tasks as it is rather time-consuming. By this method, the logging sonde is lowered into the well to a given depth and after having activated the rocks the tool is lowered deeper so that the detector will be located at the same depth where the source had been earlier. The maximum difference between depths must be kept smaller than 2 cm. Thus, the activating and measuring operations can be performed repeatedly, step-by-step.

Continuous logging is carried out by lowering the tool downwards at a speed appropriate to the half-life of the material to be activated. The maximum permissible logging speed can be determined using the approximate equation

$$\frac{a}{a_0} = \frac{2L_1 \times \lambda l}{v} \exp\left(\frac{-\lambda l}{v}\right) \tag{5.32}$$

where a is the measured specific activity, a_0 is the saturation specific activity, L_1 is the slowing down distance (see Section 5.3.1.1), λ is the decay constant, v is the logging speed, and l is the distance between source and detector.

If the distance between source and detector is increased the a/a_0 ratio decreases, and therefore a minimum distance must be chosen at which the neutron–γ effect can be neglected ($l = 1.5$–2 m).

The method can be used to detect the following elements:

— *aluminium* is detected using a Po–B neutron source, because Po–Be sources might induce the $^{28}Si(n, p)^{28}Al$ reaction, as well (the threshold energy being about 5 MeV). In principle, SiO_2/Al_2O_3 could also be determined by logging with two different neutron sources (e.g., Po–B and Po–Be), with a logging speed of about 40–80 m/h. This method is important in bauxite prospecting;

— in *copper-ores* (e.g., in calcedone), the detection of ^{66}Cu having a short half-life ($t_{1/2} = 5.1$ min) can be performed successfully using continuous logging at a speed of 30–40 m/h. This isotope is produced by activation of ^{65}Cu. Earlier, the ^{63}Cu isotope was irradiated and the ^{64}Cu thus generated ($t_{1/2} = 12.84$ h) was measured at each point.

— detection of *fluorine* is most efficiently done using Po–Be or any other sources of fast neutrons since the $^{19}F(n, \alpha)^{16}N$ reaction is of major importance with regard to this technique. The cross-section of thermal neutrons in the reaction $^{19}F(n, \gamma)\,^{20}F$ is one order of magnitude smaller than that of fast neutrons in the former reaction, and besides, the radiation emitted by ^{20}F is softer (1.63 MeV) than that of ^{16}N (6.13 and 7.13 MeV). Logging speeds as high as 500 m/h can be used with this method.

Activation analytical methods were developed to detect *oil–water contact* by determining the ^{38}Cl isotope ($t_{1/2} = 2.6$ h) of chlorine and the ^{24}Na isotope ($t_{1/2} = 15.4$ h) of sodium in the formation water as well as by determining the ^{52}V isotope ($t_{1/2} = 3.75$ min) of vanadium which is found in some oils. These methods, however, have not become widely used because they are time-consuming procedures.

5.3.2. METHODS USING NEUTRON GENERATORS

The appearance of the borehole size neutron generators contribute very much to the progress of nuclear methods in geophysics. The reaction $^3H(T) + {}^2H(D) \rightarrow He^4 + n$ is used for generating neutrons; as a result of the reaction neutrons of 14.1 MeV energy are obtained. The *borehole generators* operate in pulsed mode and yield $10^7 - 5 \times 10^8$ neutrons with a pulse width of 5–200 μs and a repetition frequency of 5–20 000 Hz, depending on the requirements of the studies at hand.

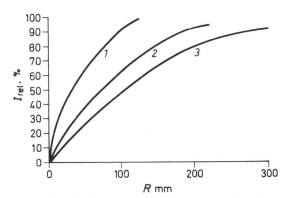

Fig. 5.35. Radial characteristics of methods using a pulsed neutron source. *1* — Oxygen activation procedure; *2* — C/O logging (sonde length: 55 cm); *3* — C/O logging (sonde length: 70 cm)

The great advantage in utilizing a *pulsed mode* is that in addition to fast neutron activation inelastic scattering (n, n') and neutron life-time logging can also be performed. Figure 5.35 shows radii of investigation determined by laboratory experiments for various methods using neutron sources.

5.3.2.1. METHODS BASED ON INELASTIC SCATTERING OF NEUTRONS (FAST NEUTRON–NEUTRON LOGGING)

Assuming that only those neutrons undergo inelastic scattering which have not been scattered previously, the following equation can be written on the intensity of γ-radiation accompanying reaction (n, n') in a homogeneous medium:

$$I_{rel\gamma} = Ai\Sigma_{n,n'} \int_V \frac{\exp[-(\Sigma_s l_s - \rho l_d)]}{16\pi^2 \, l_s l_d^2} \, dV \tag{5.33}$$

where A is the activity of the radiation source; i is the number of γ-quanta produced in the (n, n') reaction; $\Sigma_{n,n'}$ and Σ_s are inelastic and total scattering cross-sections, respectively; ρ

is the density of the medium; l_s and l_d are distances of the site of the (n, n') reaction from radiation source and detector, respectively; V is the volume.

The first approximation in solving Eq. 5.33 shows the *application possibilities* of prospecting for *oxygen, carbon, silicon* and some other elements. The method is of importance for two reasons:

— this is the only procedure usable to *detect carbon directly*;

— it makes it possible to detect *oil–water contact* in formation waters having low chlorine content (see Table 5.12).

Registration is performed by energy-selective detectors working in the ranges 3.2–4.8 MeV and 4.9–6.4 MeV, with a time-gate of 5–7 s. The γ-spectra recorded in tanks containing oil and water are shown in Fig. 5.36 to assist in selecting the proper energy range for a given measurement. For practical purposes high pulse rates should be used even though the higher level of background radiation might cause difficulties in detection.

Neutron capture spectra measured in limestone and sandstone are shown in Fig. 5.37. There are several peaks related to Si and Ca in the detection range of C and O. Therefore,

Table 5.12. Intensity of γ–γ radiation accompanying inelastic scattering of neutrons, in relative units (clean coal = 1)

Name of rock	γ-radiation intensity in (n, n') reaction in relative units, when detecting	
	carbon	oxygen
Coal, clean	1	0
Coal, 10% ash content	0.96	0.03
Coal, 25% ash content	0.84	0.09
Coal, 40% ash content	0.74	0.15
Sandstone, 0% porosity	0	0.52
Sandstone, 5% porosity oil filled	0.022	0.45
Sandstone, 10% porosity oil filled	0.045	0.43
Sandstone, 25% porosity oil filled	0.12	0.36
Sandstone, 40% porosity oil filled	0.17	0.29
Sandstone, 100% porosity oil filled	0.45	0
Sandstone, 5% porosity water filled	0	0.47
Sandstone, 10% porosity water filled	0	0.46
Sandstone, 25% porosity water filled	0	0.45
Sandstone, 40% porosity water filled	0	0.43
Sandstone, 100% porosity water filled	0	0.35

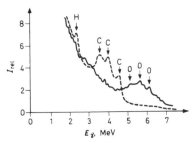

Fig. 5.36. γ-spectra measured in oil- and water-filled tank. ——— Water; – – – – Oil

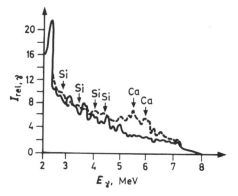

Fig. 5.37. γ-spectra measured in sandstone and limestone. —— Sandstone; – – – – Limestone

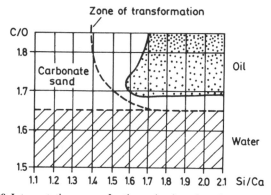

Fig. 5.38. Interpretation curves for the evaluation of sandstone reservoirs

devices are constructed for recording not only the Si/Ca ratio but the C/O ratio, as well (C/O is detected using an earlier time-gate and Si/Ca a later one).

The procedure renders it possible to determine also the lithology of rocks, in addition to a more accurate detection of oil-zones (Fig. 5.38).

Reaction (n, n′) is also utilized to detect *sulphur, magnesium, silicon* and *iron*.

5.3.2.2. ACTIVATION USING FAST NEUTRONS

In addition to the (n, n′) reaction, fast neutrons can undergo the reactions (n, p), (n, 2n) and (n, α), too, the last of these reactions also being applicable in well-logging.

Table 5.13 shows some important properties of rock-composing elements which can be activated by neutrons. Special attention should be given to oxygen which generates a particularly hard γ-radiation, useful for the detection of oxygen (see Section 4.1.4.6). As carbon does not produce γ-emitting nuclides, the C/O ratio cannot be determined by fast neutron activation. The only field of application of the method developed so far implies activation of oxygen for detecting water circulation behind a casing.

Table 5.13. Main properties of elements exposed to fast neutron activation

Stable isotope			Threshold energy, MeV	Product of reaction (radioactive isotope)			
Isotope	Relative abundance in the product, %	Type of reaction		Isotope	σ_{act}, for 14 MeV neutrons, $10^{-28}\ m^2$	Half-time, $t_{1/2}$	Energy of neutrons, MeV frequ., %
^{16}O	99.961	n, p	10.2	^{16}N	0.090	7.35 s	6.13 (76) 7.10 (6)
^{19}F	100	n, p	5.0	^{19}O	0.089	30.0 s	0.2 (96) 1.36 (54)
		n, α	3.1	^{16}N	–	7.35 s	6.13 (76)
^{23}Na	100	n, p	4.6	^{23}Na	0.034	40 s	3.0
		n, α	–	^{20}F	0.0004	11.6 s	1.64 (100)
^{25}Mg	10	n, p	–	^{25}Na	0.045	62 s	0.98 (15) 0.58 (15) 0.40 (15) 1.61
^{28}Si	92	n, p	4.4	^{28}Al	0.222	0.3 m	1.78 (100)
^{31}P	100	n, α	2.0	^{28}Al	0.146	2.3 m	1.78 (100)
^{37}Cl	24.6	n, α	–	^{34}P	0.052 1.190	12.4 s	2.1 (25) 4.0 (0.2)
		n, α	10	^{37}S	–	5.0 m	3.1 (10)
^{51}V	99.7	n, p	–	^{51}Ti	0.027	5.8 m	0.32 (96) 0.93 (4) 0.61 (1)
^{52}Cr	84	n, p	1.9	^{52}V	0.015	3.8 m	1.44 (100)
^{55}Mn	100	n, α	–	^{52}V	–	3.8 m	1.44 (100)
^{66}Zn	28	n, p	–	^{66}Cu	0.100	5.1 m	1.044 (9)

5.3.2.3. PULSED NEUTRON–NEUTRON METHOD

The pulsed neutron-neutron method is widely used. However, a predictive theory applicable to a broad range of physical conditions has not yet been elaborated. Considering a fast neutron source producing Q neutrons/s in a homogeneous field, the distribution of thermal neutrons in a medium of low hydrogen content can approximately be given by

$$n_n = f(l, t) = \frac{Q}{[4\pi(Dt+L_1)^2]^{3/2}} \exp\left[-\left(\frac{t}{\bar{t}} - \frac{l_2}{4(Dt+L_1^2)}\right)\right] \qquad (5.34)$$

where D is the diffusion coefficient, $\bar{t}$ is the average lifetime of thermal neutrons, t is the time measured from the generation of neutrons, L_1 is the slowing down distance, l is the spacing of the borehole tool.

Substituting the fast neutron source with a thermal neutron source, that is, $t > L_1^2 L/D$, Eq. (5.34) will approximate (5.35), asymptotically:

$$n_n = f(l, t) = \frac{Q}{(4\pi Dt)^{3/2}} \exp\left[-\left(\frac{t}{\bar{t}} - \frac{l^2}{4Dt}\right)\right] \qquad (5.35)$$

It follows from Eq. (5.35) that with increasing t, n_n will first increase and then decrease, the latter portion being described approximately as an exponential function of the form

$$n_n \sim n_{n,\,max} \times \exp\left(-\frac{t}{\bar{t}}\right) \qquad (5.36)$$

According to Eq. 5.34, in materials having nearly the same hydrogen content ($D_1 \sim D_2$), the following ratio can be formed:

$$\frac{n_{n,\,\bar{t}_1}}{n_{n,\,\bar{t}_2}} = \exp\left[-\left(\frac{1}{\bar{t}_1} - \frac{1}{\bar{t}_2}\right)t\right] = \exp(-at) \qquad (5.37)$$

where a is a constant, characteristic of the device.

The main feature of neutron life-time logging is expressed by Eq. (5.37): if neutron life-times do not differ too much, the ratio of neutron fluxes can be chosen rather freely by taking appropriate values for t, e.g., assuming $t_1/t_2 = 3$ and $t = 1000$ μs, $n_{n,\,t_1}/n_{n,\,t_2} = 10$ and at $t = 1600$ μs the same is about 50.

The main advantages of the method are as follows:

— the lower limit of *sensitivity* can be as low as 20–30 kg NaCl/m³, compared with a value of 80 kg/m³ obtained using the conventional method. In spite of this, the value t cannot be increased beyond a certain limit because of the rapid decrease of neutron flux;

— the method can be applied to depths much greater than that with conventional methods; the limiting value of the depth can be estimated by the formula

$$R = 2.1\sqrt{Dt} \qquad (5.38)$$

— in the case t is greater than $t_{threshold}$, the *distribution* of neutrons is independent of the construction of the well, provided that t_{rock} is greater than $t_{borehole}$—a condition most commonly fulfilled. The presence of the borehole results in a parallel shift of the curve and, considering a logarithmic scale, this is equivalent to a decrease of the intensity of the neutron source.

Table 5.14. Behaviour of rocks towards neutrons

Name of rock	D_t $\cdot$, kg/m³	Slowing down length, cm	Diffusion coefficient $D \times 10^{-9}$, m²/s	Average lifetime of neutrons, μs
Clay	2100	15.0	0.9	450
Cement	2000	12.0	1.0	330
Sandstone, compacted	2700	42.0	2.5	1300
Sandstone, oil-bearing 20% porosity	2400	16.0	1.1	630
Sandstone, water-bearing 20% porosity	2400	16.0	1.1	225
Halite	2100	—	2.0	7
Coal, 10% ash content	1300	—	—	2500
Limestone, oil-bearing 20% porosity	2400	15.0	1.0	450
Limestone, compacted	2700	35.0	2.2	700
Limestone, water-bearing 20% porosity 200 kg/m³ NaCl content	2400	15.0	1.0	195
Oil	900	12.0	0.32	190
Water	1000	12.2	0.35	207

Some important parameters of the interaction between neutrons and rocks studied most frequently in oil-recovery, are shown in Table 5.14.

Based on theoretical considerations, the *optimum spacing of the sonde* can be determined from the expression

$$l_{opt} = \frac{3}{2} \sqrt{\frac{D}{\varepsilon}} \qquad (5.39)$$

where ε is the anticipated uncertainty in the determination of $1/\bar{t}$. The optimum sonde length used in oil-recovery is 35–40 cm.

The pulsed neutron–neutron method is applied most frequently in *oil-recovery*. It is used also to trace the *movement* of *oil–water contact* in oil reservoirs during the process of recovery. Figure 5.39 shows an example of application: the log was recorded at a NaCl content of $25 \, kg/m^3$, with a sonde-length of 40 cm and a time delay of 1700 µs.

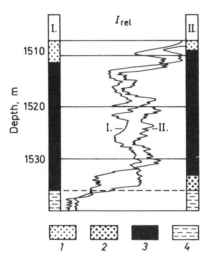

Fig. 5.39. Movement of oil–water contact during exploitation. *1* — Gas; *2* — Water; *3* — Oil; *4* — Shale. The pore content in the first (I) and seventh (II) years, respectively

The pulsed neutron method is useful also in prospecting for different *salts* ($CaCl_2$, KCl, NaCl) and in detecting other materials having great capture cross-section, such as B, Cs, Li, etc.

5.3.2.4. PULSED NEUTRON-γ METHOD

The γ-radiation field produced by a pulsed neutron generator is basically different from the field of a thermal neutron which has created it. This can be explained as follows: the γ-radiation induced by capturing neutrons; neutron travels with the velocity of light and thus the γ-radiation will leave behind slow thermal neutrons. In a homogeneous medium of infinite dimensions the field of γ-radiation formed on neutron capture can be described by the approximative formula:

$$n_\gamma = i \int\limits_V \frac{f(l,\,t)}{4\pi\,l_R^2}\exp\left(-t_\gamma l_R\right)\mathrm{d}\,V \tag{5.40}$$

where i is the intensity of quanta formed on neutron capture, $f(l,\,t)$ is the distribution of thermal neutrons (this can be determined, for example, by Eq. 5.34) and t_γ is the linear attenuation of γ-radiation at a distance l_R.

If Eqs (5.40) and (5.34) are compared and other (more exact) equations are considered, the following *conclusions* can be drawn:

— on increasing the spacing of the sonde the intensity of γ-radiation decreases more slowly than that of neutrons;

— $n_\gamma = f(t)$ approaches more rapidly the asymptote $\exp\left(-t/\bar{t}\right)$ than the $n_n - f(t)$ function;

— the optimum spacing of the sonde is about $48\varGamma60$ cm;

— the *depth of investigation* can be two times more greater than that of the neutron–neutron method.

As a result of the above factors, pulsed neutron γ-logging has grown in importance in recent years, whereas pulsed neutron–neutron logging has lost some of its importance. The only disadvantage of the former method is that the level of background radiation is an order of magnitude higher but manufacturers try to reduce this effect by providing a proper discrimination of energy. Its *field of application* in *oil-recovery* is the same as that of pulsed neutron–neutron logging. The method can be used to detect *elements having high resonance capture cross-sections*, e.g., silver, cadmium, indium, europium and gadolinium.

5.4. TRACER TECHNIQUES IN WELL LOGGING

Tracer techniques in well logging can be classified according to the phases of drilling and completing activities as follows:

— measurements carried out during drilling of the well (e.g., detection of mud-loss);

— measurements carried out during cementation (e.g., during the so-called reversed cementation when cement-milk is forced directly into the annulus (between the casing and the formation);

— testing of cement (e.g., determination of the top of the cement in the borehole);

— determination of the technical conditions of the borehole (e.g., locating any damage of casing, detection of the circulation behind the casing, etc.);

— determination of the site of the formation-perforation and qualification of the penetration;

— measurements in producing wells (recording of production logs);

— measurement carried out in boreholes to assist recovery.

From a *methodological* point of view, tracer techniques can be divided into three groups. Operations carried out

— using radioactive mud;

— in a mud (or cement-milk) traced by a radioactive isotope just before the measurement;

— in a mud traced by a radioactive isotope during logging.

From the three last methods, the first group is not recommended for with regards to health protection; the second group is, however, widely used both in measurements carried out during drilling, and in solving a number of special technical problems. The third group is applicable, in general, in cased holes when the flow rate of fluids is studied (e.g., in producing and water-injecting wells).

In the practice of borehole well logging, tracer techniques utilizing isotopes with a minimum half-life and activity are selected. The isotopes most commonly used in geophysics are summarized in Table 5.15.

Detection of mud loss, recording of the *permeability* profile log and location of the casing damage, are regarded as the most important fields of application.

A "*mud-plug*" traced by an isotope is placed above the zone of interest and is forced into the rocks *behind* the casing. Thereafter, the hole is flushed over with fresh mud and radioactive logging is performed: the location of the mud loss and other irregularities in the log appear with high amplitudes I_{rel} (Fig. 5.40).

Applying a more up-to-date technology, logging is carried out during the injection of the radioactive mud-plug. In this case, the logging tool is placed first at the top of the investigated section and so it detects the radioactive mud passing through it. Then the sonde is lowered 2–20 m deeper into the borehole and the mud injection is continued until the isotope disappears in the formation. The depth determined by this method is the location of mud-loss.

During cementing operations a radioisotope is mixed with the first dose of the cement-milk, and it is forced into the annulus between the formation and the casing. After setting, the radioactive log shows the *top of the cement sheath*. At reversed cementing operations the radioactive logging tool is placed near the bottom of the borehole, this means that the sonde is lowered to 100–200 m above the casing shoe before starting the injection of the radioactive cement. The cementing process can be followed by detecting the isotope as it reaches the detector in the outside space of the casing.

Table 5.15. Radioisotopes used in tracer techniques in well logging

Isotope	Application method	Half-life, day	Field of application	Recommended radioactive concentration (MBq) m^3 in	
				cased wells	uncased wells
^{59}Fe	Fe$_2$O$_3$ FeCl$_3$	45	A commonly used isotope with a disadvantage: in cased holes it sticks to the casing. It can be useful in making radioactive sand	80 (30)	1.4 (0.5)
^{65}Zn	ZnCl$_2$ ZnSO$_4$ Zn(NO$_3$)$_2$ dissolved in water	245	Because of its long half-life, it can be used only in special cases	160 (60)	2.7 (1.0)
^{86}Rb	RbCl Rb$_2$CO$_3$	19	Primarily at hardly accessible places, it can be used instead of iodine	810 (300)	13.5 (5)
^{124}Sb	Sb	60	It is used in boreholes having multiple casing	50 (20)	0.8 (0.3)
^{131}I	NaI, KI, I$_2$ dissolved in water	8	The most widely used isotope, also dissolved in oil	220 (80)	3.5 (1.3)
^{192}Ir	Ir salts dissolved in water	74	Used for preparation of active sand and in injection works	190 (70)	3.2 (1.2)

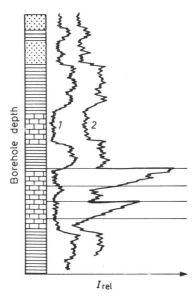

Fig. 5.40. Detecting mud-loss zone by using the tracer technique. *1* — Base log; *2* — Log detected after tracing

When control of the formation perforation is to be carried out, the isotope is mixed with the sand (adsorbed on the sand or on plastic balls).

Production and *water-injection profiles* are recorded by another technology. In this case the radioactive device is lowered into the borehole in such a manner that an isotope injector can be placed below or above it depending on the direction of the fluid stream. A few cm³ of isotopic solution is injected into the fluid and the time taken by the isotope to travel between the injection point and the detector is measured. Then the device is pulled up by 20–100 cm and the procedure described above is repeated until the whole section is measured. Applying this method, we get flow rate and flow velocity logs. The advantage of the injection method is that only small amounts of radioisotopes are required.

In addition to the method described above, there are several other applications of isotope tracer techniques in well logging. A new branch of tracer techniques used in geophysical prospecting utilizes *stable instead of radioactive isotopes*. In this method, materials containing elements of large neutron capture cross-section (B, Cd, Cl) are injected into the investigated section of the borehole (e.g., aqueous solution of $CdCl_2$) and then they are studied by neutron logging. The advantage of this technique lies in the fact that there is no radiation hazard, but it should be noted that they cost more than the other techniques described above.

REFERENCES

[5.1] Clark, S. P., *Handbook of Physical Constants*. Yale University, New Haven, Connecticut, 1951.
[5.2] Desbrandes, R., *Theorie et interprétation des Diagraphies*. Ed. Technique, Paris, 1968.
[5.3] Filippov, E. M., *(Nuclear Geophysics)* (In Russian). Zd. AN SSSR, Moscow, 1973.
[5.4] Kiss, D., *(Neutron Physics)* (In Hungarian). Budapest, Akadémiai Kiadó, 1971.
[5.5] Kobranova, V. N., *(The Physical Properties of Minerals)* (In Russian). Gostoptechizdat, Moscow, 1962.

[5.6] Komarov, S. G., *(Geophysical Handbook)* (In Russian). Vol 2, Gostoptechizdat, Moscow, 1961.

[5.7] Larionov, V. V., *(Radiometry of Logging)* (In Russian). Izd. Nedra, Moscow, 1969.

[5.8] Novikov, I. I., *(Radioactive Survey Methods)* (In Russian). Gostoptechizdat, Moscow, 1965.

[5.9] Pirson, S. I., *Handbook of Well Log Analysis.* Englewood Cliffs, N. J., Prentice-Hall, 1963.

[5.10] Tittle, C. W. and Allem, L. S., *The Theory of Neutron Logging Geophysics.* 1966. Jan.–Febr.

[5.11] Clavier, C., Hoyle, W. and Mennier, D., Quantitative Interpretation of Thermal Neutron Decay Logs. *The Journal of Petroleum Technology,* 1972.

[5.12] Schultz, W. E. and Smith, H. D., Laboratory and Field Evaluation of a Carbon/Oxygen Well Logging System. *The Journal of Petroleum Technology,* 1974.

[5.13] Lock, G. A. and Hoyer, W. A., Carbon–Oxygen Log: Use and Interpretation. *The Journal of Petroleum Technology,* 1974.

[5.14] MacKinlay, P. E. and Tanker, H. L., The Shale-Compensated Chlorine Log. *The Journal of Petroleum Technology,* 1975.

[5.15] Alger, R. P. and Conell, J. G., Progress Report on Interpretation of the Dual-Spacing Neutron Log in U. S. *The Log Analyst,* 1972.

6. RADIATION TECHNOLOGIES

Radiation technologies are practical processes for use in industry, agriculture, etc. which are based on the temporary or permanent changes in the physical, chemical or biological properties of matter under the effect of radiation.

Radiation physics processes deal with the changes in the physical properties of solids under the effect of α-, β-, γ- and X-rays, fast neutrons; accelerated electrons, protons, deuterons, as well as atoms and ions produced by nuclear reactions. These charges are, in most cases, of crystallographic character (e.g., formation of defect sites). If the lattice defect is induced by the appearance of a new element as a result of nuclear reaction, the process belongs rather to radiation chemistry. The transformation of nuclear energy to heat or light energy, as well as the increasing of the conductivity of the medium by radiation-induced ionization belong to radiation physics.

Radiation chemistry deals with chemical reactions induced by the radiation types mentioned above and by nuclear reactions. Radiation chemistry in the wider sense includes *photochemistry* treating reactions induced by visible or ultraviolet light; *electric discharge chemistry* and *plasmochemistry* studying processes occurring under the effect of various electric discharges (silent, glow or plasma discharge); and even *sonochemistry* being the science covering the chemical utilization of ultrasound. If the radiation energy exceeds about 30 eV, it causes partial ionization of matter which is the reason for it being called *ionizing radiation*. Radiation chemistry in a stricter sense includes processes stimulated by ionizing radiations only, i.e., it excludes e.g. photochemistry (Fig. 6.1).

Radiation biology involves the planned and desirable modification of the life functions of different kinds of organisms by radiation. These processes are utilized on a commercial scale by agriculture and by the food industry (stimulation, production of mutants, plant and animal protection, upgrading and conservation of agricultural products, etc.), as well as by the pharmaceutical and medical instrument industries (sterilization).

The following parameters are important for quantitative characterization of these processes from the viewpoint of radiation technologists:

— *The absorbed dose*; its different units can be converted into each other on the basis of Table 6.1.

The energy of radiation: the units can be compared with Fig. 6.1.

— *Linear energy transfer (LET)* gives the energy loss of particles (quanta) passing through matter per unit path length (its dimension is dE/dx, its commonly accepted unit is eV/nm). The characteristic *LET*-values of some radiation types are shown in Table 6.2

— *The radiation chemical yield (G)* reflects the efficiency of radiation chemical processes; it gives the number of transformed or produced molecules, atoms, ions or radicals per 100 eV absorbed energy. *G*-values can be calculated on the basis of the dose rate of the radiation (I_{abs}) and the reaction rate (dx/dt):

$$G = \frac{dx/dt}{I_{abs}} \, 100 \text{ particles per } 100 \text{ eV} \qquad (6.1)$$

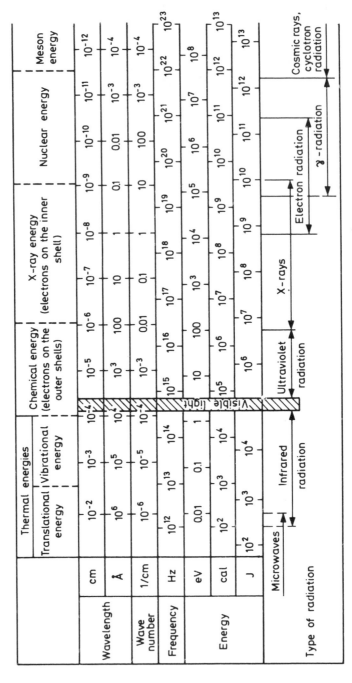

Fig. 6.1. The energy of radiation in various units

Table 6.1. Conversion table: energy equivalents

Unit	erg	$J = kg\,Gy$	kWh	cal	g rd	eV
erg	1	10^{-7}	2.78×10^{-14}	2.39×10^{-8}	10^{-2}	6.242×10^{11}
$J = kg\,Gy$	10^7	1	2.78×10^{-7}	0.239	10^5	6.242×10^{18}
kWh	3.6×10^{13}	3.6×10^6	1	8.60×10^5	3.6×10^{11}	2.247×10^{25}
cal	4.186×10^7	4.186	1.164×10^{-6}	1	4.18×10^5	2.612×10^{19}
g rd	10^2	10^{-5}	2.78×10^{-12}	2.389×10^{-6}	1	6.242×10^{13}
eV	1.602×10^{-12}	1.602×10^{-19}	4.45×10^{-26}	3.827×10^{-20}	1.602×10^{-14}	1

Table 6.2. LET-values of various radiations

Radiation	LET, eV/nm
α-Particles	
1 MeV	264
10 MeV	56
Deuterons	
2 MeV	28
10 MeV	8.2
100 MeV	1.26
1000 MeV	0.28
10000 MeV	0.21
Protons	
1 MeV	28
10 MeV	4.68
100 MeV	0.74
1000 MeV	0.22
2800 MeV	0.20
10000 MeV	0.23
Electrons	
0.01 MeV	2.32
0.1 MeV	0.42
1 MeV	0.18
10 MeV	0.20
γ-Radiation	
^{60}Co	0.2

Special care should be taken with regard to the consistent use of units: $mol/J = 9.648 \times 10^6 \times 100/eV$.

or, on the basis of the dose (D) and the conversion (dx):

$$G = \frac{dx}{D} \, 100 \text{ particles per } 100 \text{ eV} \qquad (6.2)$$

Plotting G-values as a function of the duration of irradiation or the dose rarely gives straight lines because the products of the reactions, as a rule, decrease (in a few cases, increase) the reaction rates. In view of this, it is generally more reasonable to extrapolate the reaction rates to zero conversion value. Conversion values as is usual in the practice of physical chemistry are not utilized in radiation chemistry because the overall conversion is extraordinarily low (generally below 1%), thus the consumption of the starting substance cannot be followed analytically.

In the case of *mixtures*, the ratio of absorption exerted by their individual components can be calculated on the basis of the *electron fractions* (e_A); i.e., by calculating the ratio of the number of electrons in the A component in question (E_A) to the total number of electrons in the mixture ($E_A + E_B + \ldots$):

$$e_A = \frac{E_A}{E_A + E_B + \ldots} \tag{6.3}$$

For heterogeneous systems the concept of G-values is even less unambiguous than for homogeneous mixtures. G-values can be calculated taking it into account that energy is absorbed in the whole system or, otherwise, as if it were absorbed in one phase only (e.g., in the liquid). This latter assumption is evidently incorrect, since it gives smaller G-values than the actual ones. If, however, the calculation is based on the assumption that the absorbed energy is distributed between a solid and a gas phase in proportion to their respective masses or electron fractions, the calculation obviously does not give a correct result either, because the energy absorbed by the solid phase can be—completely or partially—transferred to the gas phase. All corrections proposed until now are arbitrary.

Catalysts are irradiated simultaneously with the reactants or alone; in this latter case, the catalytic processes occur over irradiated catalysts. The following factors have to be considered when the simultaneous irradiation is applied;

— *energy transfer* from the solid to the surface reactant, i.e., the energy absorbed by the latter increases;

— *kinetic changes* due to the modification of the structure of the catalyst, as a result of irradiation (e.g., formation of defect sites). Kinetic changes may give rise to alteration of the selectivity, the reaction pathway, improvement of yields, etc.

The change of the catalytic effects may be *permanent*, i.e., the radiation effects remain after irradiation, too, as opposed to energy transfer which occurs only during the irradiation. The effect exerted on the *crystal structure* of various heterogeneous catalysts depends on the fact whether there are any nuclear transformations under the effect of

Table 6.3. Duration of various radiation chemical elementary processes in liquids

Process	Duration, $-\log t$, s
Passing of an ionizing particle through a molecule	18–17
One fluorescence vibration	15
Intramolecular vibration, instantaneous dissociation	14–12
Time between two collisions of small molecules	13
Duration of transversal motion of a small molecule due to *Brown*-motion as far as its diameter at room temperature (e.g., in the case of water molecules)	12–11
Slowing down of electrons (thermolization)	10–8
Time of excitation of permitted states	8–7
Time of reactions of ions or radicals in molar concentrations	5–4

irradiation or not. This is one of the reasons why the results obtained with neutron or γ-radiation are not the same: in the latter case the annealing of the crystals can occur, e.g., at higher temperatures. Changes in the catalytic properties are observed only if the effects caused by irradiation are not negligible compared with the pre-irradiation state.

The G-values observed are, as a rule, between 10^{-3} and 10^5; high G-values indicate chain reactions.

Radiation chemical changes followed by analysis are the results of several *partial processes*: the time required by partial processes occurring in the liquid phase is shown in Table 6.3.

6.1. RADIATION PHYSICAL PROCESSES

6.1.1 CHANGE OF MATERIAL STRUCTURE

The primary effect of radiation on metals, semiconductors and ionic salts consists in the changing of the positions of elementary particles in the lattice. Thus, the radiation physics of inorganic crystals is basically different from the radiation chemistry of crystals of organic polymers having relatively large molecular masses; in the latter case the main radiation effect is the induction of decomposition or rearrangement of covalent bonds.

The crystal structure is modified by direct impacts of energy-rich particles (quanta) as well as by further collisions. The removal of particles from their original positions in the crystal lattice is possible if the impact delivers an energy higher than a threshold value characteristic of the substance in question; the threshold is generally between 20 and 30 eV. γ-quanta and high energy electrons have a very low efficiency in this respect; measurable effects can be expected from heavy particles (α-particles, deuterons, fast neutrons). Primarily displaced particles may acquire a kinetic energy so high that they may be effective in causing secondary, etc. displacements. These particles are accommodated in interstitial (irregular) positions in the lattice giving rise to *Frenkel-type defects*. Extremely high doses can completely disrupt the crystal lattice.

Another reason of changes may be a temporary (10^{-11}–10^{-10} s) heating of regions in the crystals up to 700–800 °C, i.e., a *"thermal spike"*. Such a melting is brought about by radiations with high *LET*-values (i.e., having high specific ionization). This is the situation also with larger particles, e.g., in the case of the final stage of the path of displaced lattice elements. Molten regions solidify rapidly, but it is not the originally damaged region which serves as the centre of crystallization. Resolidified regions contain anomalies other than Frenkel-type defects: e.g., microcrystals which are oriented in a way different from the general orientation of the original crystal.

New elements are produced practically only under neutron irradiation; the properties in this case are influenced by the chemical "impurity". This may be important in semiconductor production, e.g., for telecommunication, or in catalyst manufacturing.

6.1.1.1. ELECTRIC CONDUCTORS

A decisive factor for radiation effects in *metals and alloys* is the temperature. This determines the mobility of removed particles and, consequently, whether they can be removed far enough from their original site to be accommodated in a new position. Another factor influencing radiation physical processes is the pretreatment of the metal (e.g., cold-working), because this is in close correlation with eventual lattice anomalies.

Irradiation can change, in addition to the mechanical properties of metals, their electric and thermal conductivity, optical properties, etc.

It is a general experience that irradiation of metals and alloys improves some of their properties and damages others; to a given limit, both processes are in a positive correlation with the dose. Properties of pre-tempered metals are modified to a greater extent than those of cold-worked ones. The range of changes under irradiation is between a few percents and 100–200%. No general conclusions could be drawn up to now; experiments are necessary to determine the behaviour of every structural material under the effect of irradiation.

If the irradiation does not involve transformation of elements, lattice defects are more or less annealed at higher temperatures and the original properties can be *restored*. The thermal stability of defects is, as a rule, higher as far as electric properties are concerned than in the case of mechanical properties.

Graphite lattice transformations brought about by irradiation have been studied very thoroughly, first of all, because graphite is used widely in nuclear reactors as moderator. It was shown that irradiation dislocates several carbon atoms from the lattice layers; these take inter-layer positions subsequently. As a result, a crystal contraction occurs parallel to the crystal base, together with a dilatation in a perpendicular direction (about 15%). Both the volume and the thermal conductivity changes are reversible at higher temperatures— the regeneration involves a considerable change of the heat content.

6.1.1.2. SEMICONDUCTORS

Semiconductors like *germanium, silicon, indium-antimonide*, etc. conduct electricity due to their "anomalies" (lattice defects, impurities) and the electric properties depend first of all just on these anomalies. These may be permanent or temporary: the former group contains *vacancies, interstitial* and *dislocated particles*; one of the latter group is ionization which lasts up to 10^{-7} s after the ceasing of irradiation. The classical method of producing defects is to *admix* very small amounts of contamination to the liquid substance before its solidification; or, afterwards, they are introduced by *diffusion* or *ion implantation* into the bulk of the crystal. Ionizing radiations represent a tool suitable to remove lattice elements from their original position, creating thus defects. Cavities and vacancies remaining in the lattice have electron acceptor properties (p-conductors). The effect of the defect is determined by the character of the energy anomaly or location of the barrier layer about by the vacancies and interstitial particles.

The effect of different types of radiation is qualitatively identical, but there are quantitative differences. There are also temperature effects: at lower temperatures more defects are produced, but they are completely or partially annealed by heating. No defects are produced by high temperature irradiation (e.g. at 450 °C for germanium). In many cases, impurities produced as stable (nonradioactive) products of *nuclear reactions* are important as far as semiconductor technology is concerned.

For dosimetric purposes, semiconductors are used as continuous dose rate dosimeters; their use is more and more widespread (see also Section 6.6).

6.1.1.3. INSULATORS

Glasses are important from both the technological and dosimetric points of view; some of their properties change considerably under the effect of radiation: they are discoloured, some types exhibit fluorescence. Figure 6.2. depicts the *blackening (S)* of a sodium glass as a function of the dose. Equation (6.4) gives the correlation between the blackening (of a 1 mm thick glass) and the dose:

$$S = 0.42[1 - \exp(-0.36\,D)] \tag{6.4}$$

where D denotes the dose in 10^4 Gy or in Mrd.

Blackening brought about by irradiation is completely or partially reversible as shown by Fig. 6.3, which plots various relationships belonging to different temperatures. Blackening decreases first rapidly, then the decrease slows down; the overall rate of the decolouring is higher at higher temperatures. The regeneration of some colour centres involves *thermoluminescence*, which can be measured by an appropriate instrument—thus it is possible to draw conclusions on the absorbed dose (see Section 6.6).

Borosilicate glasses (e.g. *Pyrex* types) are used in reactor technique for neutron shielding although their radiation resistance is not high: boron undergoes nuclear reactions and gives, e.g., helium and lithium in an (n, α) process. *Lead glasses* are suitable for shielding against hard γ-radiation, i.e., for radiation protection. Admixing 1–2% of ceric oxide to common glass gives *cerium glass*; this has the highest radiation resistance among all types of glasses: therefore, in cases when low discolouring is important under

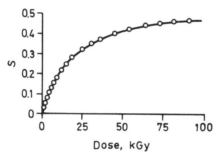

Fig. 6.2. Blackening of glass plate under the effect of radiation

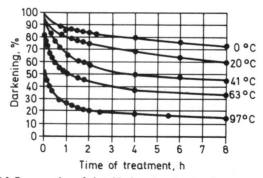

Fig. 6.3. Regeneration of glass blackened under the effect of radiation

radiation effect (e.g., for medicine ampoules to be radiation-sterilized, see Section 6.3.2), the use of cerium glass is recommended.

Oxide ceramics (aluminium oxide, beryllium oxide, titanium dioxide, etc.) are more radiation resistant than glasses, but less so than metals.

6.1.2. TRANSFORMATION OF NUCLEAR ENERGY

6.1.2.1. RADIOACTIVE LIGHT SOURCES

Luminescent substances on *luminous dial plates* are stimulated to emit light by α- or β-radiation.

In the simplest case an *α-radioactive* substance is admixed mechanically to the luminous paint, e.g., ^{226}Ra to zinc sulphide. The disadvantage of applying radium is its high energy α-radiation (4.8 MeV) and the concomitant γ-radiation, as well as its decomposition product being radon: all these facts render radium processing hazardous. The luminous efficiency of such systems (Table 6.4) cannot be high, because high LET α-radiating substances damage luminescent paints relatively rapidly; such systems remain in operating condition for 5–6 years only.

Table 6.4. Light density of luminescent paints excited by radioactive isotopes

Light density, kcd/cm²	Activity/paint mass			
	for ^{147}Pm		for ^{226}Ra	
	10⁷ Bq/g	mCi/g	10⁵Bq/g	μCi/g
0.6	1.01	0.27	0.61	1.6
0.8	1.42	0.38	0.85	2.7
1.0	1.77	0.48	1.06	2.9
1.5	2.65	0.72	1.62	4.4
2.0	3.53	0.95	2.12	5.7
2.5	4.42	1.19	2.62	7.1
3.0	5.34	1.44	3.04	8.2
5.0	10.1	2.73	5.83	15.8
10.0	23.8	6.43	15.9	43.0

β-Radiating isotopes (e.g., ^{3}H, ^{85}Kr, ^{90}Sr, ^{147}Pm, ^{204}Tl) permit one to eliminate some of the disadvantages mentioned. Luminescent systems with ^{90}Sr built in enamel are suitable for safety indicators and for lighting in mines: the necessary surface radioactivity is about 4×10^7 Bq/cm² (1 mCi/cm²).

If light is stimulated by ^{3}H or ^{85}Kr gas, objects of the appropriate shapes should be formed from thin-walled glass tubes, the inner well of which is coated with zinc sulphide. If the vessel is filled with radioactive gas and closed, the whole system will luminesce. Gaseous isotopes are not hazardous for their surrounding, because their activity is low and, advantageously, their lifetime is long (as a rule more than 10 years).

The advantage of *isotopic safety and light signals in mining*, over *emergency exits in airplanes*, etc., is that they do not require any electricity or any maintenance. If suitable isotopes (mainly ^{3}H, ^{85}Kr or ^{90}Sr) and appropriate constructions are applied, it is possible to recognize objects with isotopic illumination from a distance as great as 500 m.

EXAMPLE 6.1

What is the necessary radioactivity of ^{147}Pm or ^{226}Ra to produce 5 g of luminescent paint to give a light density of 4 kcd/cm^2?

An interpolation of the data in Table 6.4 indicates that the necessary activities are as follows: 3.8×10^8 Bq (10.4 mCi) ^{147}Pm or 2.22×10^6 Bq (60 µCi) ^{226}Ra. Although the necessary amount of ^{147}Pm is about 170 times higher than that of radium, its β-radiation is more advantageous as far as health protection is concerned, partly because the energy of the radiation is low, partly because no γ-radiation is emitted.

6.1.2.2. RADIOACTIVE ELECTRIC POWER GENERATORS

Nuclear radiation can directly be transformed into electricity: if the electrons leaving a β-radiating radioactive isotope are collected, a negative pole is obtained. The radiation source is placed on an insulated central electrode surrounded by a conducting collector system (Fig. 6.4). This arrangement produces a current intensity of 10^{-11}–10^{-10} A, at several thousand volts.

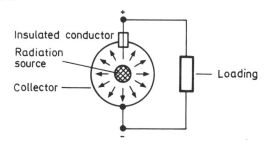

Fig. 6.4. Electric power generator based on charging by electrons

If semiconductors are irradiated by *β-radiation*, it is also possible to produce electrical voltage; suitable radiation sources have low energy β-radiation (max. 0.2 MeV, e.g., ^{3}H, ^{63}Ni and ^{147}Pm). The charge can be multiplied by germanium or silicium semiconductors with p–n barrier layer: e.g., every β-particle leaving a ^{90}Sr isotope produces about 200 000 free electrons and positive holes. The disadvantage of this device is that the radiation destroys the semiconductor completely after a few weeks. The voltage produced is low (0.2–0.3 V), but the power is relatively significant (about 0.01 mW); the efficiency is about 2.5%.

Neutron irradiation of semiconductors is also suitable to produce electricity. Boron layers or powder irradiated by thermal neutrons emit α-particles and lithium ions: voltages in the order of magnitude of millivolts can be produced in this way; the efficiency is about a few per mille. Higher effectivity can be achieved by fast neutrons: 0.5% with Si, 1% with GaAs and 8% with Pm; a performance of 1 mW/cm^2 and the temperature of 80 °C cannot be exceeded.

Photocell electricity sources are based on a different principle: if a radioactive light source provided by a luminescent paint on both sides (see Section 6.1.2.1) is built in a system of photocells, electricity is produced (Fig. 6.5). Their performance and efficiency are similar to those of the above-mentioned systems.

Thermoelectric generators supply much more power (up to 100 W) and a voltage not exceeding 100 volts. These are used as electricity sources for sea buoys, arctic

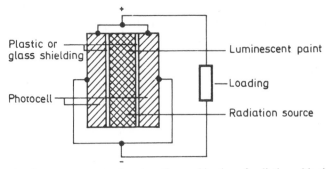

Fig. 6.5. Electric power generator based on the combination of radiation with photocells

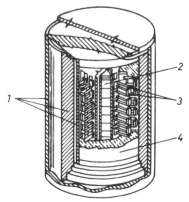

Fig. 6.6. The structure of a thermoelectric generator. *1* — Shielding; *2* — Radiation sources in holders; *3* — Thermocouples; *4* — Thermal insulation

meteorological stations, space rockets, pacemakers, etc. These devices produce electricity out of the heat of radiation of high activity (up to 10^{16} Bq or 0.1–1 MCi) by thermocouples or thermocolumns (Fig. 6.6). It is important that these devices concentrate maximum possible radioactivity in a minimum possible volume, thus the temperature in the core be high (about 500 °C). Suitable radiation sources are ^{210}Po, ^{238}Pu and ^{242}Am, as well as ^{90}SrTiO$_3$, being both heat and radiation resistant. The overall efficiency with respect to the consumer is about 6%.

EXAMPLE 6.2

What is the overall efficiency (η) of an isotopic electricity source of 10 W if it contains ^{210}Po of 2×10^{14} Bq (5400 Ci) activity?

Since the activity in becquerel is equal to the number of disintegrations per second, the given system produces 2×10^{14} disintegrations per second. Since each disintegration produces an energy of $(5.3 + 0.8) \times 10^6 = 6.1 \times 10^6$ eV (see Section 1.3), the total energy given off is:

$$6.1 \times 10^6 \times 2 \times 10^{14} = 12.2 \times 10^{20} \text{ eV/s}$$

i.e., according to Table 6.1:

$$12.2 \times 10^{20} \times 1.6 \times 10^{-19} = 195 \text{ W}$$

The efficiency is:

$$\eta = \frac{10 \times 10^2}{195} = 5.1\%$$

6.1.3. CHANGING THE CONDUCTIVITY OF THE MEDIUM

6.1.3.1. ELIMINATION OF ELECTROSTATIC CHARGES

Electrostatic charges accumulate when insulators are rubbed against each other at a high speed. This high voltage electricity *(frictional electricity)* may be disturbing or even hazardous in several industries. For example, the charging of *paper* produced continuously on a paper manufacturing mill is disadvantageous because repulsion deviates it from its path. In printing, repulsion between sheets hinders binding. Analogous phenomena can be observed with *rubber sheets* and *plastic foils*, too. In the *textile industry*, the spinning and weaving of silk, viscose. and other man-made fibres induce identical charges over neighbouring threads; therefore they repulse each other and this is harmful for processing. In order to decrease electrostatic charging, the speed of machines has to be decreased; this, in turn, sets limits to productivity and lowers the economy. At the same time, charged threads attract dust particles of the opposite charge causing thus a *contamination* which cannot even be removed completely by washing.

Where there are *inflammable* or *explosive* vapours (gasoline, benzene, etc.) or solids (dust) in the air, sparks due to electric discharges can cause fire or explosion. In surgeries, the friction of rubber or plastic gloves, foils, threads may induce sparks exploding narcotic vapours (e.g., ether). If inflammable liquids are poured from one vessel into another, both the liquid and the vessel may gather electric charge, the discharge of which can also start a fire or explosion.

Static electricity may also play a part in *corrosion*, because it produces *ozone* which can damage several structural materials (see also Section 6.2.1.1).

Radioactive isotopes are suitable for the elimination of electric charging. Radiation (especially a highly specific ionizing radiation) will ionize the air molecules; as a result, highly mobile ions are produced which increase the conductivity of the gas permitting a current to flow between the differently charged locations or objects. Thereby electric charges are neutralized.

As *radiation sources*, α- or β-radiating isotopes are suitable because of their low penetration, and from health protection viewpoints it is advantageous if they have no γ-radiation at all. Isotopes with α-*radiation* (^{210}Po, ^{226}Ra, ^{241}Am, etc.) can ionize a layer of air of 3–4 cm thickness; for ionizing thicker layers, β-*radiating* isotopes are necessary, such as ^{35}S, ^{90}Sr and ^{204}Tl. Their surface activity should not be less than 4 MBq/cm^2 (0.1 mCi/cm^2). Both types of isotopes are applied as a coating on appropriate plates (e.g., silver, gold, platinum, steel); α-radiation sources cannot be covered perfectly safely due to their low penetration (at most, a thin varnish can be applied), but the hermetic sealing of β-radiation sources is possible using a few tenths mm thick aluminium foil or a 1 mm plexiglass plate. The warranted duration of tightness should always be considered; during warranted time, as well as afterwards it is very important to check whether the system remained sealed. The advantage of ^{85}Kr is that—being a noble gas—there is no incorporation hazard even in the case of an explosion.

The removal of charges can be solved by irradiation of the air layer between the two objects in question by an isotope layer applied as a coating on an earthed plate, i.e., ions are produced in the space between the two objects (Fig. 6.7). This technique can be applied only if the object to be irradiated is radiation resistant. In the opposite case (e.g., in photographic industry) the air is ionized in an earthed small dimension cavity (which has an appropriate radiation shielding) and this *pre-ionized air* is blown over the electrostatically charged objects. The principle of this technique is analogous to that of the hair drier. This method can produce a current of about 0.1 μA.

The amount of the *charge to be neutralized* determines the activity of the radiation source. Another important factor is the position of the source. If we know the amount of electrostatic charge produced over the surface of the object per unit time, the required *activity* can be calculated (Fig. 6.8).

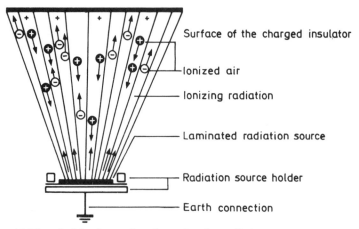

Surface of the charged insulator

Ionized air

Ionizing radiation

Laminated radiation source

Radiation source holder

Earth connection

Fig. 6.7. The principle of operation of a static voltage eliminator containing an isotope

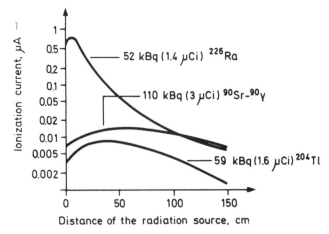

52 kBq (1.4 μCi) ^{226}Ra

110 kBq (3 μCi) ^{90}Sr–^{90}Y

59 kBq (1.6 μCi) ^{204}Tl

Distance of the radiation source, cm

Fig. 6.8. Intensity of the ionizing current as a function of the distance of the radiation source

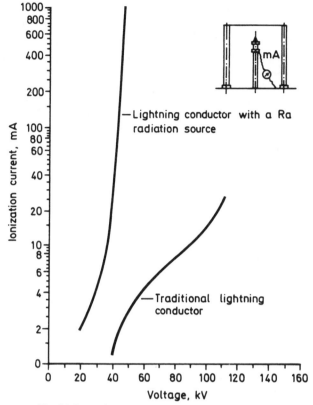

Fig. 6.9. Isotopic and traditional lightning conductor

Isotopic lightning conductors have a more or less analogous purpose. A radioactive isotope is inserted in the top or the shielding rings (^{60}Co or ^{226}Ra may be suitable). Isotopic lightning conductors are effective in an electric field weaker than necessary for traditional lightning conductors; if the field is as strong as in the case without any isotope, the current intensity is 6–15 times higher. In the case of a device shown in Fig. 6.9, the "effective height" of the isotopic device exceeds that of the traditional lightning conductor by about 50 m.

Ice formation on airplane wings is accelerated by static electricity induced by friction with air. The danger can be decreased by inserting isotopes *in critical positions.*

EXAMPLE 6.3

The velocity of paper on a printing machine is $v = 5$ m/s, its width is $a = 1$ m, its surface charge is $\sigma = 10^{-8}$ C/m^2. If we apply a ^{226}Ra source of a 50 MBq (1.4 mCi) activity, how far should it be placed to achieve elimination of all the surface charge?

Since the current intensity of the discharge (I) is

$$I = \sigma a v \tag{6.5}$$

then in our case

$$I = 1 \times 10^{-8} \times 1 \times 5 = 5 \times 10^{-8}\ \text{A} = 0.05\ \mu\text{A}$$

It follows from Fig. 6.8 that the radiation source should be about 55 cm near to the paper.

6.1.3.2. IGNITION OF ELECTRIC DISCHARGES

In discharge tubes, electron tubes, fluorescent lamps, voltage regulators, etc., electric discharges occur during operation. Since gases are insulators, the initiation of discharge by rendering them conductive requires a relatively longer period. The discharge is initiated by the presence of a few ions or electrons in the gaseous phase; these are accelerated by the electric field, collide then with other gas molecules, ionize them and an *avalanche-like* discharge is developed.

Primary ions and electrons are produced under the effect of cosmic rays and cold electron emission between the two electrodes if there is no radioactive substance in the system. The time distribution of these relatively few ions is subjected to large statistical fluctuations. Discharge cannot be started within a short interval or, if it is necessary to ignite several subsequent times, the operation of the discharge tube is poorly reproducible.

If a radioactive isotope is built into the discharge tube or electron tube, a constant concentration of primary ions can be ensured. Such isotopes may be ^{60}Co, ^{226}Ra, etc.; β-radiating ones are, however, more advantageous in many respects. Of the latter type, gases (^{3}H, ^{85}Kr, etc.) can be admixed in a homogeneous gas phase, or encapsulated in a metal tube. Other methods include absorption or powerisation of the surface. This method of ignition allows to produce pulses shorter than 1 μs—this is important in radar technique.

Radioactive isotopes can facilitate ignition in *internal combustion engines,* too; the use of such isotopes as ^{60}Co, ^{210}Po, ^{226}Ra in low amounts [e.g., 5×10^{7} Bq (1.4 mCi) ^{210}Po per sparking plug] decreases the ignition voltage considerably. The engine supplied with radioactive plugs starts with less revolution, thus the load on the battery is reduced.

The stabilizing of arc discharge is possible by adding ^{232}Th isotope to the tungsten electrode; thus the uniformity of welding in argon atmosphere can be improved. Carbon electrode gives a whiter light with such an additive.

EXAMPLE 6.4

Is it sufficient if we use a ^{60}Co radiation source with a 4 kBq (0.1 μCi) activity to ensure the ignition of a diode protecting transistor (so-called shunt gas diode limiter)?

Let us assume that the diameter of the diode is 10 mm, its length is 20 mm, it is filled with a 2.5 kPa (25 mbar) pressure argon gas and its ignition voltage is higher by 10% than the minimum ignition voltage.

The ignition process consists of two parts: the generation of the first ion pair and the developing of the avalanche. If there is at least 1 ion pair in the gas volume, the avalanche develops within less than 1μs.

The radiation source is placed as a 0.5 mm wide strip around the tube, at about its half-height. As a first, rough approximation, let us assume that a half of the emitted γ-quanta can get into the gas volume and is not absorbed either in the wall or in the radiation source itself.

There are 4×10^{3} disintegrations per second in the source of the given activity *per definitionem.* Each disintegration produces a 1.17 MeV and a 1.33 MeV γ-quantum from ^{60}Co (see Section 1.3); the concomitant β-radiation is absorbed in the wall. The dose rate in the tube is as follows:

$$I = \frac{4 \times 10^{3}}{2}(1.17 + 1.33) \times 10^{6} = 5.04 \times 10^{9} \text{ eV/(g s)}$$

The active volume of the diode tube is $0.5^{2} \times 2 \times \pi = 1.6 \text{ cm}^{3}$. The relative atomic mass of argon is 40, the density of argon at the given pressure is:

$$\varrho = \frac{40 \times 2.5 \times 10^3}{22.4 \times 10^5} = 0.045 \text{ kg/m}^3$$

The mass of argon in the tube is

$$m = 1.6 \times 4.5 \times 10^{-5} = 7.2 \times 10^{-5} \text{ g}$$

The energy absorbed per second is

$$I_{abs} = I[1 - \exp(-\mu_m \varrho l)] = I[1 - \exp(-0.0055 \times 0.045 \times 0.01)] =$$
$$= 5.04 \times 10^9 \times 2.5 \times 10^{-6} \text{ eV/(g s)} = 1.26 \times 10^4 \text{ eV/(g s)} \tag{6.6}$$

where μ_m is the mass absorption coefficient of argon with respect to the γ-radiation of ^{60}Co its value being $0.0055 \text{ m}^2/\text{kg}$ ($0.055 \text{ cm}^2/\text{g}$).

Since 26 eV is necessary to produce an ion pair in argon

$$v_{spec} = \frac{I_{abs}}{E_i} = \frac{1.26 \times 10^4}{26} /(\text{g s}) = 4.8 \times 10^2 /(\text{g s})$$

The number of ion pairs per second produced in the tube is:

$$v_{act} = v_{spec} \times m = 4.8 \times 10^2 \times 7.2 \times 10^{-5}/\text{s} = 3.5 \times 10^{-2}/\text{s}$$

Let us assume that ion pairs are consumed by recombination only, and the recombination factor (the number of ion pairs recombined per time unit) is $\alpha \approx 10^{-6}/\text{s}$. Since

$$\frac{dn}{dt} = \alpha n^2 \tag{6.7}$$

$$n = \left(\frac{dn/dt}{\alpha}\right)^{1/2} = \left(\frac{3.5 \times 10^{-2}}{10^{-6}}\right)^{1/2} = 187 \tag{6.8}$$

Thus, under the effect of 4 kBq (0.1 μCi) ^{60}Co, the steady state concentration of ion pairs is about 200 in the tube; this is sufficient to ensure ignition.

6.2. RADIATION CHEMICAL PROCESSES

6.2.1. INORGANIC SUBSTANCES

We shall confine ourselves to the radiation chemistry of gases, water and aqueous solutions, as well as ion exchangers; aqueous solutions will be treated also in Section 6.6 dealing with chemical dosimetry. Radiation-induced changes in important inorganic materials have been discussed in Section 6.1.

6.2.1.1. GASES

Ozone is produced mainly by high *LET* radiations from oxygen, in a yield as low as $G = 6$. The radiation chemical decomposition of ozone is, however, a chain reaction with $G = 10^3 - 10^4$ favoured by higher gas pressures.

Ammonia synthesis presents the difficulty that low temperature is favourable from a thermodynamical viewpoint, whereas reaction kinetics require higher temperature. The

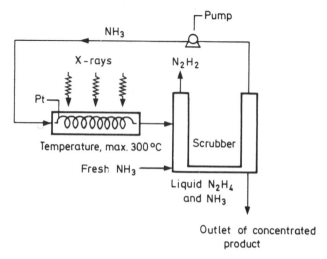

Fig. 6.10. Scheme of the radiation chemical synthesis of hydrazine

application of irradiation was intended to bridge these opposite requirements, but the *G*-values proved to be low: at room temperature, in homogeneous phase, $G = 2$; with alumina adsorbent, values of $G \approx 200$ have been reported.

Hydrazine can be produced by the irradiation of ammonia; simultaneously hydrogen and nitrogen are formed, and at higher doses hydrazine also suffers decomposition. The reaction rate can be increased by applying noble gas additives (such as krypton) and a platinum catalyst. Since hydrazine is important for *rocket propulsion*, a pilot plant has been projected for its radiation chemical synthesis (Fig. 6.10). Such a synthesis appears to be technologically advantageous; in spite of this fact, as far as we know, no commercial realization followed the pilot experiments, presumably due to economic reasons.

Radiation chemical synthesis of *nitrogen oxides* from air may be important for nitric acid production although the yield of *nitrogen fixation* carried out with fission products or with the mixed radiation of nuclear reactors is low: $G = 5{-}10$. Nitrogen oxides decompose under the effect of irradiation.

6.2.1.2. WATER AND AQUEOUS SOLUTIONS

The knowledge of the mechanism of water radiolysis is important in *nuclear reactor technique* (water is used as moderator and/or coolant, as well as in fuel element processing) and for a better understanding of *radiobiological processes*.

P. Curie observed as early as in 1901 that hydrogen and oxygen are given off from aqueous solutions of radium salts, i.e., under the effect of α-radiation. Later it was shown that oxygen is a secondary product: it is formed *via* hydrogen peroxide. The radiolysis of water is very sensitive to dissolved substances (e.g., copper sulphate, organic impurities) and is in positive correlation with the *LET* value of the radiation.

The decomposition can be characterized as follows:
(a) for low *LET* radiations:

$$\mathrm{H_2O \rightsquigarrow H, e_{aq}^-, OH^{\cdot}, H_2, H_2O_2, H_3O^+}$$

(b) for high *LET* radiations:

$$H_2O \rightsquigarrow H, e_{aq}^-, OH^\cdot, H_3O^+, HO_2^\cdot; H_2, H_2O_2$$

where e_{aq}^- denotes a *solvated electron*. This latter species is the simplest possible ion: its lifetime is as short as a few microseconds. It has a reducing nature and consists of an electron surrounded by liquid molecules. In aqueous systems it is called *hydrated electron*.

Since the processes are reversible, irradiation decomposes water up to a certain degree depending upon conditions. If ampoules totally filled with water are irradiated (in this case gases are not able to leave the liquid system) the equilibrium is established at low conversions. In turn, the presence of a gas phase over the liquid water promotes decomposition, since the hydrogen produced can leave the liquid phase (where it would be able to react by several orders of magnitude more rapidly). Thus, reverse processes are prevented. Every dissolved inorganic and organic substance which is able to react with OH' radicals and/or H-atoms accelerates the radiolysis of water because the probability of reverse processes is thus decreased.

Water in nuclear reactors suffers radiation damage under the effects of β-, γ-, proton and neutron radiation, combined with the kinetic energy of fission products, and its behaviour deserves separate treatment. In water–water type reactors water has a triple task: moderation, heat transfer and radiation protection. As long as the water is pure, reverse reactions can occur without any hindrance and water hardly decomposes. As a result of radiolysis of air and corrosion, the water in the reactor becomes more and more contaminated and decomposition *G*-values increase. The extent of decomposition is the function of temperature, too. If the water is boiling, hydrogen is removed rapidly; the remaining OH' radicals afford hydrogen peroxide which, in turn, gives water and oxygen. Thus, oxygen and hydrogen are present simultaneously over irradiated water representing an explosion hazard.

The yield of water decomposition in the vapour phase is about $G \approx 12$, in the liquid phase $G \approx 4$, in solid ice even less. The decreasing yields can be interpreted by the decreasing mobility of the free radicals, i.e., by the increasing probability of recombination reactions.

Impurities change the extent of water decomposition at very low concentrations; even tridistilled water suffers radiolysis to a much higher extent than specially purified water (the purification can be done, e.g., by preliminary radiolysis). This is an important problem also in chemical dosimetry with aqueous solutions (see Section 6.6.1). In dilute *solutions* it is assumed that the radiation interacts primarily with the solvent and primary radiolysis of the solute can be neglected. *Scavengers* are such substances which are able to react quantitatively with radicals produced from water. Such scavengers are, e.g., oxygen, diphenylpicrylhydrazine (DPPH).

The existence of H and OH' in neutral form is not proven unambiguously, but it is a fact that irradiated water exhibits simultaneously both *reducing* and *oxidizing* properties. The former indicates the presence of H-atoms, the latter that of OH' radicals. Inorganic substances with lower redox potential than 0.9 V undergo radiation chemical oxidation in aqueous solution [e.g., Fe(II) ions are oxidized to Fe(III), in the presence of sulphuric acid the redox potential being 0.75 V]; substances with a redox potential higher than 0.9 V undergo reduction [e.g., Ce(IV) ion with a redox potential of 1.44 V is reduced to Ce(III)]. Organic substances are oxidized in irradiated water.

6.2.1.3. INORGANIC ION EXCHANGERS

Several natural and synthetic inorganic substances possess ion exchanging properties. These are clay minerals or synthetic zeolites containing titanium, tin, chromium, zirconium (permutites), salts of polyvalent metals and polyvalent acids, oxides and oxide-hydrates of polyvalent metals, heteropolyacids and their salts and cyano complexes. Owing to their relatively lower ion exchange capacity and poorer mechanical stability, they are less widely used than organic ion exchangers (see also Section 6.4.3.5). Their radiation resistance, however, exceeds that of organic ion exchangers, although this also depends on their chemical composition as well as on the conditions of irradiation.

6.2.2. ORGANIC SUBSTANCES

6.2.2.1. HYDROCARBONS

Investigation of the radiation chemical behaviour of hydrocarbons has led to technological utilization in the plastics industry, and also to the development of relatively radiation-resistant structural, lubricating, etc. materials for nuclear reactors. Table 6.5 summarizes G-values of the formation of hydrogen and hydrocarbon fragments from some important hydrocarbons, indicating also the type of radiation and the state of hydrocarbons, both being significant factors influencing product composition.

The following primary reactions are important in *alkane (paraffin)* radiolysis:

(a) molecular hydrogen formation:

$$C_nH_{2n+2} \rightsquigarrow C_nH_{2n} + H_2$$

(b) C—H bond rupture:

$$C_nH_{2n+2} \rightsquigarrow C_nH_{2n+1}^{\cdot} + H$$

(c) C—C bond rupture:

$$C_nH_{2n+2} \rightsquigarrow C_iH_{2i+1}^{\cdot} + C_{(n-i)}H_{2(n-i)+1}^{\cdot}$$

$$\text{cyclo—}C_nH_{2n} \rightsquigarrow CnH_{2n}^{\cdot}$$

(d) ion–molecule interactions, e.g.:

$$C_nH_{2n+2}^{+} + C_mH_{2m+2} \rightsquigarrow C_nH_{2n+3}^{+} + C_mH_{2m+1}^{\cdot}$$

Alkanes give, in addition to hydrogen, saturated and unsaturated hydrocarbon products with lower and higher molecular mass than the parent hydrocarbon: gases, oligomers and polymers. On longer irradiation, insoluble and non-melting compounds are also formed.

The changes in the *melting points* of some alkanes under the effect of room temperature irradiation have been collected in Fig. 6.11. The initial melting point depression is attributed to the formation of hydrocarbon fragments with lower molecular mass than the starting compound. At the minimum, the compensating effect of crosslinking due to irradiation can be observed; this, at higher doses, outweighs other processes. The molecular mass increases from 10^2–10^3 up to the order of magnitude of 10^5, as a consequence of crosslinking.

Table 6.5. Informatory data on radiation chemical decomposition of hydrocarbons

Hydrocarbon	Phase	Hydrogen	Fragment*
		formation, G, molecule/100 eV	
Aliphatic alkanes			
Methane	gas	5.7	–
Propane	liquid	4.9	2.0
n-Hexane	liquid	5.0	2.1
3-Methylpentane	liquid	3.4	4.7
n-Heptane	liquid	4.9	1.9
2,2,4-Trimethylpentane	liquid	3.5	8.0
Cycloalkanes			
Cyclopropane	liquid	1.1	2.2
Cyclobutane	liquid	1.7	6.0
Cyclohexane	liquid	5.6	0.3
Methylcyclohexane	liquid	4.8	0.3
Cycloheptane	liquid	5.8	0.2
Aliphatic olefins			
Ethylene	gas	1.3	0.2
1-Butene	liquid	0.7	0.8
1-Hexene	liquid	0.9	0.6
3,3-Dimethyl-1-butene	liquid	0.4	0.6
1,5-Hexadiene	liquid	0.5	0.2
1-Octene	liquid	1.0	0.3
Cycloolefins			
Cyclobutene	liquid	0.6	3.0
Cyclohexene	liquid	1.2	0.2
3-Methylcyclohexene	liquid	1.3	0.4
1,4-Cyclohexadiene	liquid	0.9	<0.1
Cycloheptene	liquid	1.0	0.1
Cyclooctatetraene	liquid	0.02	0.02
Aromatics			
Benzene	liquid	0.04	0.02
Toluene	liquid	0.14	0.03
Naphthalene	solid	0.09	<0.02

* "Fragments" are hydrocarbon products with less carbon atoms than the parent hydrocarbon; the column contains their summarized G-values.

Heterogeneous catalysts, e.g., finely dispersed silica, alumina and aluminium silicates may increase the radiation chemical yields considerably, but several other contact substances have no such effect, and what is more, some conductors (metals) decrease the reactivity as compared with the homogeneous processes.

Alkenes (olefins) suffer polymerization as their most important radiation chemical transformation; at the same time, the G-values of gaseous decomposition products are about 1/3 of those observed for saturated hydrocarbons. This phenomenon can be interpreted on the one hand, in terms of *"physical protection"*, i.e., the loss of radiation energy without chemical reaction, on the other hand, by *reactions between free radicals and double bonds*, i.e., by scavenging processes. Mass spectrometric studies have demonstrated the role of very rapid ion–molecule interactions, especially at low temperatures.

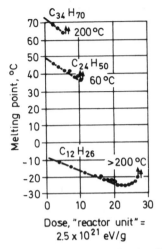

Fig. 6.11. Melting points of irradiated *n*-alkanes

Acetylene gives cuprene at room temperature and ambient pressure as the main product, apart from 15–20% of benzene. The *G*-value of acetylene decomposition is about 70, that of benzene formation is $G \approx 5$.

Aromatic hydrocarbons are distinguished for their very high radiation resistance; this statement is especially valid for polycyclic ones such as terphenyls: their G_{H_2} values are, as a rule, lower than 0.05.

Mixtures containing alkene or aromatic components exhibit physical and chemical "protection": the compounds mentioned "protect" more radiation sensitive components (e.g., alkanes) according to the mechanisms mentioned for alkenes. "Physical protection" in mixtures is achieved by transferring radiation energy from higher ionization potential (so-called *donor*) components to molecules of lower ionization potential, practically without chemical reaction. These latter (so-called *acceptor*) molecules may react chemically, or the energy taken up is dissipated as heat or light.

Alkylaromatic homologues are less radiation resistant than those without side alkyl groups, but the aromatic ring exerts an intramolecular "protecting" effect on the side chain for both gas and polymer formation.

In the foregoing the radiation chemistry of hydrocarbons at room temperature has been discussed. In practice, however, several hydrocarbons or petroleum products are subjected to irradiation at elevated temperatures. *Lubricants* in nuclear power stations are kept at about 100°C, but *reactor coolants and moderators* must operate at about 400°C without *damage* exceeding permissible limits (see also Section 6.4.3.1). High temperature irradiation can have also *useful* effects (apart from harmful ones): in principle, radiation chemical *cracking, alkylation and hydrogenation* at about 300–400°C and radiation chemical *isomerization* at about room temperature can produce engine fuel in better yields than traditional processes.

6.2.2.2. PREPARATIVE PROCESSES

Radiation chemical oxidation of technical paraffin proceeds at a high rate at such a low temperature where there is no reaction at all without irradiation and, what is more, the chain reaction initiated by an irradiation of appropriate duration is continued at nearly identical rate after stopping the irradiation. Radiation chemical oxidation permits one to avoid some undesirable side reactions, as well as the contamination of the product with a catalyst. The experience of a pilot plant indicates that solid paraffin should be irradiated for 1.5 h at 120°C with a dose rate of 55–110 mGy/s (20–40 krd/h); this leads to a product with an acid number as high as 60 mg KOH/g; refining this technical fatty acid gives a product with an acid number of 160 mg KOH/g.

Chlorination processes are important as far as insecticide and herbicide manufacturing is concerned. For example, *Gammexane* (γ-isomer of 1,2,3,4,5,6-hexachlorocyclohexane) and *octachlorocyclopentene* can be produced in this way. It has been stated that photochemical and radiation chemical chlorinations occur by the same mechanism, but the latter process is more economical.

Chlorination of toluene can lead to substitution in the side chain or addition of the chlorine atoms to the aromatic ring. Visible or ultraviolet light does not induce this latter reaction, only high energy (e.g., γ) radiation does.

Manufacturing of ethyl bromide was the first radiation chemical synthesis realized on a commercial scale. This was reported by Dow Chemical Company (USA) in 1963. The process having a chain character is nearly quantitative. The irradiating reactor lies 1.3 m deep underground and is supplied by a ^{60}Co γ-radiation source of an activity of about 10^{14} Bq (several thousand Ci). The sectional view of the apparatus is shown in Fig. 6.12, the process chart in Fig. 6.13. The pump circulates liquid ethyl bromide in the system through an appropriate heat exchanger (at a circulation rate of 6–9 m^3/h); the feed of hydrogen bromide and ethylene is continuous and so is the unloading of the product. Raw ethyl

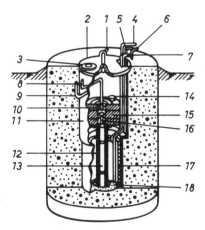

Fig. 6.12. Reactor for the radiation synthesis of ethyl bromide. *1* — Locking bar; *2* — Cover plate; *3* — Moisture sight glass; *4* — Outer N$_2$ purge inlet; *5* — Inner N$_2$ purge outlet; *6* — Reactants in; *7* — Product out; *8* — Inner N$_2$ purge inlet; *9* — Outer N$_2$ purge outlet; *10* — Closure plug with N$_2$ line; *11* — Steel silo jacket; *12* — Source well in vessel; *13* — Source carrier; *14* — Locking dog; *15* — Shielding plates; *16* — Lift rod; *17* — Reaction vessel; *18* — Distribution ring

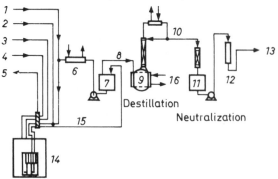

Fig. 6.13. The scheme of radiation synthesis of ethyl bromide. *1* — HBr gas; *2* — C_2H_4 gas; *3* — N_2 purge inlet; *4* — N_2 purge outlet; *5* — Control instruments; *6* — Cooler; *7* — Overflow tank; *8* — Product overflow; *9* — Distillation column; *10* — Product outlet; *11* — Neutralizing column; *12* — Drier; *13* — Ethyl bromide to storage; *14* — Reactor; *15* — Ethyl bromide solution; *16* — Steam in- and outlet

bromide is purified by distillation, neutralization and drying. The process is a chain reaction ($G = 10^5$); the yield is 98–100%.

Sulphoxidation of hydrocarbons is carried out by their irradiation in the presence of SO_2 and O_2; the process results in the formation of liquid and solid sulphonic acids, thiosulphonic esters, sulphinic acids, etc. C_6—C_{16} n-paraffins and cyclohexane can be sulphoxidated at a high efficiency: the yields are high ($G = 3000$–7000). The process may have a potential importance in detergent manufacturing because its products are biodegradable which is advantageous from the point of view of environment protection.

Sulphochlorination is less economical than sulphoxidation, although its radiation chemical yield is higher ($G = 10^3$–10^6). Saponified sulphochlorides are used as detergents, like the products of direct sulphoxidation (both giving sulphonates). There exists a pilot plant in the USSR using radiation chemical sulphochlorination process.

6.2.2.3. POLYMERS

Radiation effects on polymers are similar to those exerted on analogous monomers. Crosslinking (dimerization, oligomerization, polymerization), chain rupture (giving gaseous products of low molecular mass), dehydrogenation and hydrogenation are the most important processes. Of these reactions, crosslinking and chain rupture have particular importance. It is possible to carry out radiation polymerization in the *solid phase*, too.

Crosslinking leads to the increase of the molecular mass; as a result, insoluble and non-melting products are obtained which may represent one single giant molecule; on the other hand, chain rupture decreases the molecular mass. Both processes—like in the case of low molecular mass organic substances—occur simultaneously and the net result of two opposite reactions is observed. Accordingly, if the number of ruptures of the main chain per unit time is higher than the number of formation of crosslinks, the polymer suffers *degradation*; in the opposite case it undergoes *crosslinking*. Those polymers which contain at least one hydrogen atom on each carbon atom in their main chain are usually prone to crosslinking. Those containing at least one quaternary carbon atom per monomer unit, suffer degradation (Table 6.6). Stereochemical aspects are also important. It is another rule for orientation that the tendency for crosslinking is higher with polymers which have a high heat of polymerization.

Oxygen influences not only the rates of crosslinking and degradation, but may also change the direction of the resulting process. Whereas, e.g., PVC, polystyrene, polypropylene, undergo crosslinking if irradiated in vacuum, they suffer degradation in the presence of oxygen. This effect is especially important with plastic objects of high specific surface, such as films, powders, chips, and it is not significant for spheres. *Additives* may exert a considerable effect.

Graft copolymerization renders possible the colouring of polymers which could be coloured only poorly otherwise (e.g., polypropylene). If an appropriate polymer is

Table 6.6. Radiation chemical transformations of polymers (at room temperature, without oxygen)

Compounds undergoing mainly crosslinking	Compounds undergoing mainly degradation

Buna
Natural rubber
Polyacrylamide:
 	
$-CH_2-CH-CH_2-CH-$
      $\overset{|}{C}ONH_2$ $\overset{|}{C}ONH_2$
 	
Polyacrylate:
 	
$-CH_2-CH-CH_2-CH-$
      $\overset{|}{C}OOR$ $\overset{|}{C}OOR$
 	
Polyacrylonitrile
Polyamide
Polyester
Polyethylene:
 	
$-CH_2-CH_2-CH_2-$
 	
Polypropylene:
 	
$-CH_2-CH-CH_2-CH-$
      $\overset{|}{C}H_3$ $\overset{|}{C}H_3$
 	
Polysiloxane
Polystyrene:
 	
$-CH_2-CH-CH_2-CH-$
      $\overset{|}{C}_6H_5$ $\overset{|}{C}_6H_5$
 	
Polyvinyl alcohol
Polyvinyl chloride:
 	
$-CH_2-CH-CH_2-CH-$
      $\overset{|}{C}l$ $\overset{|}{C}l$
	Cellulose
	
	Cellulose derivatives
	
	Poly-α-methylstyrene:
	
	      CH_3 CH_3
	      $\overset{|}{}$ $\overset{|}{}$
	$-CH_2-C-CH_2-C-$
	      $\overset{|}{C}_6H_5$ $\overset{|}{C}_6H_5$
	
	Polycarbonates
	Polyfluorochloroethylene
	Polyfluoroethylene
	Polyisobutylene:
	
	      CH_3 CH_3
	$-CH_2-C-CH_2-C-$
	      $\overset{|}{C}H_3$ $\overset{|}{C}H_3$
	
	Polymethacrylamide:
	
	      CH_3 CH_3
	$-CH_2-C-CH_2-C-$
	      $\overset{|}{C}ONH_2$ $\overset{|}{C}ONH_2$
	
	Polymethacrylate:
	
	      CH_3 CH_3
	$-CH_2-C-CH_2-C-$
	      $\overset{|}{C}OOR$ $\overset{|}{C}OOR$
	
	Polyvinylidene chloride:
	
	      Cl Cl
	$-CH_2-C-CH_2-C-$
	      $\overset{|}{C}l$ $\overset{|}{C}l$
	
	                Thiokol

irradiated in the presence of a monomer, free radicals produced from the monomer initiate the polymerization and the polymer formed is attached to the originally added polymer. If molecules of monomer B are incorporated into the main chain of the polymer A, a *block polymer* is obtained:

$$\ldots -AAAA- \ldots \overset{+nB}{\leadsto} \ldots -AAAA-BBB-AAA- \ldots$$

If B is attached as a side group, a *graft copolymer* is formed (apart from low molecular mass radicals):

$$\ldots -AAAA- \ldots \overset{+nB}{\leadsto} \ldots -AAAAA- \ldots +R$$

$$\begin{matrix} | \\ B \\ B \\ B \\ \vdots \end{matrix}$$

Radicals may induce *homopolymerization* of the monomer:

$$R^{\cdot}+nB \leadsto \ldots -BBBB- \ldots$$

The three following methods are used mainly for radiation grafting:

(a) the polymer is irradiated in vacuum and subsequently the monomer is added; grafting occurs on the free radicals of the polymer;

(b) the polymer is irradiated in the presence of air and subsequently the monomer is added; grafting is initiated by heating when the peroxides and hydrogen peroxide during irradiation decompose into free radicals;

(c) the polymer and monomer are irradiated simultaneously.

Grafting of a polyethylene foil with styrene gives *ion exchange membranes*; the polystyrene is sulphonated to produce ion exchanging groups. Such membranes can be built in electrolysers and they allow the passage of either positive or negative ions only, thus they are suitable to effect special electrodialysis.

The grafted copolymer obtained from chlorinated polyethylene and a mixture of acrylonitrile and styrene is very suitable for *injection moulding*; similar substances can also be used as *glues* and *synthetic fibres*. Grafting of ethylene with carbon monoxide gives a cheap polymer having many advantageous properties. This substance cannot be produced by any other method. Very hard *moulded pieces* can be made from PVC–styrene copolymer if the latter is prepared at low temperatures.

The first radiation chemical process in the *textile industry* was the grafting of polyester–cotton mixed fibre to give a permanent pressing; this can be done on the ready textile. The process improves the colouring of man-made fibres, also its water absorptivity and several other properties of cellulose-based fibres. At the same time, decreased uptake of contamination by the textile is another advantageous result.

Plastic–wood combinations can be produced if the appropriate wood is first vacuum-dried, then impregnated with a monomer (such as methyl methacrylate, vinyl acetate, styrene). The impregnated system should be irradiated with a dose of 10^4–10^5 Gy (a few Mrd). The hardness, compression, bending and shearing strength of these wood–polymer combinations are greatly increased. The products have improved chemical, tropical and dimensional stability and lower water absorption. Better machine workability represents a special advantage in the wood industry; further the product looks nicer than wood. *Paper–plastic* combinations can be prepared similarly.

There have been experiments to produce *concrete–polymer* and *glass chips–polymer* systems by irradiation. Polymerization of styrene, methyl methacrylate, etc., in cavities of such systems leads to materials having a high resistance against sulphate ions, hot sea water, freezing and melting and, what is more, tubes manufactured in this way are cheaper than coated or enamelled ones. These products have extraordinarily high compression strength.

Hardening of paints and varnishes ("curing", "drying") can be carried out by electron accelerators, as opposed to the γ-irradiation used for producing plastic–wood and concrete–plastic combinations. The necessary energy is only about 0.2–0.5 MeV instead of 1–3 MeV used commonly. This method of *surface treatment* permits one to carry out a faster and cheaper fixation of thin paint layers and other coatings on metal, wood, paper, etc. surfaces than possible by the traditional hardening in ovens. The omission of solvents represents a great advantage from the viewpoint of energy saving and environment protection. Polyester monomers are the most suitable for this purpose; these can be applied by hardly any other technique.

Crosslinked polyethylene was the first and has remained the most important radiation chemical product. It is usually produced from polyethylene by its subsequent irradiation. The radiolysis of gaseous ethylene gives a product with relatively low molecular mass; polyethylene with the usual properties can only be obtained in the pressure range 4–10 MPa (40–100 bar) and somewhat above room temperature. The properties of polyethylene produced by radiation are between those of high and low pressure polyethylenes: e.g., its density is 940 kg/m^3, its crystallinity being 77%. It contains no traces of metallic catalysts, therefore its insulating properties are excellent.

When protected against oxidation, polyethylene crosslinked by irradiation withstands high temperatures; e.g., a crosslinked polyethylene film is not damaged by molten soldering metal even at 165–180°C. At room temperature, its mechanical properties are practically the same as those of traditional polyethylene (i.e., it is not brittle). Above 100°C, keeping partly its strength, it behaves elastically, like rubber. Non-irradiated polyethylene flows above 100°C, but a product irradiated with a dose of 100 kGy (10 Mrd) has a tensile strength of 1 MPa/cm^2 (10 kp/cm^2) even at 160°C. Below the melting point the crystalline structure, above it crosslinking ensure the mechanical strength. Crosslinking also influences gas permeability; besides the irradiation, this also depends on other experimental conditions (solvent, temperature, etc.). The product has no melting point if 10–20% of it has become insoluble.

Crosslinked polyethylene is suitable to protect wires, cable end insulations and electric coils against the heat effects of electric discharges lasting not too long. This polymer can also protect the axles of heavy duty rotors (e.g., in electric locomotives) during high temperature impregnation. From this respect, poor adhesion of polyethylene represents a particular advantage; thereby the polyethylene which has become damaged, softened, welded (but not molten) during discharge, can be easily taken off from the protected coil after having fulfilled its protecting duty, and replaced by a new one if necessary. This sort of polyethylene is used, for example, in *cable industry* for the insulating of high temperature cables (e.g., for geophysical purposes), but it is also suitable for corrosion-resistant gaskets.

Memory effect is a particularly interesting property of slightly irradiated polyethylene. This is utilized in up-to-date electrotechnics and packaging. Polyethylene is irradiated in the required shape; it is then expanded with warming up. After cooling, the cable end or wire to be insulated is inserted, or the plastic is filled with the material to be packed (generally a liquid, plastic or granulated material). When warmed up again above its softening temperature, the polyethylene resumes the original size and shape "prescribed" during irradiation, as required for the *insulating* or *packaging* purpose.

When polyvinyl chloride (PVC) is made by the irradiation of vinyl chloride, the polymer segregates from the monomer, thus the process is heterogeneous. Primarily produced PVC is a white powder; further irradiation, however, induces the releasing of hydrogen chloride and a characteristic discolouration. If the radiation polymerization is carried out at $-20°C$, the transition temperature to the glassy phase is higher than that of common PVC.

Polyvinyl chloride has an intermediate position between crosslinking and degrading polymers. The main radiation chemical reactions are the formation of hydrogen chloride and double bonds. The evolution of HCl can be used for dosimetric purposes; it can be followed analytically or spectrophotometrically; the latter requires special preparations. *Dose indicators* made on this basis are also of importance (see also Section 6.6.5).

Polyacrylonitrile is produced in a heterogeneous reaction because the polymer segregates from the monomer as a solid. Radiation polymerization gives hard, ceramics-like materials, which can be worked mechanically well. If the dose rate is too high (above 14 Gy/s, i.e., 5 Mrd/h) polymerization proceeds like an explosion; even the reactor may burst.

Polymethyl methacrylate (Plexiglass, Lucite) becomes slightly yellow when irradiated with even relatively small doses ($1-5 \times 10^4$ Gy; 1–5 Mrd). This discolouring can be rendered characteristic by means of appropriate additives: this is how *Perspex-type chemical dosimeters* work (see Section 6.6.4); spectrophotometrical measurement is necessary to obtain information about the absorbed dose.

Small gas inclusions formed during irradiation expand in the soft polymer to 6–8 times of their original volume when the material is warmed up; in this way *plastic foams* are obtained, which are useful for thermoinsulation. There is an inverse correlation between the dose and the necessary temperature: e.g., with 290 kGy (29 Mrd) 70°C, with 46 kGy (4.6 Mrd) 140°C, and with 4.6 kGy (0.46 Mrd) 200°C is necessary to start foaming.

The radiation chemical preparation of *polystyrene* is a homogeneous reaction, because polystyrene remains dissolved in the monomeric styrene. This material has the best radiation resistance among all pure polymers; its copolymers which contain less aromatic rings are more sensitive to radiation.

The radiation resistance of *silicons* depends on their structure: those containing benzene rings are almost as resistant as aromatic hydrocarbons. With decreasing aromatic content the G-values increase. The crosslinking G-values of dimethyldiphenylsiloxanes also depend on the ratio of the components: with 5% phenyl content, we have about $G = 1$; with 25 mol-% phenyl groups, $G < 0.1$.

Natural rubber, being a polyisoprene, belongs to the group of crosslinking polymers. The elastic modulus is proportional to the degree of crosslinking to a certain degree of vulcanization; above this it remains constant. Literature data suggest that there is no significant difference between the products of chemical and radiation vulcanizing.

EXAMPLE 6.5

A sample of 8.8 cm^3 of n-octane is irradiated with ^{60}Co γ-radiation at a dose rate of 3.5 Gy/s (1.25 Mrd/h). After 30 min, 0.53 cm^3 gas is given off. The pressure in the room is atmospheric, the temperature is 20°C. What is the value of G_{H_2}, if the gas contains 98 mol-% of hydrogen?

The normal volume of the gas produced during 30 min is as follows:

$$V_{gas} = 0.53\frac{273}{293} = 0.49 \text{ cm}^3$$

of this, 98% is hydrogen:

$$V_{H_2} = 0.49 \times 0.98 = 0.48 \text{ cm}^3$$

Using *Avogadro*'s number we can calculate the number of hydrogen molecules produced:

$$n_{H_2} = 0.48 \times \frac{6.02 \times 10^{23}}{22\,400} = 1.30 \times 10^{19}$$

The rate of hydrogen formation is then

$$\frac{dn_{H_2}}{dt} = 1.30 \times 10^{19}/0.5 = 2.6 \times 10^{19} \text{ molecules per hour.}$$

Since the density of n-octane is $703 \text{ kg/m}^3 = 0.703 \text{ g/cm}^3$ at 20°C, the mass of the irradiated n-octane is:

$$8.8 \times 0.703 = 6.19 \text{ g}$$

The energy absorbed per unit time by the irradiated substance will be:

$$mI_{abs} = 6.19 \times 3.5 \text{ g Gy/s} = 21.7 \text{ g Gy/s} = 7.8 \text{ g Mrd/h}$$

and this, according to Table 6.1 is equal to

$$7.8 \times 10^4 \times 6.24 \times 10^{15} \text{ eV/h} = 4.87 \times 10^{20} \text{ eV/h}$$

Since $G = \dfrac{dx/dt}{I_{abs}}$ 100 particles per 100 eV [see, Equation (6.1)], in the given case

$$G_{H_2} = \frac{2.6 \times 10^{19} \times 10^2 \text{ particles/h}}{4.87 \times 10^{20} \times 100 \text{ eV/h}} = 5.3 \text{ molecules per 100 eV}$$

this is the radiation chemical yield of hydrogen evolution from n-octane.

EXAMPLE 6.6

Let us attempt to calculate the G-value of nitrogen formation from N_2O if the irradiation is carried out by γ-radiation, in the presence of uranium-impregnated silica gel.

The experimental data are as follows: amount of catalyst 1 g (13.2% U, 86.8% SiO_2); volume of reaction vessel: 3 cm^3; volume of the solid phase: 0.4 cm^3; gas volume $3-0.4$ $=2.6$ cm^3; pore volume: $0.9-0.4=0.5$ cm^3; amount of N_2O: 2.25×10^{-3} mole; total absorbed dose: 100 kg Gy (6.24×10^{20} eV); amount of nitrogen produced: 26×10^{-6} mole $= 1.56 \times 10^{19}$ molecules; radiation chemical yield of nitrogen formation in the homogeneous phase: $G_{N_2, hom} = 8$.

(a) Let us assume that the solid adsorbs the whole amount of nitrogen.

The radiation energy is distributed according to the ratio of electrons in each phase: 2.25×10^3 mole of N_2O contains $2.25 \times 10^3 [(2 \times 7) + 8] = 4.95 \times 10^{-2}$ electron moles; the solid phase contains $\dfrac{132 \times 10^{-3}}{238} 92 + \dfrac{868 \times 10^{-3}}{28 + (2 \times 16)} (14 + 2 \times 8) = 48.5 \times 10^{-2}$ electron moles per gram.

Thus, the energy absorbed by the gaseous phase is:

$$6.24 \times 10^{20} \times \frac{4.95 \times 10^{-2}}{(48.5 + 4.95) \times 10^{-2}} \text{ eV} = 5.8 \times 10^{19} \text{ eV}$$

Consequently,

$$G_{N_2, \text{het}, a} = \frac{1.56 \times 10^{19}}{5.8 \times 10^{19}} \, 10^2 = 26.9$$

(b) Let us assume that no N_2O is adsorbed by the solid:

The pore volume is 0.5 cm^3, the free volume between the grains—i.e., outside the pores—is $2.6 - 0.5 = 2.1$ cm^3. The process in the pores is heterogeneous, because the mean free path and pore diameter are comparable. Between the grains the situation is the opposite: here the mean free path is lower by about 5–6 orders of magnitude than the dimensions of the vessel.

It was calculated above that the gaseous phase absorbs an energy of 5.9×10^{19} eV and a part of it reacts according to a homogeneous, another part according to a heterogeneous mechanism. The energy share of homogeneous phase will be:

$$5.8 \times 10^{19} \times \frac{2.1}{2.1 + 0.5} \text{ eV} = 4.7 \times 10^{19} \text{ eV}$$

Since $G_{N_2, \text{hom}} = 8$, the number of nitrogen molecules produced in the homogeneous phase will be $4.7 \times 10^{19} \times 8 \times 10^{-2} = 3.76 \times 10^{18}$.

This has to be deducted from the total number of nitrogen molecules. The result gives the number of nitrogen molecules produced in heterogeneous reactions: $1.56 \times 10^{19} - 3.76 \times 10^{18} = 1.18 \times 10^{19}$.

The energy share of heterogeneous reactions is:

$$5.8 \times 10^{19} - 4.7 \times 10^{19} = 1.1 \times 10^{19} \text{ eV}$$

Thus,

$$G_{N_2, \text{het}, b} = \frac{1.21 \times 10^{19}}{1.1 \times 10^{19}} \times 10^2 = 110$$

(c) Comparing the two calculations with each other and the homogeneous process, it is obtained that the radiation chemical yield of nitrogen formation from N_2O irradiated by γ-radiation in the presence of uranium-impregnated silica gel should be between $G_{N_2, \text{het}, a} = 26.9$ and $G_{N_2, \text{het}, b} = 110$, i.e., it is 3–13 times higher than the value of $G_{N_2, \text{hom}} = 8$.

6.3. RADIOBIOLOGICAL PROCESSES

Whereas earlier emphasis was put on the harmful effects of radiation and the necessity of radiation protection (see also Chapter 8), nowadays the biological effects of radiation are applied—in addition to medicine—in many fields of national economy:

— *in agriculture:* in plant improvement, animal breeding, as well as in plant, animal and environmental protection;

— *in the food industry:* for preservation, for upgrading, and against parasites;

— *in the pharmaceutical industry and in the manufacturing of medical tools:* for the sterilization of the products.

Table 6.7. Doses necessary to achieve various biological effects with β- and γ-radiation

Effect	Necessary dose	
	Gy	rd
Stimulation of plants and animals	0.01–10	1–1000
Plant breeding by mutation	10–500	10^3–5×10^4
Destruction of insects by the sterile male method	50–200	5×10^3–2×10^4
Sprouting inhibition (potato, onion)	50–400	5×10^3–4×10^4
Destruction of insects and their eggs	250–10^3	2.5×10^4–10^5
Disinfection (radicidation)	10^3–10^4	10^5–10^6
Pasteurization of food (radurization)	10^3–10^4	10^5–10^6
Sterilization of pharmaceuticals and therapeutic equipment	$(1.5–5) \times 10^4$	$(1.5–5) \times 10^6$
Sterilization of food (radappertization)	$(2–6) \times 10^4$	$(2–6) \times 10^6$
Inactivation of viruses	10^4–1.5×10^5	10^6–1.5×10^7
Inactivation of enzymes	2×10^4–10^5	2×10^6–10^7

The essence of radiobiological effects consists in the modification (damage) of cells sensitive to radiation. As the dose absorbed by a cell increases, the damage can attain the lethal effect. Table 6.7 gives illustrative data about the doses necessary for biological purposes.

The first theory in radiobiology explaining cellular damage was the *theory of direct target*: if a high-energy particle or quantum impacts against a vitally important molecule in the sensitive part of a cell (e.g., in a chromosome), it alters and eventually kills the cell. Later research has shown, however, that a considerable part of radiobiological effects is due to *indirect* processes. These indirect effects are related to the radiolysis of water molecules (see Section 6.2.1.2). Free radicals formed from water molecules can oxidize vitally important molecules, they can cause chain rupture, disintegration of colloid systems, etc.; as a final result, an abnormal metabolism and the death of the cell will ensue.

6.3.1. PROCESSES IN AGRICULTURE AND THE FOOD INDUSTRY

The individual processes will be treated in the order of increasing dose requirements:

Stimulation requires small doses and causes non-inheritable alterations. Irradiations with doses between 0.1 and 10 Gy (10–1000 rd) may have a stimulating effect, e.g., they may result in a richer harvest. This technique is promising especially in the case of small-seed plants (e.g., tobacco, grass, clover, cotton, poppyseed, tomato, paprika), because the volume of the seed necessary for large cultivated areas is relatively small. The costs of irradiation are, as a rule, directly proportional to the volume of the substance to be irradiated (see also Section 6.5.2).

Mutations are abrupt changes which are inheritable; the individuals carrying mutations are the mutants. Mutations may be spontaneous or induced by man. Under given conditions, mutations can cease to exist or can be reversed.

Induced (artificial) plant mutants can be produced by chemicals, by radiation (electromagnetic, e.g., γ-radiation, fast or thermal neutrons, electron irradiation, ionic radiations and, by π-mesons of much higher doses than necessary for stimulation), or by the *combined effect* of chemicals and radiation. The purpose of irradiation breeding is to produce plants which are resistant to winter cold, diseases (often against a given special disease), to ensure earlier ripening, to enhance straw strength (to facilitate mechanical harvesting), to provide male sterility (for the hybridization of self-pollinating plants), to promote hybridization etc.

Mutations can also be induced in *animals and microorganisms*: the investigation of lower moulds, fungi and bacteria have shown that their mutants can be utilized in the pharmaceutical industry. For example, penicillin is sometimes used to cure cows. If, however, penicillin gets into the milk, it kills lactobacilli and this is disadvantageous for cheese and butter manufacturing. By irradiation it was possible to develop penicillin-resistant lactobacilli.

Tubers and bulbs (potato, onion) *will not sprout* after irradiation; inhibition of sprouting was the first legally approved method of radiation utilization in agriculture and in the food industry. Irradiation of *potato* and *onion* has obtained—general or limited—official permission from the health authorities of many countries. The doses to be used are between 50 and 400 Gy (5–40 krd).

Sterilization of males is a more economic tool for plant or animal protection than killing the parasites, because a dose as low as 50–200 Gy (5–20 krd) brings about such modifications in the male sperm that offsprings cannot be generated, although the ability to mate remains intact. The precondition of the success of this method is that the insects be confined to a more or less localized area; the method is promising on islands being far away from the mainland, or in the case of a strict and concerted international cooperation. The male insects to be radiation-sterilized are bred in the laboratory. After irradiation they must be dispersed uniformly. It is very important that their number should exceed the number of males in the natural population by several orders of magnitude; even in this case the process should be repeated with several breeding cycles.

Irradiation can also be used for *insecticidal, vermicidal* and *acaricidal* purposes. This method has been developed for cereals, because their loss owing to such pests is large: about 5% of the world harvest, but under tropical conditions it may be as high as 25–50%. The damage can be eliminated by applying a dose of 300–750 Gy (30–75 krd), if the re-infection of the irradiated product can be avoided, e.g., by storage in closed silos. Killing insects in dried *vegetables*, dried *fruits*, and dried smoked *fish* by radiation is also effective; they must be irradiated after packaging. The *quarantine* irradiations of plants are also important; a similar task is the killing of insects in stored products. A peculiar field of application is the killing of worms in wood; this requires, according to presently available data, a dose of $5–30 \times 10^2$ Gy (50–300 krd). The killing of insects in *museum exhibits* is based on the same principles.

Radicidation is a radiation disinfection: *Salmonella* species can be eliminated from frozen meat, egg yolk, poultry, fish meal, feed premixes by applying a medium (1–10 kGy or 0.1–1.0 Mrd) dose. The number of *Salmonella* cells (causing paratyphoid) can be decreased by seven log cycles by applying 3 kGy (0.3 Mrd) for liquid egg, and 7 kGy (0.7 Mrd) for powdered egg. As a rule, 3–5 kGy (0.3–0.5 Mrd) irradiation causes no organoleptic changes; also the off-flavour brought about by higher doses disappears during spray drying or frying. It is important that the radiation disinfection of egg is possible also under frozen conditions.

Irradiation of *sewage* and *waste* aims at killing pathogenic microorganisms, as well as facilitating coagulation of the colloid system; this latter promotes filtration.

Radiation pasteurizing is called *radurization*. Relatively low doses (2–6 kGy; 0.2–0.6 Mrd) are sufficient to remove a considerable fraction of the microflora present in food-stuffs; this prolongs the shelf-life of the product. However, killing of the remaining 0.1–1% of the microflora (to ensure sterility) requires incomparably higher doses.

The radiation resistance of *vegetables* and *fruits*, as well as *juices*, is very different. For instance, tomato and orange juices are very sensitive, whereas apple pulp is radiation resistant. The irradiation of champignon mushroom and strawberry with doses up to 2.5 kGy (250 krd) has been permitted in some countries, partly to a limited extent.

Spices are irradiated generally with higher doses (3–15 kGy; 300–1500 krd).

In the case of *meat*, the irradiation or pre-packed slices is the most reasonable. There are combined processes preventing the discolouration of even the very sensitive beef. The problems are much less with poultry: simple irradiation is sufficient for this purpose requiring 3–6 kGy (300–600 krd).

Fish has to be conserved by irradiation due to the continuous increase of the operating distance between the shore and the fishing fleet; the radiation sources are installed on the ships themselves. If cooling is combined with irradiation, the time of preservation of marine products can be increased by 2–5 times.

Wine—as far as experiments up till now suggest—cannot be conserved by irradiation without a considerable and disadvantageous change of its flavour.

Nutritive materials (proteins, carbohydrates, fats) suffer so little chemical changes under the effect of irradiation that they can be followed analytically only in cases of doses exceeding several times those necessary for radiation preservation. On the other hand, *vitamins* behave very differently: their radiation resistance depends very much on the conditions of irradiation and the composition of the food.

Radappertization is the radiation sterilization of food: the dose should be selected here to be so high that all bacteria, moulds, fungi and yeast be killed (i.e., their number be decreased by 10^{12} times). The necessary doses are high: 25–56 kGy (2.5–5.6 Mrd). A disadvantage of the process is that enzymes in tissues and toxins cannot be inactivated even by these high doses. In addition, high dose requirements make the process very expensive and most food-stuffs are damaged as far as their flavour and aroma are concerned. This technique works with some sorts of food and with some combinations, but practically it is limited to the purposes of military supplies and space research.

Milk can be sterilized by considerably lower doses than meat (5–15 kGy, i.e., 0.5–1.5 Mrd), but it is a problem that doses as low as 100–200 Gy (10–20 krd) give rise to off-flavour and taste. Milk products are similarly sensitive to irradiation. If irradiation is combined with vacuum distillation, the organoleptic changes can be eliminated: undesirable materials are eliminated and even a dose as high as 20 kGy (2 Mrd) causes no unpleasant flavour.

One of the main problems in the radiation sterilization of food is the *change of organoleptic properties* (flavour, aroma) upon irradiation. These determine an upper dose limit for most food-stuffs. The critical dose is thus different for each food and depends on the conditions or irradiation. Radappertization can be used for the conservation of *bacon, minced meat, canned ham, pre-packaged pork fillets, ready-made dishes, spices* (e.g., paprika), as well as for decontamination of *enzyme preparations*.

Combination of radiation pasteurization with other preservation methods enhances the effects without undesirable side phenomena. Most obvious is the combination of irradiation and refrigeration: storage even at 5 °C has a selective and advantageous effect on the microflora of food-stuffs because refrigeration prevents the multiplication of pathogenic and radiation resistant germs; on the other hand, food stored in refrigerated state before irradiation contains cold-resistant germs which are not radiation resistant. If irradiation is carried out in the *frozen state*, most of the water content of cells freezes and the critical doses causing organoleptic changes are usually much higher than with room temperature irradiations. On the other hand, the radiation resistance of vegetative microorganisms is increased by freezing. The optimum irradiation temperature will be a compromise between these two opposite factors.

Mild heat treatment before irradiation inactivates radiation-resistant enzymes and, in addition, considerably decreases the initial number of viable microorganisms. Thus the required dose decreases. The product of such a combined method will have a better organoleptic quality than when irradiation is used alone. The combination is more effective from the microbiological point of view if heat treatment follows irradiation.

The purpose of *additives* is to prevent or decrease the undesirable organoleptic changes or to enhance the antimicrobial effect (antibiotics, etc.). Ascorbic acid and its derivatives are suitable additives: if 0.5% ascorbic acid is added to minced meat, radiation sterilization brings about no organoleptic changes.

Improvement of quality can also be achieved in food-stuffs by irradiation: the *controlled ripening of fruit and vegetables* will be possible. *Dried fruit and vegetables* irradiated with 3–6 kGy (0.3–6.0 Mrd) can be cooked more easily (e.g., prunes, green beans, peas, carrot, cabbage, onion and potato). The *taste* of canned strawberry is *improved* upon irradiation. Irradiation of raw coffee beans and cocoa beans (with 10–90 kGy, i.e., 1–9 Mrd) leads to a shorter roasting time, as well as to better *aroma, flavour* and *colour*. In alcoholic beverages (such as brandy, whisky) a *faster aging* can be achieved.

The utilization of radiation in the food industry and agriculture spreads rather sluggishly. This can be attributed to the slow procedure of premission granting by health authorities—which can be understood, but may not be always justified—rather than to technological reasons. In 1983 the specialized organizations of the United Nations Organization (FAO, IAEA, WHO) recommended for the member states to authorize the irradiation of any food by an average dose not higher than 10 kGy (1 Mrd).

6.3.2. STERILIZATION OF PHARMACEUTICALS AND MEDICAL INSTRUMENTS

Radiation obtains more and more importance in sterilization of medical instruments, in addition to generally accepted high temperature and gas sterilization by ethylene oxide. The fact that irradiation is a *cold process* allows to sterilize preparations and packaging materials sensitive to heat, permitting thus the industry to introduce new products.

Radiation sensitivity is the average dose necessary to kill a given fraction of germs (e.g., a dose which decreases the number of microorganisms to $1/e$ times as compared with the original):

$$\ln \frac{n}{n_0} = -k_{1/e}D \qquad (6.9)$$

where n_0 is the number of microorganisms in the system before irradiation, n is their number after irradiation, D is the dose and $k_{1/e}$ denotes a constant characteristic of the radiation resistance, the value of which is the same within a wide dose range but depends on the temperature, the age of the microorganisms, the temperature, the presence of

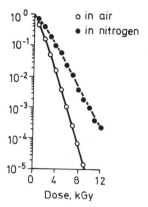

Fig. 6.14. Radiation damage of *Bacillus pumilus* E. 601 suspended in a phosphate buffer. The two lines represent radiation effects with and without oxygen

oxygen or water, the pH, etc. The radiation sensitivity of various microorganisms decreases in the following order: vegetative forms > moulds > yeasts > bacterial spores > viruses.

The exact determination of the *dose* applied is very important because a complete sterility should be achieved even if the substance was contaminated with germs prior to irradiation (Fig. 6.14); an overdose, on the other hand, involves chemical pyrogenic decomposition and the substance becomes pyrogenic. The exact dose should be checked by simultaneous irradiation of an ampoule containing radiation resistant bacteria, e.g., *Bacillus pumilus* (see Section 6.6). Since the dose necessary for sterilization is inversely proportional to the viable cell count in the product to be sterilized, no general sterilizing dose values can be given, and it is recommended to determine the permissible maximum of the number of microbes before irradiation.

Sometimes *pyrogenic decomposition*—which can be judged on the basis of the change of the chemical composition followed analytically—as well as discolouration of the product (or a change in some other respect of the appearance) occurs under the effect of even lower doses than necessary for sterilization; in these cases radiation sterilization cannot be applied.

The radiation sterilization of *solutions*, suspensions and emulsions is more difficult than that of solid substances, because the solution or the *colloid system* itself may suffer damage. Several materials give off gas. Some additives—added irrespective of the irradiation—can increase or decrease the extent of radiation damage (e.g., sugars have a protecting effect).

The advantages of radiation sterilization compared to thermal sterilization are as follows:

— previously hermetically closed packages can be sterilized;
— in some cases more suitable packing materials can be used;
— sterilization can be adjusted to continuous manufacturing technology;
— the sterility is more certain.

At the same time, the *costs* of radiation sterilization at present exceed those of thermal sterilization and, in some cases, even those of ethylene oxide sterilization; therefore it should be considered in such cases only when this method must be applied, due to the chemical composition of the substance to be sterilized.

Inocula (serums and antiserums), *microbal and viral vaccines*, as well as *microbe-antigen complexes* are obtained mostly by inactivation. In the case of smaller viruses larger doses are necessary.

Sterile tissues for transplantation of biological organs (skin, nerves, blood vessels, heart valves) can be radiation sterilized in the frozen state, at temperatures as low as -80 to $-50\,°C$, without the appearance of microscopic or other changes.

Bone transplantation and the creation of humane *"bone banks"* requires a particular sterility requirement. Irradiation is important although a radiation-sterilized bone expecting transplantation suffers slight changes even when stored in a deep frozen state.

Enzymes are, as a rule, more radiation resistant; thus their total inactivation requires very large doses, especially if they are irradiated in dried state. Much lower doses are sufficient to inactivate them—either instantaneously or in a delayed manner—in aqueous solutions.

Of antibiotics, salts of penicillin and streptomycin can be radiation sterilized either in the dry state or in oily suspensions. The dose necessary to decrease an initial viable cell count of 10^{10} down to 1 damages maximum 0.1% of the penicillin.

As far as other *pharmaceuticals* are concerned, eye drops, ointments, capsules, pills can be radiation sterilized and also the radiation treatment of some *cosmetics* is spreading.

Radiation sterilization of *therapeutical* and *medical tools* has even greater importance, among others, to fight infectious hepatitis. Bandages, surgical rubber gloves, scalpels, various vessels, instruments for taking and transfusing blood, injection needles, catheters, disposable syringes (of these, some hypodermic ones; the syringes may be filled with a pharmaceutical), plastic skin surrogates, etc. belong to commercially radiation sterilized products in several countries. Radiation sterilization also increases assortment: bandages treated with heat-sensitive additives can only be radiation sterilized. Catgut thread is successfully radiation sterilized; its protein content is very sensitive to heating, especially in the presence of water. All these products are sterilized in the factory: attempts to re-sterilize used devices in *hospitals* proved to be unsuccessful in practice. At the same time, the importance of radiation sterilization of *artificial kidney* and similar equipment joined directly to the organism is increasing.

Of all radiobiological processes, radiation sterilization of therapeutical equipment proved to be the most suitable for industrial purposes.

EXAMPLE 6.7

Compare the radiation resistance constant $k_{1/e}$ of *Bacillus pumilus* E. 601 in air and in nitrogen.

As shown by Fig. 6.14, the number of surviving microorganisms after an irradiation with a dose of 8 kGy (0.8 Mrd) is 7×10^{-5} times less in air and 4×10^{-3} times less in nitrogen than the starting value.

According to Eq. (6.9):

$$k_{1/e} = -D \ln \frac{n}{n_0} \qquad (6.10)$$

Using Eq. (6.10) the result is as follows:

$$\frac{k_{1/e,\,\text{air}}}{k_{1/e,\,\text{nitrogen}}} = 1.73$$

Hence, the radiation resistance constant of *Bacillus pumilus* in air is lower by about 70% than in pure nitrogen.

6.4. RADIATION RESISTANCE OF STRUCTURAL MATERIALS

By a careful and proper selection of structural materials it is possible
— to decrease considerably (or even eliminate) radiation damage;
— to decrease the time necessary for maintenance.

The radiation resistance of organic materials, e.g., polymers, is, as a rule, lower than that of inorganic ones. This, together with the increasing importance of plastics as structural materials, explains why an overwhelming majority of the literature in this field deals with problems connected with polymeric materials. Table 6.8 demonstrates the limits of radiation resistance of various substances. It can be seen that some materials do not lose their excellent physical and chemical properties even under the effect of high doses.

Table 6.8. Radiation resistance of some structural materials

Material	Limits of application*		Remark
	MGy	Mrd	
Transistors	10^{-3}	0.1	
Glass	10^{-3}–0.01	0.1–1	As a transparent material
	$> 10^4$	$> 10^6$	As an insulator
Teflon	10^{-3}–0.1	0.1–10	
Plexiglass	0.1–1	10–100	
Water	$0.01(> 10^4)$	$1–(> 10^6)$	Depending on the purpose
Natural and butyl rubber	0.1–1	10–100	
Graphite	$0.1–(> 10^4)$	$10–(> 10^6)$	Depending on the purpose
Liquid organic substances	1–10	$100–10^3$	
Polyethylene	0.1–10	$10–10^3$	
Phenoplasts with mineral fillers	1–10	$100–10^3$	
Hydrocarbon-based oils	1–10	$100–10^3$	
Polystyrene	10–100	$10^3–10^4$	
Special polymers	$100–10^3$	$10^4–10^5$	
Ceramics	$100–10^3$	$10^4–10^5$	
Carbon steel	$10–(> 10^4)$	$10^3–(> 10^6)$	Depending on the purpose
Stainless steel	$10^3–(> 10^4)$	$10^5–(> 10^6)$	
Aluminium alloys	$10^4–(> 10^4)$	$10^6–(> 10^6)$	

* Informatory value permitting considerable variations depending on the purpose of application.

6.4.1. INORGANIC MATERIALS

Sections 6.1.1 and 6.1.2 deal with processes induced by radiation in inorganic structural materials. Only a few additions should be made here.

6.4.1.1. AIR

Ozone and nitrogen oxides are the chemically most active substances formed from air under the effect or irradiation. An air of 60% relative humidity at room temperature will have its dew point at about 10–15 °C. The condensing water interacts with nitrogen oxides and as a result extremely corrosive nitric acid is formed. This—especially in the presence of ozone—very rapidly corrodes metal structures. Therefore the concentrations of these corrosive products should be limited, and it is a practical method to decrease corrosion by coating the metal walls and structures with acid resistant paints or coatings.

6.4.1.2. WATER AND AQUEOUS SOLUTIONS

With high LET radiations (see Table 6.2) or in the case of high dose rates, the *G*-values of the radiolysis of water are relatively small, which is advantageous from the point of view of *nuclear reactor technique*.

Purity has a paramount importance in water radiolysis; it can be checked by measuring the electric resistivity. To reduce decomposition, the system must always be charged by de-ionized (or distilled) and de-gassed water; corrosive products should be continuously removed from closed systems, e.g., by ion exchange. If possible, it is advantageous to dissolve small amounts of *hydrogen* in the purified water.

The value of pH should be at its optimum. Alkaline medium reduces corrosion; for *steel* structures the best *pH* is between 9 and 11. The corrosion resistance of *copper* and its alloys is not particularly good: slightly alkaline medium (pH = 8–9) is recommended for them. *For aluminium* and its alloys, a slightly acidic medium is necessary, the optimum *pH* being about 6.5.

Inhibitors may decrease analytically observable corrosion caused under irradiation by some ions; nevertheless, they may cause increased damage by poisoning the ion exchanger purifiers, decreasing their capacity and their efficiency.

Other aspects of radiolysis of water and aqueous solutions are treated in Section 6.2.1.2 in detail.

6.4.2. POLYMERS

Radiation effects on polymers are of importance in the technologies of nuclear reactors, high energy accelerators, high activity radiation sources, fuel element reprocessing plants and spaceships.

General aspects of radiation damages in polymers have been discussed in Section 6.2.2.3; the next section will treat mainly the consequences of irradiation.

Table 6.9 summarizes data with respect to the radiation damages in some polymers.

6.4.2.1. MAJOR TECHNOLOGICAL CONSEQUENCES
OF RADIATION DAMAGES

In addition to changes in the chemical composition (and in connection with it), radiolysis modifies mechanical, electrical, optical, etc. properties.

Of the mechanical properties, crosslinking increases the softening point, *Young's* modulus and the tensile strength; at the same time, solubility and ultimate elongation are decreased. Apart from this, gas evolution and brittleness may result. With decomposition as the main process, reverse effects are observed.

The electrical properties will also differ during and after irradiation from those of the non-irradiated sample. Direct current conductivity increases considerably during

Table 6.9. Radiation resistance of plastics

| Plastic | Approximate limit of application | | | | | |
| | Almost no damage | | Often acceptable | | Limited | |
	MGy	Mrd	MGy	Mrd	MGy	Mrd
Acryl rubber	0.08	8	0.8	80	8	800
Aniline–formaldehyde resins	1	100	2	200	>40	>4000
Aromatic epoxy resins	20	2000	>40	>4000	–	–
Aromatic polyamides	0.8	80	8	800	>40	>4000
Cellulose	0.02	2	0.06	6	0.4	40
Cellulose acetate	0.03	3	0.09	9	0.9	90
Cellulose nitrate	0.02	2	0.05	5	0.5	50
Ethyl cellulose	0.02	2	0.07	7	0.7	70
Furan resin	2	200	20	2000	>40	>4000
Natural rubber	0.05	5	0.1	10	3	300
Neoprene rubber	0.1	10	1	100	>40	>4000
Nylon	0.1	10	0.2	20	10	1000
Phenoplast (cast)	0.3	30	0.8	80	20	2000
Phenoplast (paper-filled)	0.03	3	0.09	9	8	800
Polyacrylonitrile	0.1	10	0.5	50	8	800
Polyamide	0.04	4	0.07	7	0.4	40
Polycarbonate	0.02	2	1	100	2	200
Polyester (glass-filled)	10	1000	>40	>4000	–	–
Polyester (unfilled)	0.01	1	0.1	10	0.5	50
Polyethylene	0.05	5	0.1	10	20	2000
Polyformaldehyde	0.003	0.3	0.008	0.8	0.05	5
Polyimide	1.5	150	10	1000	>40	>4000
Poly(methyl methacrylate)	0.02	2	0.06	6	0.8	80
Poly(methyl vinylsiloxane)	0.02	2	0.04	4	1	100
Polypropylene	0.01	1	0.02	2	0.1	10
Polystyrene	0.5	50	1	100	>40	>4000
Polytetrafluorethylene	0.002	0.2	0.007	0.7	0.06	6
Polyurethane	10	1000	>40	>4000	–	–
Poly(vinyl acetate)	0.3	30	0.8	80	6	600
Poly(vinyl alcohol)	0.03	3	0.08	8	0.8	80
Poly(vinylcarbazole)	1	100	3	300	>40	>4000
Poly(vinyl chloride)	0.15	15	0.4	40	9	900
Poly(vinylidene chloride)	0.1	10	0.2	20	3	300
Silicon (glass-filled)	10	1000	>40	>4000	–	–
Silicon (unfilled)	1	100	10	1000	>40	>4000
Styrene–butadiene rubber	0.3	30	1	100	7	700
Urea–formaldehyde resin	0.1	10	0.2	20	8	800

irradiation due to the charged particles formed; the conductivity of epoxy resins, polyethylene, poly(methyl methacrylate), polytetrafluoroethylene (Teflon), etc. increases rapidly to a saturation value and remains then unchanged under further irradiation. The dose rate and temperature determine the actual value of conductivity of a given irradiated polymer.

If the irradiation ceases, the conductivity of the polymer decreases again; its initial value is reached (or approached) after a certain time, depending upon the nature of the polymer. The conductivity of most non-polar polymers decreases abruptly after stopping the irradiation.

Some electrical properties undergo permanent change as a result of irradiation; these can be attributed to permanent, radiation-induced chemical changes. The deterioration of mechanical properties, as well as gas evolution may lead to electrical breakdown in many cases. The failure of electrical insulation due to radiation effects can primarily be attributed to the loss of mechanical strength and formation of gas inclusions and not to the alteration of the dielectric properties as a whole.

The optical properties—first of all the colour of the polymer—are changed by higher doses, depending on the conditions (vacuum, air, temperature, dose rate), on the size of the sample, on its chemical composition, etc. Additives in small amounts (e.g., plasticizers) may alter the colour effects—this has been utilized by dose indicators (see Section 6.6.5).

The colour of most polymers turns yellow to brown (eventually black) under irradiation. With a dose of 100 kGy (10 Mrd) of γ-radiation, under atmospheric conditions, polyvinyl chloride becomes dark green, poly(methyl methacrylate) (Plexiglass) yellow-green, polystyrene light yellow and polyethylene remains unchanged. Most epoxy resins become dark brown under the effect of 10^7 Gy (1000 Mrd).

Whereas the discolouration of silicate glasses can be reversed by heating (see Section 6.1.1.3), polymers behave in a different way. Radiation induces the formation of unsaturated bonds in polymers and these contribute to discolouration since, as a rule, conjugated double bonds are responsible for the colour of organic compounds. Since these double bonds do not disappear on heating, it is obvious that an elevated temperature cannot reverse the darkening of the polymers; it will even deepen the colour by promoting further decomposition.

6.4.2.2. FACTORS INFLUENCING DECOMPOSITION

The effect of chemical composition

Crosslinking polymers give off mainly hydrogen under irradiation; those undergoing *degradation* give off such gases which contain cracked products (e.g., hydrocarbons) pointing to the splitting of bonds in the vicinity of quaternary carbon atoms. The absorbed dose and the temperature also influence the composition of the gases.

The presence of *halogens* decreases—more or less—the stability of plastics against irradiation.

A rigid molecular structure is generally more radiation resistant than a flexible one. Thus, thermosetting resins are more radiation resistant than thermoplastics or elastomers. Of thermoplastics, hard, glassy types (e.g., polystyrene and polyimide) are the most resistant.

The degree of crystallinity has also an influence on the ratio of crosslinking and decomposition. Crystalline regions stop (or at least hinder) the movement of the polymeric radicals formed under the effect of irradiation in the amorphous regions and, by doing so, disturb recombination processes. In addition, polymeric radicals are produced within the crystals; they are practically unable to react with each other, because they do not have the sufficient mobility. These "frozen" radicals have a long lifetime and may lead to subsequent changes; their stability is, of course, a function of the temperature.

The number of crystalline regions decreases upon irradiation. Above a certain dose, not only their effect disappears, but also the polymeric structure is destroyed. This influences, of course, all the physical properties.

The method of production has also an effect on the radiation resistance; e.g., low and high pressure polyethylene behave differently.

The effect of conditions

Temperature has hardly any effect on the radiation chemistry of polymers, especially under the transition temperature. The radiolysis of most kinds of thermoplastics cannot be influenced by changing the temperature, under the temperature of formation of the glassy polymer. At the same time, most polymers exhibit an increased temperature sensitivity in radiolysis above this temperature. This can be attributed, to all probability, to the enhancement of decomposition chain reactions as opposed to competing crosslinking.

In spite of rough similarities in heat and radiation resistance, polymers which are resistant to high temperatures are not necessarily also radiation resistant: e.g., the thermally very stable polytetrafluoroethylene is very sensitive to radiation.

The stretching of the target can influence the mechanical properties after irradiation; e.g., some stretched elastomer gaskets are shortened to 10% of their original size. Internal stretching causes an increased damage in polymers. This is due to the enhanced crystallization ability of elastomers under stretching, at room temperature. Under pressure, radiation effects are less expressed than without it.

In the presence of oxygen, also the *dose rate* has an effect on free radical reactions. In the case of polyethylene and polypropylene, if the bulk is irradiated in air with a high dose rate of 300 Gy/s (100 Mrd/h), the result is similar to that obtained with an irradiation in vacuum. On the other hand, with a low dose rate (1 Gy/s or 400 rd/h), the same polymers undergo oxidation in the presence of oxygen.

If the polymers are irradiated in the *presence of air*, the aggressive gaseous products formed from air (ozone and nitrogen oxides, see Section 6.2.1.1) also attack them. In addition to the direct radiation effects, this also gives rise to various decomposition products, as indicated, e.g., by the behaviour of rubber: its double bonds are sensitive to ozone, therefore, it decomposes if irradiated in oxygen, but no decomposition occurs if the irradiation is carried out in vacuum. At the same time, polymers resistant to ozone do not become brittle even if they are irradiated in air.

The lifetime of free radicals is especially long in *crystalline systems;* hence, when the oxygen of air penetrates into the target subsequent to irradiation, it reacts continually with "frozen" free radicals and the mechanical, optical, etc. properties of the sample keep on changing.

Humid atmosphere has a considerable influence on radiolysis, especially if—as mentioned in Section 6.4.1.1—the condensation of water occurs owing to lowering of the temperature. The extent of sorption of condensed water depends on several physical properties of the material, e.g., on its porosity. Of course, the radiolysis of sorbed water gives hydrogen and oxygen; these gases cause stresses in the sample, accelerating thus mechanical damage. The presence of water in the sample may lead to undesirable chemical processes: e.g., some elastomers undergo hydrolysis and this decreases their radiation resistance further. This effect is especially significant if polyurethane is irradiated in the presence of water; this increases its radiation sensitivity ten times as compared to the value observed in air.

The effect of additives

Low concentrations of *organic protecting agents* may exert an advantageous influence on the ratio of crosslinking and decomposition. For example, if such substances are added to natural rubber, they inhibit chain rupture and/or hinder crosslinking.

So far the mechanism of effect of such additives has remained unknown. The general opinion is that those substances may act as anti-radiation additives in elastomers, which have the same properties as antioxidants: e.g., β-naphthol and pyrogallol act equally in both ways in natural rubber. α-Naphthylamine is a good anti-radiation additive in styrene–butadiene rubber and nitrile rubber; quinhydrone decreases radiation damages in Hypalon rubber. The radiation resistance of epoxy resin adhesives increases upon the addition of 2,5-diphenyloxazole.

Inorganic fillers—besides ensuring a lower price for the product—may enhance the radiation resistance of polymers. As opposed to organic additives, relatively large amounts of mineral fillers are necessary. Although no general rules could be recognized until now, the favourable results may be due to the formation of a solid structure, or to the fact that a fraction of the radiation energy is transformed to heat without chemical reaction in the inorganic additives. The addition of alumina to polyester or epoxy resin, and the addition of silica to rubber can be mentioned as examples.

Radiosensitizers have an opposite effect: they accelerate the decomposition of polymers even in low concentrations. These substances increase the flexibility of polymeric systems (e.g., aliphatic flexibilizers in polyester and epoxy resins) and decrease the price (e.g., in the case of phenoplasts filled with paper).

6.4.3. VARIOUS ORGANIC MATERIALS

6.4.3.1. LUBRICANTS AND ORGANIC MODERATORS

Up to 50 kGy (5 Mrd) lubricating oils and greases suffer practically no damage under the effect of radiation, i.e., their viscosity does not increase above 125% of the original value. The technologically decisive interval is between 5×10^4 and 10^8 Gy (5 and 10^4 Mrd), within this, first of all, above 10^6 Gy (100 Mrd). If the lubricated machine is shielded against radiation, the change of lubricants may be performed after a longer period, but this renders the availability and maintenance of the device more difficult.

The radiation resistance of *lubricating oils* is generally measured by the viscosity change of a sample irradiated by a dose of about 10^6 Gy (100 Mrd), eventually by plotting the viscosity changes as a function of the time or the dose. In agreement with the statements of Section 6.2.2.1, paraffinic lubricants are more rapidly damaged, i.e., irradiation for the same period causes a more severe damage in them than in aromatic ones.

Radiation stability is determined mainly by the chemical nature of the basic oil. Table 6.10 contains summarizing data related to an inert atmosphere, i.e., to irradiation in nitrogen or argon gas, or in vacuum. Oxidation decreases the lifetime of lubricants drastically or, in other words, irradiation accelerates oxidation processes.

Some analogy can be observed with the radiation chemistry of polymers discussed in Section 6.4.2. Lubricants based on alkylaromatic compounds (ethers or petroleum products) are, e.g., very resistant to radiation. The limit of their use is about 5 MGy (500 Mrd), but for special products this value may reach 100 MGy (10^4 Mrd). On the other hand, lubricants containing phosphorus, fluorine or chlorine and fluorine are not resistant to radiation: their properties are changed even by a dose of 10 kGy (1 Mrd). Of conventional lubricants, esters, silicones and polyglycols are utilizable up to a dose of about 6×10^5 Gy (60 Mrd).

Table 6.10. Radiation resistance of lubricants without additives

Lubricating oil type	Approximate limit of application					
	Almost no damage		Often acceptable		Limited	
	Gy	Mrd	Gy	Mrd	Gy	Mrd
Fluorinated compounds	7×10^3	0.7	9×10^4	9	10^6	100
Chlorofluorocarbons	7×10^3	0.7	8×10^4	8	7×10^5	70
Phosphates	10^4	1	2×10^5	20	3×10^8	300
Silicones	2×10^4	2	3×10^5	30	7×10^6	700
Polyglycols	5×10^4	5	6×10^5	60	9×10^6	900
Esters	9×10^4	9	10^6	100	10^7	10^3
Petroleum-based oils	5×10^5	50	2×10^7	2×10^3	10^8	10^4
Ethers	2×10^6	200	7×10^7	7×10^3	10^8	10^4
Alkylaromatics	8×10^6	800	7×10^7	7×10^3	10^8	10^4

As a rough approximation, it can be stated that heat resistant lubricants are also radiation resistant: e.g., polyphenyl ethers possess excellent thermal stability, they can be applied up to 400 °C, and the same substances also withstand the highest doses of radiation without considerable damage. At the same time, there are some basic controversies, too: the thermal stability of fluorine compounds is excellent, but their radiation resistance is poor.

If the lubricant contains *additives*, the effects of radiation are not unambiguous: both the hydrocarbon and the additive suffer simultaneous and—more or less—independent radiation chemical degradation. Additives decompose generally more rapidly than the basic oil, consequently they lose their effect within a relatively short time.

Lubricating greases contain petroleum products and gelling agents. They suffer more radiation damage than lubricating oils; both the oil and gelling agent may change— interacting with each other, but basically independently of each other. The same is true also for additives. This is the reason why the direction of the changes is not unambiguous, especially for soap-based greases. If some properties are plotted as a function of the dose, curves with extreme values are obtained.

Of synthetic lubricants, good experiences have been obtained with the following oil types: octadecylbenzene and di-(2-ethylhexyl) sebacinate containing additives, as well as with alkylbenzene type greases containing didodecyl selenide (e.g., *Du Pont*'s "Estersil") or a terephthalate gelling agent (of a relative molecular mass of about 250). These substances preserve their advantageous properties even above 50 MGy (5×10^3 Mrd), and what is more, they possess optimum properties above 30 MGy (3×10^3 Mrd). Some data for APL (Atomic Power Lubricants) oils and greases of the *Shell* company have been summarized in Table 6.11.

Dry lubrication involves different problems. Graphite, molybdenum sulphide and their mixtures are suitable lubricants even at temperatures as high as 260–320 °C. It has been shown that a dose of 1 MGy (100 Mrd) decreases the lifetime of the lubricated machine part in some cases, but lubrication improves again above 26 MGy (2.6×10^3 Mrd).

Moderators in nuclear reactors slow down fast neutrons produced during nuclear fission in order to make them suitable to maintain the chain reaction. In most cases hydrogen (sometimes deuterium) is used for this purpose, mainly as water (heavy water). Experience in nuclear technology pointed to some disadvantage of water as moderator (and coolant), therefore investigations were carried out to replace water by hydrocarbons.

Table 6.11. Radiation-resistant lubricants

(a) Lubricating oils

Product	Viscosity, 10^{-6} m²/s (cSt)		Setting point, below °C	Maximum permissible neutron flux, 10^{18} neutron/cm²	Site of lubrication	Remark
	50°C	100°C				
Shell APL 710	171	16.6	−18	1.6	Bearings and drives	—
Shell APL 719	562	35.8	−6	1.6	Drives	—
Shell APL 729	29.5	6.9	−6	<1.0	CO_2 circulators and as turbine oil	Low vapour pressure; with anti-rust and antioxidant additives
Shell APL 731	70.6	10.7	−26	1.0	General lubricating oil (e.g., for machine tools)	With anti-rust and antioxidant additive
Shell APL 734	116	13.1	−18	1.6	Like Shell APL 710	With anti-rust additive
Shell APL 742	20.3	4.5	−40	1.0	Hydraulic fluid	With anti-rust, anti-oxidant and anti-foam additives

(b) Lubricating greases

Product	Penetration (ASTM D-217), at 25°C	Pour point, above °C	Maximum permissible neutron flux, 10^{18} neutron/cm²	Site of lubrication	Remark
Shell APL Grease 700	265/295	300	1.8	Rolling and other bearings operating in CO_2, fuel element lift mechanisms	—
Shell APL Grease 701	265/295	300	<1	Control devices for long-time operation	Very low vapour pressure, good oxidation stability, CO_2 compatibility
Shell APL Grease 702	265/295	300	<1	Heavy duty, low velocity, high temperature sleeve bearings, etc.	Like Shell APL 701 but with graphite additive

Table 6.12. Radiolysis of polyphenyls at high temperature, under the effect of the mixed radiation of nuclear reactors

Compound	$G_{total gas}$			$G_{polymer}$		
	300 °C	350 °C	400 °C	300 °C	350 °C	400 °C
Diphenyl	0.112	0.159	–	0.82	1.13	–
o-Terphenyl	0.080	0.108	–	0.63	0.70	–
m-Terphenyl	0.069	0.081	–	0.58	0.64	–
p-Terphenyl	0.062	0.073	0.156	0.47	0.54	1.32
Santowax-R	0.064	0.080	0.153	0.51	0.59	1.30

Of hydrocarbons, paraffins have most advantageous moderator effect because the hydrogen/carbon ratio is the highest in them; but of all hydrocarbons, paraffins have the poorest thermal and radiation resistance. Aromatic hydrocarbons—e.g., diphenyl, terphenyl and their homologues, methylphenyl and mono-(isopropyl)-biphenyl—have highest thermal and radiation stability; they and their mixtures can be applied the most reasonably as moderators (Table 6.12) although their moderator effect is not so good, due to their lower hydrogen content. In practice, therefore, one has to be satisfied with a compromise.

6.4.3.2. TEXTILES

Textile reinforcement is often used for flexible cable joints and for other types of electrical insulation.

Synthetic fibres are, as a rule, more radiation resistant than *natural* ones: the most sensitive is cotton and the most resistant are aromatic polyamides. The highest permissible dose for natural fibres is 50 kGy (5 Mrd), for synthetic ones, it is higher by an order of magnitude (500 kGy or 50 Mrd), except for aromatic polyamides (e.g., Nomex) which can withstand a dose as high as 1 MGy (100 Mrd) (see Table 6.9).

If possible, glass fibre reinforcement should be applied.

6.4.3.3. PAINTS

The radiation resistance of paints is determined first of all by the medium. In addition, the state of the material as well as the technique of painting (the temperature of treatment, atmospheric conditions during drying) are also important. Table 6.13 summarizes the doses corresponding to the limits of radiation resistance; as can be seen, most advantageous are polyurethane type paints which can be used up to a dose of 10 MGy (10^3 Mrd).

Several types of damages are caused by irradiation in organic paints: they are blown up by gas evolution, they become porous, their permeability increases, cracks are formed because of the degradation of the polymeric structure, they peel off due to worsening adhesion, etc.

Table 6.13. Radiation resistance of paints

Medium of the paint	Approximate limit of application	
	Gy	Mrd
Cellulose esters	10^4	1
Poly(methyl methacrylate)	5×10^4	5
Neoprene	5×10^5	50
Chlorinated rubber	5×10^5	50
Sulphochlorinated polyethylene	$5 \times 10^5 - 10^6$	50–100
Poly(vinyl chloride)	$5 \times 10^5 - 10^6$	50–100
Silicons	10^6	100
Polyesters	$5 \times 10^6 - 10^7$	$500 - 10^3$
Polyurethanes	$5 \times 10^6 - 2 \times 10^7$	$500 - 2 \times 10^3$
Phenol derivatives	$7 \times 10^6 - 2 \times 10^7$	$700 - 2 \times 10^3$
Melamine resins	$7 \times 10^6 - 2 \times 10^7$	$700 - 2 \times 10^3$
Epoxy resins	$5 \times 10^6 - 5 \times 10^7$	$500 - 5 \times 10^3$

6.4.3.4. ADHESIVES

Adhesives are damaged by radiation in several ways: e.g., crosslinking increases their fragility, the decomposition of molecules contributes to gas formation, decreases' the adhesiveness and elasticity. In addition, the danger of corrosion increases to an extent depending on the chemical nature of the adhesive; the resistance against solvents decreases, cold flow increases and also thermal properties change.

The following groups can be distinguished from the point of view of radiation resistance:

— some heteroatomic polymers, e.g., polybenzimidazole and its copolymers keep their excellent thermal and oxidation properties up to 10^8 Gy (10^4 Mrd);

— adhesives withstanding a dose of 10^7 Gy (10^3 Mrd) or higher are radiation resistant; such are epoxy-phenol, vinyl-phenol, modified nylon-phenol, polyurethane and polyimide type adhesives;

— the following systems can be used up to 5×10^6 Gy (500 Mrd): epoxy adhesives, epoxy-thiokol, nitrile rubber and nitrile rubber–epoxy-phenol systems;

— neoprene-phenol systems can be used up to 10^6 Gy (100 Mrd); neoprene-rubber–nylon-phenol systems can withstand even lower doses: up to 5×10^5 Gy (50 Mrd).

Phenol derivatives containing also mineral fillers are, as a rule, more radiation resistant: e.g., filled epoxy and silicone preparations and mixed rubber–resin types have a medium position as far as resistivity is concerned.

6.4.3.5. ION EXCHANGE RESINS

The properties of organic ion exchangers change linearly or abruptly as a function of the dose, above a certain threshold value: their functional groups suffer chemical decomposition, their mechanical resistance decreases; as a result, their most important property: the ion exchange capacity is deteriorated. In addition, their colour and swelling are altered, but this has no technological importance. The different properties can undergo modification into opposite directions, too (Fig. 6.15). The influence of conditions may also

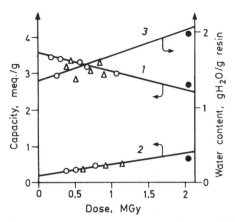

Fig. 6.15. Radiation damage in a polystyrene sulphonic acid ion exchange resin (Katex S) in water (^{60}Co-γ). 1 — Change of the ion exchange capacity on the basis of determination of sulphonic groups; 2 — Change of the capacity on the basis of titration of the functional groups; 3 — Water content: ○ — 0.13 Gy/s (47 krd/h); △ — 0.23 Gy/s (83 krd/h); ● — 0.44 Gy/s (158 krd/h)

Table 6.14. Radiation resistance of ion exchange resins

Matrix	Functional group	Commercial name of the resin	G_0, ion/ 100 eV	Approximate limit of application	
				Gy	Mrd
I. Irradiated in aqueous medium					
(a) Cation exchangers					
Phenol condensate	$-SO_3H$	Katex FN (Duolite C 3)	0.8	2×10^6	200
Polystyrene	$-SO_3H$	Amberlite IR-120	1.6	1.9×10^6	190
		Dowex-50Wx12	1.4	8×10^5	80
(b) Anion exchangers					
Polystyrene	$-N^+(CH_3)_3$	Amberlite-IRA-400	1.6–3.2	9×10^6	900
		Dowex-1x8	1.7–3.4	2.7×10^7	2700
Polystyrene	$-N^+R_2'R''$	Amberlite-IRA-410	1.7–3.4	9×10^6	900
II. Irradiated in air, in dry state					
(a) Cation exchangers					
Phenol condensate	$-SO_3H$	Duolite C 3	0.3	3.2×10^5	32
Polystyrene	$-SO_3H$	Amberlite-IR-120	1.6	1.9×10^6	190
		Dowex-50Wx12	1.4	2.7×10^6	270
(b) Anion exchangers					
Phenol or polyamide condensate	$-NR_2$ $-NHR$	Amberlite-IR-4B	3.3	2.2×10^7	2200
Polystyrene	$-N^+(CH_3)_3$	Amberlite-IRA-400	3.6	2.2×10^7	2200
		Dowex-1x8	13	7.6×10^5	76
Polystyrene	$-N^+R_2'R''$	Amberlite-IRA-140	4.0	2.2×10^7	2200

be of importance, e.g., that of the *pH* or the fact whether the irradiation is carried out with a dry resin, in air or in an aqueous system (Table 6.14).

Cation exchangers containing carboxyl functional groups are the most radiation sensitive.

The following resins were found to be the most suitable for use under the conditions of radiation exposure (e.g., for isotope production or purification of radioactive waste): polyfunctional polycondensed sulphonic acid resins, polymerized and polycondensed mono- and bifunctional resins containing phosphite groups and anion exchangers containing a pyridine group. They can be used up to a dose of about 10^7 Gy (10^3 Mrd).

6.4.4. COMPOSITE SYSTEMS

Technological practice often requires composite systems consisting of various materials: e.g., cables and coils containing metals and insulators, laminated metallic–organic systems (e.g., condensers), glass windows and TV cameras with polymeric adhesives. The lifetime of such systems is determined, as a rule, by the radiation resistance of the organic substance.

In organic materials, used as glues for glass or metal pieces, suffering some radiation damage, the equipment becomes unserviceable when maintained or repaired even if it could have served some more time under normal conditions.

6.4.4.1. CABLES

PVC can be used for cable insulation up to 20 kV. Its radiation resistance is acceptable from the mechanical point of view up to 1 MGy (100 Mrd).

Thermoplastic polyethylene has a low dielectric loss even at high frequency, therefore it is suitable for insulating radiofrequency cables and power cables up to 50 kV. From the mechanical point of view, it can be used at 20 °C up to 1 MGy (100 Mrd).

Crosslinked polyethylene has better mechanical properties than thermoplastic polyethylene: it can be used at room temperature up to 5 MGy (500 Mrd).

Polytetrafluoroethylene has excellent electrical properties, even at high frequency and elevated temperatures but its radiation resistance is poor: its mechanical properties remain acceptable up to 10 kGy (1 Mrd) only.

Mylar, polycarbonates (Makropol) and polyimide (Kapton film) have good electrical and thermal properties; e.g., Kapton resists high temperatures up to 400 °C, whereas polycarbonate can insulate cables up to 300 kV. The radiation resistances are different: polycarbonates can be used up to 2, Mylar up to 5, polyimide at least up to 20 MGy (i.e., 200, 500 and 2000 Mrd, respectively) as far as their mechanical properties are concerned.

6.4.4.2. ELECTRONIC PARTS

Table 6.15 gives informative data on the radiation resistance of electronic parts against mixed γ- and fast neutron as well as electron irradiation; these values are influenced also by the circumstances.

Main radiation effects on electronic devices are as follows:

— *in condensers*, the insulating resistance decreases, dielectric loss increases, the capacity is altered and some organic materials become defective due to gas evolution in the dielectric;

Table 6.15. Radiation resistance of electronic parts

Type of parts	Approximate limit of application					
	Almost no damage		Often acceptable		Limited	
	Gy	Mrd	Gy	Mrd	Gy	Mrd
Semiconductors	1	10^{-4}	10^4-10^9	$1-10^5$	10^{11}	10^7
Organic insulators	10^2	0.01	10^4-10^8	$1-10^4$	$>10^{13}$	$>10^9$
Resistors	10^2	0.01	10^7-10^9	10^3-10^5	10^{13}	10^9
Transducers	10^3	0.1	10^4-10^{11}	$1-10^7$	10^{13}	10^9
Electron tubes	10^3	0.1	10^6-10^9	$100-10^5$	$>10^{13}$	$>10^9$
Condensers	3×10^3	0.3	10^7-10^8	10^3-10^4	$>10^{13}$	$>10^9$
Magnetic materials	10^6	100	10^8-10^{10}	10^4-10^6	$>10^{13}$	$>10^9$
Piezo-electric crystals	3×10^6	300	10^{10}	10^6	$>10^{13}$	$>10^9$
Inorganic insulators	10^7	10^3	$10^{10}-10^{12}$	10^6-10^8	$>10^{13}$	$>10^9$

— the natural frequency of *oscillator crystals* is changed: the extent of this change depends on the direction of crystal cleavage and the construction;

— *gas-filled electron valves* will have altered conductivity and gas ionization; other electron tubes exhibit an accelerated aging of the cathode, due to secondary electron emission. An altered impedance and increased stray currents on the insulating and glass parts are observed, together with an increased tendency of glass parts to break or crack;

— in *inorganic insulators* stray currents increase and eventually chemical changes also occur;

— the same is the case with *organic insulators;* chemical changes, as a rule, cause brittleness or softening, together with gas evolution;

— in *magnetic materials*, coercivity, remanent magnetism and permeability all change;

— the stray current grows over *resistivities* causing the decrease of their nominal value;

— in *semiconductors*, the stray current increases; in *transistors*, amplification decreases, saturation voltage increases; operating voltage data are changed in *silicon rectifiers;*

— in *transducers*, the stray current increases, disturbing noise is generated and the calibration curve is altered.

6.5. THE TECHNOLOGY OF IRRADIATION

6.5.1. RADIATION SOURCES

The structures of irradiating facilities are determined first of all by the nature of the task to be performed and the character of the radiation. Either electron beams or γ-radiation are used for technological purposes.

Electron irradiation is carried out on a commercial scale with *accelerators*, but the utilization of an *isotopic β-irradiation* with ^{90}Sr is also investigated because this isotope, being in equilibrium with its ^{90}Y daughter element, can be obtained relatively economically from spent fuel elements.

Electrons are accelerated by "generators"; such are, e.g., linear accelerators, *van de Graaff* generators, cascade generators and resonance transformers. The maximum energy of practically applied accelerators is 8 MeV; the power is maximum 200 kW.

The advantage of an electron beam is—owing to its high dose rate—that it can be readily applied in continuous technologies (Fig. 6.16), e.g., for the treatment of polymers, conservation of food, or to produce special technical rubber products. Its disadvantage is the rather low penetration, which is only a fraction of that of a γ-radiation having the same energy. This disadvantage can often be eliminated by irradiating the target from both sides; in this way the dose absorbed by the material as a whole is much more uniformly distributed. In some cases, e.g., for the surface treatment of polymers, the inhomogeneity of the dose rate can be advantageous: if the energy is correctly selected, the bulk of the polymer remains undisturbed, at the same time, the surface layer can be grafted. Another disadvantage of electron accelerators from the point of view of operation is that they require several and highly qualified persons for maintenance; their advantage is that after switching off there is no remaining radiation hazard at all.

The γ-irradiation facilities are loaded as a rule, with ^{60}Co (less often ^{137}Cs); their maximum activity is about 10^{17} Bq (i.e., 2–3 MCi). The design of the irradiation chamber is, in principle, analogous to those of radiographic laboratories (see Section 7.2.1), but the radioactivity, being higher by several orders of magnitude, requires a much more elaborate safety system. Owing to the high penetration of γ-radiation, it is essential to provide an optimum geometrical arrangement of the radiation sources and the target (Fig. 6.17) in order to ensure the economical utilization of the dose.

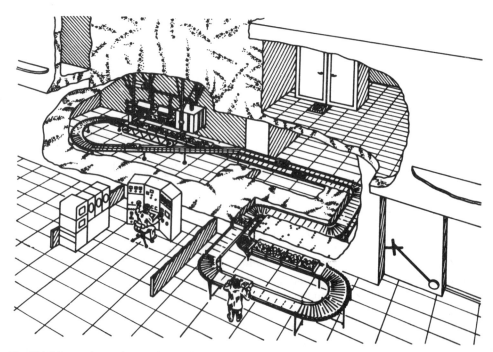

Fig. 6.16. Scheme of a continuous electron irradiator. *1* — Accelerator; *2* — Instrument rack; *3* — Switchboard; *4* — Shielding

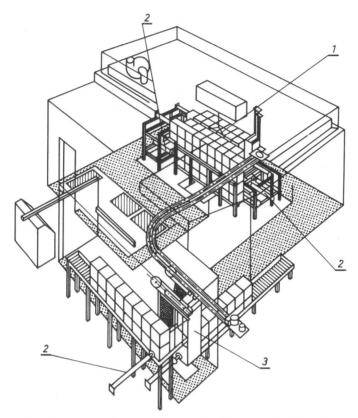

Fig. 6.17. The operation scheme of a stationary continuous γ-irradiator. *1* — Plate-like radiation source system;
2 — Pneumatic plungers for box moving; *3* — Hanging frame

The optimum arrangement of the *transportation system* in the irradiation chamber requires a very high level of engineering skill in the case of both electron and γ-irradiation stations. Undisturbed operation requires a careful construction, radiation-resistant lubrication, etc. *Radiation shielding* may be either concrete—completed with iron or sand—or water of an appropriate depth (5 m).

In addition to *stable facilities, mobile irradiation systems* gain more and more in importance, first of all in the food industry and agriculture. They may be installed on railway waggons, trucks (Fig. 6.18) or ships. Overland mobile irradiators can be used, e.g., for sprout inhibition of potato and onion, or deworming of cereals. Radiation sources on sea vessels are used for fish conservation on commercial scale. Another promising economic use is the killing of insects in agricultural products during their transport from tropical areas (tropical products being highly contaminated by insects). Thereby the importation of such products becomes permissible without any further quarantine. Whereas the most suitable isotope for stable sources is ^{60}Co, it is more economical to use ^{137}Cs for mobile irradiators. This latter isotope has a lower energy γ-radiation, therefore the mass of radiation shielding sufficient for the necessary safety is lower. This is advantageous from the point of view of both investment and transportation costs.

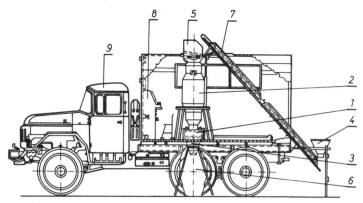

Fig. 6.18. Mobile continuous irradiation station (installed on a truck trailer). *1* — Truck; *2* — Switchboard; *3* —
Feeder; *4* — Spoon conveyor; *5* — Irradiator; *6* — Conveyor; *7* — Filling funnel; *8* — Holder; *9* — Bag

Irradiation with *spent fuel elements* of nuclear reactors was expected in the early 1960s
to have a bright future. More thorough studies have revealed, however, that the use of such
irradiation sources is disadvantageous from both the technical and economical aspects.

Nuclear reactors are suitable for irradiations with chemical or biological purposes, but
their principal disadvantage is that the neutron-irradiated targets may become
radioactive. Nuclear reactors have been constructed for expressly radiation chemical
purposes; they utilize the kinetic energy of fission products in chemical processes,
improving thus the economy of the process (e.g., synthesis of nitrogen oxides; see also
Section 6.2.1.1).

Reactor loops may combine the advantages of nuclear reactors and radioactive
isotopes. They use molten metals or alloys with high activation cross-section (e.g., indium–
gallium, magnesium, sodium). These metals can be activated rapidly in the active core of
the reactor and they give off their radiation energy as γ-radiation in the outer half of the
loop ensuring thus a high efficiency for the chemical reactor. Although their utilization on
commercial scale has not been reported so far to our knowledge, it is possible that under
appropriate investment conditions (e.g., if the nuclear reactor and the chemical plant are
built on a common site) loops will become an economical type of radiation source,
especially if the reactor provides also the necessary heat.

6.5.2. ECONOMIC FACTORS

The capacity of radiation chemical technologies can be calculated from the following
formula:
$$k = 3.7 GM \boldsymbol{I} \boldsymbol{m} \, 10^{-7} \tag{6.11}$$

where *k* denotes the product produced per unit time, t/h; *G* the radiation chemical yield of
the product, molecule per 100 eV; *M* the relative molecular mass of the product; *$\boldsymbol{I}$* the
applied dose rate, Gy/s, and *$\boldsymbol{m}$* is the quantity of the irradiated material, t.

The specific energy requirement is also determined by the *G*-value and the relative
molecular mass.

The specific cost of radiation energy depends, in addition, on the physical properties of
the target and of the radiation source. The nomogram depicted in Fig. 6.19 helps to
estimate the costs of irradiation.

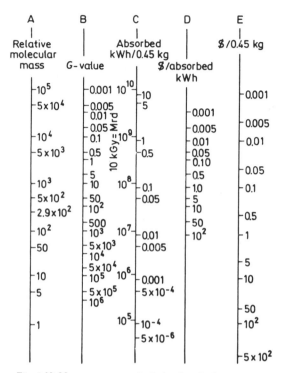

Fig. 6.19. Nomogram tor calculating irradiating costs

It is very difficult to suggest anything definite in connection with the economy of commercial scale irradiation technologies, because the available data on both the investment and operation costs are mostly approximate and often contradictory. It should also be remembered that many countries have a monopoly of isotope production, hence they may determine the costs of isotopic products more or less arbitrarily. This may have a paramount importance as far as apparent economy is concerned. At the same time geographical circumstances (e.g., the availability of hydroelectric energy) determine the costs of electric energy and, thereby, that of radiation energy produced by electron accelerators.

There are cases, however, where economy is of no decisive importance because there is no substitute for radiation energy: such is the killing of *Salmonella* species in, e.g., frozen eggs (see Section 6.3.1).

6.6. THE CONTROL OF IRRADIATION

The purpose of dosimetry is to check regularly the dose absorbed by a system during the period of a technological process or during a given time. This is basically another concept than dosimetry in the sense of health physics (see Chapter 8), although practical wording often mixes the two.

The most precise method of *physical dosimetry* is calorimetry providing direct data on the value of the radiation dose absorbed by a given system. Physical dosimetry also includes glass dosimetry, thermoluminescent dosimetry (see Section 6.1.1.3) and dosimetry with semiconductors (see Section 6.1.1.2). Special instruments or devices belong to the methods of physical dosimetry; they are used according to the instructions issued by the manufacturers, therefore they will not be discussed here in details.

Chemical dosimetry is based on the determination of chemical changes brought about by irradiation, i.e., the energy absorbed can be followed by the radiation chemical change occurring in the system. These methods are of secondary character, because they have to be calibrated on the basis of physical dosimetry (e.g., by ionization chamber or calorimeter), but they are well reproducible and relatively simple. The best known dosimeters suitable for practical purposes are: the *Fricke* dosimeter using iron(II) sulphate solution, the modified *Fricke* dosimeter using iron(II) and copper(II) sulphate, the cerium(IV) sulphate, the oxalic acid, the alcohol–chlorobenzene, the film and the Perspex-type dosimeters; of these four procedures (*Fricke*, ceric sulphate, chlorobenzene and the Perspex methods) will be discussed in detail. Approximative chemical dosimeters, called dose indicators (monitors), will also be described shortly.

Biological dosimetry draws conclusions about the absorbed dose on the basis of the radiation damage of bacterial, etc. cultures [see Eq. (6.9)]. It is used generally as a test method, i.e., radiation resistant bacteria are irradiated under the same circumstances as the target (product) to be sterilized; if the bacteria are killed, the target can also be regarded as sterile.

6.6.1. FERROUS SULPHATE *(FRICKE)* DOSIMETRY

The iron(II) sulphate dosimetry elaborated by *Fricke* and co-workers can be regarded as a secondary standard for dosimetry. It is suitable to determine the energy absorbed from γ- and electron radiation both in static and pulse type operation.

It is based upon the oxidation of iron(II) ions to iron(III) ions under the effect of radiation. The dosimetric solution consists of iron(II) sulphate in air-saturated aqueous solution containing also 0.4 M sulphuric acid:

$$OH^{\cdot} + Fe^{2+} \xrightarrow{rapid} OH^- + Fe^{3+}$$

$$H_2O_2 + Fe^{2+} \xrightarrow{slow} OH^{\cdot} + OH^- + Fe^{3+}$$

If the water contains dissolved oxygen, HO_2 is also produced with high efficiency. Dissolved oxygen reacts with H atoms:

$$H + O_2 \xrightarrow{rapid} HO_2$$

$$H^+ + Fe^{2+} + HO_2 \xrightarrow{rapid} Fe^{3+} + H_2O_2$$

The latter two equations demonstrate that a good oxygen supply is of paramount importance for the *Fricke* solution: some methods use saturation with oxygen in order to extend the limit of the measurement.

Organic compounds present in the system change the yield of Fe^{3+} ions because of the occurrence of the

$$OH^{\cdot} + RH \rightsquigarrow H_2O + R^{\cdot}$$

process competing with the oxidation of iron(II) ion by $OH^{\cdot}$ radicals. The resulting $R^{\cdot}$ organic radical

— either reacts with oxygen and gives $RO_2^{\cdot}$ radical which, in turn, oxidizes Fe(II) ions, i.e., the value of $G(Fe^{3+})$ increases, e.g.:

$$R^{\cdot} + O_2 \rightarrow RO_2^{\cdot}$$

— or reduces Fe^{3+} ions, i.e., the value of $G(Fe^{3+})$ decreases:

$$R^{\cdot} + Fe^{3+} \rightarrow Fe^{2+} + R^{+}$$

The presence of organic impurities can be checked by adding chloride ions (e.g., a 1 mM NaCl solution); in this case the following reaction also occurs:

$$OH^{\cdot} + Cl^{-} \rightarrow OH^{-} + Cl$$

The chlorine atom produced oxidizes a Fe(II) ion similarly to the $OH^{\cdot}$ radical, therefore the value of $G(Fe^{3+})$ remains unchanged:

$$Fe^{2+} + Cl \rightarrow Fe^{3+} + Cl^{-}$$

Since chlorine atoms react more rapidly with Fe^{2+} ions than organic radicals, the change of $G(Fe^{3+})$ observed upon adding chloride ions indicates the presence of organic impurities.

The radiation chemical yield is: $G(Fe^{3+}) = 15.6$ ion per 100 eV.

The scope and limits of application are the following:

— *Dose*: a linear correlation exists between the absorbed dose and the chemical transformation between 40 and 400 Gy (4–40 krd).

— *Dose rate*: the accuracy of determination of γ-doses does not change up to 40 Gy/s, i.e., 150 kGy/h (4 krd/s, i.e., 15 Mrd/h); with electron pulses, the corresponding values are: 2 Gy per pulse $= 2$ Gy/10^{-6} s (200 rd per pulse $= 200$ rd/10^{-6} s).

Temperature: the variation of temperature between 10 and 50°C has no effect on the accuracy of the measurement.

— *Energy dependence*: the method is energy independent between 0.1 and 1.6 MeV.

The procedure is as follows: The solution is prepared with water purified by ion exchange, then distilled from potassium permanganate and potassium bichromate. The concentrations are: 1 mM for iron(II) sulphate, 1 mM for sodium chloride and 400 mM for sulphuric acid. The solutions should be kept in dark bottle, exposure to sunshine must be avoided. In this case they can be preserved for several months. Complete purity of the vessels, the reagent and the water is essential. The ampoules should be open and their inner diameter must be at least 8 mm to ensure an appropriate diffusion rate of oxygen. If the oxygen in the solution is used up during irradiation, the data will be faulty.

The concentration of Fe^{3+} ions produced by irradiation is determined spectrophotometrically. The absorbance (optical density) of the solution is measured against 400 mM sulphuric acid at 302 nm in a 1 cm thick quartz cuvette.

The Lambert–Beer law gives the absorbance of the solution as follows:

$$A = \varepsilon c l \tag{6.12}$$

and

$$c = \frac{A}{\varepsilon l} \tag{6.13}$$

where ε denotes the molar absorption coefficient ($212 \text{ m}^2/\text{mol}$); c the molar concentration, A; and l the thickness of the cuvette, m.

With a 1 cm thick cuvette, the absorbed dose will be as follows (at 20°C):

$$D = 0.285A \ kGy = 0.0285A \ Mrd \tag{6.14}$$

For the purpose of checking, several measurements with different exposures should be carried out in the same selected point of the irradiation chamber. The dose data are correct if the dose values plotted as a function of the time of exposure give a straight line passing through the origin. The slope of the straight line gives the dose rate. A straight line can be obtained even if the dosimetric solution contains organic impurities but, according to general experience, this straight line does not pass through the origin, but gives a positive intersection with the ordinate. In this case, the slope gives a faulty value for the dose rate.

The exact description of the method can be found, e.g., in the US standard ASTM-D 1671-63.

The accuracy of the method is ± 1–2%.

EXAMPLE 6.8

The dose rate is determined by *Fricke* dosimetry in a small volume of the irradiation system. At the same time, the purity of the distilled water is checked by carrying out parallel measurements with *Fricke* solutions with and without NaCl. An appropriate spectrophotometer should be used for the measurements.

The results have been tabulated in Table 6.16. The absorbance of the *Fricke* solution containing also NaCl is before the irradiation 0.038, at 22°C, against 200 mм H_2SO_4.

Table 6.16. Experimental results obtained by *Fricke* dosimetry

No.	Time of irradiation, min	$A_{304 \ mn}$	Absorbed dose	
			Gy	krd
1	15	0.175	49.2	4.92
1*	15	0.178	50.0	5.00
2	30	0.355	99.8	9.98
2*	30	0.361	101.4	10.14
3	45	0.533	149.8	14.98
3*	45	0.557	156.5	15.65
4	60	0.720	200.2	20.02
4*	60	0.730	200.5	20.05

* Sample with added NaCl

The calculated dose is at 22°C:

$$D_{22} = 0.285(A - 0.038)kGy = 0.0285(A - 0.038) \ Mrd$$

where A denotes the absorbance of the irradiated solution.

Since the molar extinction coefficient is given for 20°C and has a rather large temperature coefficient ($+0.7\%$ per centigrade), the D_{22} value should be recalculated to 20°C:

$$D = \frac{D_{22}}{1 + 0.007 \times 2}$$

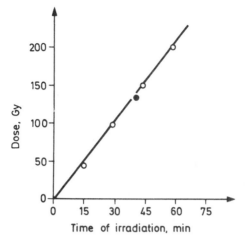

Fig. 6.20. Dose determination by means of *Fricke* dosimetry

Plotting these doses as a function of the time of irradiation, Fig. 6.20 is obtained; the slope gives the dose rate value:

$$tg\ \alpha = \frac{200\ Gy}{1\ h} = 200\ Gy/h = 0.056\ Gy/s = 20\ krd/h$$

6.6.2. CERIUM SULPHATE DOSIMETRY

Cerium sulphate dosimetry is used mainly for calibration purposes, for the determination of doses absorbed in liquid systems from static and pulse type electron or electromagnetic irradiation.

The method is based on the fact that intermediates produced from water under the effect of ionizing radiation reduce cerium(IV) ions to cerium(III) in a reversible process taking place in aqueous solutions of cerium(IV)–cerium(III) systems:

$$Ce^{4+} + H \rightsquigarrow Ce^{3+} + H^+$$

$$Ce^{3+} + OH^\cdot \rightsquigarrow Ce^{4+} + OH^-$$

The above equation shows that the process requires no oxygen.

Various organic impurities—even if present in very low amounts—have an extremely great influence on the chemical processes and their G-values. Therefore reproducible results can only be obtained under very strictly controlled conditions, after a lengthy preparatory work; a large practice is required. The procedure should exclude the possibility of getting organic impurities into the system, either from the air or from the glass vessels.

Another precondition of the application of this method is the complete purity of the water used; this can be checked in a separate sample by adding NaCl, similarly to the *Fricke* method.

The yields of $G(Ce^{3+})$ have been reported to be between 2.0 and 3.4 ion per 100 eV; the great dispersion can be explained by the extremely strong influence of organic impurities, which increase the radiation chemical yield even if present in a concentration as low as $10^{-5}\%$.

The scope and limits of application are as follows:

— *Dose:* 0.5–40 kGy (0.05–4 Mrd).

— *Dose rate:* the accuracy remains unchanged for γ-radiation up to 1 MGy/s $\approx 4 \times 10^9$ Gy/h, i.e., 100 Mrd/s $\approx 4 \times 10^5$ Mrd/h and, for electron pulses, up to 1 Gy per pulse $= 1$ Gy/1.1×10^{-6} s (100 rd per pulse $= 100$ rd/1.1×10^{-6} s).

— *Temperature:* its effect is within the error of the method between 5 and 50 °C.

— *Energy dependence:* the accuracy is energy independent between 0.1 and 10 MeV.

As mentioned, when effecting the measurement, complete purity is essential. Thus the water has to be tridistilled, the Pyrex glass vessels heated for 8 h at ca. 530 °C, and afterwards rinsed with the dosimetric solutions three times, etc. To avoid contamination by direct contact with the atmosphere, it is necessary to apply traps cooled with liquid nitrogen.

One of the methods is described in the US standard ANSI/ASTM D 3001–71 (Reapproved 1977).

Recently there has been developed a less sensitive procedure. In this method an aqueous solution of both Ce^{4+} and Ce^{3+} ions is used, the latter ions are produced by the irradiation of an aqueous $Ce(SO_4)_2$ solution. The irradiation products not only to Ce^{3+} ions, but simultaneously also oxidizes the impurities in the solution: thereby the necessary purity of the system is established. Since this method is not so sensitive to impurities, it is sufficient to use monodistilled instead of tridistilled water, it is possible to omit the baking of the glass vessels, the use of cooled traps, etc.

The concentration of the solution should be selected such that no less than 10% and no more than 80% of the Ce^{4+} ions are reduced during the irradiation. The dosimetric solutions are prepared with a 100 mM $Ce(SO_4)_2$ stock solution: 20.2165 g of $Ce(SO_4)_2 \times 4H_2O$ is dissolved in about 250 cm^3 of 400 mM H_2SO_4 and completed to 500 cm^3 by dilution with 400 mM H_2SO_4. For the determination of doses between 0.5 and 5.0 kGy (50–500 krd) the cerium ion concentrations are about 1.0–1.5 mM Ce^{4+} and 2.0–2.5 mM Ce^{3+}; between 5.0 and 40 kGy (0.5–40 Mrd), these values are 8–10 mM Ce^{4+} and 8–10 mM Ce^{3+} The Ce^{3+} ions are produced by irradiation of the dosimetric (not of the stock) solution.

The concentration of Ce^{4+} remaining after irradiation is determined spectrophotometrically. In order to improve the accuracy of the measurement, both the irradiated and non-irradiated solutions are diluted to 0.04–0.2 mM with 400 mM sulphuric acid. Photometry is carried out in a 1-cm quartz cuvette at 320 nm, against 400 mM sulphuric acid solution. The molar absorption coefficient of the Ce^{4+} ion is 561 m^2/mole at room temperature, in 400 mM sulphuric acid.

The dose can be calculated from the following equation

$$D = 8.23 \frac{\Delta A f}{G} \text{ kGy} = 0.823 \frac{\Delta A f}{G} \text{ Mrd} \qquad (6.15)$$

where ΔA denotes the change of the absorbance (20 °C), f is the ratio of dilution and $G = 2.04$ Ce^{3+}/100 eV.

Cerium sulphate dosimetry is used mainly for the measurement of high doses (above 400 Gy or 40 krd, representing the upper limit of *Fricke* dosimetry), as a comparative method, because its range of application is wide and its error is small. Because of its very high sensitivity to impurities—mentioned several times above—this method has not gained widespread acceptance for routine measurements.

The accuracy of the method is about $\pm 3\%$.

6.6.3. CHLOROBENZENE DOSIMETRY

The only considerable disadvantage of *Fricke* dosimetry is its unsuitability for measuring doses above 400 Gy (40 krd). At the same time, the cerium sulphate method, suitable for dose checking in a much wider dose range, is very sensitive to impurities.

Although the accuracy of alcoholic chlorobenzene dosimetry does not reach that of the *Fricke* method or cerium sulphate dosimetry, but it is much more applicable for routine measurements owing to its simplicity, relative insensitivity to impurities and wide range of measurement.

The mechanism of the basic radiation chemical process has not been clarified yet; the following reactions of chlorobenzene are suggested:

$$C_6H_5Cl \rightsquigarrow C_6H_5^{\cdot} + Cl^{\cdot}$$

$$2\,C_6H_5^{\cdot} \rightarrow C_{12}H_{10}$$

The product to be measured in the alcoholic chlorobenzene system is basically hydrochloric acid formed from chlorine atoms. HCl is radiostable in solution even when present in high concentrations (e.g., 500 mM). Ethyl alcohol, acetone and water are added in order to stabilize the chlorine atoms produced in the free radical reaction and to promote their transformations to chloride ions. Ethyl alcohol serves also as an inhibitor of a chain type oxidation reaction and it is a very good solvent of hydrochloric acid.

The radiation chemical yield in an alcoholic solution containing 25% of chlorobenzene is $G_{Cl} = 5.5$ ion per 100 eV.

The scope and limits of application of chlorobenzene dosimetry are the following:

— *Dose*: there is a linear correlation between the dose and the chemical transformation between 500 and 4×10^5 Gy (0.05–40 Mrd).

— *Dose rate*: the accuracy of determination remains unchanged between the dose rates 0.045 and 25 Gy/s $= 160$–9×10^4 Gy/h, i.e., between 4.5–2.5×10^3 rd/s $= 0.016$–9 Mrd/h.

— *Temperature*: the method is temperature independent between 20 and 90 °C.

The solution necessary for dosimetry contains 24% of chlorobenzene, 4% water and 0.04% acetone and alcohol. In general, analytical grade reagents and distilled water should be used.

Correct dose data are obtained even if the irradiated ampoules are kept in darkness for several years and evaluated afterwards.

The titration of hydrochloric acid produced by the radiolysis of chlorobenzene can be based on the determination of hydrogen ions in the presence of Bromophenol Blue indicator:

$$KHCO_3 + HCl = KCl + H_2O + CO_2$$

or on the determination of chloride ions in the presence of diphenylcarbazone indicator:

$$Hg(NO_3)_2 + 2HCl = HgCl_2 + 2\,HNO_3$$

The value of the absorbed dose can be calculated from the following formula:

$$D = 2.19 \, c \, MGy = 219.5 \, c \, Mrd \tag{6.16}$$

where c denotes the hydrogen chloride concentration in the irradiated sample.

The absorbed dose can also be determined by oscillometry. *Oscillometry* belongs to the group of conductivity measurements and is carried out by means of a high frequency alternating current. Its considerable advantage is that the measurement can be carried out without any galvanic contact with the solution to be studied, thus valuable information can be obtained on the composition of the sample even without opening the ampoule.

A calibration curve can be constructed by measuring the high frequency conductivity for several ampoules containing chlorobenzene solutions irradiated with known doses; the values obtained at a given frequency are plotted. It must be remembered that not only hydrochloric acid, but also other products of irradiation may contribute to the alteration of conductivity and capacity of the solution. Therefore the calibrating samples should be produced by irradiation, and the chloride ion concentration in each ampoule should be determined after oscillometric measurement by titration with mercury(II) nitrate. This measurement gives the value of absorbed dose for each calibration ampoule.

It is essential that ampoules with identical diameter and wall thickness should be used both for calibration and measurement. For checking, about 50 ampoules of each series of production should be measured in every sensitivity range; these, when filled with non-irradiated chlorobenzene solution should give practically the same deviation in a given range of the instrument. The structural change of the ampoules (discolouring) has no effect on the accuracy of the measurement.

After having measured the calibrating series and plotting the values of absorbed dose as a function of the scale reading, we obtain a calibration curve. The ampoules for dosimetry are then measured, and the absorbed dose values are read from the calibration curve.

Systems with various electron densities have different properties as far as their γ-radiation absorption is concerned. Therefore dose values measured with different systems should be recalculated on the same basis.

If the *Fricke* dosimetry is selected as a secondary standard (D_F), then the dose obtained by alcoholic chlorobenzene dosimetry (D_{cb}) should be corrected:

$$D_F = 1.01 \, D_{cb} \tag{6.17}$$

The accuracy of the measurement is $\pm 5\%$.

The irradiated ampoules can be preserved in darkness for at least three years without change in their high frequency conductivity.

EXAMPLE 6.9

The absorbed dose is determined by alcoholic chlorobenzene dosimetry at a given spot of an irradiation system by both the oscillometric and titration methods.

The sensitivity of the instrument is adjusted in a way that the full range can be utilized, i.e., the ampoule with the highest and the lowest conductivity can be measured within the same range. The dose values shown in Table 6.17(a) and Fig. 6.21 are obtainable by titration.

First the doses corresponding to the contents of the ampoules to be investigated are determined by oscillometry [Table 6.17(b)]; the ampoules are then opened and their contents titrated [Table 6.17(c)].

Table 6.17. Experimental data for chlorbenzene dosimetry

(a) Data of calibration (see Fig. 6.21)

| No. of ampoule | Dose | | Instrument reading |
	kGy	Mrd	
231	15.1	1.51	94
139	14.1	1.41	91
217	12.4	1.24	86
212	12.0	1.20	85
149	10.7	1.07	78
232	9.1	0.91	72
228	8.9	0.89	70
121	8.4	0.84	68
237	6.7	0.67	59
151	6.3	0.63	54
143	5.6	0.56	49
161	5.3	0.53	47
183	5.1	0.51	45
140	4.9	0.49	44
255	4.5	0.45	42
229	3.3	0.33	33
266	3.0	0.30	31
218	2.0	0.20	24

(b) Data measured by oscillometry

| No. of ampoules | Instrument reading | Instrument range | Dose | |
			Gy	Mrd
1	25	4	22.7	2.27
4	84.5	8	12.0	1.20
10	40	4	27.3	2.73
12	36	4	25.0	2.50

(c) Measured by titration

| No. of ampoule | Volume of the sample, cm^3 | 5mM KHCO$_3$ solution consumed, cm^3 | 10 mM Hg(NO$_3$)$_2$ solution consumed, cm^3 | Absorbed dose | |
				kGy	Mrd
1	1.00	1.14 1.15 1.16	1.18 1.18 1.18 } 1.18	23.5	2.35
4	1.00	0.62 0.63 0.63	0.65 0.65 0.65 } 0.65	12.9	1.29
10	1.00	1.28 1.26 1.28	1.38 1.34 1.35 } 1.36	27.0	2.70
12	1.00	1.26 1.27 1.28	1.32 1.32 1.33 } 1.32	26.3	2.63

The doses are calculated on the basis of the equation (6.16) considering also the factor of the titrating 5 mM $Hg(NO_3)_2$ solution (being equal to 0.971, related to 10 mM NaCl).

A comparison of the titration data with those obtained by oscillometry shows that the deviation exceeds $\pm 5\%$.

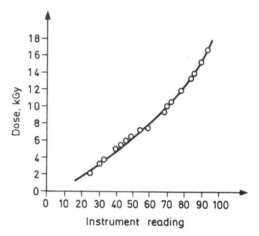

Fig. 6.21. The calibration curve of the oscillometer for chlorobenzene dosimetry

6.6.4. POLYMERIC DOSIMETERS

Films and pieces of polymers are more and more often used for the chemical dosimetry of both γ- and electron radiations. Of these, poly(methyl methacrylate) (PMMA) dosimeters are relatively new. The *red Perspex* dosimeter should be mentioned here, used mainly for routine checking in γ-irradiation technique.

The essence of the method is the irradiation of PMMA foils or cylinders containing appropriate dyestuffs; these change their colour in a close correlation with the dose (see also Section 6.4.2.1). The absorbancy changes can be measured at, e.g., 656.5 nm and the dose can be determined by means of a calibration curve.

The limits of application are:
— *Dose*: it should be between 5×10^3–4×10^4 Gy (0.5–4.0 Mrd) if a calibration curve is used.

— *Dose rate*: 0.14–2.8 Gy/s $\approx$ 500–10^4 Gy/h (14–280 rd/s $\approx$ 0.05–1.0 Mrd/h) are the limits of dose rate values where the colour change does not depend on the dose rate.

— *Temperature*: the absorbance change is temperature independent between 10 and 32 °C.

— *Energy dependence*: the method can be used for energies up to 4 MeV in the case of both γ- and electron radiations.

The method should be carried out according to the instructions for use of the evaluating spectrophotometer: the colour changes give information about the absorbed doses.

The accuracy of the measurement is ± 2-3% if the evaluation is carried out instantaneously after irradiation. The absorbancy changes rapidly even at room temperature; there are appropriate correction curves permitting one to make recalculations, however, not without non-negligible errors.

6.6.5. DOSE INDICATORS (MONITORS)

As radiation sterilization gains ground on the commercial scale, efforts to produce handy radiation indicators becomes more and more actual. These should indicate the fact of irradiation at a first glance.

Every dose indicator makes use of the discolouring of polymers under the effect of irradiation. They utilize the evolution of hydrochloric acid from PVC owing to the radiation chemical reaction; this is detected by a coloured acid–base indicator admixed to the polymer. Such radiation indicators are sold as tags or adhesive tapes, eventually in printed form.

Dose indicators have a ± 20–30% accuracy in the dose range suitable for sterilization, e.g., most indicators can be used between 2 and 30 kGy (0.2 and 3 Mrd) with the accuracy mentioned, independently of the dose rate. Good dose indicators are insensitive to visible and ultraviolet light, to heat, humidity and chemical reagents and remain unchanged for a long period.

REFERENCES

[6.1] Broda, E. and Schönfeld, T., *Die technischen Anwendungen der Radioaktivität*. Verlag Technik, Porte Verlag, Berlin–München, 1957.

[6.2] Mohler, H., *Chemische Reaktionen ionisierender Strahlen*. Verlag Sauerländer, Aarau–Frankfurt am Main, 1958.

[6.3] Charlesby, A., *Atomic Radiation and Polymers*. Pergamon Press, Oxford–London–New York–Paris, 1960.

[6.4] Swallow, A. J., *Radiation Chemistry of Organic Compounds*. Pergamon Press, Oxford–London–New York–Paris, 1960.

[6.5] Lind, S. C., *Radiation Chemistry of Gases*. Reinhold, New York, 1961.

[6.6] Kirchner, J. F. and Bowmann, R. E., *Effects of Radiation on Material and Components*. Reinhold, New York, 1961.

[6.7] Haïssinsky, M., *La chimie nucléaire et ses applications*. Masson et Cie, Paris, 1957.

[6.8] Bolt, R. O. and Carroll, J. G., *Radiation Effects on Organic Materials*. Academic Press, New York–London, 1963.

[6.9] Vereshchinskij, I. V. and Pikaev, A. K., *(Introduction to Radiation Chemistry)* (In Russian). Izdatelstvo Akademii Nauk SSSR, Moscow, 1963.

[6.10] Charlesby, A., *Radiation Sources*. Pergamon Press, Oxford–London–Edinburgh–New York–Paris–Frankfurt, 1964.

[6.11] *Ionising Radiation and the Sterilization of Medical Products*. Taylor.and Francis, London, 1965.

[6.12] Rexer, E. and Wuckel, L., *Chemische Veränderungen von Stoffen durch energiereiche Strahlen*. Deutscher Verlag Grundstoffindustrie, Leipzig, 1965.

[6.13] Henglein, A., *Einführung in die Strahlenchemie mit praktischen Anleitungen*. Verlag Chemie, Weinheim, 1969.

[6.14] Van de Voorde, M. H., *Effects of Radiation on Materials and Components*. CERN, Geneva, 1970.

[6.15] *(Modern Aspects of Radiation Chemistry)* (In Russian). Izdatelstvo "Nauka", Glavnaja redaktsija fiziko-matematicheskoj literatury, Moscow, 1972.

[6.16] Swallow, A. J., *Radiation Chemistry*. Longman, London 1973.

[6.17] Braginskij, R. P., Finkel, E. E. and Leshchenko, S. S., *(The Stabilization of Radiation-modified Polyolefines)* (In Russian). Izdatelstvo "Khimija", Moscow, 1973.

[6.18] Frumkin, M. L., Kovalskaja, A. P. and Grelfand, S. Ju.: *(Technical Bases of Radiation Processing in the Food Industry)* (In Russian). Izdatelstvo Pishchevaja promyshlennost, Moscow, 1973.

[6.19] Gaughran, E. R. L. and Goudie, A. J. (Eds), *Sterilization by Ionizing Radiation.* Multiscience Publ. Co., Montréal–Québec, 1974.

[6.20] *Requirements for the Irradiation of Food on a Commercial Scale.* Proc. Panel International Atomic Energy Agency. IAEA, Vienna, 1975.

[6.21] Pikaev, A. K., *(Dosimetry in Radiation Chemistry)* (In Russian). Izdatelstvo "Nauka", Moscow, 1975.

[6.22] Spinks, J. W. T. and Woods, R. J., *An Introduction to Radiation Chemistry* (2nd ed.). John Wiley, New York, 1976.

[6.23] *Radiation Chemistry of Major Food Components.* Elias, P. S. and Cohen, A. J. (Eds), Elsevier, Amsterdam, 1977.

[6.24] *Food Preservation by Irradiation.* Vols I and II. Proc. Symp. International Atomic Energy Agency. IAEA, Vienna, 1978.

7. INDUSTRIAL RADIOGRAPHY

The nondestructive testing (NDT) methods which use ionizing radiations (X-ray, γ-source, etc.) to detect defects of nontransparent materials or apparatus or to discover their internal construction belong to industrial radiography. The attribute *industrial* distinguishes them from the medical application which utilizes similar methods. In the following we will restrict the term radiography to its industrial application.

There are *additional NDT methods* such as ultrasonic, magnetic or eddy current NDT method or crack detection using penetrants. The different NDT methods may also be applied in combinations. Radiography is *combined* frequently *with ultrasonic methods* to find defects of welded joints. Frequent checking of the quality of welded joints is performed by *fluorescent liquids,* by *magnetic particles* or by *soap solutions* since radiography is not very suitable for crack detection.

X-ray diffraction methods do not belong to the armoury of NDT methods because they give information on the lattice. Bearing in mind the radioisotopic theme of our book, we shall only deal with radiography carried out with radioactive sources. This *procedure is analogous to radiography using X-ray apparatus or accelerators,* apart from their specific operational properties.

7.1. CLASSIFICATION OF RADIOGRAPHIC METHODS

The methods of radiography are open to arbitrary classification.

7.1.1. CLASSIFICATION ACCORDING TO TYPE OF RADIATION

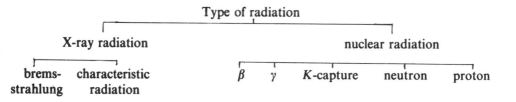

The high energy bremsstrahlungs of accelerators (e.g., betatron, linear accelerators) also belong to *X-ray radiations. Characteristic X-ray radiation* can be generated by X-ray tubes or by electron capture (K-capture, see Section 1.2) isotopes and can be used for X-ray diffraction methods as mentioned above: for X-ray fluorescent analysis (see Section 4.1.4), for continuous testing of coatings (see Section 2.2.3).

Only the high energy electromagnetic radiations (bremsstrahlung, γ-radiation) are generally accepted in industry. These two methods have similar techniques—apart from the radiation sources. At one time, industrial radiography used only X-ray radiation and even today it does so to a great extent, though for testing thick slabs, welded joints and concrete structures accelerators, mainly industrial betatrons are, for example, now being used more and more.

Gamma-radiography *(gamma-defectoscopy)* is based on the application of γ-radiating radio isotopes and it has the following *advantages* over X-ray sources:

— *the radiation source has small dimensions* thereby enabling it to be placed almost anywhere in a complex construction;

— *the radiation is homogeneous in every direction* which enables it to be used for panoramic exposition; a single exposure is sufficient instead of many individual collimated ones;

— *high energy* radiating isotopes enable the testing of greater wall thicknesses than does X-ray radiation;

— *external parameters* have no influence on the radiation of the isotopes (this means that the method has considerable stability);

— *the method needs no special supplies* such as electric energy, cooling water, etc.; this makes it cheaper and simpler to use, particularly in outdoor testing;

— *the cost of radiation sources* is relatively low.

Gamma-radiography also has some *disadvantages* in that γ-sources:

— have a *smaller dose rate* than X-rays, therefore longer exposure time is necessary (which influences economy);

— with *smaller wall thickness* the radiograms are of poorer quality (the image contrast is worse);

— *their geometrical size* may sometimes be larger than that of the focal point of the X-ray tubes, therefore the outer (geometrical) unsharpness of the radiograms is increased;

— *they need to be exchanged* from time to time due to radioactive decay, which operation involves special experience;

— *they radiate continuously*, therefore radiation protection has to be carefully considered;

— *they decay continuously*, therefore efficient time utilization is an important economic factor.

Apart from γ-radiation, other radioactive radiations are also used for radiography— though within a fairly limited field. *β-Radiation* is suitable to penetrate thin sheets of low density materials *(rubber, plastics)* because of its limited penetrability (β–R Beta-radiography).

Samples containing elements with different thermal neutron attenuation (H, B, Cd, Eu, Gd, Li, Sm) are investigated with *neutron sources. Neutron radiography* has found only special applications because the sources are not easily available. This is even more true of *proton radiography*—though this has some other advantages (see Section 7.5).

7.1.2. CLASSIFICATION ACCORDING TO THE ENERGY OF RADIATION SOURCES

In radiography the spectral distribution of the radiation energy of the source is of little interest; the source can either be bremsstrahlung or it can originate from the nuclear energy transitions. The only important factor is the *peak energy* of the radiation (for X-ray)

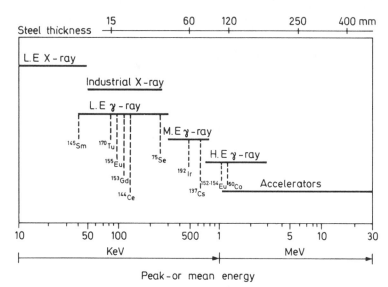

Fig. 7.1. Energy ranges of different radiation sources and the corresponding steel thickness ranges which can be tested economically

or the *prevailing (average)* energy (for γ-radiation), since it is these that characterize the penetrability.

The energy ranges and the optimum thickness of a steel sheet for radiography are shown in Fig. 7.1.

7.1.3. CLASSIFICATION ACCORDING TO THE DETECTION OF DEFECTS

The detection of defects by ionizing radiations can be carried out by different methods:

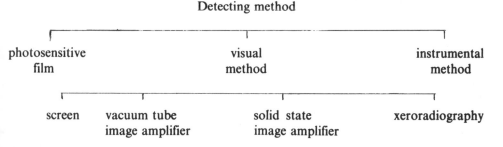

X-ray film (less frequently X-ray paper) is the most commonly used defect detector so we shall first deal with this.

If we put a film against the side of a sample opposite to the source we obtain an intensity distribution which corresponds to the thickness distribution and gives the internal structure of the sample (Fig. 7.2).

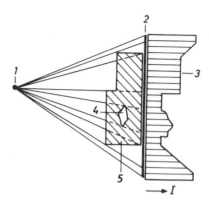

Fig. 7.2. Change in γ-radiation intensity passing through samples which contain inclusions. *1* — Radiation source; *2* — Film; *3* — Distribution of intensity at the film; *4* — Inclusion; *5* — Sample being tested

As a result of interaction between the ionizing radiation and the sample to be examined in the presence of the amplifying screen, a *latent image* will develop in the photographic emulsion which can be made *visible* by photographic processing (developing, fixing, etc.).

The intensity of the primary radiation decreases exponentially with the penetration depth in the material, but at the same time the intensity of the *scattered (Compton-) radiation*, which originates from the interaction between the radiation and the material increases and will be added to the actual value of the primary radiation (Fig. 7.3). As a result of the interaction *secondary electrons* form whose number is proportional to the radiation intensity at a depth larger than the mean range of the electrons. Where the layer thickness is smaller than their mean range the electrons leave the material. This phenomenon can be used to reduce the exposure time: the film is placed between two metal foils (amplifier foils) whereupon the electrons leaving the foils produce a latent image in a shorter time (see Section 7.3.2.2).

Visual defect detection has two advantages: it is rapid and it saves film costs—which comprise a considerable part of the whole cost. The disadvantages are: its application has

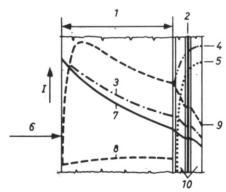

Fig. 7.3. Dose-distribution in the arrangement of γ-radiography. *1* — Steel-thickness; *2* — Film; *3* — Gross-radiation; *4* — Gross secondary electrons; *5* — Secondary electrons rising in lead-screens; *6* — Direction of radiation; *7* — Primary radiation; *8* — Scattered radiation; *9* — Secondary electrons rising in steel and screens; *10* — Lead screens

limitations, defect evaluation is more difficult, it gives no documentation of the results. For visual investigation mostly X-ray sources are used. The most important methods are the following:

The (invisible) X-rays will be transformed to visible radiation *on the screen*. The *luminescent* (generally ZnS) screen shows a magnified picture which must not be viewed directly but through a lead glass sheet or in a mirror. The image on the screen can also be amplified and transmitted to a television set. The defects appear on the screen—opposite the film—as brighter spots.

Vacuum tube image amplification aims at reducing the dose necessary for image formation. Instead of a luminescent screen a particular vacuum tube is applied which has an electron emitter layer on its front surface. The electrons generated by the X-ray radiation will be accelerated and focused on a fluorescent screen by means of an electron-optical lens and the screen can be directly observed. It is a great advantage that the defects can be detected using only a 1/100 dose rate.

With a solid state image amplifier the X-radiation reaches an electron emitter layer. The emitted electrons form an image with the help of a semiconductor layer. The necessary radiation intensity is only 1/3000 of the direct method.

Xeroradiography detects defects by semiconductors. Instead of a film a charged semiconductor layer is applied on which a charge loss corresponding to the dose rate takes place. The dose rate follows the shape of the sample. The image becomes visible by blowing powder particles on to the layer. The powder distribution and density depend on the local charge.

The major disadvantage of *instrumental non-destructive testing* (which uses radiation detectors) is that it presents no visible image of defect distribution and size but gives only numerical values. This limits its applicability. Theoretically it agrees with the method of thickness measuring described in Section 2.4.4. The information presented by these NDT instruments is often processed by *computer*. The radiation intensity data are compared with those of a standard sample and defective specimens will be discarded automatically.

7.2. TECHNIQUES OF GAMMA-RADIOGRAPHY

There are numerous *control methods*, but as far as samples are concerned, there are two main groups according to whether the samples are transportable or not. These two groups need different equipment, radiation protection and techniques. If transportation costs are not too high the transportable samples should be tested in a laboratory where suitable radiation protection can be ensured. Local testing is more common and it requires that great attention be paid to protection.

7.2.1. LABORATORIES FOR RADIOGRAPHY

Gamma-radiography of samples which can be *transported economically*, is carried out in special laboratories under the safest conditions. The basic types of laboratories for gamma-radiography are shown in Figs 7.4 and 7.5.

Among the laboratories with panoramic exposures large *multipurpose* chambers are fairly common (Fig. 7.4(a)): they have dimensions of 10×10 m; they are equipped with a suitable portal crane, with transport way and have a traffic maze. They generally have a

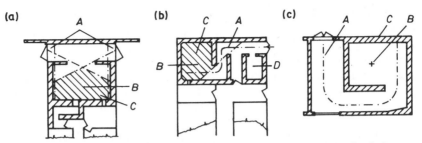

Fig. 7.4. Basic laboratory types for gamma-radiography. *A* — Crane track; *B* — Site of radiation source; *C* — Exposure room; *D* — isotope store; (a) — Laboratory with transporter and maze for staff; (b) — Isotope store and laboratory with entrance for staff and for transport; (c) — Simple laboratory

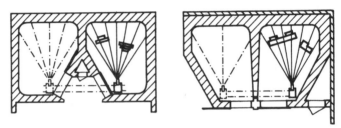

Fig. 7.5. Twin laboratories with collimated radiation source

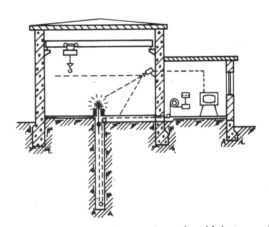

Fig. 7.6. Laboratory for gamma-radiography with isotope well

so-called *isotope well* (tile hole) type radiation source (Fig. 7.6). The isotope well is an approximately 10 m deep tube with closed bottom which contains the radiation source and the necessary electromechanical parts for moving the source. The correct localization of the source for irradiation is carried out by means of an expansion tongs and for its observation either a periscope or a television system can be used. All doors are locked and should there be unauthorized opening the radiation source is immediately directed back into the well, in order to avoid any accidents.

Figures 7.4(b) and 7.4(c) show smaller, more simple laboratories but still with panoramic exposures (without service rooms and photographic laboratory; Fig. 7.7).

Laboratories for gamma-radiography generally have a *maze* to diminish the γ-radiation; this maze ensures that the intensity of the scattered radiation decreases to a permissible level. *Protecting doors* are large and heavy, thus they are used *only if no maze can be built.*

Among the laboratories with oriented radiation beams, twin laboratories are very convenient for controlling small items, first of all castings (Fig. 7.5). The two laboratories have a common, portable gamma-radiographic unit (isotope container for radiography) which irradiates with a collimated beam in one of them while the samples and films are prepared in the other. In such cases only a part of the radiation solid angle, about 50–90° can be used.

Laboratory isotope containers may be of different construction. Some of the most common types are shown in Figs 7.8–7.12.

Isotope containers with a vertical frame (Fig. 7.8) reduce the time-loss. Samples and films are placed on the platform of a small trolley which is then pushed under the gamma-radiographic unit which has a radiation beam directed downward. While exposure is taking place the next trolley is prepared. The radiation beam is enabled or prevented with the help of a remotely controlled tong.

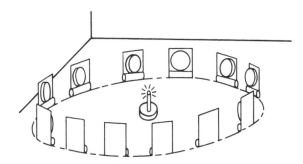

Fig. 7.7. Panoramic exposure for simultaneous radiography of several samples

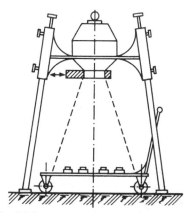

Fig. 7.8. Isotope container with vertical frame

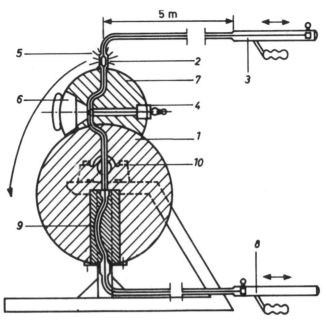

Fig. 7.9. Combined laboratory γ-radiographic unit, *1* — Storage container; *2* — Radiation source; *3* — Upper moving tool and bowden; *4* — Fork lock; *5* — Panoramic final position; *6* — Collimator stopper; *7* — Upper collimator head; *8* — Lower moving tool and bowden; *9* — Inner maze channel; *10* — Stopper

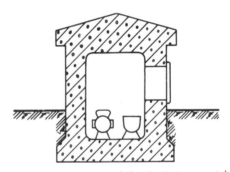

Fig. 7.10. Concrete storage shelter for isotope container

The combined laboratory γ-radiographic unit (Fig. 7.9) has two heads. The lower one *(1)* is a storage container for the radiation source *(2)* when it is not in use. For exposure the source can be lifted up to the arrester *(4)* by the upper trigger unit and cable *(3)* or to the final position *(5)*. If the arrester is locked the source is held in the centre of the funnel shaped collimator *(6)* and it irradiates an oriented beam if the funnel plug is taken off. If the plug is replaced by one with a smaller hole a very fine beam can be collimated and used, for example, for instrumental NDT. Opening the arrester the source can rise above the top of the upper head *(7)* and can be used for panoramic exposures in about 3π st (steradian, 270°) enabling several samples to be irradiated simultaneously. After exposure the source is returned to the container with the help of the lower trigger unit *(8)*. In order to prevent

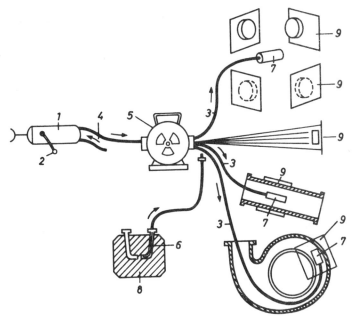

Fig. 7.11. Telemechanical γ-radiographic unit and its application. *1* — Control mechanism; *2* — Crank; *3* — Tube for carrying the radiation source; *4* — Cable for moving the source; *5* — Isotope container; *6* — Changeable source; *7* — Exposure head; *8* — Multichannel isotope container

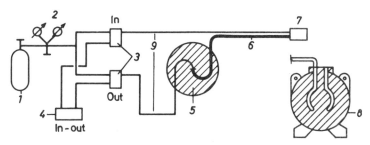

Fig. 7.12. Pneumatically remote controlled gamma-radiographic unit. *1* — Steel flask for compressed air with a pressure of 120 bar; *2* — 250/160 press reducer; *3* — Electromagnetic air ventiles, *4* — Flow indicator; *5* — Isotope container with tungsten shield and with maze exits, *6* — Flexible tube for leading out the radiation source; *7* — Exposure terminal for the source; *8* — Multichannel isotope container

radiation escape the inner channel *(9)* is convoluted. The radiation source is changed by taking off the plug *(10)* which contains the source holder, too. The combined gamma-radiographic unit can be turned around a horizontal axis and can be fixed in any position. The equipment can contain 2×10^{11} Bq (5 Ci) ^{60}Co isotope. Its mass is 250 kg and it is easily portable on a two wheeled trolley.

It is advantageous to have *industrial control laboratories* joined to the technological line. This ensures homogeneous product quality and decreases transportation costs.

Production quality control can be performed *inside the workshop* but care has to be taken of radiation protection. However as the area of the workshop is of a high value and the need to keep sufficient distance between the radiation source and the workers, as a usual protection method, cannot be applied, it is thus, a more economical solution to separate a part of the building with a wall *built of concrete blocks*. This has the advantage that it can easily be demolished or rebuilt in another form if necessary. For one-storey buildings no protecting roof is needed and even big samples can be moved by a crane from above. Personnel movement takes place through a small maze. For unusually-shaped samples, *special structures* like tunnels, shelters, pits should be built. The thickness of the protecting wall or the size of the separated area can be reduced considerably if a shield reduces a part of the radiation solid angle with suitably formed (conic or panoramic) collimators (Fig. 7.13).

Dimensioning of wall thickness for radiation protection of laboratories and containers is given in Section 8.1.4.2.

7.2.2. GAMMA-RADIOGRAPHIC FIELD-TESTS

Radiographic field-tests are essential for the safety of fixed steel constructions like pipelines, vessels, steel frames or buildings. Local radiographic tests require special protection measures. For this reason the isotope containers and the radiographic technique differ from those of laboratories in the ways given below.

Field tests need easily transportable containers which can be controlled from a large distance. The radiation sources can be moved from the *mobile containers* to the required position by mechanic, pneumatic, hydraulic or electrical equipment. Present-day containers (γ--radiographic units) not only have a *lead protection shield* but also a *tungsten or uranium* one which results in a considerable weight reduction and in easier transportation. (Uranium with reduced ^{235}U content is subject to the control of the Nuclear Non-Proliferation Treaty (NPT); it has therefore to be registered in countries which have signed the treaty.)

Multichannel storage containers are able to contain 4–5 radiation sources simultaneously: the required source can be directed to a less protected, remotely controlled *single channel working container* under completely safe conditions. The radiation protection of the storage container has to ensure a surface dose rate less than 20 µSv/h (2 mR/h) as opposed to working containers, for which 100 µSv/h (10 mR/h) is allowed. In a suitably protected area the source can be stored in the working container too. The shelter for the portable single channel working container (Fig. 7.10) must be closed with a safety lock. It is advisable that an area of 5×5 m around it be protected with a 2 m high fence.

An underground waterproofed conrete pit closed with a safety lock can be used as a simple (temporary) storage.

Remote controlled gamma-*radiographic units* are used mainly for local radiographic testing. There are several constructions for this purpose and two of them are described below. The usual *telemechanical* gamma-radiographic unit (Fig. 7.11) has two flexible tubes, a larger diameter one *(3)* and a smaller diameter one *(4)* containing a flexible cable to move the source, and join to the working container *(5)* covered with a uranium or a tungsten shield. At one end of the larger diameter tube there is a small case *(7)* with or without collimators. The smaller diameter tube has, at one end, an apparatus (1) with a crank-arm (2). From the apparatus a steel cable (4) can be wound out, the other end of which is fixed to the source (6). When making a radiogram, the container is transported close to the position where it will be used.

Having removed the protecting shields of the pipe junctions the flexible pipes are fixed onto them. The exposure head on the end of the larger diameter tube, either the panoramic one or that with the collimator will be located to where it will be most effective in terms of radiography. The transport container will be taken as far as allowed by the tube. The remote control mechanism *(1)* should stay in the direction of the axis of the larger diameter tube. The source is moved by the control mechanism into the exposure head *(7)* and after the exposure back into the working container *(5)*. At the end of a particular task the source *(6)* has to be replaced by remote control into the storage container *(8)*.

The pneumatically remotely controlled gamma-radiographic units (Fig. 7.12) operate by means of compressed air originating from a steel flask *(1)*. The pressure is controlled by a reducer *(2)*. The path for the compressed air is determined by magnetic valves *(3)* which are controlled electrically from a distance of 3–5 m. The air current can be controlled by means of the flow indicator *(4)*. The spherical source having a diameter of 3.5 mm can be moved by the compressed air and the valves, either from the tungsten shielded working container *(5)* through a flexible tube *(6)* into the exposure position *(7)* or into the opposite direction. Signals show the actual position of the source.

The length of the outgoing tube can change from 6 to 30 m, the only difference being caused is in the air consumption. The tube must be resistant to twisting, bending and external pressure. After the termination of the preselected exposure time the source is directed back to the working container *(5)* by a timer. At the end of the day's shift the exposure head *(7)* is removed, the tube *(6)* is joined to the multichannel storage container *(8)* and the source is moved into it. An about 1T Bq (30 Ci) ^{192}Ir radiation source is used. The exposure head *(7)* can be partially shielded if necessary. The mass of the whole apparatus is about 32 kg. It works either by, e.g., a silver–zinc battery or on mains current.

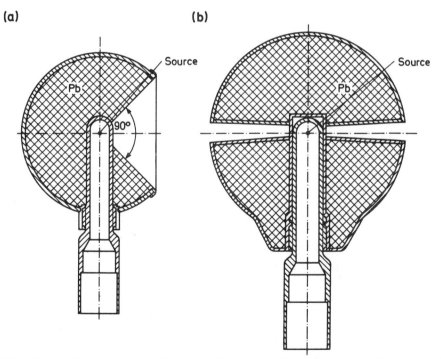

Fig. 7.13. Collimators for remotely controlled gamma-defectoscopes. (a) — For oriented radiation; (b) — For annular arrangement

Panoramic exposures can be prepared without collimators and an oriented radiation beam can be formed by use of collimators (Fig. 7.13 (a), (b)).

For the testing of *welded joints* of self-propelled gamma-radiographic units, so-called *radioisotope pipeline crawlers* are the most economical ones. The source is in the centre of the pipe thus the radiation passes through the wall of the tube only once. This reduces the exposure time 20 to 30-fold and gives better image quality. Without using pipeline crawler the introduction of the source into the tube is a complicated process. To simplify the examination of annular welded joints self propelling pipeline crawlers have been developed with or without wire control.

The remotely controlled radioisotope pipeline crawlers (Fig. 7.14) generally consist of 4 main parts: the radiating head *(1)*, the vehicle *(2)* with drive and control, the energy supply *(3)* and the control apparatus which operates from outside the pipe *(4)*. As a result

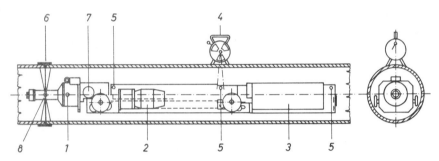

Fig. 7.14. Remotely controlled automotive gamma-radiographic unit (pipeline crawler). *1* — Radiating head; *2* — Vehicle with drive; *3* — Batteries for energy supply; *4* — Isotopic control apparatus; *5* — GM-tubes for the γ-relays; *6* — Film; *7* — Apparatus for moving the radiation source

of an external signal the pipeline crawler advances inside the pipe from one welded joint to the next where it waits for the signal of the gamma-*relay* (Section 2.2.1.1) with an accuracy of ± 4–5% as a function of the pipe diameter. A film *(6)* is placed on the welded joint outside and on a next signal a retarded (10–20 s) driver *(7)* releases the ^{192}Ir radiation source for an exposure time of between 3 and 999 s. After exposure the source will pushed back to be container whereupon the crawler moves automatically to the next joint or, at the end of the day's shift (if need be moving backwards) it leaves the pipeline, running 1–2 km distance. The signals are relayed by a ^{137}Cs radiation source of about 200 MBq (6 mCi) to the instrument. The instruments are manufactured in two sizes: e.g., for 200–400 mm and for 300–800 mm pipe diameter. There is a possibility for broadening the testing range up to 1500 mm dia by applying an additional instrument. The respective masses of the instruments are 40 and 70 kg. The radiation source is up to 4 TBq (100 Ci) ^{192}Ir isotope. The pipeline crawler moves at a speed of 8–10 m/min. The maximum gradient allowed is 55–65%. If the maximum gradient exceeded this, the four driven wheels may skid even on a dry inner surface of pipeline. The mean range of the machine is 1.5 to 4 km per battery charge.

Collimators (Fig. 7.13) limit the solid angle of the radiation to a useful beam around the exposure head of the remotely controlled gamma-radiographic unit. Collimators are prepared either for forming oriented radiation beam (Fig. 7.13 (a)) or annular-shaped radiation beam (Fig. 7.13 (b)). Their radiation protection is generally not enough to enable one to approach them (as in the case of the working containers), but the radius of the evacuated zone can be reduced significantly (see Section 8.4.4.2). This is particularly important for exposures in places where there are likely to be many people (residential areas, factories, etc.). By using a collimator, the work and traffic around the radiographic workplace is less limited.

7.2.3. PREPARATION FOR RADIOGRAPHY

As far as the *quality of a radiogram* is concerned, it is not the external conditions (outdoor or indoor testing, kind of the applied isotope container) but rather the previous preparation for radiography that is important, though the external conditions are of course important from the viewpoint of radiation protection of persons.

7.2.3.1. SELECTION OF RADIATION SOURCE

The selected radioactive isotope should have the best radiation parameters (first of all the energy) from the viewpoint of the *material quality* and the *wall thickness* (Fig. 7.1). If many different radiation sources are available for the given thickness range, the advantages and disadvantages described in Section 7.1.1. have to be considered. The detailed data of γ-radiation sources are given in Section 1.4.2.3.

Thicker walls have to be irradiated by a radiation source of *higher energy* in order to achieve a measurable dose rate on the other side. Higher energies have higher dose constants (Section 1.3) therefore a more economic exposure time also means the application of higher energy, as opposed to *the quality* needs of a radiogram due to contrast.

An isotope can be considered to be *optimum* if the thickness of the investigated sample is *between two and fourfold* of its attenuation half value thickness for the given material. In the case of steel: for the range of 16–65 mm thickness a ^{192}Ir, for 20–90 mm a ^{137}Cs, for 40–150 mm a ^{60}Co source can be recommended. For samples thinner than 16 mm either X-ray radiation or soft γ-radiation (^{170}Tm, ^{144}Ce, ^{155}Eu, ^{169}Yb, etc.) should be used. A wall thicker than 150 mm can be examined mostly with high energy accelerators (betatrons, linear accelerators). If the wall-thickness is less than the optimum, the quality of the radiogram worsens; if it is more, the exposure time is uneconomically long.

7.2.3.2. PLANNING THE ARRANGEMENT
FOR RADIOGRAPHY

The direction of the beam and the distance between the source, the sample and the film also determine the quality of the radiogram. In the stage of selecting (see Sections 7.2.1 and 7.2.2) the number of exposures and the positions have to be determined, as an optimum.

One should get maximum information from the lowest possible number of radiograms. For a sample with different wall thicknesses the exposure time has to be calculated *either for the dominant or for the critical thickness*, because on the other parts of the sample a limited detection of the defects might also be satisfactory. If each part is essential, a number of radiograms have to be prepared with different exposure times.

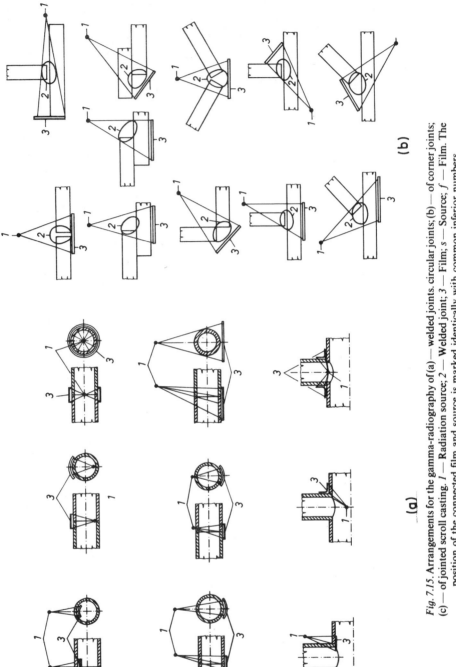

Fig. 7.15. Arrangements for the gamma-radiography of (a) — welded joints. circular joints; (b) — of corner joints; (c) — of jointed scroll casting. 1 — Radiation source; 2 — Welded joint; 3 — Film; s — Source; f — Film. The position of the connected film and source is marked identically with common inferior numbers

Fig. 7.15. (cont.)

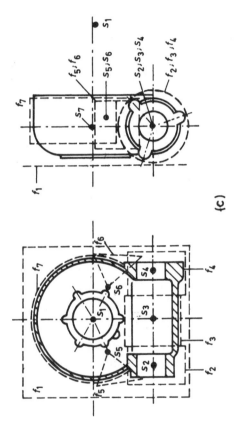

Fig. 7.15. Arrangements for the gamma-radiography of (a) — welded joints, circular joints; (b) — of corner joints (c) — of jointed scroll casting. *1* — Radiation source; *2* — Welded joint; *3* — Film; *s* — Source; *f* — Film, The position of the connected film and source is marked identically with cimmon inferior numbers

The radioactive radiation sources radiate in all direction. This can be utilized covering as great part of the irradiated area with samples and films as possible. For this purpose in a hollow sample the source is placed inside or in the case of more samples to be tested a panoramic arrangement has to be made. In this second case samples of different wall thicknesses can be tested by a single exposure because the attenuation differences can be equalized by selecting various distances.

In order to obtain the best possible detection of a defect the correct *direction for radiography* is essential. Defects on the film-side of the sample can be recognized more easily. With the thicker walls of *hollow castings* the external side of the casting cools down faster thereby causing e.g., shrinkage cavity defects mostly at the inner third of the wall-thickness: therefore in critical cases instead of the more economical inside exposure it is more reasonable to place the source outside the casting. Cracks and other nearly *two dimensional defects* can be detected if their plane deviates not more than about 10° from the direction of radiation.

In the case of *double wall samples* (tubes, closed vessels) the exposures through both walls might be avoided. The defects of the wall at the source-side cannot be seen on the radiogram and the quality of the latter is poorer because of the scattered radiation. It sometimes helps to bore a hole and to pass a small source (e.g., 0.5 mm dia) through it into the sample, and subsequently to seal the hole.

Figures 7.15 (a), (b), (c) show some special arrangements for radiography.

7.2.3.3. DETERMINATION OF EXPOSURE TIME

Exposure time depends on the following: the kind of radiation source, its total and specific activity (its sizes), the quality and thickness of the sample, the type of film and intensifying screen, the absorbancy (optical density, blackening), the distance between the source and the film, the processing techniques, etc. These effects can be taken into consideration partly on the basis of mathematical relations and partly on the basis of empirical data.

The use of *exposure nomograms* is the simplest way to determine the exposure time. The nomograms fix some of the parameters and generally give the exposure-time, i.e., the product of activity by time in GBq.h (Ci.h) vs. the wall-thickness. The trajectories of the nomograms show the film-densities at definite source-film distances or inversely. If commercially produced nomograms are not available they can be prepared by making radiogram with standard steel sheets of different thickness.

The exposure time can also be calculated using the dose values in the equation of the radiation absorption (2.13):

$$D_f = I_0 t \exp(-\mu x) \tag{7.1}$$

where D_f is the dose needed for the required film-density in units of µGy in air (where 1R $= 8690$ µGy); I_0 is the dose-rate without any absorber sheet at the site of the film, in µGy/h in air (R/h); μ is the linear absorption coefficient, m^{-1}; x is the thickness of the sample, m; t is the exposure time, s;

In another form

$$I_0 = \frac{AK_\gamma}{l^2} \tag{7.2}$$

where A is the activity of the source, Bq (Ci); K_γ is the dose constant $\mu Gy \cdot m^2/(GBq \cdot h)$ $[R \cdot m^2/(Ci \cdot h)]$; l is the distance between the source and the film, m.

It is known that

$$\exp(\mu x) = 2^{x/x_{1\,2}} \qquad (7.3)$$

where $x_{1/2}$ is the half value thickness, m.

The exposure time is

$$t = \frac{D_f \times l^2 \times 2^{x/x_{1\,2}}}{A \times K_\gamma} \qquad (7.4)$$

The rough values of D_f for some films and isotopes are shown in Fig. 7.16, but these can easily be determined experimentally for other films or conditions too.

On the basis of such calculations a *disc* has been constructed to determine the exposure time (Fig. 7.17), for steel samples and for a ^{192}Ir source. Using the logarithmic circular nomogram, the inner disc *(1)* is moved till the mark *(5)* reaches the required film dose value. In this position it is fixed by a screw *(2)*.

The required source–film distance value on the middle (mobile) disc *(3)* is fitted by means of a pin *(6)* to the corresponding source activity value on the inner disc *(1)*. In this position the exposure time for any steel wall thickness (see the outer scale of the middle disc) can be read on the time-scale of the outer disc *(4)*.

For *materials other than steel* (copper, aluminium, etc.) the exposure time has to be determined experimentally, because it depends on the composition and density of the alloys. Table 7.1 gives some attenuation half value thickness which can be used to calculate exposure times.

EXAMPLE 7.1

Calculate the exposure time for the following conditions: steel wall thickness: $x = 100$ mm; radiation source: ^{60}Co [300 GBq (8 Ci)]; film type: Agfa-Gevaert D7; distance: $l = 0.7$ m; intensifying screens: lead of 0.15 mm before the film and of 0.45 mm behind the film; required film-density: $S = 2$.

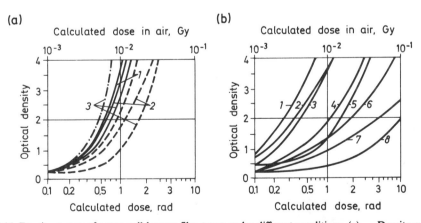

Fig. 7.16. Density curves of some well-known film types under different conditions. (a) — Density curve of Gevaert D7 film exposed behind 40 mm thick steel sheet. Developer G 209 A; temp. 20 °C; time 5 min. *1* — ^{60}Co; *2* — ^{137}Cs; *3* — ^{192}Ir.———— 0.15/0.45 Pb; — — — Without foil; -.-.– M 200 foil. (b) — Density curves of different film types for ^{192}Ir isotope and 40 mm steel sheet. *1* — Agfa-Gevaert D10; *2* — Ilford G; *3* — Ferrania ID; *4* — Kodak Industrex D; *5* — Agfa-Gevaert D7; *6* — ORWO Texo; *7* — Ilford CX; *8* — Agfa-Gevaert D4

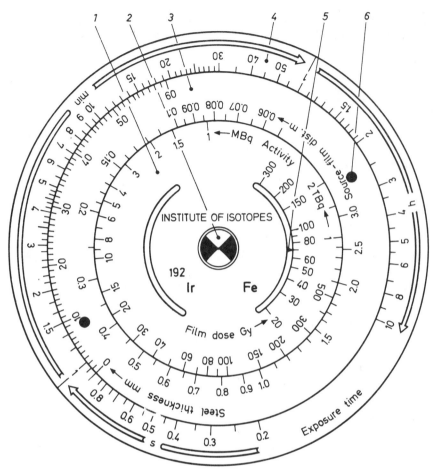

Fig. 7.17. Disc for determination of exposure time. *1* — Inner fixed disc; *2* — Fixing screw; *3* — Mobile disc; *4* — Outer fixed disc; *5* — Mark for film factor; *6* — Pin

Table 7.1. Approximate half value thicknesses of attenuation

Metal	Radiation source		
	^{60}Co	^{137}Cs	^{192}Ir
	thickness, mm		
Steel, cast iron	20	15	14
Non-ferrous metals (copper, bronze)	18	13	12
Light metals (Al) alloys)	58	43	40

The following data should be used: $D_f = 3.4$ mGy (0.39 R) (from Fig. 7.16 (a)); $x_{1/2} = 20$ mm (from Table 7.1); $K_\gamma = 305$ µGy · m²/(GBq · h) [1.3 R · m²/(Ci · h)].

On the basis of Eq. (7.4) using SI units:

$$t = \frac{3400 \times 0.7^2 \times 2^{100/20}}{300 \times 305} = 0.58 \text{ h} = 35 \text{ min}$$

in conventional (non-SI) system:

$$t = \frac{0.39 \times 0.7^2 \times 2^{100/20}}{8 \times 1.3} = 0.59 \, \text{h} = 35 \, \text{min}$$

EXAMPLE 7.2

At what distance should the $x = 42$ mm thick castings be placed around an $A = 2.10^{11}$ Bq (5.4 Ci) ^{192}Ir source to obtain a film-density $S = 2$ on a Ferrania ID film after a $t = 16$ h exposure time?

Data: $D_f = 3.9$ mGy (0.45 R from Fig. 7.16 (b)); $x_{1/2} = 14$ mm (from Table 7.1); $K_y = 117 \, \mu\text{Gy} \cdot \text{m}^2/(\text{GBq} \cdot \text{h})$ [0.5 R.m^2/(Ci.h)].

Expressing the distance from Eq. (7.4) in SI-units:

$$l = \left[\frac{A \times K_y \times t}{D_f \times 2^{x/x_{1\,2}}} \right]^{1/2} = \left[\frac{200 \times 117 \times 16}{3900 \times 2^{42/14}} \right]^{1/2} = 3.46 \, \text{m}$$

in the conventional (non-SI) system:

$$l = \left[\frac{5.4 \times 0.5 \times 16}{0.45 \times 2^{42/14}} \right]^{1/2} = 3.46 \, \text{m} \, .$$

7.2.3.4. COMPLEMENTARY PROCESSES OF RADIOGRAPHY

During the *preparation of the exposure*, first of all the sample has to be *fixed* so that it is *vibration-free and cannot be displaced*. After that an image quality indicator, e.g., a series of steel wires (I.Q.I.; see Section 7.3.1), corresponding to the quality and thickness of the sample, are placed at a convenient site (they will help later to qualify the radiogram). To determine the correct location of the defect, *certain marks or a series of lead figures* are used. Lead figures are also used to mark the location and the serial number of the film whose shadow remains on the film. If possible the sample should also be marked by means of marking tool. The position of the source has to be determined correctly and finally the waterproof film—marked with a serial number—is placed on the corresponding side of the sample. The films can be fixed with magnets, with rubber bands, with adhesive tape or may only need to be supported.

For outdoor work the films must be protected from sunshine or from high temperature: therefore, if temperature isolation is not possible, it is reasonable to work at night in order to avoid the melting of the emulsion. *The danger zone has to be marked* with warning signs and closed off with great care; only after that may the source be directed into the exposure position.

During exposure the operators must remain at a reasonable distance or behind a shielding. For this period nobody is allowed to pass the safety zone (see Section 8.4.4.2). The size of the safety zone depends on the kind and activity of the source and can be calculated on the basis of Fig. 7.18. The dose rate should decrease to 20 μSv/h (2 mR/h) on the border of the zone. The relatively long distances can significantly be reduced by applying collimators or by taking advantage of the shielding effect of buildings, other constructions or even of the sample itself. In such cases the border of the zone must be determined instrumentally.

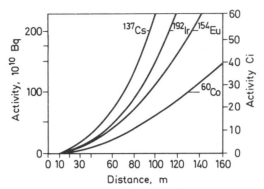

Fig. 7.18. Protection distances corresponding to the 20 µSv/h (2 mR/h) dose rate, for different source activity
values

For radioisotope pipeline crawlers no constant safety zone can be determined. In such
cases audible and visual signals remind people of the radiation danger. In cases of on-site
testing this generally causes no problem.

After the exposure the source is directed back into the container and the films will be
processed as described in the following section.

7.2.3.5. PROCESSING AND HANDLING OF FILM

The detector material is a cellulose acetate based *double coated X-ray film* sensitized
for ionizing radiations. The silver content of its emulsion is about 40% higher than in usual
photographic materials.

In the emulsion of the *exposed film* the silver bromide decomposes as a result of the
ionizing radiation, a latent image is formed which can be developed and fixed by *usual
photographic processing* techniques. Great care and purity are needed because of the
danger of artifacts which may result in wrong decisions. The usual processing technique
will not be described but attention needs to be paid to some aspects of the handling, viz. *the
age of the developer solution* considerably influences the quality of the radiogram. The
storage possibilities under different conditions are as follows:

— in a full, closed bottle for 6 months;
— in a half filled, closed bottle for 3 months;
— in a developer tank for 1 month;
— in a developer tank over 24°C for 2 weeks;
— in open air for 1 day.

These figures should be treated as approximations.

Temperature is an extremely important factor and is closely linked with the developing
time. This *time* is given by the producer: it is generally 5–6 min at 20°C. Over 25°C only
tropical developers should be used. During developing, or at least in the first 30 s the film
should be moved to remove gas bubbles and to obtain a homogeneous chemical effect.

Before fixing the developed film, it is rinsed for 20 s then allowed to fall in drops for a
few s. The *overexposed* film also has to be developed for the prescribed time but later it will
be reduced with a special solution. With *underexposed* film the developing time can be
doubled without any damage to the image quality.

During developing the solution will be partly exhausted, and partly removed from the vessel on the film (about $1/3$ l/m^2 film), on the developer frame, etc. This loss should be made up with the same volume of so-called *replenisher* solution. One litre of the developer solution is generally enough for one m^2 of film but 10 m^2 can be developed in the same quality if replenisher is used.

The film should be moved for 30 s in the *fixing bath* which contains twice as much as needed for the disappearance of the yellowish colour of the emulsion. But this time depends very much on temperature, on the grade of depletion and on the composition of the fixer solution (acidic, hard, or normal). The fixing time is generally 6–15 min. The temperature does not play such an important role as in the case of developing, but a similar temperature is desirable in order to avoid the "marbling" effect of the film.

The film is *washed* in running water: 40 min at 5–12°C; 30 min at 13–25°C, or 20 min at 26–30°C. After washing the water *drips off* for 5 min, but this time can be reduced to 2 min, on applying hydrophobic solutions.

The film should be *dried* in a clean air-flow not warmer than 35°C. Scratches and breakage of the film should be avoided.

7.3. RADIOGRAPHIC SENSITIVITY OF IMAGES

The measure of the detectability of the defects is called radiographic sensitivity. The reliability and the economy of a construction increases with increasing sensitivity, i.e., with the decrease in the size of the detectable defects (e.g., cracks).

Sensitivity depends partly on the image projected by the radiation, and partly on the quality of the film and on its handling. When judging the image quality one should ignore the presence of the film (like a photographer, when he controls the sharpness of the image). The factors which determine the radiographic sensitivity, are summarized in Table 7.2; the details are given in Sections 7.3.2 and 7.3.3.

7.3.1. MEASURE OF THE RADIOGRAPHIC SENSITIVITY

The quality of radiograms, made in order to detect internal defects, depends on the radiographic sensitivity. In other cases, when, for example, the *structure of constructions* or the *location of the iron framework* inside concrete pieces needs to be discovered, the

Table 7.2. Factors determining radiographic sensitivity

Image quality		Film quality	
Contrast	Contour sharpness	Contrast	Contour sharpness
Differences in thickness and/or the ratio of the dose reduction	Geometrical conditions (outer sharpness and magnification)	Type and age of the film	Thickness of the emulsion layer
Energy of radiation	Application of intensifying screens	Type and age of the developer	Turbidity of the emulsion
Intensity of the scattered radiation	Secondary radiation inside the material	Conditions and measure of the development	Size and distribution of the silver halide grains

sensitivity is negligible compared to the accurate, proportional projection of the details. The radiographic sensitivity may have different interpretations but it is expressed in all cases as the visibility of a special image-quality indicator on the film.

The indicators are either *series of wires* with increasing diameter or *stepholes or plaques with increasing thickness.* The sensitivity is given as the serial number (the thickness or the diameter, etc.) of the smallest (thinnest) element of the indicator recognizable on the radiogram. It can be expressed also as a percentage value related to the material thickness. *Image quality indicators (I.Q.I.)* are used either according to the recommendations of the International Standards Organization *(ISO) standard* or in other forms and may be different in various countries.

Radiographic sensitivity is *not identical with flaw sensitivity,* which relates to undetermined defect sizes and therefore it cannot be defined accurately.

Before making a radiogram the indicators have to be *fixed* to the source-side of the sample. If this is impossible (e.g., in the case of a pipe). the film-side can also be used, but this fact should be noted in the record. In the case of international deals it may happen that not only the standardized indicator of the producer country but that of the buyer country also has to be visible on the radiogram.

The wire type indicator consists of wires *(needles)* whose diameter changes according to the completed number series R 10 (the series coefficient is $10^{0.1} = 1.2589$). In the various countries a smaller or a wider range is taken from the wire diameters (0.032, 0.040, 0.050, 0.063, 0.080, 0.100, 0.125, 0.160, 0.200, 0.250, 0.320, 0.400, 0.500, 0.630, 0.80, 1.00, 1.25, 1.60, 2.00, 2.50, 3.20, 4.00 mm); moreover the different diameters may get a serial number (code number) too. Not only is it possible that the range of wire diameters and the serial numbering may change in different countries but it is also possible that their sequence direction may change too. According to British BS 3971-1966, number 1 is a wire of 0.032 mm diameter and it increases up to number 21 and 3.20 mm. The USSR GOST 7512-75 has no serial number and the range is from 0.08 up to 4.0 mm. According to the recommendations of the DIN 54109/1,2-1964 FRG, TGL 10646/4 (GDR), MSZ 15963 (Hungary), IIS/IIW 340-69 (International Institute of Welding), serial number 1 belongs to a diameter of 3.2 mm and the diameters decrease with increasing numbers. Number 16 relates to 0.100 mm. The tolerance zone of the wire diameters according to ISO/R 1027-1969 (E) are given in Table 7.3.

The wire type indicator (Fig. 7.19) has to be *embeaaea* in flexible materials (such as PVC), whose radiation absorption is insignificant, and which are oil-, petrol- and water-resistant. The *distances between the axes of the wires* must not be smaller than three times the diameter but at least 5 mm. The *length of the wires* is 25 mm and they are made of unalloyed iron (Fe) for testing iron and steel. For testing aluminium and its alloys pure Al wires are used. For copper, zinc and their alloys pure Cu is applied. The following have to be embedded into the elastic material also using lead number and letters: the number of

Table 7.3. Tolerance zone of the wire diameters

Wire diameter d, mm	Tolerance, mm
$0 < d \leq 0.125$	± 0.005
$0.125 < d \leq 0.5$	± 0.01
$0.5 < d \leq 1.6$	± 0.02
$1.6 < d \leq 4.0$	± 0.03

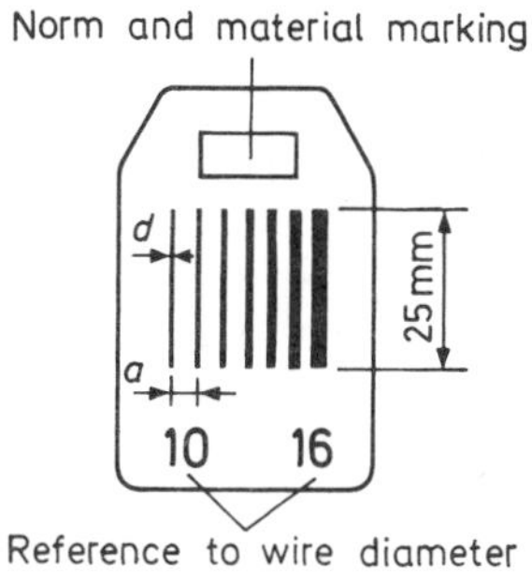

Fig. 7.19. Wire series type indicator (MSz 15963-67)

the particular standard, the kind of material tested (for example, Fe) and the serial number of the thinnest and thickest wire according to the national standard.

The measure of image quality used to be given earlier (and in certain countries even today) as the percentage of the diameter of the thinnest wire visible on the film related to the thickness of the sample. The interpretation of this non-standard measure was sometimes disputable.

At present, the visibility of a defined, thinnest wire (and naturally all the thicker ones) is demanded for a certain sample thickness range. The required image quality can vary depending on the kind of radiation source (X-ray, ^{192}Ir, ^{60}Co) and on the sample thickness according to the standards ISO 2504 and BS 3971-1966. Other standards (DIN 54109)1, 2, TGL 10646/4, MSZ 15963, etc.) do not differentiate according to source, but have descriptions of radiogram quality classes I and II. The expert has to determine what kind of source should be used in order to fulfil the quality requirements.

According to GOST 7512-75 for wire type indicators, the sensitivity K is equal to the diameter of the thinnest recognizable wire expressed in mm. The sensitivity can be determined also as a percentage of thickness. In the USSR the obligatory sensitivity of a testing is half of the allowed maximal defect size for pores and inclusions, expressed in mm.

Because of the different practice of different countries the *test* record must contain the standard to which the wire series (or other indicators) correspond, and also the serial number of the recognizable thinnest wire. According to the recommendations of the ISO, if a wire of 1.25 mm diameter (serial number 11) is recognizable: I.Q.I. ISO/R 1027, 11 wires, 1.25 mm.

Stephole type indicators, corresponding to ISO R 1027-1969 E recommendations, are used in different forms and size mainly in France (AFNOR-04-304, IIS/IIW 340-69) (Fig. 7.20). These standards are bodies (prisms or hexagonal bodies having triangular or quadratic steps) consisting of stepwise bored metal pieces whose thickness steps increase according to the geometrical progression of R-10. The diameters of the boreholes are the same as the thicknesses of the steps. The thickness of the steps increases according to the following series in mm: 0.125, 0.16, 0.20, 0.25, 0.32, 0.40, 0.50, 0.63, 0.83, 1.0, 1.25, 1.6, 2.0, 2.5, 3.2, 4.0, 5.0, 6.3. The steps (and holes) have the serial numbers 1–18 in the British Standard BS 3971-1966. The steps of the standard bodies of different shape and size overlap according to the French AFNOR standard.

The two thickest steps of a standard are as thick as the thinnest steps of the neighbouring one. The following code numbers involve the step ranges as follows:

Code number	Step range, mm
H 3	0.125–0.40
H A	0.32 –1.0
H B	0.8 –2.5
C	2.0 –6.3

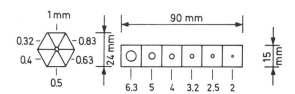

Fig. 7.20. Stephole type indicators (AFNOR)

For the steps 0.63 mm or less two holes are bored and the visibility of both of them is required. The upper four indicator ranges cover the testing of samples 4–300 mm thick.

The tolerance zones of the step thicknesses and of the borehole diameters (d) are given in Table 7.4 according to ISO/R 1027-1966 (E).

For judging the image quality the number of visible boreholes and the diameter of the thinnest visible borehole has to be determined. These two data give the percentual sensitivity

$$E_e = (d_{min}/l) \times 100\% \tag{7.5}$$

and the detectability of the borehole

$$N = a - b \tag{7.6}$$

Table 7.4. Tolerance zones of the steps and holes

Step thickness or borehole diameter d, mm	Tolerance, mm
$0 < d \leq 0.5$	± 0.015 0
$0.5 < d \leq 1.0$	± 0.020 0
$1.0 < d \leq 2.5$	± 0.025 0
$2.5 < d \leq 5$	± 0.030 0
$5 < d \leq 10$	± 0.036 0

where d_{min} is the diameter of the thinnest visible borehole, mm; l the exposed wall-thickness, mm; a the number of visible boreholes; b the number of boreholes with the same or larger diameter than 5% of the wall-thickness.

In some countries, among them the USSR and the USA, other indicator types are also standardized besides the acceptance of the image quality indicator recommended by the ISO.

The USSR standard GOST 7512-75 allows the use of *notch indicators* in addition to the wire series indicators. Notch indicators are made in three different sizes with notches increasing according to arithmetical progression (Fig. 7.21, Table 7.5).

Notch indicators give an excellent basis for judging the dimensions of notch-like defects (cracks) in depth. The K sensitivity for notch indicators is

$$K = \frac{h_{min}}{h_s + h_i} 100, \% \tag{7.7}$$

where h_{min} is the smallest depth of the notch still detectable, mm; h_s is the thickness of the sample, mm; h_i is the thickness of the indicator, mm.

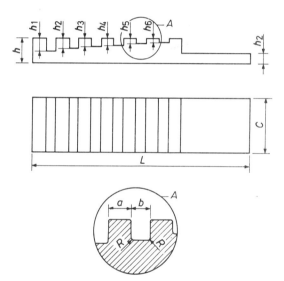

Fig. 7.21. Notch indicator (GOST 7512-75)

Table 7.5. Sizes of notch indicators in mm, with the code numbers given in Fig. 7.20 (according to GOST 7512–75)

Serial number of the indicator	Notch depth						Tolerance zone of notch depth	Main sizes of the indicator					
	h_1	h_2	h_3	h_4	h_5	h_6		R	a +0.1	b +0.1	c	h	L
1	0.6	0.5	0.4	0.3	0.2	0.1	−0.02	0.1	2	0.5	10	2	30
2	1.75	1.5	1.25	1.0	0.75	0.5	−0.05	0.2	2.5	1.5	12	4	45
3	4.0	3.5	3.0	2.5	2.0	1.5	−0.1	0.3	3.0	3.0	14	6	60

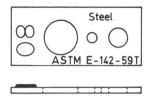

Fig. 7.22. Bored plaque type indicators (ASTM)

Bored plaque-type indicators (Fig. 7.22) have been standardized in the USA (ASTM E-142 59 T) and are called *penetrameters*. These indicators are rectangular metal plates whose thickness (T) is about 2% of the samples, and they contain mostly boreholes with diameters of $4T, 1T, 2T$. The mark of the standard, the kind of material and the code number expressing the thickness in mil units are given on the indicators (e.g., the code number 10 means 0.010 inch = 0.254 mm). The indicators with code numbers as given above have to be used in the following wall thickness ranges (measured in inches): Code No. 5 up to 1/4″ (6.35 mm) Nos 7, 10, 12, 15, 17 and 20 up to 1″ (25.4 mm) with 1/8″ steps, and it continues as follows:

Code number	Upper thickness limit	
	in inches	in mm
25	1 1/4	31.75
31	1 1/2	38.1
35	2	50.8
40	2 1/2	63.5
45	3	76.2
50	4	101.6

etc., up to 8 inches (203.2 mm).

Two test levels are given for the evaluation: the level $2 - 2T$ means, that at least the borehole of $2T$ diameter must be visible; according to level $1 - 1T$ the hole of $1T$ diameter (i.e., all three holes) must be recognizable. The latter one naturally means more strict image quality requirements.

The GOST 7512-75 standard of the USSR also allows the use of bored plaque-type indicators. The plaques have the following sizes: 10×25, 12×35 and 14×45 mm, and have the same 16 thickness values (I) as the notch indicators. Each plaque contains two boreholes: one of them has the diameter of the plaque thickness, the other is twice as big. The determination of the sensitivity is the same as given in the case of the notch indicators.

The *judgment of the image quality* has subjective and objective requirements.

Subjective requirements: the person making the evaluation needs good eyes; his vision has to be checked annually. The use of glasses or a 3–4 fold magnifier might be obligatory. He must be able to read without error from 40 cm distance the 0.5 mm high letters given on the 4th side of the ISO 2504-1973 E recommendation.

Before commencing evaluation the person's eyes must accommodate to the darkness of the room. On coming from sunshine, the waiting time is 10 min; it is 30 s if the direct light of the illuminator (negatoscope) blinds the person.

The *objective requirements* are connected with the illuminator. The transilluminating light intensity cannot be less than 30 cd/m² but, if possible, it should be 100 cd/m² or more. The light intensity of the illuminator should be fitted to this requirement according to the following:

Film density	Light intensity, cd/m²
1	300
2	3000
3	30000

7.3.2. INFLUENCE OF IMAGE QUALITY ON RADIOGRAPHIC SENSITIVITY

The quality of the projected image appears in two parameters. They are the contrast and the contour-sharpness.

7.3.2.1. FACTORS DETERMINING IMAGE CONTRAST

Radiographic image contrast is the ratio of dose rate values corresponding to the sample thickness and/or to the material density differences:

$$k = \frac{\text{highest dose rate}}{\text{lowest dose rate}} = \frac{I_{max}}{I_{min}} \tag{7.8}$$

To investigate the effect of *sample thickness differences* steplike samples have to be taken a part of which is of $4x_{1/2}$, the other part being $2x_{1/2}$ thickness (Fig. 7.23a; $x_{1/2}$ is the attenuation half value thickness for the given material at an energy of radiation E_1). The

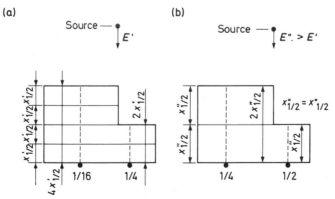

Fig. 7.23. Influence of sample thickness and of radiation energy on image contrast

radiation intensity coming through the thicknesses $4x_{1/2}$ and $2x_{1/2}$ decreases according to the ratios $1:2^4 = 1/16$ and $1:2^2 = 1/4$, respectively, and the value of the image contrast is:

$$k_1 = \frac{1/4}{1/16} = 4$$

The half value thickness obviously differs for materials of *different densities* and this also influences the image contrast. The greater the difference of the material thickness (the size of the defect) the better the contrast, or the density difference between the basic material and material being in the defect is a *gas pore* better recognizable than a *sand* or a *slag inclusion.*

To study the *effect of radiation energy* the sample described above should be exposed with a source whose radiation energy is $E' > E$. Its attenuation half value thickness is $x''_{1/2} = 2x'_{1/2}$ (Fig. 7.23(b)). In such a case the contrast is given by the equation:

$$k_2 = \frac{1/2}{1/4} = 2$$

The comparison of k_1 and k_2 shows that k_2 is half of k_1 only because of the higher radiation energy. As the image quality increases with increasing contrast, the *smallest radiation energy possible which is still just enough for exposure of the sample should be used.* The exposure time increases uneconomically with decreasing radiation energy and also results in worse contour sharpness; in view of this a *compromise* generally has to be found.

Scattered radiation increases the background and *reduces the contrast.* Let us assume that scattered radiation changes the intensity as much as one attenuation half value thickness in the sample of Fig. 7.23(a). In this case the contrast will be

$$k_3 = \frac{\dfrac{1}{2^2 + 2^1}}{\dfrac{1}{2^4 + 2^1}} = 3$$

instead of the original $k_1 = 4$ value, which means 25% reduction. In order to reduce the intensity of the backscattered radiation, the *use of collimated radiation beam, a larger laboratory* and *lead shields* put *behind the film* is advisable.

7.3.2.2. FACTORS DETERMINING CONTOUR SHARPNESS

The *image quality* is much better if the contrast change occurs along a sharp edge, i.e., if the *half shadow* of the radiation is as small as possible. This is characterized by the *geometrical or outer unsharpness* (Fig. 7.24).

If the defect of the sample is at a depth l_d, the geometrical unsharpness U_{gd} is

$$U_{gd} = \frac{d \times l_d}{l_s - l_d} \tag{7.9}$$

where d is the diameter of the radiation source, cm; l_d is the distance between the defect and the film, cm; l_s is the distance between the source and the film, cm.

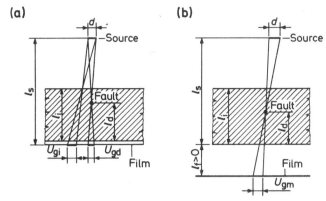

Fig. 7.24. Scheme of outer unsharpness for (a) — normal (b) — and magnified defectoscopy

11 $I_d = I_i$ which means that the distances between the defect and the film (I_d) as well as between the indicator and the film (I_i) are equal to the sample thickness, the geometrical unsharpness of the indicator U_{gi} is

$$U_{gi} = U_{g\,max} = \frac{d\,I_i}{I_s - I_i} \tag{7.10}$$

If the film is at a distance $I_f > 0$ from the sample, the *image is enlarged*. In this case the geometrical unsharpness (U_{gm}) increases according to Eq. (7.10) using the terms of Fig. 7.24(b):

$$U_{gm} = \frac{d(I_f + I_d)}{I_s - I_d} \tag{7.11}$$

The magnification (m) is

$$m = \frac{U_{gm}}{U_{gd}} = \frac{I_f + I_d}{I_d} \tag{7.12}$$

The geometrical unsharpness can be reduced by using *small radiation sources or larger source–film distances.* Based on experience the source–film distance must not be smaller than that corresponding to the $U_{gi} = 0.4$ mm value.

The application of an *intensifying screen* results in an image of better quality and in a more economic exposure time (see Section 7.1.3). *Lead screens* are most frequently used and are placed on both sides of the film. The exposure time compared with that without an intensifying screen can be reduced by a factor of 2 by applying the optimum foil thickness (about 0.1 mm). The specific effect increases with decreasing radiation energy.

Fluorescent materials as intensifying-*screens* reduce also the exposure time: the *salt-intensifying screen* placed beside the film transforms γ-radiation to visible or UV light. Fluorescent screens can be made of *zinc sulphide, cadmium sulphide* (with copper and silver activator) and of *zinc and cadmium silicate* (with manganese activator), but mostly *calcium tungstate* screens are used. As the films are double coated, two screens are applied like a sandwich. In favourable cases the amplification can be as much as an order of magnitude. The worsening of the image quality limits the application of fluorescent screens because the film- or inherent unsharpness (see Section 7.3.3.2) increases due to its grainy property.

The contour sharpness decreases very much if the contact between the film and the screens is not perfect. Secondary radiation develops also as a result of different interactions between the γ-radiation and the absorbing material (see Section 7.1.3.), and it can be scattered in different directions. Because of the scattering the decrease of the contour-sharpness will be the greater the larger the distance between the defect and the film: for this reason the wire series indicator must be placed on the side of the sample opposite the film.

7.3.3. INFLUENCE OF FILM QUALITY ON RADIOGRAPHIC SENSITIVITY

The quality and handling of the film influence the radiographic sensitivity through the film-contrast and the contour sharpness of the image.

7.3.3.1. FACTORS DETERMINING FILM CONTRAST

Film contrast—contrary to image contrast—depends on the type and age of the *film*, on the type, age and level of oxidation of the *developer solution*, and on the time and temperature of the *developing process*—as described in details in Section 7.2.3.5.

7.3.3.2. FACTORS DETERMINING CONTOUR SHARPNESS OF THE FILM

The appropriate type of film and its careful, not too long, storage result in good contour sharpness. The choice of film determines the main parameters: first of all the layer thickness and the turbidity of the emulsion, the size and distribution of the silver-halide grains.

The *film or inherent unsharpness* (U_f) is the measure of the contour sharpness: it is double the mean range of the photoelectrons produced by the γ-radiation from the emulsion and from the intensifying screens. It can be determined experimentally in the following way: a radiation beam of small, known diameter (0.01 mm) is radiated onto the film and the diameter of the image is determined. The usual values of the film unsharpness are:

0.2 mm for a film without intensifying screens;
0.3 mm with fine intensifying screens;
0.4 mm with salt-intensifying screens;
0.7 mm for conventional fluoroscopy (screen without film).

Film unsharpness cannot be avoided but an effort should be made so that the geometrical unsharpness is not greater than it. The same is also valid for motion unsharpness which is caused by removing the radiation source, object and/or the film.

Equation (7.10) of the *total unsharpness* (U_T) describes the summarized effect of the geometrical and film unsharpness

$$U_T = (U_{gi}^3 + U_f^3)^{\frac{1}{3}} \tag{7.13}$$

7.4. EVALUATION OF RADIOGRAMS

Evaluation requires the great experience of an expert, because he has not only to recognize the image of different defects of castings or of welded joints on the films, but he must also be able to draw conclusions about the influence of these defects on the strength of a construction. He has to consider circumstances originating from the technology and from the later use of the product and then he must give an unambiguous decision about its state and use. *Standard prescriptions* help one to judge the effects of defects and to fix the conditions of the acceptance of a product.

7.4.1. DETERMINATION OF WELDING DEFECTS

The inner welding defects detected on the radiograms of butt joints made by fusion welding technique have to be determined according to their *type, size and frequency.* For the qualification of welded constructions the tolerance limits of other defects of the butt joints (deviations from the mechanical requirements, misfits, the slope and meltings-out of the joint edges, the height of reinforcements, etc.) must also be taken into consideration. Only the defects detected on the radiograms are not enough for the qualification of joints. The preparation of radiograms is prohibited if a welding has visible outer defects (detected by means of a magnifying glass or penetrant).

The defects detected by radiography can be evaluated as follows:

— according to their *characteristic defect size* and frequency;

— on the basis of their radiographic *defect grade calculated* from the size and frequency data;

— on the basis of the *largest defect size* or defect area.

The classification of welding defects is described in national standard based on recommendation of international organizations (International Institute of Welding, CMEA, etc.). The defect classification system (RS 4670–74) valid in the CMEA countries describes six main types of defects similar to the IIW document (IIS/IIW 340–69). *The six main types of welding defects* are marked *in capital letters, and the subtypes are marked in lower case.*

A   *gas pore*
    Aa spheroidal gas pore;
    Ab tubular porosity;
    Ac linear porosity;
    Ad group of gas pores;
    Ae elongated gas pore

B   *slag or metallic inclusion*
    Ba spheroidal slag inclusion;
    Bb elongated slag inclusion;
    Bc metal inclusion

C   *lack of fusion*

D   *root or concavity*
    Da root;
    Db lack of root fusion;
    Dc lack of penetration

E   *crack*
    Ea longitudinal crack;
    Eb transverse crack;
    Ec radial crack

F   *surface (shape) and other defects*
    Fa excessive penetration;
    Fb irregular surface;
    Fc undercut

Further surface defects (distortion of the cross-section, high reinforcement, etc.) should be described in the record.

The size and frequency of the defects are characterized by *index numbers*. The number and the area of the defects are related to a so-called normal range of the radiogram. The length of the normal range N (mm) depends on the main wall-thickness s (mm), which is generally equal to the thickness of the elements to be welded. For different elements it is equal to the thickness of the thinner element. The main wall-thickness s determines the length of the normal range N as follows:

$$
\begin{array}{lll}
\text{if} & s \leq 10, & N = 100 \text{ mm} \\
\text{if} & 10 < s \leq 30, & N = 10 \text{ mm} \\
\text{if} & s > 30, & N = 300 \text{ mm}
\end{array}
$$

With thin sheets the main wall thickness is equal to the sheet-thickness plus half of the reinforcement thickness. The latter value is limited according to the following:

Sheet thickness, mm	1/2 Reinforcement thickness, limit, mm
1–4	≤ 2
5–6	≤ 1
over 6	0

The index number of the defect size has two parameters: the characteristic defect size depends on the type of defect, whereas the relative characteristic defect size W does not. If the two parameters are different, the index number is determined from the greater one. The characteristic defect size x is either h, the depth of the defect (the largest size of the defect perpendicular to the surface of the joint) or b, the *width of the defect* (the largest measurable defect perpendicular to the longitudinal axis of the defect).

The characteristic defect size is the depth for defects with indices of the type C, Da, Db, Dc, Fc. This is determined by means of density measurement on the parts of the radiogram exposed from the perpendicular and inclined directions. From the latter a modified thickness can be calculated ($X = h$). For three dimensional defects which have the type indices Aa, Ab, Ac, Ad, Ae, Ba, Bb, Bc, the width is characteristic ($X = b$). The relative characteristic defect size, $W\%$, is the ratio of the characteristic defect size x and of the main material thickness

$$W = \frac{X}{s} \times 100 \tag{7.14}$$

The index number of the defect size has to be determined on the basis of Table 7.6.

The index number of the defect frequency—determined on the basis of occurrence frequency $n\%$—is the ratio of summarized defect length, L, mm (i.e., the sum of the length values of defects of the same type measured along the normal range of a radiogram) and normal range, N, mm, expressed as a percentage:

$$n = \frac{L}{N} \times 100 \tag{7.15}$$

Table 7.6. Determination of the index number of the defect size

Method of the determination		Index number of the defect size				
		1	2	3	4	5
According to $W\%$ independently of the type of defect		$W \leq 10$	$10 < W \leq 20$	$20 < W \leq 30$	$30 < W \leq 50$	$W > 50$
According to X, mm, depending on the type of defect	A and B	$X \leq 3$	$3 < X \leq 5$	$5 < X \leq 7$	$7 < X \leq 9$	$X > 9$
	C and D	$X \leq 1.5$	$1.5 < X \leq 2.5$	$2.5 < X \leq 3.5$	$3.5 < X \leq 5$	$X > 5$
	Fc	$X \leq 0.5$	$0.5 < X \leq 1.0$	$1.0 < X \leq 1.5$	$X > 1.5$	–
	Ea, Eb, Ec, Fa, Fb	It does not have to be determined				

The defect frequency has the following index numbers:

index number
1   if $n \leq 10\%$
2   if $10\% < n \leq 20\%$
3   if $20\% < n \leq 30\%$
4   if $30\% < n \leq 50\%$
5   if $n > 50\%$

For defect frequency calculations the defects of types Aa and Bb have to be considered identical and their frequencies have to be summarized. The length of parallel defects and of those lying close to each other is the distance between their most distant points. The frequencies of defects of types Ea, Eb, Ec, Fa, Fb are to be ignored when determining the index number.

In order to have *uniform interpretation* of the effect of defects in welded joints, the joint defects need to be categorized into radiographic grades (R1–R5) as shown in Table 7.7, on the basis of their types, sizes and frequencies (e.g., Aa 11, Da 13, etc.). The first figures of the two figure numbers in Table 7.7 are the index numbers of the defect sizes, the second ones are those of the defect frequencies.

Table 7.7. Categorization of joint defects into radiographic grades (according to the standard RS 4670-74)

Defect types	Radiographic defect grades				
	R1	R2	R3	R4	R5
Aa, Ad, Ba, Bc	11, 12	13, 21, 22	14, 23, 31, 32	15, 24, 33, 41, 42	All other index
Ab, Ac, Ae	11	12, 21	13, 22, 31	14, 23, 32	numbers which
Bb	–	11, 12	13, 21, 22	14, 23, 31, 32	are not listed
C	–	–	–	11, 12	in R1-R4
Da	–	11, 12, 13, 14	15, 21, 22	23, 31, 32	
Db, Fc	–	–	11	12, 21	
Dc	–	–	11, 12	13, 14, 21, 22, 23	
Ea, Eb, Ec	–	–	–	–	
Fa, Fb	Listing according to additional prescriptions				

EXAMPLE 7.3

Two steel sheets, one of which is 22 mm, the other 24 mm thick, are welded together on one side, having a 2 mm joint reinforcement height. In the 10 × 36 cm size radiogram of the joint, 3 slag strips can be seen each of which is 30 mm long and 4 mm wide. The distances between them are 20 and 40 mm, respectively. In which radiographic defect grade should the sample be categorized?

Solution: elongated slag inclusion; mark Bb. Main material thickness $l = 22$ mm (the thickness of the thinner sheet). The height of the joint does not have to be taken into consideration.

Length of the normal range: $10 < l = 22 < 30$

$$N = 10 \, l = 220 \text{ mm}$$

For defects of type Bb the characteristic defect size is equal to the width of the defect: $x = b = 4$ mm. The index number of the defect size according to x is 2 (from Table 7.6). The relative characteristic defect size: $W = (x/l) \, 100 = \dfrac{4 \times 100}{22} = 18.2$. The index number of the defect size according to W is also 2 (from Table 7.6). The summarized defect length: $L = 3.30 = 90$ mm. The distances between the defects are $20 + 40 = 60$ mm. The sum $90 + 60 = 150$ is less than the normal length, therefore they should be evaluated together.

The defect frequency is $n = (L/N) \, 100 = \dfrac{90 \times 100}{220} = 40.9\%$. The index number of the defect frequency is 4

$$(\text{because } 30\% < n = 40.9\% < 50\%)$$

The defect is Bd 14 which should be categorized as radiographic defect grade R4 (see Table 7.7).

7.4.2. CLASSIFICATION AND EVALUATION OF CASTING DEFECTS

The classification of the casting defects seen on radiograms is complicated because of their great variety, therefore no comprehensive standards exist for judging them. Any particular standard classifies the casting defects according to their kind, size and frequency.

Marking of different *types of defects*:

A: gas pores;
B: sand and other non-metallic inclusions;
C: porous, slack, spongy structures;
D: hot and cold cracks;
E: cracks (heat crack, cold tear) and cold lap;
F: surface, shape and size defects.

The size of defects is characterized by numbers 1–4 on the basis of the percentage of defect size related to wall-thickness, according to Fig. 7.25(a). With varying wall size the thickness of the thinner wall is considered.

(a)   (b)

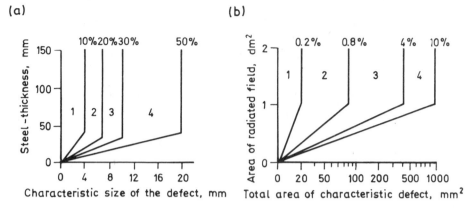

Fig. 7.25. Measures of (a) — the size and (b) — frequency of casting defects. *1* — Steel-thickness; *2* — Characteristic size of the defect, mm; *3* — Area of radiated field, mm²; *4* — Total area of characteristic defect, mm²

Table 7.8. Classification groups of castings

Mark of the defect group	I	II	III	IV	V
Kind of defect	Numbers characterizing the size and frequency of defects				
A	11	12 21	13 31 22	23 32	
B	–	11	12 21	13 22	According
C	–	11	12 21	13 22	to special
D	–	–	– –	–	agreement
E	–	–	12 21	13 22	
F	According to the prescriptions of the drawing of the casting				

The frequency of defects is also characterized by numbers 1–4, taking the percentage value of the *characteristic defect areas* (Q) related to the exposed areas (A) of 1 dm², according to Fig. 7.25(b). The characteristic defect area is the sum of the areas of each defect. If the tested area is larger than 1 dm², then 1 dm² part of the film has to be considered which has the highest defect content.

For classification the defects can be categorized into five groups according to Table 7.8. The first figures of the numbers give the defect size, the second ones the defect frequency.

In the presence of different defect types the classification occurs according to the size and frequency index numbers of the largest characteristic type.

7.5. NEUTRON AND PROTON RADIOGRAPHY

Radiations containing heavy particles (neutron, proton) extended the radiography to such areas where classical X-ray or gamma-radiography was not applicable.

The main advantage of *neutron radiography* is that it differentiates in many cases neighbouring elements (e.g., Co and Fe); moreover, it differentiates between different isotopes of the same element (for example ^{10}B and ^{11}B, ^{235}U and ^{238}U) because the absorption coefficient for the neutrons is independent of density but it is determined by the characteristics of the nucleus.

For neutron radiography mainly *thermal neutrons* (0.02 eV $< E < 0.3$ eV) are used as they have greater absorption differences. Recently epithermic neutrons (0.3 eV $< E < 10$ keV) have been applied for the testing of nuclear heating elements. Neutron radiography applies mainly two methods: direct exposure and the transfer technique.

With the direct exposure method a converter material (foil) and an X-ray film are placed behind the sample. In the foil either prompt (n, γ) reactions, or electron emitting nuclear reactions take place. These secondary radiations cause a proportional optical density of the film. In most cases *foils of large activation cross-section* (e.g., 12.5–25 mm thick gadolinium foils), or a neutron scintillator (e.g., ^{6}Li and ^{10}B powder mixed with ZnS/Ag "phosphor") are applied. The latter emits visible light.

With the transfer technique (*autoradiographic transferring* method) no film is present during the irradiation, only *transferring foils* made of materials having large activation cross-section and radioactive isotopes with a proper half life (Au, Cd, Dy, Ga, In, Rh) are applied. After the irradiation the image-carrying foil is placed on X-ray film, and it is handled as in case of regular autoradiography (see Section 7.6). The method has great importance for investigating nuclear fuel elements, when photosensitive materials cannot be used because of the high γ-background-radiation level.

As radiation sources, mostly nuclear reactors, neutron generators or radioactive neutron sources (Am-Be, Cm-Be, Po-Be, Pu-Be, ^{252}Cf) are used (see Section 1.4.2.4). The integrated flux levels needed for the method are the following: 10^{19} neutron/m^2, if a scintillator is applied; 3×10^{10} neutron/m^2 for fast X-ray film with metal foil; this value can increase to 3×10^{13} neutron/m^2 with a high resolution, single coated, fine-grained (Kodak R) film.

For a high quality radiogram a neutron beam collimated in the ratio 1 : 500 is needed. With smaller neutron sources 1 : 30 or 1 : 20 ratios can also be used, but the quality will be worse. A well collimated, high flux beam can originate only from a reactor, which is not mobile and not easily available.

For proton radiography radiation sources (cyclotrons or other *high-energy accelerators*) are needed which are even much less available. The protons, similarly to other charged particles, penetrate only to a defined depth in the material. At 90% of the mean range the intensity decreases considerably. At this depth very small differences of the thickness (0.05%) are also detectable. A further advantage of this method is that the beam of charged particles can be well collimated electromagnetically.

The techniques of detection are similar to that of neutron radiography. One difference is that *polaroid and colour films* are also regularly used, and Cd, Cu, Pr, Se, or Zn converter foils are applied for (p, n) reactions. The electronic image transferring methods are well applicable in both cases as are scanning detectors and solid state track detectors (see Section 7.6.5).

7.6. INDUSTRIAL AUTORADIOGRAPHY

Autoradiography is a method for detecting the *distribution* of radioactive materials in the solid state and for *quantitative determination of their local concentration* (under suitable conditions) both for macro- and for micro-structural testing. The distribution of radioactive tracers contains informations concerning the structure of the base material.

Whereas in the case of γ-radiography an outer radiation source helps to detect the defects in the macrostructure of the material, in autoradiography *no outer source is utilized* because it is the radioactive tracer content of the sample itself which photographs (see also Chapter 3). (This, as is known, was the means by which *Becquerel* discovered the natural radioactivity of uranium in 1896.)

Autoradiograms show the radioactive tracer distribution in the form of *blackening differences in photosensitive films* whose optical density originates from the photographic effect of the nuclear radiation (see Section 7.2.3.3).

In recent years, *solid state track detectors* (mainly in the form of different plastic foils) became commonly used, their application being based on the interaction of heavy charged particles, e.g., α-particles, with the material of the detector. This method enables the localization of distribution of such stable, mostly light elements, which cannot be located by the classical silver halide emulsions because of the lack of suitable radioactive isotopes. Solid-state detectors fix the tracks of particles developed in nuclear reactions, e.g., (n, α) reactions. In this method no radioactive tracer is needed (see Section 7.5.5). It obviously has a disadvantage, this being that it needs a neutron source. The method itself is generally known as *induced autoradiography*.

Autoradiography has two main fields of application:
— investigation of *inorganic solid state* materials;
— investigation of *biological* samples.

The techniques used for these two fields differ. For industrial purposes mostly *metals and alloys, semiconductors or ceramics* are studied, therefore our description will be limited to them. For the autoradiography of biological specimens see Section 3.6.2.

7.6.1. EFFICIENCY OF THE AUTORADIOGRAPHIC METHOD

The parameters expressing the efficiency of autoradiography are less exact than are other experimental methods. Similarly to radiographic sensitivity (see Section 7.3.1) the efficiency depends on
— *subjective factors* (e.g., the skill and experience of an expert);
— *objective factors* (e.g., the parameters, such as density, chemical activity of the material to be investigated, the radiation properties of the radioactive isotope, the properties of the emulsion).

As far as objective factors are concerned, the efficiency to be expected for a certain system can be calculated and *optimized*. The efficiency of autoradiography can be described by its resolution power and by its sensitivity.

7.6.1.1. RESOLUTION POWER OF AUTORADIOGRAPHY

The resolution power of autoradiography means the *smallest distance, for which two radiation sources can be separately detected*. This value depends first of all on the broadening of the autoradiograph as compared to the diameter of the radiation source (Fig. 7.26).

The differences between the peak and the saddle point of the blackening curve have to be measurable. Effects increasing the broadening of the optical density decrease the resolution power. Such effects are:

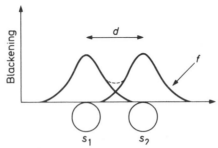

Fig. 7.26. Resolution power of the autoradiogram. S_1 and S_2 — radiation sources; d — resolution power; f — density curve

— thickness of the radiation source,
— distance between the emulsion and the radiation source,
— thickness of emulsion,
— radiation energy.

Microautoradiography is a high resolution method which reflects fine details of the radioactive material distribution of the sample. The method can be realized by carefully controlling the techniques.

The distance between the radiation source and the emulsion has the greatest influence on the resolution power therefore attention has to be paid to the *good contact* between the sample and the emulsion.

Liquid emulsion and the stripping film give a perfect contact, but this contact has sometimes to be reduced artificially. When investigating metallic samples which are able to reduce the silver halides—this takes place in the case of most metals—an *insulating layer* has to be placed between the emulsion and the metal samples and the contact will thereby be worsened artificially (see Section 7.2.2). The thickness of the insulator layer is generally about a few tenths of a µm.

The emulsion thickness can be decreased to a monolayer, which is realized in the case of *electronmicroscopic autoradiography*; the commercially available films utilized in practice have a lower thickness limit of 5–10 µm which means about 15–30 monolayers. Below this size the treatibility and the sensitivity of the film decreases rapidly.

Radiation energy cannot be influenced, but it is to be considered that, because of the decreasing mean range, smaller radiation energy means better resolution power.

In autoradiographic practice, mostly β-radiators (e.g., ^{14}C, ^{32}P, ^{35}S, ^{45}Ca, ^{59}Fe, ^{185}W) are used. Their mean range in the sample is in the order of 10–100 µm. Because preparation of thinner samples presents great difficulties in laboratories with normal equipment it is pointless to attempt this. The increase in sample thickness may result in an increase in the background; this is caused either by γ-radiation or by the bremsstrahlung originating from the β-particles; there is, however, a considerable difference between the linear deceleration (*LET*-value, see Chapter 6) of the electromagnetic and corpuscular radiation, therefore the small, homogeneous increase of the first one does not affect the evaluation of the autoradiographs.

There is the realistic aim of reaching a resolution power between 3–10 µm with present materials and with the usual equipment of an applied laboratory.

7.6.1.2. SENSITIVITY OF AUTORADIOGRAPHY

The sensitivity of autoradiography can be defined either by the *number of photons or charged particles needed to make a grain developable,* or by *the blackening caused by a certain radiation dose.* As it depends on the energy loss per unit length of the radiation in the emulsion, i.e., on the linear deceleration, the sensitivity varies with the type and energy of radiation.

Both the grain size and the grain density have an influence on the *energy loss*: their increase increases the sensitivity. Naturally, larger grain size means smaller resolution power of the emulsion. Relatively sensitive autoradiographic emulsions with high resolution power are fine-grained and of high grain density. They can be well evaluated microscopically if exposed with 10^{12}–$10^{14}\beta$-particles/m^2. X-ray films, which are more sensitive but have worse resolution power, need 1–2 orders of magnitude less exposure. With X-ray or γ-radiation 10–100 fold exposure is necessary because of the much smaller linear deceleration of these radiations. As shown above, sensitivity depends very much on the conditions, therefore—first of all in the case of quantitative evaluation—preparation, exposure and developing process should be kept constant.

Underexposed pictures can be intensified by *double autoradiography* if the necessary exposure cannot be ensured otherwise. The underexposed, but developed autoradiogram should be soaked into a ^{35}S containing aqueous solution, e.g., into a thiocarbamide solution. Ag_2S will be formed on the silver grains and the dried film can be used as an identical sample for preparing another autoradiogram. Such intensified pictures cannot be evaluated quantitatively but they give excellent qualitative information.

7.6.1.3. ERROR SOURCES OF AUTORADIOGRAPHY

The quality and the evaluability of autoradiography is influenced by different parameters:

— *chemical effects* (e.g., the reduction of the silver halide) transform unexposed grains rendering them developable *(chemogram)*;

— *mechanical effects* (pressure, scratch) result in developable grains *(mechanogram)*;

— *too long or improper storage of the film* can cause a background;

— *during too long exposure* a fading effect takes place (the latent image fades away) and some exposed grains lose their developability. The *danger of fading increases with increasing relative humidity:* 30% relative humidity has a negligible effect. Under conditions of higher relative humidity the approximate density decrease over a period of 6 weeks is:

rel. humidity	density decrease
60%	50%
80%	75%

7.6.2. TECHNIQUE OF AUTORADIOGRAPHY

The advantages of autoradiography become obvious immediately on use. No special conditions are needed apart from those normally available in a chemical laboratory for tracer application. The only special material required is the photosensitive detector film which can easily be bought. Manuals are available giving details of the method; furthermore, the suppliers give prescriptions for the processing of the films. In the following we give the main steps of autoradiographic investigation.

7.6.2.1. SAMPLE PREPARATION

For good contact between the specimen and the emulsion layer a smooth *flat surface* is needed on the sample. This can be prepared by metallographic methods. Preparation of very thin samples to increase the resolution power is generally unnecessary as mentioned already, because the mean range of β-particles is in the same range as the thickness of a thin sample prepared with usual equipment.

The flat surface of the sample is covered directly with the film in the case of the contact method, but if wet methods (e.g., *stripping film*) are applied, an insulating layer might be needed.

For insulation any hydrophobic material can be used (e.g., polyvinylalcohol, paraffin, Canadian resin) which does not react with the emulsion. A diluted solution of a few percent concentration has to be prepared with a volatile organic solvent. The sample is dipped into the solution and dried. A $0.1–1$ μm layer will be formed which hinders the electron exchange between the metal sample and the emulsion. For ceramics and for metals which do not reduce silver halides, insulators are not necessary.

7.6.2.2. LABELLING THE SAMPLE WITH RADIOACTIVE ISOTOPES

The material to be investigated has to be labelled with a radioactive tracer *before industrial treatment*. The methods can be diffusion, melting, sintering, etc. (see Section 3.1.1.1).

If the matrix consists of atoms which have isotopes of short half life (such as silicon or aluminium), the tracer atoms can be added in inactive form. After the necessary technological treatments (mechanical deformation, annealing, etc.) which result in the material distribution to be investigated by autoradiography, *the sample will be activated in a reactor*. After a calculated cooling period the matrix no longer has a considerable effect on the autoradiogram. This method offers great advantages if large amounts of material are to be labelled under industrial conditions and the direct application of radioactive isotopes would be dangerous.

7.6.2.3. CHOICE OF EMULSION AND ITS HANDLING

Nowadays, industrial autoradiography applies practically only the contact and the stripping film methods.

The contact method is generally used if small resolution power is needed. *A single coated X-ray film or an autoradiographic film* is pressed gently on to the polished, flat surface of the sample. A soft, elastic material (e.g., a rubber sheet) is desirable between the film and the jaw of the press in order to avoid inhomogeneous pressing which can cause mechanograms (Fig. 7.27).

The film should be exposed *in a closed container*. The quality of the autoradiogram can be improved (particularly with a long exposure time), if exposure takes place *at low temperature* (around 5 °C) and under *low relative humidity* conditions, and applying some drying chemicals.

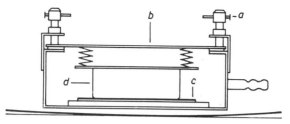

Fig. 7.27. Preparation of contact autoradiogram. a — Clamp jaw; b — Elastic pressure equalizer sheet; c — Film; d — Sample

The optimal exposure time may be calculated either from the activity and radiation energy of the sample and from the sensitivity of the film or determined experimentally. After exposure the film should be processed and dried as in the case of radiography (see Section 7.2.3.5).

The stripping film method is the most frequently used high resolution method of autoradiography. Its popularity is based on the uniform quality, on the photographic sensitivity and on the relatively simple handling. The film is delivered on glass sheets and before use it should be stripped off the glass base and taken over on to the sample. The minimum thickness of the sensitive layer is 5 μm strengthened with a 10 μm thick gelatine support. The structure of the film is shown in Fig. 7.28.

A suitable piece of film is cut with a sharp knife, stripped off the glass base and placed in water with its sensitive side down. After 1–2 min of immersion in water the sample is taken out so that the film covers the polished surface to be investigated (Fig. 7.29). During drying the film shrinks and makes excellent contact with the sample.

The exposure time will be determined—similarly to gamma-radiography—mostly by the quality of the film and by the activity of the sample: a good picture needs 10^{10}–10^{14} radiation particles/m². Increasing activity results in shorter exposure time but the activity of the sample is limited by the specific activity of the radioactive tracer, by the efficiency of the labelling process and by the radiation danger during the sample preparation. Exposure times are usually between a few hours and several weeks.

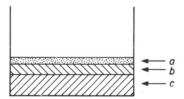

Fig. 7.28. Structure of the stripping film. a — Photosensitive layer; b — Gelatine support layer; c — Glass carrier sheet

Fig. 7.29. Handling technique of the stripping film

7.6.2.4. FILMS FOR AUTORADIOGRAPHY

For industrial autoradiography two basic types of films are utilized:
— *films with X-ray emulsion* for short exposure time, with poor resolution power;
— *films with nuclear emulsion* for longer exposure time, with good resolution power.
Many well-known firms dealing in photographic material produce films for autoradiographic purposes (Table 7.9).

Table 7.9. Characteristic data of some photosensitive autoradiographic films

Product firm	Commercial name	Type and thickness of emulsion, μm		Average grain size, μm
Kodak Ltd	AR-50	X-ray	12	1.2
Kodak Ltd	AR-10	nuclear	5	0.2
ORWO	K-102	nuclear	5	0.11
ORWO	K-106	nuclear	5	0.12–0.16
Ilford	G-5	nuclear	(Liquid emulsions,	0.27
Ilford	K-5	nuclear	but also available	0.20
Ilford	L-4	nuclear	in the form of stripping films	0.15

7.6.3. QUANTITATIVE AUTORADIOGRAPHY

The quantitative evaluation of an autoradiogram is based on the proportionality between the local concentration of a radioactive isotope in a sample and the optical density caused by its radiation in a photosensitive film. The accuracy of the evaluation depends on the practical realization of the conditions for this proportionality.

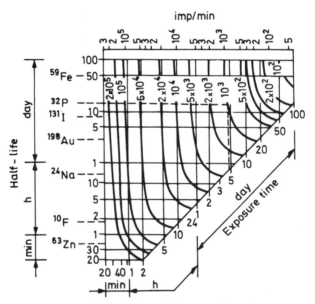

Fig. 7.30. Exposure diagram for autoradiography

A main point of the quantitativity is a *well-defined exposure time*. When applying an isotope with a short half life, the decay during the exposure has to be taken into consideration. The exposure time can easily be determined using the diagram of Fig. 7.30. It shows, for example, that if the sample is labelled with ^{32}P (half-life 14 days), and the surface activity of the sample was 1000 imp/min measured with a GM-end-window counter, an exposure time of 4.5 days is needed applying highly sensitive X-ray films.

The exposed *stripping film is processed on the sample itself*. After processing it will be taken off under water and replaced on a glass sheet (usually a microscope slide) where it dries. It can be studied by microscope and can also be compared with the metallographic picture.

If the handling conditions (preparation, film selection, exposure, processing) were appropriate, the autoradiogram can be evaluated quantitatively, in most cases *by densitometry* and sometimes *by track counting*.

With a high resolution film the resolution power of the microdensitometer has to be comparable with that of the film, i.e., *absorbance (optical)* density, blackening measurement is to be carried out on a surface area of a few μm diameter. The average optical density of a larger area leads to "saturation" as a function of dose or time, because the background of the unexposed areas increases only slowly, and on the exposed areas even a several-fold overexposure does not result in an optical density increase.

Microdensitometers work basically in two modes: in the *flying spot mode* a light spot scans the sample. In most constructions *the autoradiogram moves* in X–Y directions over the light source and a slit whose dimensions can be controlled.

7.6.4. APPLICATION OF AUTORADIOGRAPHY

Industrial research and development utilizes autoradiography in several fields. Some of the most important applications are mentioned below.

7.6.4.1. INVESTIGATION OF DIFFUSION PROCESSES

The study of diffusion is probably the most frequent application of autoradiography. A radioactive tracer deposited on the surface of a sample diffuses into the matrix during the heat treatment. After a certain annealing time the sample is cut in the direction of the diffusion and an autoradiogram will be made on the new surface. The density curve is

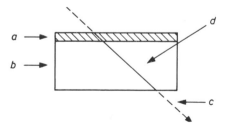

Fig. 7.31. Sample for investigation of diffusion. *a* — Radioactive layer; *b* — Matrix of the sample; *c* — Direction of cut after anneal; *d* — New surface for autoradiography

proportional to the tracer penetration profile and enables one to calculate the diffusion coefficient. A cut which is not parallel with the diffusion direction but at an angle to it magnifies geometrically the diffusion path and makes the density measurement more accurate (Fig. 7.31).

Not only can a penetration profile be observed on autoradiograms but also *local differences*. This is of importance, for example, for polycrystalline materials because of grain boundary diffusion and metallographic structure is as illustrated in Fig. 7.32.

(a)   (b)

(c)

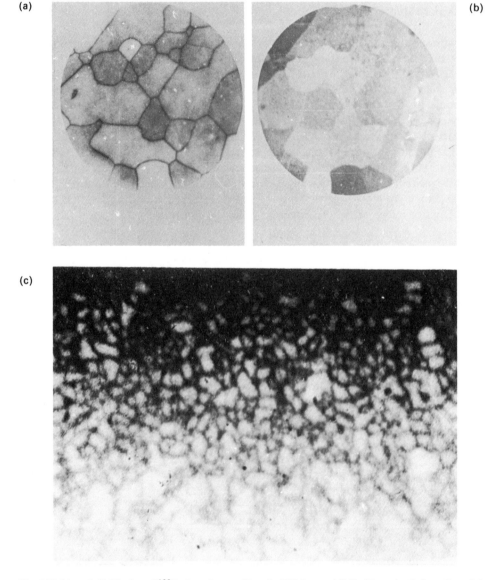

Fig. 7.32. (a) — Self diffusion of ^{123}Sn in polycrystalline tin (166 hours, 449 K, 40 μm depth from the original surface). The enhanced diffusion along the grain boundaries and the orientation dependence of the volume diffusion in the tetragonal lattice can be observed. *M*: 4 × ; (b) — Metallography of the sample; (c) — Diffusion of ^{59}Fe in polycrystalline tungsten [0.05 rad (3°) cut angle, 5 hours, 1881 K]. *M*:80 ×

7.6.4.2. INVESTIGATION OF SEGREGATION

During the solidification of *multicomponent melts* some components concentrate locally and will distribute inhomogeneously in the solid state. Such inhomogeneities are not always desirable and can be influenced by certain treatments. Figure 7.33 provides an example of segregation during the dendritic solidification of a melt.

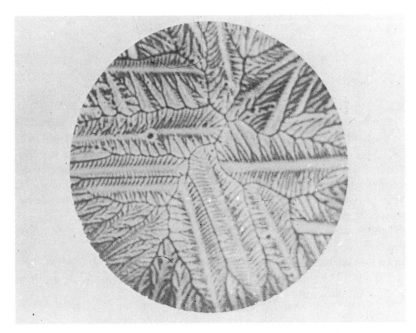

Fig. 7.33. Segregation of ^{111}Ag in tin during dendritic solidification. *M*: 4 ×

7.6.5. SOLID STATE TRACK DETECTORS
IN AUTORADIOGRAPHY

Autoradiography based on photosensitive films is not applicable if the element to be investigated has no suitable radioactive isotope. This represents a limitation primarily in the case of elements with a low atomic number. But even the light elements have such nuclear reactions which result in the emission of heavy, charged particles [e.g., (n, α) or (n, p)]. This helps us to locate the reacting element by means of a *hot-atom chemical reaction* between the charged particle and a suitable detector material, the so-called solid state track detector. The method is analogous to the reversed labelling technique (see Sections 3.1 and 3.8.8).

Solid state track detectors are mostly cellulose-based thin foils which are known under different commercial names (Table 7.10). The detector types differ in the minimum α-particle energy needed to form a developable track.

The particles cause radiation damage to the detector: the destroyed material becomes soluble in strong alkaline solutions and will be visible under a microscope. The damaged part corresponds to the site of the atom taking part in the nuclear reaction, because of the

Table 7.10. Solid state track detectors

Producer firm	Commercial name	Material	Layer thickness, μm
Bayer	Macrofol-E	Polycarbonate	200–500
Bayer	Macrofol-G	Polycarbonate	15–200
Kodak-Pathé	LR-115 I, II	Cellulose nitrate	6, 13
Kodak-Pathé	CN 85	Cellulose nitrate	100
ORWO	T-Cellite	Cellulose acetate	30, 100
Pershode Ltd.	CR 39	Allyl-diglycole-carbonate	10–1000
MOM	MAND/α	Allyl-diglycole--carbonate	1000
	MAND/p		

very short mean range of the particles. As for the necessary radiation damage, at least the mass of an α-particle is needed; smaller particles (protons, neutrons, electrons) or photons do not cause a background, therefore the sample covered with the detector can be put directly into the suitable channel of a *reactor or other neutron source* where the nuclear reaction takes place. In order to eliminate the tracks of other reactions, e.g., those of recoiled nuclei of the detector material, it is expedient for comparison to put an empty detector foil into the reactor too.

Recently, encouraging experiments have been carried out to produce *proton*-sensitive detectors too.

The tracks formed in the irradiated foils are developed in 2–6 M KOH or NaOH solutions, at 50–70 °C, within 20–60 min. The process is interrupted several times and *microscopic observation* informs us when it should be finished.

REFERENCES

[7.1] Clauser, H. R., *Practical Radiography for Industry*. Reinhold, New York, 1952.

[7.2] Rumyancev, S., *Industrial Radiology*. Foreign Language Publ. House, Moscow, 1969.

[7.3] Schnittenheim, R., *Siemens Zeitschrift.*, **29**, 483 (1955).

[7.4] *(The Technical Informations of Gamma-instruments)* (In Russian). "POLON", Warsaw, 1970. 1970.

[7.5] Sipos, T., Fekete, Z. and Hirling, J., *Development and Use of Radioisotope Pipeline Crawlers in Hungary*. Conf. on "Defectoscopy '77". Varna, Bulgaria, 1977.

[7.6] Halmshaw, R., *Industrial Radiology Techniques*. Wykeham Publications (London) Ltd, London–Winchester, 1971.

[7.7] Sharpe, R. S., *Research Techniques in Nondestructive Testing I–II–III*. Academic Press, London–New York, 1970–73/77.

[7.8] Berger, H., *Neutron Radiography—Methods, Capabilities and Applications*. Elsevier, Amsterdam, 1965.

[7.8a] Müller, C. H. F., *Automation in the Field of X-ray Inspection*. 8th World Conf. on Nondestructive Testing. Cannes, 1976.

[7.9] GEVAERT: *Industrial Radiography*. (Handbook).

[7.10] *Gamma defektoskopi i prinadlezhnosti dlja promishlennoj radiografii*. Atomizdat, Moscow, 1970.

[7.11] Sauerwein, K., "Prinzip und Anwendung der Gamma-Radiografiegeräte Typ *GAMMAT für die Materialprüfung*". Berichte der Deutschen Keramischen Gesellschaft e.V. **43**, 457–462 (1966).

[7.12] Westhäusser, R. and Sauerwein, K., "Ein selbfahrendes Isotopendurchstrahlungsgerät für Zentralaufnahmen von Rohrrundschweissnähten." *Technische Überwachung*, **10**, 345–349 (1969).

[7.13] Hirling, J., *Isotopentechnik*, **1**, 159 (1962).

[7.14] Horowev, V. N., Kulescsov, A. V., Majorov, A. N. and Sulkin, A. G., *(Gamma-defectoscopy and Equipment for Radiography)* (In Russian). CMEA, Zakopane, 1970.

[7.15] Hungarian Standard No. MSZ 4310.
[7.16] Hungarian Standard No. MSZ 15963.
[7.17] Hungarian Standard No. MSZ 62–78.
[7.18] Hungarian Standard No. MSZ 5–36. 3505.
[7.19] Standard of USSR No. GOST 7212–75.
[7.20] French Standard No. AFNOR–04–304.
[7.21] USA Standard No. ASTM E 142–59 T.
[7.22] International Standard No. ISO/R 1027–1969 (E).
[7.23] International Standard No. ISO 2504 Proposal.
[7.24] British Standard No. BS 3971–1966.
[7.25] FRG Standard No. DIN 54109/1.2–64.
[7.26] GDR Standard No. TGL 10646/4.
[7.27] Hungarian Standard No. MSZ 4310/5–77.
[7.28] International Institute of Welding No. IIS/IIW 340–69 document.
[7.29] Herz, R. H., *The Photographic Action of Ionizing Radiations*. Wiley-Interscience, New York–London–Sydney–Toronto, 1969.
[7.30] Rogers, A. W., *Techniques of Autoradiography*. Elsevier Publ. Co., Amsterdam–London–New York, 1969.
[7.31] Fleischner, R. L., Price, P. B. and Walker, R. M., *Nuclear Tracks in Solids*. University of California Press, Berkeley 1975.
[7.32] Benes, J., *Fundamentals of Autoradiography*. Iliffe Books Ltd., 1966.
[7.33] Fischer, H. A. and Werner, G., *Autoradiographie*. de Gruyter Co., 1971.
[7.34] Boyd, G. A., *Autoradiography*. Academic Press, New York, 1955.

8. RADIATION PROTECTION

8.1. GENERAL ASPECTS OF RADIATION PROTECTION

8.1.1. THE ROLE OF RADIATION PROTECTION

The application of radioactive materials is always accompanied by an amount of risk. This risk appears in two forms: the exposure of humans to ionizing radiation increases the probability of some detrimental somatic and hereditary effects ("stochastic" effects) and relatively high doses may cause harm to particular tissues or organs ("non-stochastic" effects).

In addition, deleterious effects other than detriment to health may also occur, such as the restriction of some areas or products from use.

The risk may be reduced by appropriate measures but, in general, it cannot be entirely eliminated. Penetrating radiation around radioactive sources and radioactive contaminants in consumer products contribute, even if at an almost negligible level, to the exposure of individuals coming into contact with them. Careless handling of radioactive materials or their abnormal release into the *environment* (i.e., *pollution*) may lead to accidental exposure of individuals. However, the risk associated with the application of radioactive materials—as has been proved by experience gained during the last few decades—is not higher than the risk of industrial activities in general [8.1].

It should also be taken into consideration that the number of persons likely to be effected is relatively small as is the probability of serious harm to health. Despite this, the hazard arising from the use of radioactive materials should not be underestimated, but all possible practicable measures should be taken to *minimize risk and harm*. Before reviewing the methods and devices of protection, the detrimental effects of ionizing radiation are outlined.

The deleterious health effects of ionizing radiation are developed on a complex biological mechanism which cannot be discussed here, but just to outline the consequences, it is mentioned that high doses *damage the cells and tissues* and impede the normal function of that organ; moreover, apart from local injury the whole organism will be affected. For example, irradiation of the blood forming tissues (e.g., bone marrow) leads to a depletion of blood cells which will obviously inhibit the normal functioning and protective ability of the organism as a whole.

Higher doses lead to *somatic detriment to individuals,* mostly to local injuries. Mild forms of somatic injuries appear in temporary erythema, loss of hair and opacity of the lens; severe forms lead to ulceration and/or disorders of internal organs. The severity of these so-called *non-stochastic effects* depends on the magnitude of the dose received, and there may be a threshold dose below which no detrimental effects can be observed.

The effects of small doses near background level or below it are hardly predictable. Although the relationship between dose and biological response is not yet fully understood, it is now accepted that exposure of individuals or the population at large increases the *risk of hereditary* and of some somatic (malignant) diseases. These effects are *stochastic,* thus the risk is proportional to the dose received. However, it is clear at present that risk associated with the application of radioactive materials is much lower than the risk due to other man-made and natural sources of radiation. Despite this, the application of radioactive methods is advisable only in those cases when the benefit is obvious and there is no other non-radioactive and so less harmful method available for the given purpose.

The aim of radiation protection is threefold:

— to establish *safety standards* which ensure that the individual deleterious effects are socially acceptable;

— to take *preventive measures* in order to limit the exposure of individuals to a reasonably achievable low level—also bearing in mind the risk of incidents and accidents;

— to carry out *monitoring and surveillance* in order to make protection and safety measures effective.

The special feature of the industrial application of radioisotopes from the aspect of radiation protection is that the place of work is generally not an adequately equipped and controlled laboratory and persons specialized in isotope technique are available only temporarily and in a limited number.

For this reason, the technical protection of such work should be planned and controlled with a higher degree of safety—as in special laboratories.

Because this chapter deals mainly with the health physics problems arising from nuclear radiation utilized in industry, radiation protection problems concerned with laboratories specially designed and equipped for the regular use of radiation are excluded.

8.1.2. DOSE CONCEPTS

The definitions of radiation quantities and their units are formulated by the International Commission on Radiation Units (ICRU) [8.2], partly based on the recommendations of the International Commission on Radiological Protection (ICRP) [8.1].

Radiation protection needs such definitions of dose that characterize the radiation field through directly measurable quantities and that account for the interaction of radiation with biological objects, too. Such conditions cannot be fulfilled by one dose definition because of the variety of interactions and biological consequences.

Exposure, X. The SI unit is C/kg; the special non-SI unit is the roentgen, R, where $1 R = 2.58 \times 10^4$ C/kg (see Section 1.2). The exposure is applicable to a wide energy range of X- and γ-radiation and it is more or less proportional to the biological effect. Nowadays, the use of exposure is not recommended: it should be replaced by absorbed dose in air.

Absorbed dose, D. The SI unit is joule/kg, a special name for this is the gray, Gy, where $1 Gy = 1$ J/kg. The non-SI unit is the rad (radiation *a*bsorbed *d*ose), its symbol is rad or rd where $1 Gy = 100$ rd (see Section 1.2). The absorbed dose characterizes the physical interaction of radiation with matter even for *corpuscular radiation*, e.g., for α, β and neutron radiations. It should be noted that the dose absorbed in a material placed in a radiation field depends also on the material (density, elemental composition, size). The absorbed dose by itself is insufficient as a means of predicting the severity or probability of deleterious effects for all kinds of radiation and irradiation conditions in general. For radiation protection purposes it has been found convenient to introduce a further quantity—dose equivalent—which reflects better the harmful health effects.

Dose equivalent, H. The SI unit is the sievert, symbol Sv where $1 Sv = 1$ J/kg (if $Q \times N = 1$). The non-SI unit is the rem (see Section 1.2), $1 Sv = 100$ rem.

$$H = DQN \tag{8.1}$$

Table 8.1. Relationship between linear energy transfer: L_∞ in water and quality factor (Q); and approximate values of $\bar{Q}$ related to various types of primary radiation

L_∞ in water (keV/m)	Q	Type of radiation	$\bar{Q}$
3.5 (and less)	1	X-rays, γ-rays and electrons	1
7	2	Thermal neutrons	2.3
23	5	Neutrons, protons and other singly-	
53	10	charged particles of unknown energy	10
175 (and above)	20	α- and other multiply-charged particles of unknown energy	20

Note: Interpolated values for Q vs. L_∞ are given, e.g., in Ref. [8.4]. Average quality factor

$$\bar{Q} = \frac{1}{D} \int_0^\infty Q(L_\infty) \times \frac{dD}{dL_\infty} \times dL_\infty$$

where Q is the quality factor depending on the *linear energy transfer* of a given radiation (see Section 1.2, Chapter 6 and Table 8.1). The average value of Q (see Table 8.1) can be applied if the distribution of L_∞ is not known in detail; N is the product of modification factors taking into account the space and time distributions of absorbed dose. According to the present standpoint of the ICRP [8.1] the value of N is taken as 1.

The use of detectors calibrated directly in dose equivalent is advantageous in radiation protection practice because the effects of different types of radiation can be summed up.

There are some other dose definitions too, such as effective dose equivalent, dose commitments, collective doses, etc. but they are not very often used yet in practical radiation protection for industrial radioisotope applications. However, one of them, the effective dose equivalent, which accounts for the total detriment, whatever the dose distribution in the body is, should be mentioned here.

Effective dose equivalent, H_E. The SI unit is the sievert, Sv.

$$H_E = \sum_T w_T H_T$$

where H_T is the mean dose equivalent in tissue T and w_T is a weighing factor representing the proportion of the detriment from stochastic effects resulting from tissue T to the total detriment from stochastic effects when the body is irradiated uniformly.

The values of w_T are specified by ICRP [8.1] and are:

Tissue	w_T	Tissue	w_T
Gonads	0.25	Thyroid	0.03
Breasts	0.15	Bone surfaces	0.03
Red bone marrow	0.12	Remainder	0.30

Dose index (absorbed dose index, D_1, or dose equivalent index, H_1) is the maximum absorbed dose or dose equivalent referred to a given point within a 30 cm diameter sphere consisting of tissue equivalent material with unit density and centred at this point. The unit of dose index is J/kg. The 30 cm diameter tissue equivalent sphere is a model simulating the trunk of the body. (Note that the centre of the sphere cannot be closer than 15 cm from a surface or a source!) To make provision for radiation which is strongly attenuated in the sphere (i.e., soft γ-, X- and β-rays) the sphere is divided into 2 shells and a core. The outer shell (0.07 mm) reflecting the basal layer of the epidermis with radiation effects negligible, is ignored; the next shell extends to a depth of 1 cm; the inner core has a radius of 14 cm. Thus the 2 maxima in the intermediate shell and in the core are termed the shallow and deep absorbed dose or dose equivalent indices, with symbols $H_{1,s}$ and $H_{1,d}$, respectively. Dose indices are not yet widely used in health physics, but it is expected that in the future they will be used in setting and controlling external radiation levels.

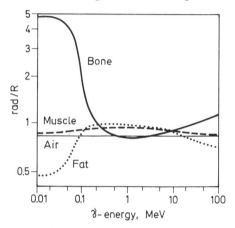

Fig. 8.1. Relationship between absorbed dose (rad) and exposure (R) for air and certain body tissues vs. γ-ray energy

Relationship between dose concepts. Although we are interested primarily in the dose equivalent delivered to the whole-body or a certain tissue, we often measure *air dose* in the place of the body, and tissue dose is established by calculation if necessary. Fortunately— at least for γ-radiation of usual energy—the absorbed dose in air is a good approximation of both soft and bone tissue doses (see Fig. 8.1). For ^{137}Cs and ^{60}Co, e.g., in soft tissue $D(\text{Gy}) = 1.1 D$ in air (Gy). This consideration is more or less true also for the non-SI exposure unit, R, as 1 R = 0.869 rd (or 8.69 mGy) in air.

Therefore, if instead of dose equivalent rate for γ-radiation at about 1 MeV energy, absorbed dose rate in air (or even exposure rate) is considered, the error introduced can be neglected for practical protection purposes. On the other hand, if the dose rate of soft X-, β- or neutron radiation is established from the reading of a radiation monitor calibrated only for γ-radiation, the result may be quite misleading.

8.1.3. THE DOSE LIMITATION SYSTEM

Radiation protection standards are incorporated in the national regulations on the basis of the International Commission on Radiological Protection (ICRP) recommendations. These standards are revised from time to time by the ICRP in the light of up-to-date scientific achievements. The following text is based on the latest ICRP recommendations [8.1], [8.3] and [8.14] appearing with some delay in other international recommendations (e.g., International Atomic Energy Agency) [8.4], and in national regulations. Where it is necessary, concepts from earlier ICRP recommendations—still in practice in many countries—are also mentioned.

The population, from the viewpoint of radiation protection, is divided into two categories:

Category A: all persons concerned with work involving ionizing radiation: e.g., persons who are engaged in tracer experiments, radiographic work or in mounting radioactive sources of gauges.

Category B: members of the public who may be exposed to radiation due to different applications of radioactive materials; e.g., persons living or working in the vicinity of nuclear facilities but not directly involved in radiation work, generally belong to this category.

The radiation dose (see Section 8.1.2) is the most important factor determining the biological effect: the dose corresponding to the acceptable risk for the individual is the dose equivalent limit (DEL), its former name is the *maximum permissible dose (MPD). The dose limit is not a tolerance dose* because, to our present knowledge, *even the smallest dose corresponds to a certain risk.* However, it is reassuring to know that radiation injuries could only be observed at several times higher doses than the dose limits as proved by many years of experience. *Higher doses may cause acute radiation sickness:* a single irradiation of the total body with a dose of 2 Sv (200 rem) is highly likely to be lethal.

The biological effects depend to some extent on *dose rates.* With lower *dose rates* or *protracted irradiation* a higher dose is needed to cause the same injury, e.g., skin erythema, lens opacification compared with higher rates, and even a dose of 2 Sv (200 rem) to the total body will not lead to severe injury if it is delivered uniformly over the occupational lifetime, i.e., about forty years. However, the long-term effects of radiation (cancer) have to be considered because in the case of these late effects there is no threshold dose and the risk is proportional to the dose.

The ICRP recommendations [8.1], when establishing the dose equivalent limits, takes the stochastic and non-stochastic effects into account separately.

Table 8.2. Annual dose equivalent limits

Limits	For occupational exposure of workers		For exposure of members of the public[a]	
	mSv	rem	mSv	rem
Effective dose equivalent limit	50[b]	5	5[c]	0.5
Dose equivalent limit				
— for eye lenses	300	30	50	5
— for other tissues and organs	500	50	50	5

[a] Resulting from justifiable and optimized practices, and not to be fully exploited by any single source or practice.

[b] Long continued exposure of a considerable proportion of workers near the dose equivalent limits would only be acceptable if it is justified by a careful cost-benefit analysis.

[c] If the exposure of the same individuals approaches this limit over many years, an average annual effective dose equivalent of 1 mSv (0.1 rem) should be maintained.

The dose equivalent limit for *non-stochastic effects* is relatively high, equal to 0.5 Sv (50 rem) except the lens for which the limit is 0.3 Sv (30 rem). The dose equivalent limit recommended for *stochastic effects* makes allowances for organs and tissues subject to special risk in such a way that the risk of uniform exposure of the whole body should be equal to the risk of non-uniform exposure:

$$\sum_T w_T H_T = H_{WB,L} \qquad (8.2)$$

where w_T is the risk weighting factor of the tissue; H_T is the annual dose equivalent in tissue; $H_{WB,L}$ is the annual dose equivalent limit for uniform irradiation of the whole body, namely 50 mSv (5 rem).

The dose equivalent limits (DEL) are applied to the sum of external and internal irradiation.

No persons *under the age of 18* shall be allowed in radiation Working Condition A, in which exposures might exceed three-tenths of the basic dose limits.

Workers or students *between the ages of 16 and 18* are allowed to work in Working Condition B for training purposes, where it is most unlikely that the annual exposure will

Table 8.3(a). Derived limits for soluble compounds of selected radioisotopes, corresponding to occupational exposure of workers (Category A)

Radioisotope	Organ	$t_{1/2,\,eff}$, days	MPBB, μCi	MPC$_{air}$, kCi/cm^3
^{3}H	Total body	12	1000	5×10^{-6}
^{14}C	Fat	35	3000	4×10^{-6}
^{24}Na	G. I.* tract	0.6	7	1×10^{-6}
^{32}P	Bone	14	6	7×10^{-8}
^{35}S	Testes	76	90	3×10^{-7}
^{51}Cr	G. I.* tract	27	800	1×10^{-5}
^{55}Fe	Spleen	400	1000	9×10^{-7}
^{59}Fe	G. I.* tract	40	20	1×10^{-7}
^{60}Co	G. I.* tract	10	10	3×10^{-7}
^{64}Cu	Total body	0.5	10	2×10^{-6}
^{65}Zn	Total body	220	60	1×10^{-7}
^{82}Br	Total body	1.3	10	1×10^{-6}
^{90}Sr	Bone	6400	2	3×10^{-10}
^{111}Ag	G. I.* tract	3	20	3×10^{-7}
^{131}I	Thyroid	8	0.7	9×10^{-9}
^{137}Cs	Total body	140	30	6×10^{-8}
^{140}Ba	G. I.* tract	12	4	1×10^{-7}
^{140}La	G. I.* tract	1.7	9	2×10^{-7}
^{170}Tm	Bone	107	9	4×10^{-8}
^{192}Ir	G. I.* tract	16	6	1×10^{-7}
^{198}Au	G. I.* tract	2.7	20	3×10^{-7}
^{203}Hg	Kidney	8	4	7×10^{-8}
^{204}Tl	G. I.* tract	5	10	6×10^{-7}
^{226}Ra	Bone	16000	1.1	3×10^{-11}
^{239}Pu	Bone	72000	0.04	2×10^{-12}
^{241}Am	Kidney	23000	0.05	6×10^{-11}

* G. I.: gastrointestinal

Note: MPBB: Maximum permissible body burden
MPC: Maximum permissible concentration

exceed three-tenths of the *DELs*, i.e., 15 mSv (1.5 rem) annual effective dose equivalent. *Under the age of 16* years no person is allowed to work as a radiation worker.

The occupational exposure of *women* of reproductive capacity shall be permitted only under such conditions that:

— the resulting dose equivalent shall not exceed the dose limits provided that exposure is received at an appropriately regular rate, if pregnancy is not known;

— the women continue to work in Working Condition B only, if pregnancy is known.

The dose equivalent limits, as recommended by the ICRP [8.1] are summarized in Table 8.2.

The dose equivalent limits given in Table 8.2 are often converted to derived limits, such as for example μSv/week (mrem/week) and μSv/h (mrem/h). Basic derived values for planning purposes are 25 μSv/h (2.5 mrem/h) and 1 mSv/week (100 mrem/week). These values are calculated with the annual dose limit, 50 mSv/year (5 rem/year), and 40 working

Table 8.3(b). Annual limits of intake and derived air concentrations for selected radionuclides

Nuclides	Oral ALI, Bq		Inhalation ALI, Bq			DAC, Bq/m³*		
	a	b	c	d	e	f	g	h
³H	3×10^9		3×10^9			8×10^5	2×10^{10}	
²⁴Na	1×10^8		2×10^8			8×10^4		
³²P	2×10^7		3×10^7	1×10^7		1×10^4	6×10^3	
³⁵S	4×10^8	2×10^8	(i) 6×10^8	8×10^7		3×10^5	3×10^4	
			(ii) 5×10^8			2×10^5		
⁵¹Cr	1×10^9	4×10^9	2×10^9	9×10^8	7×10^8	7×10^5	4×10^5	3×10^5
⁵⁹Fe	3×10^7		1×10^7	2×10^7		5×10^3	8×10^3	
⁶⁰Co	2×10^7	7×10^6	6×10^6	1×10^6		3×10^3	5×10^2	
⁶⁴Cu	4×10^8		1×10^9	9×10^8	8×10^8	5×10^5	4×10^5	3×10^5
⁶⁵Zn	1×10^7		1×10^7			4×10^3		
⁸²Br	1×10^8		2×10^8	1×10^8		6×10^4		
⁹⁰Sr	1×10^6	2×10^7	7×10^5	1×10^5		3×10^2	6×10^1	
¹¹¹As	3×10^7		6×10^7	3×10^7		2×10^4	1×10^4	
¹³¹I	1×10^6		2×10^6			7×10^2		
¹³⁷Cs	4×10^6		6×10^6			2×10^3		
¹⁴⁰Ba	2×10^7		5×10^7			2×10^4		
¹⁴⁰La	2×10^7		5×10^7	4×10^7		2×10^4	2×10^4	
¹⁷⁰Tm	7×10^7		3×10^7			1×10^4		
¹⁹²Ir	4×10^7		1×10^7	1×10^7	8×10^6	4×10^3	6×10^3	3×10^3
¹⁹⁸Au	4×10^7		4×10^7	6×10^7	6×10^7	2×10^4	2×10^4	2×10^4
²⁰³Hg	9×10^7		(i) 5×10^7	4×10^7		2×10^4	2×10^4	
			ii) 3×10^7			1×10^4		
²²⁶Ra	7×10^4		2×10^4			1×10^1		
²³⁹Pu	2×10^5	2×10^6	2×10^2	5×10^2		8×10^{-2}	2×10^{-1}	
²⁴¹Am	5×10^4		2×10^2			8×10^{-2}		

* for 40 h/week

Note: *ALI*: Annual Limit of Intake; *DAC*: Derived Air Concentration; *a–h*: refer to specific compounds as given in Table 8.3 (c) and in Refs [8.3, 8.4 and 8.13] for each radionuclide. (*i*) and (*ii*) see Table 8.3 (c). The relationships between DAC and *ALI* is: DAC = *ALI*/2000 × 60 × 0.02 = *ALI*/2.4 × 10³ Bq/m³ where: 0.02 m³ is the volume of air breathed at work by "Reference Man" per minute under working conditions of "light activity" [8.5]

Table 8.3 (c). Compounds or physical form of radionuclides listed in Table 8.3 (b) with reference to classes a–h

Nuclides	Oral		Inhalation		
	a	*b*	*c ≡ f*	*d ≡ g*	*e ≡ h*
^{3}H	HTO		HTO	Elemental	
^{24}Na	All		All		
^{32}P	Dietary		All other than in d	Phosphates	
^{35}S	All inorganic	Elemental	(*i*) Sulphides, sulphates other than in *d* (*ii*) Vapours: SO_2, H_2S or CS_2	Elemental sulphides of Sr, Ba, Ge, Sm, Pb, As, Sb, Bi, Cu, Ag, Au, Zn, Cd, Hg, Mo, W, sulphates of Ca, Sr, Ba, Ra, As, Sb, Bi	
^{51}Cr	Cr^{6+}	Cr^{3+}	All other than in d, e	Halides, nitrates	Oxides hydroxides
^{59}Fe	All		All other than in d	Oxides, hydroxides, halides	
^{60}Co	Inorganic in tracer quantities, oxides and hydroxides	Organic, all inorganic with carrier, except oxides, hydroxides	All other than in d	Oxides, hydroxides halides, nitrates	
^{65}Zn	All		All		
^{82}Br	All		All bromides of H, Li, Na, K, Rb, Cs, Fr	All other, bromides	
^{90}Sr	Soluble	$SrTiO_3$	All soluble except $SrTiO_3$	All insoluble	
^{111}Ag	All		Elemental, all other than in d, e	Nitrates, sulphide	Oxides, hydroxides
^{131}I	All		All		
^{137}Cs	All		All		
^{140}Ba	All		All other than in d	Oxides, hydroxides	
^{170}Tm	All		All		
^{192}Ir	All		All other than in d, e	Halides, nitrates, elemental	Oxides hydroxides
^{198}Au	All		All other than in d, e	Halides, nitrates	Oxides, hydroxides
^{203}Hg	All inorganic		(*i*) Sulphates (*ii*) vapour	Oxides, hydroxides, nitrates, sulphides, halides	
^{226}Ra	All		All		
^{239}Pu	All other than in b	Oxides, hydroxides	All other than in d	PuO_2	
^{241}Am	All		All		

Table 8.4. Physiological data of Reference Man

Weight of total body: 70 kg (male)

Litres of air breathed for Reference Man (adult man)

8 h working "light activity"		9600
8 h nonoccupational activity		9600
8 h resting		3600
	Total	2.3×10^4 l

Water intake for Reference Man (adult man), ml/day

Milk	300
Tapwater	150
Other	1500
Total fluid	1950
In food	700
By oxidation of food	350
Total	3000 ml/day

hours per week. However, when an individual receives a few mSv (a few hundred mrem) during a week or even during a shorter period, it is not regarded as an overexposure if the dose does not exceed the annual dose limit.

The dose limit for individual *members of the public*, Category **B**, is equal to 5 mSv/year (0.5 rem/year). This dose limit concerns all man-made sources of radiation, medical applications excepted. The permissible levels for all special activities (e.g., nuclear power industry, industrial application of radioisotopes) are determined by the competent national authority after a *risk-benefit evaluation* of that given activity.

Internal doses due to the incorporation of radioisotopes (radionuclides) cannot be determined directly in health physics practice; their calculation is also difficult. In view of this, derived quantities are introduced which can substitute the dose limit values and can be measured directly.

The incorporated radioisotope may be accumulated in certain organs, but finally is excreted both from the organ and body. The process is governed by the biological half-life which, in turn, depends on metabolic processes. Parallel the radioactivity decreases due to physical decay. *The effective half-life* of these two processes is:

$$t_{1/2,\, \text{eff}} = \frac{t_{1/2,\, \text{phys}} \times t_{1/2,\, \text{biol}}}{t_{1/2,\, \text{phys}} + t_{1/2,\, \text{biol}}} \tag{8.3}$$

where $t_{1/2}$ is the half-life.

The secondary limits for internal exposure are the *annual limits of intake (ALI)* by inhalation or ingestion related to adult Reference Man, (see Tables 8.3 and 8.4 Refs [8.3–8.5]). The incorporation of the activity of the *ALI* value leads to the occupational dose equivalent limit in a person. Appropriate *ALI* values are also defined for different groups of population. For practical radiation purposes, *derived limits* are defined: e.g., *derived air concentration (DAC)*, this represents that concentration of a radionuclide in air (Bq/m^3), which breathed by Reference Man for a working year of 2000 h (50 weeks at 40 h per week) under conditions of "light activity", would result in the *ALI* by inhalation. The *DAC* equals the *ALI* (of a radionuclide) divided by the volume of air inhaled by Reference Man in a working year (that is, 2.4×10^{13} m^3). These concepts are intended to replace the former limits, the so-called *maximum permissible levels*:

— *the maximum permissible body burden, MPBB,* is the activity delivering the maximum permissible dose to the critical organs if the burden is continuous, (see Table 8.3),

— *the maximum permissible concentration, MPC,* is calculated in such a way that the dose commitment to the critical organs will be equal to the *MPAD* when the contaminated air or water is continuously consumed by Reference Man at the permissible levels, (see Tables 8.3b, c and 8.4). Appropriate *ALI* values are also defined for different groups of population.

The activity, *inhaled* via contaminated air may be calculated using radioactive concentration data which are easily measurable.

Similar considerations can be applied for the *ingested* activity but the process here is more complex due to the complexity of nourishment. However, food is usually not directly contaminated, thus water has primary importance in establishing the dose burden of the population.

The maximum permissible surface contamination levels are determined on the basis of incorporation hazard due to ingestion or inhalation probabilities of contaminants. These levels are applied only at radioisotopic laboratories where the handling of unsealed radioactive materials is a matter of routine; no contamination is allowed on surfaces of equipment, floors, etc. during or following the industrial application of sealed radioactive sources.

With industrial tracer experiments using unsealed radioactive materials, the permissible contamination levels of products, wastes and materials affected by the experiment have to be determined in advance. It is very unusual that a contaminated product can be followed precisely: how it comes into contact with individuals or groups of the population. For this reason the *MPC* (or *DAC*) values have to be used together with appropriate safety factors.

EXAMPLE 8.1

Mercury of an electric cell is traced with ^{203}Hg. The mass of mercury is 1500 kg and the tracer activity is 200 GBq (5.4 Ci).

The average concentration of mercury on the premises is 10^{-10} kg/l air. What will the expected ^{203}Hg concentration in the air be, and what will the body burden of workers be after eight hours work?

The specific activity of ^{203}Hg

$$\frac{200\,\text{GBq}}{1500\,\text{kg}} = 133\,\text{MBq/kg} \quad \left(\frac{5400\,\text{mCi}}{1500\,\text{kg}} = 3.6\,\text{mCi/kg}\right)$$

which, after the equilibrium of evaporation has been reached, results in a ^{203}Hg concentration in air of 10^{-10} kg/l $\times$ 133 MBq/kg = 1.33×10^{-2} Bq/l (= 3.6×10^{-10} µCi/ml). The MPC_{air} value for ^{203}Hg and for Category **B** is 2.6×10^{-4} Bq/ml = 0.26 Bq/l ($7 \cdot 10^{-9}$ µCi/ml) (see Table 8.3), therefore the dose burden of personnel working on the given premises, due to the continuous incorporation of radioactive mercury is

$$\frac{1.33 \times 10^{-2}\,\text{Bq/l}}{0.26\,\text{Bq/l}} \times 100 = 5.1\%\ \text{of the MPD}$$

The activity of incorporated ^{203}Hg, during eight hours work, using Reference Man data will be: 1.33×10^{-2} Bq/l $\times 10^4$ l = 133 Bq (3.6×10^{-10} µCi/ml $\times 10^7$ ml = 0.0036 µCi).

This activity is only

$$\frac{133 \text{ Bq}}{148\,000 \text{ Bq}}\,100 \quad \left(\frac{0.0036\,\mu\text{Ci}}{0.4\,\mu\text{Ci}} \times 100\right) \leq 1\%$$

of the $MPBB$ value, see Table 8.3 (a) for Category **B**, and even a smaller fraction of the ALI (see Table 8.3(b) and (c)).

8.1.4. CALCULATION OF DOSE RATE

8.1.4.1. DOSE RATE IN AIR

The distribution *of radiation flux density and dose rate around a point source* is spherically symmetric and if there is no absorber it is proportional to $1/R^2$, where R is the distance from the source:

$$\Phi_R = \Phi_{R_0}\frac{R_0^2}{R^2} \tag{8.4a}$$

$$I_R = I_{R_0}\frac{R_0^2}{R^2} \tag{8.4b}$$

Equations are valid precisely in a vacuum and approximately in most practical cases, in air.

The flux density of a point source at a source distance R metre is

$$\Phi_P = \frac{An}{4\pi R^2}\exp(-\mu R) \quad 1/(\text{m}^2\,\text{s}) \tag{8.5}$$

where A is the source activity, Bq; n is the number of particles or photons per decay, μ is the linear attenuation coefficient; $\exp(-\mu R)$ is the transmission in air, is taken as unity for γ-radiation but is <1 for β-radiation.

The dose rate from a point source at a distance of R, is

$$I_P = K\frac{A}{R^2}\exp(-\mu R) \tag{8.6}$$

where A is the source activity; μ is the linear attenuation coefficient; K is the dose constant which is the dose rate measured at unit distance from a given radioisotope having unit activity (see Section 1.2). If the radiation is complex the partial dose rates have to be calculated and summed up. The unit of dose is determined by the unit of K chosen. The flux density and the dose rate Eqs (8.5) and (8.6) can be generalized for interposed absorbers.

$$I_P = K\frac{A}{R^2}\exp(-\mu l) \tag{8.7}$$

where l is the thickness of the absorber; μ is the linear attenuation coefficient, which may be interpreted for narrow or broad beams, and also for complex absorbers.

The range of α-*radiation* is only a few cm in air, therefore in radiation protection practice there is no need for external dose rate calculations.

In the case of β-radiation K_β may also be introduced but Eq. (8.6) loses its validity in practical cases, because of self-absorption in the source. An approximative rule may be applied here

$$I_{P,\beta} = 45.7 \; \bar{E}_\beta \frac{\mu}{\rho} \frac{A}{R^2} \exp\left(-\mu R\right) \quad \text{mGy/h} \tag{8.8a}$$

$$I_{P,\beta} = 169 \; \bar{E}_\beta \frac{\mu}{\rho} \frac{A}{R^2} \exp\left(-\mu R\right) \quad \text{rd/h} \tag{8.8b}$$

where A is the source activity, MBq (mCi*); $\bar{E}_\beta$ is the mean energy of beta particles, MeV, this is approximately equal to one-third of the maximum β-energy; R is the distance from the source, cm; μ is the linear attenuation coefficient of the medium (air), 1/cm; it can be

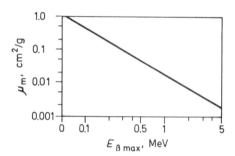

Fig. 8.2. Mass absorption coefficient of β-radiation, μ_m, cm²/g, vs. maximum energy of the β-particle

obtained by multiplying ρ (density of the medium, g/cm³) by μ_m, cm²/g obtained from Fig. 8.2.

If $E_{\beta max} > 0.5$ MeV, then instead of Eq. (8.8) the following expression serves well for practical purposes:

$$I_{P,\beta} \approx 0.8 \quad A \; \text{mGy/h} \tag{8.9a}$$

$$I_{P,\beta} = \approx 3 \; A \quad \text{rd/h} \tag{8.9b}$$

where A is the activity, MBq (mCi*). This expression gives the dose rate at a distance of 10 cm from the source, neglecting self-absorption and absorption in the air.

For larger distances, even at 1 m, the expressions based on Eqs (8.4 a, b) are not valid even approximately (Fig. 8.3).

Equation (8.6) gives a good approximation for γ-radiation if the medium is air. Taking into account the absorption, the dose rate for a narrow beam will be:

$$I_{P,\gamma} = K_\gamma \frac{A}{R} \exp\left(-\mu R\right) \quad \mu\text{Gy/h} \tag{8.10a}$$

$$I_{P,\gamma} = K'_\gamma \frac{A}{R^2} \exp\left(-\mu R\right) \quad \text{R/h} \tag{8.10b}$$

where A is the activity, Bq (mCi*); R is the distance from the source m (cm*); $K_\gamma (K'_\gamma)$ is the γ-dose constant,** $\mu\text{Gy} \cdot \text{m}^2/(\text{GBq} \cdot \text{h})$ [R $\cdot$ cm²/(h $\cdot$ mCi)*[; μ is the linear attenuation coefficient 1/m (1/cm*).

---

* applied to the equation in brackets
** $K_\gamma[\mu\text{Gy} \cdot \text{m}^2/(\text{GBq} \cdot \text{h})] = 23.5 \; K'_\gamma[\text{R} \cdot \text{cm}^2/\text{Ch} \cdot \text{mCi})] = 235 \; K'_\gamma[\text{Rm}^2/(\text{h} \cdot \text{Ci})]$

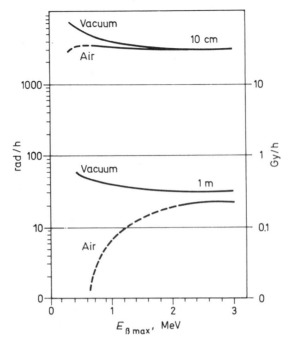

Fig. 8.3. Dose rate of a 37 GBq (1 Ci) β-point source

The γ-dose constant is strongly energy-dependent, (see Fig. 8.4).

The values of K_γ, for the most important radioisotopes, are tabulated in Table 1.1.

When K_γ is unknown and the energy of γ-radiation is in the range of 0.3–3 MeV, the dose rate in air at 1 m from the source is

$$I_{P,\gamma} \approx 0.15\, AE \quad \mathrm{nGy/h} \tag{8.11a}$$

$$I_{P,\gamma} \approx (0.54\, AE \quad \mathrm{rd/h}) \tag{8.11b}$$

where A is the activity in GBq (Ci*), E is the total γ-energy per decay in MeV.

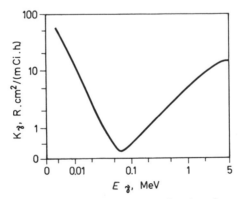

Fig. 8.4. The γ-ray dose constant, K_γ, as a function of γ-energy

$$E = \sum_i n_i E_i \tag{8.12}$$

where n_i is the number of photons with energy E_i per decay. The values of n_i and E_i are given in Table 1.3.

Eq. (8.6) is also a good approximation for *neutrons*:

$$I_{P,n} = K_n \frac{S_n}{R^2} \quad \mu Sv/h \tag{8.13a}$$

$$I_{P,n} = K' \frac{S_n}{R^2} \quad mrem/h \tag{8.13b}$$

where S_n is the neutron yield of the source s^{-1}, $K_n (K' *)$ is the dose constant, $\mu Sv \times m^2 \times s/h$ (mrem $m^2 s/h*$), R is the distance from the, $1/s$, source.

Table 8.5. Dose constants of neutron point sources, $\mu Sv\ m^2 \cdot s/h$ (mrem $m^2 \cdot s/h$). Neutron fluence rate (flux density), $1/(cm^2 s)$ corresponding to 10 $\mu Sv/h$ (1 mrem/h); and the average quality factor as a function of neutron energy* [8.4, 8.13]

Neutron energy, MeV	$K'_n \times 10^6$ (mrem $\cdot m^2\ s/h$) $K_n \times 10^5$ ($\mu Sv \cdot m^2 \cdot s/h$)	1/(cm · s) 10μ Sv/h	Average quality factor, $\bar{Q}$
Thermal	0.030	260	2.3
0.005	0.035		2.0
0.02	0.071	170	3.3
0.1	0.25	48	7.4
0.5	0.66	14	11
1	1.1	8.5	10.6
5	1.1	6.8	7.8
10	1.17	6.8	6.8
14	1.34		
20	2	6.5	6.0

* For unidirectional broad beams of monoenergetic neutrons at normal incidence

The rounded values of $K_n (K'_n)$ are given in Table 8.5. This table gives the neutron fluxes corresponding to the dose equivalent rate of 10 μSv/h (1 mrem/h), too. The following expression is useful for practical purposes giving the distance in m from an unshielded neutron source at which the dose equivalent rate is 10 μSv/h (1 mrem/h)

$$R \approx 0.65 \sqrt{A} \quad m \tag{8.14a}$$

$$I_{P,n} \approx R \approx 2.5 \sqrt{A} \quad m \tag{8.14b}$$

where A is the ^{226}Ra-Be equivalent activity in GBq (Ci*) (see Section 1.4.2.4).
From Eqs (8.14) and (8.4b)

$$I_{P,n} \approx 4.2 \frac{A}{R^2} \quad \mu Sv/h \tag{8.15a}$$

$$\left(\approx 15.6 \frac{A}{R^2} \quad mrem/h \right) \tag{8.15b}$$

where R is in m and A is the ^{226}Ra-Be equivalent activity in GBq (Ci**).

* Applied to equations (8.11b) and (8.13b)
** Ci applied to equation (8.14b) and (8.15b)

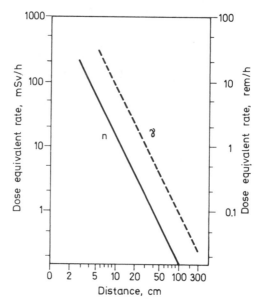

Fig. 8.5. γ- and neutron dose rates, mSv/h (rem/h), for a ^{226}Ra-Be source from Ref. [8.5]

Figure 8.5 gives the neutron and γ-dose rates of a ^{226}Ra–Be source as a function of distance from the source. The calculation of neutron dose attenuation due to absorption tends to be complicated because of the strong energy dependence of the dose equivalent.

Expressions describing space distribution of fluxes from *extended sources* are usually very complex even if the absorption is neglected; in practice such expressions are utilizable only with the help of computers. Figure 8.6 shows some cases for which calculation is comparatively simple.

Line source [Fig. 8.6 (a)]. Flux density from a line source, having a length of L cm and an activity of A Bq, at a point taken at distance d cm from the straight line through the source ($d=0$ is excluded) and neglecting the absorption in air is given by

$$\Phi_L = \frac{An}{4\pi Ld}\left(\arctan\frac{L-x}{d} + \arctan\frac{x}{d}\right) \quad 1/(cm^2s) \tag{8.16}$$

where n is the number of particles or photons per decay, x is the distance between the perpendicular through the given point and the end of the line source, cm.

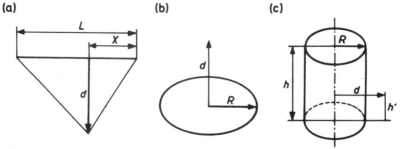

(a)   **(b)**   **(c)**

Fig. 8.6. Parameters for extended sources. (a) — Line source; (b) — Disc source; (c) — Cylindrical source

When x equals to $L/2$ (the point is on the perpendicular through the centre of the source) the flux density will be

$$\Phi_L = \frac{An}{2\pi Ld} \arctan \frac{L}{2d} \; 1/(\text{cm}^2\text{s}) \qquad (8.17)$$

The equation for the dose rate is

$$I_L = 10^4 \, K \frac{A}{Ld} \left(\arctan \frac{L-x}{d} + \arctan \frac{x}{d} \right) \qquad (8.18)$$

and if $x = L/2$:

$$I_L = 2 \times 10^4 \, K \frac{A}{Ld} \arctan \frac{L}{2d} \qquad (8.19)$$

where A is the total activity, Bq (or mCi) of the source of L m (or cm); K is the dose constant; d is the distance in m (or cm) on the normal from the source, Bq or mCi and m or cm to be used appropriate to the unit of K, which also determines the unit of dose.

Disc source [Fig. 8.6(b)]. The flux density on the axis at a distance of d cm of a disc source having a radius of R cm and activity of A Bq.

$$\Phi_D = \frac{An}{4\pi R^2} \ln \left(\frac{R^2}{d^2} + 1 \right) \quad 1/(\text{cm}^2\text{s}) \qquad (8.20)$$

and the dose rate

$$I_D = K \frac{A_2}{R} \ln \left(\frac{R^2}{d^2} + 1 \right) \qquad (8.21)$$

All units should correspond to the unit of K used here.

Cylindrical source [Fig. 8.6(c)]. The dose rate from a cylindrical source with a radius of R cm and height of h cm at a point d cm from the axis of the cylinder is

$$I_c = 2K \frac{A}{R\pi} \left[G(h', R, d) + G(h-h', R, d) \right] \qquad (8.22)$$

where *A is the source activity in* Bq, $G(h, R, d)$ is a tabulated function given in Ref. [8.6].

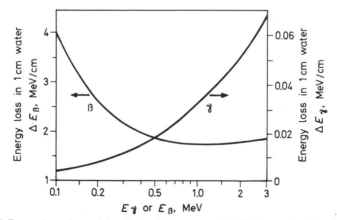

Fig. 8.7. Energy loss of γ- and β-radiation in 1 cm water (MeV/cm) as a function of energy

The absorbed dose rate can be calculated from the flux density data, too. In the case of soft tissue ($\rho \approx 1$), for β- and γ-radiation

$$I_{\beta,\gamma} = 5.75 \times 10^{-4} \Phi \, (\Delta E/\text{cm}) \quad \text{mGy/h} \tag{8.23a}$$

$$I_{\beta,\gamma} = 5.75 \times 10^{-5} \Phi \, (\Delta E/\text{cm}) \quad \text{rd/h} \tag{8.23b}$$

where Φ is the flux density, $1/(\text{cm}^2 \cdot \text{s})$; and the values of $\Delta E/\text{cm}$, in MeV/cm can be taken from Fig. 8.7.

The absorbed dose rate in air (or the exposure rate) of γ-radiation in the energy range of 0.03–3 MeV is given by

$$I_\gamma = 1.62 \times 10^{-2} E \Phi \quad \mu\text{Gy/h} \tag{8.24a}$$

$$I_\gamma = 4.8 \times 10^{-10} E \Phi \quad \text{C/(kgh)} \quad (\sim 2 E \Phi \quad \mu\text{R/h}) \tag{8.24b}$$

where E is the photon energy, MeV, Φ is the flux density, $1/(\text{cm}^2 \text{s})$.

Dose rates of neutrons and γ-radiation corresponding to unit flux density, are given in Figs 8.8 and 8.9.

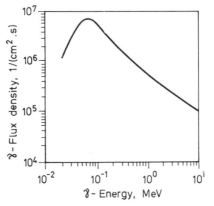

Fig. 8.8. γ-Flux density equivalent to $\sim 10\,\text{mGy/h}$ in air (1R/h) as a function of γ-energy

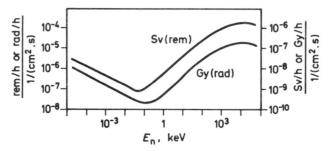

Fig. 8.9. The ratio of flux density, $1/(\text{cm}^2 \cdot \text{s})$ and dose rate, $\mu\text{Gy/h}$ or $\mu\text{Sv/h}$ (rad/h or rem/h) of neutrons for tissue as a function of neutron energy. The applied values of quality factors are 2.5 for slow and 10 for fast neutrons

EXAMPLE 8.2

The activity of a point source is 70 MBq (1.9 mCi), 1 particle is emitted in each decay, absorption is neglected. What is the flux density at 50 cm from the source?

From Eq. (8.5)

$$\Phi_P = \frac{7 \times 10^7}{4\pi \times 50^2} = 2228 \text{ particles/(cm}^2\text{s)}$$

EXAMPLE 8.3

Consider a line source of length 30 cm and activity of 70 MBq (1.9 mCi). Then n is taken as unity and there is no absorption. What is the flux density at a point located 50 cm on the perpendicular through the centre of the source?

From Eq. (8.17)

$$\Phi_L = \frac{7 \times 10^7}{2\pi \times 30 \times 50} \arctan \frac{30}{2 \times 50} = 2164 \text{ particles/(cm}^2\text{s)}$$

EXAMPLE 8.4

Consider a flux density of 1 MeV γ-radiation of 2000 photons/(cm^{2}s). What is the absorbed dose rate in soft tissue?

From Fig. 8.7, ΔE/cm is approximately 0.03 MeV/cm, using Eq. (8.23).

$$I_\gamma = 5.75 \times 10^{-4} \times 2000 \times 0.03 \text{ mGy/h} = 34.5 \text{ } \mu\text{Gy/h} \quad (\approx 3.5 \text{ mrd/h})$$

What is the absorbed dose rate in air or the exposure rate at the same point?

Using Eq. (8.24)

$$I_\gamma = 1.62 \times 10^{-2} \times 1 \times 2000 = 32.4 \text{ } \mu\text{Gy/h}$$

$$I_\gamma = 4.8 \times 10^{-10} \times 1 \times 2000 \approx 9.6 \times 10^{-7} \text{ C/(kg h)} (\approx 3.2 \text{ mR/h})$$

8.1.4.2. SHIELDING CALCULATIONS

The limitation of personal exposure simply by remaining at an appropriate distance is usually impossible or uneconomic and in many cases even not safe. Before building a shield, it is necessary first to determine
— the reduction factor for the given *source distance*, and then
— the *thickness of a shielding material* that reduces the dose rate to the required level.

The first step in determining the *reduction factor* is to estimate the distance that will exist between the source and the body or hand of the worker. The typical mean or smallest distance between an installed radioactive source and members of the public, i.e., persons working or living in its neighbourhood should also be estimated. Then the dose rate I_0 is calculated for this estimated distance and, taking into account the suitable dose limit (*DL* or *MPD* for a year or week or other period as appropriate) and the occupancy time, t(e.g., yearly or weekly working hours), the necessary reduction factor will be:

$$F = S \frac{I_0 t}{DL} \tag{8.25}$$

where I_0 is the dose rate for unshielded source, t is the occupancy time, h, for example μSv/h working hours per year (or week), S is the safety factor, its usual value is 2 or 3 if not otherwise prescribed by regulation, DL is the dose limit for the same period as for t, μSv/year or μSv/week.

Rearranging Eq. (8.25) the denominator can be written as $\dfrac{DL}{t}$, which is the maximum permissible dose rate at places occupied by radiation workers or members of the public, depending on the values chosen.

Assuming whole body exposure their values will be for Category **A** and for 36 working hours per week $\dfrac{1000\ \mu Sv/week}{36\ h/week} \approx 28\ \mu Sv/h$ (2.8 mrem/h); for Category **B**, i.e., for members of the public including non-radiation workers, this value is e.g., $\dfrac{100}{40} = 2.5$ $\mu Sv/h$ (0.25 mrem/h).

The required *thickness of a shielding material* is determined by using the equation expressing the attenuation of a narrow beam

$$I = I_0 \exp\left(-\mu l\right) \qquad (8.26)$$

where l is the thickness, cm; μ is the linear absorption coefficient, 1/cm which is dependent on energy of radiation and on the atomic number of the absorber.

Broad beams are more common in practice than narrow ones and due to scattering in the absorber an increase in the intensity—with respect to a narrow (collimated) beam—is observed. This can be taken into consideration using the *build-up factor, B*:

$$I = B(l)\ I_0 \exp\left(-\mu l\right) \qquad (8.27)$$

The build-up factor depends on radiation energy, and on the type and size of material applied, its value is determined from calculations, graphs or by using the tables in Refs [8.7 and 8.8]. The build-up factor can be built in the exponent

$$I = I_0 \exp\left(-\mu l\right) \qquad (8.28)$$

where μ depends on thickness l, too.

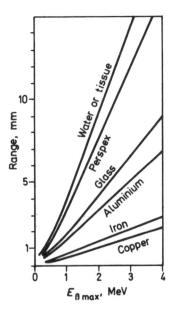

Fig. 8.10. Maximum range of β-particles in some selected materials as a function of their maximum energy

α-Radiation does not need shielding because α-particles are totally absorbed within a few cm of air or in the dead layer of the skin.

The absorption of *β-particles* more or less follows the exponential law but the protection against β-radiation is facilitated by its finite range. This range is energy- and medium-dependent. For a given radioisotope the maximum range of β-particles is governed by the maximum energy of β-decay (Fig. 8.10). For the most important β-energies in practice, i.e., 2–3 MeV, the maximum range is about a few metres in air but a thickness of 10–15 mm of a plastic (e.g., Perspex) is quite sufficient to absorb them fully.

Bremsstrahlung due to absorption of β-radiation is a more penetrating radiation, therefore a few mm thick plastic will not significantly reduce its intensity. The bremsstrahlung yield is low, its intensity is lower by three or four orders of magnitude than the primary β-radiation, so it becomes important only when sources have higher than 0.5 GBq (10–20 mCi) activity. For example, the dose rate from a 350 MBq (10 mCi) ^{32}P source at a distance of 10 cm, after absorbing the β-radiation in a light absorber, is approximately 10 μGy/h (1 mR/h) in air.

Absorption of *γ-rays* follows well the exponential law (see Eq. 8.10).

Quick estimations can be made using some aids:

— *the half-, or tenth-value thicknesses* as a function of energy can be obtained from tables and graphs, e.g., from Fig. 8.11;

— knowing the prescribed reduction factor, graphs and tables give the necessary thickness for various energies [8.9] and [8.10].

Quick methods are also useful for radioisotopes emitting more than one photon with different energies when the transmission (reduction factor) for the given radioisotope is given [8.9].

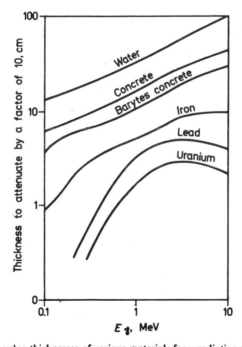

Fig. 8.11. Tenth-value thicknesses of various materials for γ-radiation (narrow beam)

Other methods of calculation are:

— *use of tables* containing the necessary thicknesses as a function of energy and reduction factor [8.6(b)]

— *graphical method* applied for broad beam. The first step is to form an auxiliary value, U:

$$U = \frac{R^2}{A} \times \frac{DL_{weekly}}{t_{weekly}} \tag{8.29a}$$

$$U = 3.7 \frac{R^2}{A} \times \frac{MPD_{weekly}}{t_{weekly}} \tag{8.29b}$$

where R is the distance, m; A is the activity, MBq (mCi*); $DL(MPD)_{weekly}$ is the weekly dose limit, μSv/week (mrem/week); t_{weekly} is the number of working hours per week. An occupancy factor may also be used here.

The desired lead thickness for a given radioisotope can be determined from Fig. 8.12 using the predetermined value of U. Conversion factors, when using materials other than lead, are given in Table 8.6.

Protection against *scattered radiation* is also required especially with provisional shields which are sufficient to attenuate the direct beam only. Intensity of scattered radiation behind the shield due to the scattering of γ-radiation on the walls and even in the air at relatively high activity GBq-TBq(Ci-kCi) sources may exceed by orders of magnitude the intensity of the primary but attenuated radiation.

In such cases, protection has to be extended against scattered radiation, too. *The calculation of scattering and shielding is complex* because the energy of the scattered beam

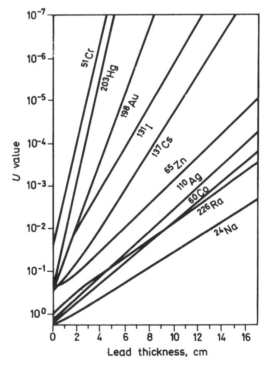

Fig. 8.12. Diagram for lead shielding thickness calculation

* applies to equation (8.29b)

Table 8.6. Multiplication factors of lead equivalent thickness calculated for various materials

Materials	Radioisotopes							
	^{203}Hg	^{51}Cr, ^{131}I, ^{198}Au	^{137}Cs, ^{192}Ir	^{65}Zn, ^{89}Rb	^{22}Na, ^{59}Fe, ^{60}Co	^{42}K, ^{56}Mn, ^{124}Sb	^{226}Ra	^{24}Na
Iron	4.5	4	2.7	2.2	2.1	2	1.9	1.9
Barytes concrete	13	10	8	5.7	5	4.8	4.7	4.7
Ordinary concrete	18.5	14.5	10	8.1	7.3	6.7	6.5	6.8

decreases due to scattering in the shield and the initially discrete energy spectrum becomes continuous. The flux density of the scattered beam depends on the angle of scattering, on the distance, on the size and composition of the absorber and of course on the flux density and energy of the primary radiation. Detailed calculations are available in the literature, e.g., Ref. [8.7].

The energy of *neutrons* depends on the type of source: neutron generators produce ~ 14 MeV neutrons, sources utilizing the (α, n) reaction give neutrons with a mean energy of 4–5 MeV, (γ, n) sources emit neutrons of about 0.1 MeV, (see Section 1.4.2.4 and Tables 1.4 and 4.3, respectively).

Fast neutrons passing through a medium gradually lose their energy due to elastic and inelastic collisions and after reaching thermal equilibrium they are captured by nuclei. As the energy loss of neutrons is the greatest using shields of small atomic number, mostly those rich in hydrogen, the commonly used shielding materials are water, concrete, paraffin and polyethylene. Fast neutrons cannot be absorbed efficiently, but neutrons slowed down are easily captured by, say, boron or cadmium.

The attenuation of neutron flux density is calculated by expressions similar to Eq. (8.10) but with neutrons, attenuation of the dose equivalent rate is more important because of the strong energy dependence of the quality factor hence the dose equivalent.

Exact calculation requires large computational and measuring apparatus and it is usually impractical in industrial applications. Radiation transmitted through a shield having a thickness common in practice (about 100 cm) will mainly consist of neutrons with energy of 0.5 eV–100 keV and the dose contribution of thermal neutrons will be negligible.

The dose attenuation factor is:

$$R_n = \frac{I_{n,0}}{I_{n'}} \tag{8.30}$$

Values of R_n for 14.2 MeV neutrons can be found in Fig. 8.13.

Attenuation of the fast neutron flux density of a Po-Be source is given in Fig. 8.14.

With neutron sources emitting γ-radiation too, it is effective to line the light element neutron absorber with iron or lead; this results in more efficient neutron shielding and it also saves space.

Thermal neutrons are fully captured by an 0.5–1 mm thick Cd foil.

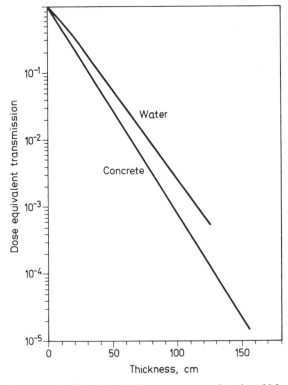

Fig. 8.13. Dose attenuation coefficient R_n, of a 14 MeV neutron source plotted vs. thickness of shielding material

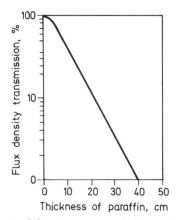

Fig. 8.14. Fast neutron flux density of a ^{210}Po-Be source vs. thickness of a paraffin absorber expressed as a percentage of primary flux density

EXAMPLE 8.5

What is the required wall-thickness of a working container made of lead or iron containing a ^{60}Co source of 130 MBq (3.5 mCi) activity if the permanent place of work is 1 m from the source?

From Eq. (8.29) determine the ratio U, assuming that the DL (or MPD) is 100 μSv/week and $t_{weekly} = 48$ h:

$$U = \frac{37}{130} \times \frac{10}{48} = 6.10^{-2}$$

from Fig. 8.12 the lead thickness is 6 cm and using the multiplication factor from Table 8.6 this corresponds to $6 \times 2.1 = 12.6$ cm thick iron.

EXAMPLE 8.6

The yield of a 14 MeV neutron generator is 10^9 n/s.

How high is the dose rate at a distance of 2 m from the source behind a 1 m thick concrete wall?

From Eq. (8.13), using Table 8.5

$$I_0 = 1.34 \times 10^{-5} \frac{10^9}{4} = 3350 \ \mu\text{Sv/h} \quad (335 \ \text{mrem/h})$$

The dose attenuation factor R_n for 1 m concrete taken from Fig. 8.13 is ~ 150, therefore according to Eq. (8.30)

$$I = \frac{I_0}{R_n} = \frac{3350}{150} \approx 22.3 \ \mu\text{Sv/h} \quad (2.2 \ \text{mrem/h})$$

which level is permissible for personnel.

8.1.5. MEASUREMENT OF DOSE AND DOSE RATE

A dosimeter has to measure a quantity which is proportional to the physical or biological dose.

This requirement is not fulfilled by all types of dosimeters and cannot be met in all conditions. The types of *personal dosimeters* are outlined in Section 8.3.1. In radiation protection practice the dose rate measurement is also important and the industrial application of radioisotopes needs portable, easy-to-handle dose rate meters (see Section 8.3.2.1). The main requirements of such equipments are:

— energy independence;
— better than 20% accuracy;
— independent measurement of different types of radiation;
— maximum response at dose rates higher than the upper range of the monitor;
— small detector and instrument size;
— battery life of at least 8 h;
— high stability and reliability.

The dose rate of *α-radiation* cannot be measured by simple devices.

The dose rates of *β-radiation*, except for those of very low energy, can be measured by a thin-walled ionization chamber but the measured value tends to be informatory only. The determination of β-radiation flux densities is more evident in health physics routine than of the dose rates.

The exposure rate of *γ-radiation*, above 100 keV is approximately equal to the dose rate absorbed in soft tissue. The exposure rate can be well established through the measurement of *ionization chamber* current.

The measurement of low dose rates (around 1 μSv/h (0.1 mrd/h)—important in radiation protection practice—can be carried out only by relatively large, about 1 litre volume ionization chambers. Such large chambers are inconvenient for portable monitors and disadvantageous for narrow beam measurement. Radiation monitors with GM-counters (see Section 2.1.1.3.) are the most widespread; their response is nearly energy independent for γ-radiation in the range of 0.3–3 MeV. The *scintillation detector* (see Section 2.1.3) is also very sensitive but it is strongly energy dependent. These two latter types of detectors can usefully be applied for contamination surveys.

Table 8.7. Classification of neutrons and detectors used for their measurement

Neutron	E_n	Detector
Slow	<0.5 eV	BF$_3$ counter of ZnS–B or LiI(Eu) scintillator
Intermediate	0.5 eV– 0.5 MeV	ZnS–B scintillator + moderator
Fast	>0.5 MeV	ZnS(Ag) polystyrene scintillator or proton recoil counter
Thermal — 10 MeV		"rem" counter

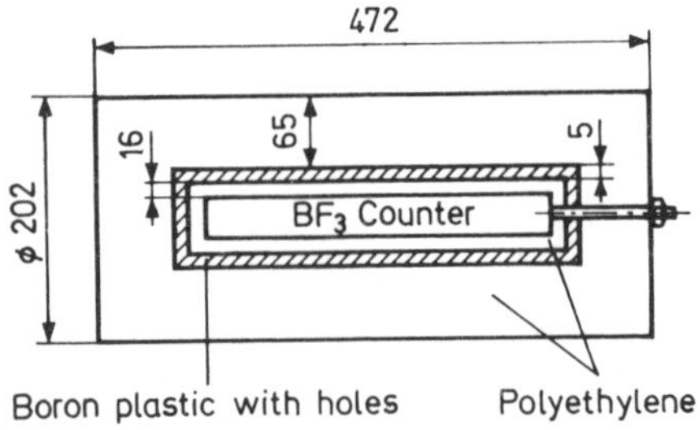

Fig. 8.15. Detector for neutron dose equivalent rate measurement (Anderson–Braun type "rem" dosimeter)

Neutrons can be measured only indirectly through the detection of ionizing particles produced by them. The dose rates are measured in practice by detectors, as proportional counters or scintillation detectors, operated in pulse counting mode but these can only be used in specific energy ranges (see Table 8.7). If a thermal neutron counter is covered by a suitable *polyethylene–boron* layer (see Fig. 8.15), its sensitivity will correspond to the dose equivalent from 0.025 eV to 10 MeV. Using a 20 or 25 cm diameter paraffin sphere, the energy dependence is adequate up to 10 or 14 MeV, respectively. With a suitable boron layer the moderator mass can be reduced by a factor of two.

When simple neutron counters are calibrated by a standard neutron source, it should be kept in mind that the calibration is valid precisely only for the given neutron spectrum.

8.2. METHODS OF PREVENTION

The objective of prevention is to limit the individual dose of all people who are directly involved in the application of radioisotopes or who live in the vicinity as well as to keep population doses to a safe level.

8.2.1. TECHNICAL PROTECTION

The objectives of the technical protection are
— limitation of *external radiation* by maintaining a distance and by absorption or attenuation of radiation by shields (manipulators, containers, lead-, concrete-, etc. walls);
— prevention of dispersion of radioactive materials, i.e., contamination, which may lead to *incorporation* (source encapsulation, hermetization, etc.).

Technical protection sometimes also needs special means; these will be discussed in Sections 8.4 and 8.5.

8.2.2. ADMINISTRATIVE AND ORGANIZATIONAL MEASURES

The prescriptions and *regulations* contain the rules for safe working and instructions for handling the protective devices. First of all the related acts, standards, code of practices have to be applied but special instructions have also to be prepared for certain special operations.

For further details, see Sections 8.4 and 8.5.

Training. Persons, working with radioactive materials (Category **A**), should be qualified for such work by special training prescribed by the competent authority. For persons not directly involved in tracer experiments or in the installation of instruments containing radioactive sources (Category **B**), special instructions given in advance are satisfactory.

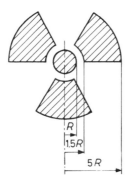

Fig. 8.16. International symbol of radiation hazard (the shaded part is red against a yellow background)

Authorization, licensing. The use of radioactive materials in most countries must be licensed by the competent authority. Radioactive materials, instruments or equipment containing radioactive sources may be delivered to the user only if such a licence is provided. This is the case when radioactive products are intended to be *borrowed* from other institutions, too. The *transport* of radioactive materials might also need authorization.

However, the application of a radioisotope with activity below the so-called "*exempted quantity*" can be carried out without requesting a licence.

Radiation hazards can be *indicated* by barriers, notices, and audiovisual signs together with the international radiation hazard symbol (Fig. 8.16).

A health physicist (radiological officer, etc.) is required to be appointed in all laboratories or at other premises where work with radioactive materials is carried out. Generally it is advisory to designate a person as a factory (facility) health physicist (whose duties are listed below), where instruments with radioactive sources are permanently used, even if the licence does not contain such a requirement.

With regard to temporary tracer experiments with open radioactive materials the necessary protection measures are fulfilled by the health physicist belonging to the institution or firm carrying out the experiment.

The terms of reference of the *factory health physicist* include

— arranging any necessary licences for using radioactive materials or equipment;

— taking part in any installation work;

— obtaining necessary operational and safety instructions and technical data relating to radioactive sources;

— keeping records on radioactive sources. A record must first of all contain the technical data of sources (name, activity, date of production, description, serial number, etc.), the date and place of installation, and all relevant later events (checks, damage, replacements, etc.);

— regularly inspecting the integrity of sources, the conditions of containers or installations and that the barriers and signs are adequate;

— keeping survey instruments needed for monitoring at hand;

— taking part in all preventive and counter-measures concerning incidents (accidents, breakdowns) and reporting all radiation accidents to the health authority and to the manufacturer or institution that installed the equipment.

8.2.3. PACKAGING AND TRANSPORT OF RADIOACTIVE MATERIALS

8.2.3.1. PACKAGING

Packaging means an assembly of components necessary to ensure compliance with the packaging requirements of transport regulations. It is usually known as a container, which is correctly a component of the packaging, or it is a freight container. The packaging has to comprise the radioactive material, to attenuate the radiation and to protect the source against mechanical effects and fire. A packaging is generally required to *consist* of:

— receptacle (for liquid or gas it must be able to be hermetically sealed);

— absorbent material to damp vibrations and to absorb liquids if they spill;

— radiation shielding to attenuate radiation;

— devices for absorbing mechanical shock and thermal insulation;

— an external covering.

The main requirements to be applied:

— the packaging has to be securely closed with a positive fastening device such that it will remain closed during transportation even if subjected to tossing, etc.;

— the maximum dose equivalent rate is limited: it may not exceed 2 mSv/h (200 mrem/h) on the surface and 0.1 mSv/h (10 mrem/h) dose equivalent rate at 1 m from the surface of the package at any moment during the transport;

— the maximum activity of radioisotopes allowed to be put in one package depends on the type of package and on the toxicity of the radioactive material, e.g., the maximum activity of ^{192}Ir in special form transportable in packaging type **A** is 740 GBq (20 Ci) and for packaging type **B**, which is constructed in such a way that it is able to withstand more severe traffic accidents, the activity is unlimited but clearly specified in its approval document;

— radioactive contamination is limited on the package surfaces;

— the package has to retain its containment and shielding integrity during transport, i.e., to the extent required by specified tests;

— the package has to bear specified labels (including symbol in Fig. 8.16) and any others required by transport regulations.

8.2.3.2. TRANSPORT

People and radiosensitive goods, e.g., undeveloped films affected by the transport of radioactive material, should be protected against external radiation, and radioactive contamination has also to be avoided. Transport may be by regular transport means: road, rail, or by special conveyance, e.g., by the firm's car. In all cases, regulations of packaging, loading, etc. must be followed, but other regulations concerning transport of goods must be complied with as well.

Instruments, manufactured articles such as clocks, electronic tubes or apparatuses having radioactive material as a component part *shall be exempted*, provided that these items are securely packed and the conditions such as stated below are fulfilled:

— the dose equivalent rate at any point on the external surface of the package shall not exceed 5 μSv/h (0.5 mrem/h);

— the dose rate at 10 cm from the surface of any unpacked equipment shall not exceed 0.1 mSv/h (10 mrem/h);

— the activity of an instrument or article and per package is limited, see Table V in [8.11].

In cases *when the* above mentioned *requirements are not met the source must be dismounted and transported separately* or the whole consignment should be considered as a radioactive package.

Main transport and loading regulations:

— the number of packages allowed to be put into one vehicle or aircraft is generally limited;

— the dose rate shall not exceed 2 mSv/h (200 mrem/h) on the surface of the vehicle;

— safe segregation distances must be kept from places occupied by people; undeveloped photographic films can be exposed to a maximum of 0.1 mGy (10 mrad);

— the radioactive shipment is required to be indicated on the outer surface of the transport vehicle by the prescribed signs.

With a full load, i.e., the consignor has the sole use of a vehicle, all the above mentioned regulations are required to be kept except the requirements of loading and the surface dose rate of the package is allowed to exceed the value of 2 mSv/h (200 mrem/h) up to a maximum of 10 mSv/h (1 rem/h). However, in such cases all initial, intermediate and final loading and unloading has to be carried out in accordance with the directions of the consignor or consignee.

The contamination of the vehicle should be checked after final unloading.

8.2.4. HANDLING OF WASTES AND CONTAMINATED PRODUCTS, DECONTAMINATION

In *tracer experiments* using unsealed radioactive substances all procedures for handling and disposal of contaminated products, materials and wastes should be planned in advance.

If the radioactive concentration is higher than the limit given in Section 8.1.3 or as stated by any other regulations, then the products before their commercial use should be *stored until their concentration decreases* to the prescribed level. For products whose route can be followed and whose ultimate destination is definitely known the permissible concentrations should be determined separately.

If these levels are exceeded the product or *material should be kept separately from the production process* for the necessary period.

If it cannot be expected that the activity of the product will decrease to the permissible level within a reasonable time then *it should be handled as waste and should be disposed of.* Discharge of low activity or low concentration waste into the public sewage system or into the air is regulated by the relating standards and codes of practice.

Radioactive waste is also produced by the *decontamination* of working surfaces, having been contaminated during the normal operations of a tracer experiment, or either, accidentally. It should be noted that sealed sources, if they are damaged, may also contaminate their environment.

Great attention and experience is needed in carrying out decontamination because of the higher risk of incorporation and spreading of contamination. The main rules are:
— selection of the proper and safe procedure (dust cleaning, washing with dissolvers or with detergents, scrubbing the surfaces, etc.);
— the proper order of actions;
— the use of protective clothes and devices;
— continuous monitoring;
— safe collection of contaminated wastes.

Wet decontamination can be started by washing using diluted solutions of household washing agents or detergents. In the case of metal surfaces the composition of one of the recommended solutions is 20–30 g sodium- or ammonium-citrate and 1–2 g of detergent per litre; in the case of glazed pottery, glass, plastic and rubber, a mixture of 5 g acetic acid, 10 g oxalic acid, 10 g sodium-hexa-meta-phosphate and 2 g detergent per litre water can be applied. If these solutions do not prove to be efficient enough, then 0.1–2 M solution of sulphuric acid or nitric acid can be applied. In certain cases special chemicals give the best results, e.g., the water solution of Na-EDTA or citric-acid in concentrations of a few thousandths per mille in most cases eliminates cations but for anions the best decontamination agent needs to be found experimentally.

8.3. SUPERVISORY METHODS

Checking the efficiency of protection includes:
— determination of personal doses;
— medical supervision;
— radiological investigation of working places and their surroundings;
— record keeping.

8.3.1. PERSONAL DOSIMETRY

The most adequate method would be the detection of *biological changes* produced by radiation but the applicability of such methods is limited to higher doses only. Moreover, the direct measurement of dose absorbed in the body or in a particular organ should be substituted by estimation based on indirect measurements. The dose can be determined in the case of

external irradiation:
— by individual dosimeters on the body surface;
— through the measurement of dose rate at the work place and estimating residence time (see Section 8.3.2);
internal irradiation is determined through the measurement of activity incorporated:
— direct measurement of radiation emerging from the body;
— excretion analysis.

8.3.1.1. PERSONAL DOSIMETERS

The condenser chamber (pocket ionization chamber, pen dosimeter) is a small ionization chamber which is charged from time to time. The chamber potential decreases as the charge produced by ionizing radiation discharges the chamber. The decrease of chamber potential is proportional to the dose delivered to it or to the exposure at the location of the chamber. The momentary chamber potential is measured directly by a built-in electroscope or by an external reader.

One of the desired features of personal dosimeters is energy independence. This can be achieved by using an air-equivalent chamber wall of appropriate thickness.

These requirements can be met theoretically for the γ-energy range of 0.05–3 MeV but many existing constructions do not realize this and they are significantly energy dependent at energies lower than 0.2 MeV. For low energies, i.e., for soft γ- or X-radiation, only special thin-walled chambers can be used.

The self-reading condenser chamber consists of a built-in electroscope (Fig. 8.17): the dose is read directly with the help of an optical system. The chambers most commonly used in practice have a range of about 2.0 mGy in air (200 mR) but other types with higher dose ranges up to about 10 Gy in air ($\sim$1 kR) are also used mainly as *accidental dosimeters.*

The use of *film badges* is based on the blackening of silver-bromide emulsion due to ionizing radiation (see Section 8.1.3) so this blackening determined, e.g., by absorbance (*optical density*) measurement can be a measure of the dose. The sensitivity of film badges

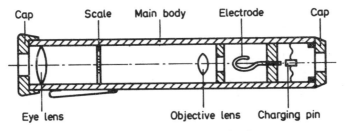

Fig. 8.17. Self-reading condenser chamber

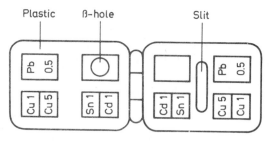

Fig. 8.18. Film badge

depends on the γ-*ray* energy but correction for this can be done at the evaluation if the radiation quality has been determined by a suitable filter system. The films, cut and packed to size by the producer, should be used in a plastic badge (Fig. 8.18). The common film badges cover a γ-dose range of 100 μGy–5 Gy in air ($\sim$ 10 mR-500 R). Film badges under a suitable window measure also the β-*doses* if the energy is higher than a few tenths of an MeV. The dose of *thermal neutrons* is determinable if the badge contains Cd and Sn filters; in this case the difference between the readings under these two filters will be proportional to the thermal neutron dose as γ-radiation causes equal absorbancy (blackening) under both filters. The dose determination of *mixed radiation* is also possible by using appropriate filter combinations.

It is a time-consuming as well as a delicate task to develop, evaluate and calibrate film dosimeters, so it is usually centralized. One of the advantages of film badges is that developed films can be preserved as documents.

Luminescent dosimeters utilize the luminescence phenomena also exhibited by many natural substances, when after exposing them to ionizing radiation they emit visible light if they are stimulated by ultraviolet radiation or heating. The former is called *radio-photo-luminescence*, e.g., on silver activated phosphate glass, the latter is called *thermolumines-cence*, usually on activated (Mn, Dy, Ti, etc. doped) LiF, CaF_2, $CaSO_4$, $Li_2B_4O_7$, BeO and Al_2O_3 crystals. The luminescence intensity is measured by a reader consisting of a stimulating device (UV light source or heater), photomultiplier and auxiliary electronics. The light intensity vs. time or i.e., temperature, in the case of heating can be recorded by a plotter (glow curve), and/or it can be integrated in fixed intervals and displayed directly as electric charge (integrated photomultiplier current) or converted to dose according to calibration.

Nowadays, thermoluminescence is a widely used method due to the highly sophisticated readers and good quality dosimeters available on the market.

Luminescent dosimeters are small, applicable in wide dose and energy range and their evaluation is simple and quick. The measurable dose ranges from μGy up to kGy (from 10^{-4} up to 10^5 rd) and besides β- and γ-radiation they may be used for neutron dosimetry as well.

Track detectors are suitable for neutron fluence measurement. If an appropriate nuclear emulsion is used the tracks of protons knocked out by fast neutrons may be counted by a microscope, but the dose range of this method is narrow: only 1–20 mSv (0.1–2 rem). Solid state track (or etch-pit) detectors utilize the damage caused in certain materials (mica, plastics) by the penetrating particle. The tracks become visible and countable after etching. Their important advantage is the lack of γ-background.

Accidental doses of fast and slow neutrons are generally measured by activation methods, using, for example, P, Rh or S and Au or In foils, respectively. The dose is calculated by the use of the neutron spectrum but this is rather complicated because the activation foils are sensitive only to certain parts of the spectrum.

8.3.1.2. DETERMINATION OF INCORPORATED RADIOACTIVITY

Direct measurement of radiation emitted from the body can be done in the case of γ-*emitters* and with a strongly limited sensitivity at harder *β-radiation* through bremsstrahlung measurement. The composition and amount of incorporated radioactive material can be determined from the γ-energy spectrum usually taken by scintillation detectors. In some cases a well-shielded detector is put close to the *organ or part of the body* where the given radioisotope is concentrated most, e.g., radioiodine measurements at the thyroid gland. It is a more complex but usually a more sensitive method when the whole body together with the detector are put into a large and well-shielded room—into the so-called *whole body counter*. The shield of such chambers (Fig. 8.19) is usually made of iron free from radioactive contamination. The sensitivity of whole body counters is high, a small fraction of the *MPBB* (or ALI) can be determined.

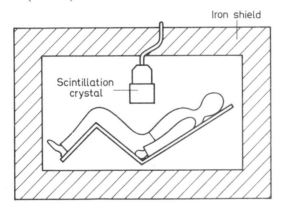

Iron shield

Scintillation crystal

Fig. 8.19. Whole body counter

Excretion analysis is based on the fact that incorporated radioactive materials are excreted through metabolic processes from the body. The body burden can be estimated from the radioactive concentration of the excreta (urine, faeces, expired air, etc.) assuming that the fate of the radioactive material in the body is also known. These methods are important especially for α- and soft β-emitters (^{3}H, ^{14}C) because such nuclides cannot be measured directly. The method is useful also for γ-emitters to obtain additional information or maybe the only method when the whole body counter is not available.

The results of an excretion analysis, under the conditions of the industrial application of radioisotopes, may be used to establish or exclude the fact of an incorporation, but the burden of a particular organ can only be roughly estimated because the metabolic processes are not fully understood.

In case of significant incorporations observed whole body counting and/or exerction analysis should be continued to determine individual parameters for more accurate organ dose estimation.

8.3.2. WORK PLACE MONITORING

Work place monitoring generally consists of:
— *dose rate measurement* in the close proximity of radioactive sources;
— *radioactive contamination measurements* on the surfaces of equipment and in rooms, in the laboratory air, in wastes (sewage water, etc.), and in industrial products.

8.3.2.1. DOSE RATE MEASUREMENT

The main aim is the measurement of dose rates of γ, sometimes β- or neutron radiation from a few μSv/h (a few tenths of mR/h) up to a few mSv/h (few hundreds of mR/h). Here only portable monitors are discussed.

Monitors with *ionization chambers* must have quite a large chamber in order to reach the desired sensitivity (see Section 8.1.5). Such monitors can be used in not very inhomogeneous fields even to measure soft γ- or β-radiation if their wall is thin enough. Their great advantage is their energy independence.

Monitors with *proportional counters* are, in principle, similar to ionization chamber monitors except that their sensitivity at small detector volume, is increased by the gas amplification due to the high voltage applied to the counter. The proportional chamber can be used both in current and pulse (counting) mode.

The most common radiation monitors use *GM-counters*; many versions of such monitors are commercially available. With suitable GM-counters α-, β-, γ-radiation and neutrons can be detected but as a dose rate meter it is good only for γ-radiation in a limited energy range, i.e., 0.2–3 MeV, and with an error which cannot be disregarded. For practical purposes it works well, it is cheap and the desired accuracy in practice can be achieved.

Care should be taken in the use of GM-monitors when soft γ-radiation is measured because *the sensitivity of the GM-counter becomes strongly energy dependent, and also at high intensities, e.g. when searching for a source which has been dropped and has rolled away, many types of monitors will give a false response or no response at all.*

Among the *scintillation detectors*, sodium iodine, activated with Tl [NaI(Tl)] and, more rarely, plastic scintillators have found a use. Scintillation detectors are more sensitive than the detectors discussed previously: a dose-rate of a few tens of nGy/h (μR/h) is measurable but the upper range is also low, around a few tens μGy/h(mR/h). Their further drawback is the strong energy dependence and insensitivity below 0.05–0.1 MeV. For fast neutron flux density measurements ZnS(Ag)-plastic scintillators are widely used but these have to be calibrated with known neutron sources, or more precisely with a known neutron spectrum.

8.3.2.2. SURFACE CONTAMINATION MONITORING

For direct measurement the detector sensitive to the particular radiation is brought close to the surface and the particle flux density is counted. From this measurement the contamination can clearly be located and if the detector is calibrated, the contamination can be expressed in units used by the regulations (e.g., in kBq/cm^2 or μCi/cm^2). However, the numerical value of such calculation is uncertain even if the radioisotope is known and appropriate calibration has been done because it may be uncontrollably altered by the *surface quality.* In some cases, especially in industrial conditions where any contamination should be completely removed, it is sufficient to measure particle flux and decide whether the surface is still contaminated or not.

α-*Emitters* could be measured by endwindow GM-counters, if the window thickness is less than 20 g/m^2 (2 mg/cm^2), but ZnS(Ag) scintillators are more frequently used.

Hard β-*emitters*, when the energy is higher than 0.3–0.5 MeV, may be detected by most of the portable GM-monitors. Special, portable β-contamination meters or detectors connectable to ratemeters or counting devices, are also produced. Such equipments usually have a large sensitive surface using several GM-counters or plastic scintillators.

β-emitters, when the energy is around 0.06–0.5 MeV, could be monitored by endwindow type GM or proportional counters. The measurement is more convenient and efficient if larger surfaces are used; such larger surfaces can be achieved by putting more detectors together or by using a multiwired detector having a common gas atmosphere.

β-*emitters* with energy less than 0.05 MeV, basically this refers to tritium, cannot be detected even with such kinds of detectors. Recently, windowless proportional and special scintillation counter probes are available for direct tritium monitoring. However, the ^{3}H concentration of air exhausted from the surface can be directly measured by flow chambers or counters but this is, in principle, a sampling method.

There are only a few pure γ-*emitters* among the radioisotopes therefore the accompanying β- or α-radiation is measured to which the monitor is more sensitive. A β-radiation monitor is sensitive to γ-rays, too. When there is no accompanying charged particle or it is too weak or contamination is fixed to a deeper layer, the scintillation monitors discussed previously are useful for contamination detection.

General purpose meters detect all types of contamination and most of the dose rate measurements could be carried out by them.

With regard to procedures based on *sampling*, the surfaces are *wiped* with some soft material and the activity of this sample is measured. Scraped samples may also occur but they are rarely used for quantitative evaluation.

The material used for *wipe tests* is usually a filter paper, cotton-wool or textile which may be dry or wetted with water or special dissolving agent. For quantitative evaluation the size of the tested surface and the efficiency of sampling (wiping) has to be known. The sampling efficiency is variable and uncertain, it depends on the force and speed of wiping as well as on the type of dissolver, isotope, surface quality etc. For a given case, sampling efficiency can be estimated if sampling is repeated.

Smear test samples can be measured by portable monitors but for more precise determination or analysis laboratory instruments are needed (see Section 2.2).

8.3.2.3. AIR MONITORING

The continuous measurement of radioactive concentration of *gas or vapour contaminants*, such as ^{3}H, ^{14}C, ^{41}Ar, ^{85}Kr, ^{203}Hg, and their compounds may be carried out by a *flow type ionization chamber*. In order to reach the desired sensitivity a chamber volume of about a few or a few tens of litres is required. The current produced in such a chamber is measurable by an electrometer (Fig. 8.20), e.g., the sensitivity of a chamber with 20 l volume is 25 mBq ^{3}H/cm^3 ($\sim 10^{-6}$ μCi/cm^3) which corresponds to one-fifth of the MPC_{air}. Recently, more sensitive flow type proportional counters have become commercially available; their sensitivity is around 0.25 mBq ^{3}H/cm^3 ($\sim 10^{-8}$ μCi ^{3}H/cm^3).

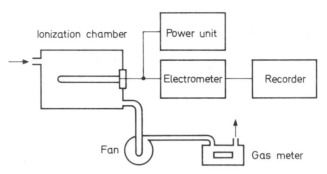

Fig. 8.20. Air monitor with ionization chamber

The most common method used for *aerosol monitoring is to filter out* and then to measure the activity of the aerosols. Using a filter moving slowly in front of the detector the air contamination can be controlled continuously assuming that the air flow-rate is known. The use of such highly sophisticated laboratory instruments under industrial conditions is not reasonable because of their cost and, furthermore, the average concentration may be determined by discontinuous sampling, too.

A discontinuous sampler consists of a sample holder, a flow-rate meter and an air blower. If the flow-rate is known the anemometer may be omitted. A vacuum pump or a suitably modified vacuum cleaner can be utilized as an air blower. Activity, filtered on glass fibre, plastic or cellulose filters, should be measured by sensitive laboratory instruments because of the usually low activities.

8.3.2.4. INDUSTRIAL PRODUCTS AND WASTES

Air contamination monitoring of *gas products* may be carried out by the methods discussed above.

The control of *liquid* and *solid* end- or by-products or wastes does not usually require special instruments because most of the radioisotope applications are carried out by sampling, and instruments used for sample measurement can be used for health physics purposes, too. In other cases, activity of a solid waste can be determined from dose rate

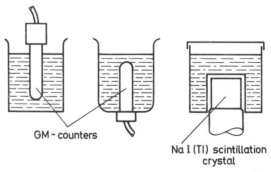

Fig. 8.21. Detector arrangement for activity measurement of liquids

measurements performed on the surface of a vessel or at a given distance from it. The direct measuring method of a product should be specified separately.

The activity of *sewage water* or other liquid wastes can be controlled directly by *immersion or bell type detectors* (Fig. 8.21). If higher sensitivity is required the sample has to be concentrated by evaporation, chemical precipitation, ion exchange enrichment or by some other way before its measurement.

8.3.3. RECORDS

Record keeping (accountancy) specified by regulations gives information on the amount, location, movement, etc. of radioisotopes and may thus be considered as a *direct control method*. For example, a measurement made around a container and showing a dose rate lower than permissible does not mean a reduced radiation hazard if the reason for it is the loss of a part or the whole source.

All other kinds of recording are regarded as indirect methods, these are any which have a bearing on safety.

The usually recorded data are:
— approval of investigation or instrument installation;
— transport licences, transport documents;
— radioisotope orders;
— records of radioisotopes (shipping documents, registration book, storage register);
— instructions concerning the investigation or installation;
— minutes relating to investigations or installations;
— documents relating to special training materials;
— notifications (on loans, damage, etc.).

8.4. USE OF SEALED SOURCES

8.4.1. DEFINITION OF SEALED SOURCE AND LEAKAGE TEST

A sealed source is a radioactive material embedded in a material and/or *permanently sealed* in a capsule in such a way that under normal conditions no leakage can occur (see Section 1.4.2). In this case only the outcoming radiation is utilized, therefore first of all the hazard from external radiation has to be considered but the possibility of contamination due to fracture of the capsule should not be disregarded.

Capsules, e.g., of welded steel for sources of *penetrating radiations* (γ, neutron), should resist all expectable thermal, mechanical and chemical effects. Experience has shown that a proper level of safety can be achieved only by using *double encapsulations* therefore nowadays sealed sources with higher activity are prepared in such a way (see Section 1.4.2).

α- and β-sources, because of the *strong absorption* of these radiations need only be *covered with thin foils* (see Section 1.4.2). These foils, such as beryllium, aluminium, etc., under normal handling ensure the integrity of sources but they are very fragile, can easily be perforated and are less resistent to corrosion. For this reason radioisotopes are frequently *embedded into glass, ceramic or glaze*. Using such sources contamination of the environment can be avoided or reduced and uptake of contaminants to the body may be prevented even if the protective foils are damaged.

Before use of a source all data relating to the *design and construction materials* should be requested from the manufacturer or supplier. Leakage testing of the source must be done not only during production but *regularly* because, especially in industrial conditions, it is liable to be subjected to thermal effects, shock, vibration and to corrosive atmosphere. If the producer does not supply instructions for leakage testing and for its frequency, it is advisable to *check the source by wiping (smear test) or soaking* according to the existing regulations with a frequency depending on the circumstances.

If the smear test (see also Section 8.3.2.2) is used, the source surface should be wiped with a tampon wetted in a dissolver, e.g., water, toluene, dioxane. With the soaking method the source should be kept for eight hours in dissolver at 50 °C or for thirty minutes in boiling dissolver. The dissolver used should not attack the capsule. Sources may be regarded as sealed if the activity of the tampon or dissolver is less than ~ 2 kBq (~ 50 nCi). Sources covered with thin foils should be wiped very carefully; such tests should be done solely by experts in a suitably equipped laboratory.

Leakage testing of a source built into a *working container* may be done, by an adhesive tape put on the radiation exit if this is not an obstacle to its normal use. The *adhesive tape* should be changed from time to time, its activity checked, and incidental contamination revealed without dismounting the source. However, this method cannot substitute the regular test discussed before by which damage of the capsule may be discovered at an early stage.

8.4.2. CONTAINERS FOR SEALED SOURCES

Radioisotopes can be transported to the place of installation in *transport containers* (i.e. packagings) or in instruments in which they are incorporated (see Section 8.2.3). Construction of a transport container is such that 2 mSv/h (200 mrem/h) of surface dose rate is allowed according to the IAEA recommendation (see Section 8.4.4.1) but the permissible dose rate on surfaces of working and storage containers is lower, according to some current national regulations, e.g., max. 0.2–1 mSv/h (20–100 mrem/h). If the dose rate on the surface of a transport container exceeds the limit prescribed for storage conditions, the source immediately after its arrival

— must be put in a working or storage container; or

— the shielding of the transport container must be strengthened by distance-keeping or by additional shielding.

Storage containers are needed if radiation sources are used temporarily, under supervision, but without complete shielding, e.g., portable devices, radiography. The containers should be *lockable and should bear identification labels and the radiation symbol* (see Fig. 8.16). Transport or working containers may be used as storage containers if the dose rate on their surfaces does not exceed the prescribed level. Generally, it is reasonable to have a storage container constructed as an addition to the working container instead of replacing the source occasionally into a separate holder. It is practicable to determine experimentally the "capacity" of a storage container with those amounts of radioisotopes which, if they are put into the container, give the maximum permissible surface dose rate. The values so obtained should be indicated on the identification label and noted in the records.

With high activity sources (irradiation or radiographic facilities), the storage container is substituted by the installation itself (Figs 6.15, 6.17 and 7.6).

The working container is a device which contains the radioactive source during its use and fulfils the following functions:

— it attenuates the radiation, except for the useful beam;

— it protects the source against mechanical and corrosion effects;

— it prevents unauthorized access to the source.

The maximum permissible surface dose rate of a working container is, e.g., 0.2 mSv/h (20 mrem/h); in some special cases it is 1 mSv/h (100 mrem/h) according to the field of application (see Section 8.4.4.1). The dose rate at a distance of 1 m is also limited because without such a limitation the dose rate would be significantly high even for large distances, because there is no restriction on container size.

Construction of a working container depends on the type of source and on the conditions of application. Containers for *α- and β-sources* are usually made of aluminium or steel, sometimes plastic. A thickness of a few mm, fulfilling mechanical requirements is usually sufficient to absorb fully the primary radiation. Most common containers of *γ-sources* are made of lead or steel; tungsten or depleted uranium have also come into use. Typical working container designs are shown in Figs 2.1, 7.8 and 7.13.

Iron vessels filled with paraffin usually serve as a *neutron source* container. A layer of paraffin of about 20–30 cm is sufficient to thermalize fast neutrons. Sometimes boron is added to the paraffin to capture thermalized neutrons. If the source also emits intensive γ-radiation as in the case of ^{226}Ra–Be, an additional lead shield should be used around the source. The necessary thickness of this shield can be calculated as shown in Section 8.1.4.2.

8.4.3. GENERAL ASPECTS

8.4.3.1. METHODS OF PROTECTION

When using sealed sources, hazards arise mainly from external radiation. *Protection* has three direct forms:

— *distance keeping*: in most cases the working container, mode of installation, the equipment itself all ensure a safe distance between the source and personnel. Safe distances needed for an unshielded source (e.g., replacement of source) may be estimated in accordance with Section 8.1.4.1. Touching the source with bare or gloved hands could be very harmful even with very low activities;

— *use of shields*: containers, shielding slabs constructed from materials selected and designed according to the type and activity of radiation source (see Section 8.1.4.2) reduce radiation to the prescribed level;

— *time shortening*: needlessly remaining near to the source should be avoided. Careful planning, preparation and training of personnel must be carried out in advance of every dangerous operation.

Indirect methods of protection are regarded as those preventive and control measures which ensure efficiency of protection. For details, see Sections 8.2 and 8.3.

Incorporation hazards when using sealed sources do not arise normally but leakage testing should be done regularly, because, neglecting the *possitibility of contamination*, may lead to severe injury to people or damage to goods. The importance of leakage tests is emphasized by the fact that sealed sources frequently contain radioisotopes which are very toxic and have long half-lives. The maximum permissible body burdens for Category **B** and for continuous intake (see Section 8.1.3), are, for example 7.4 kBq (0.2 µCi) of ^{90}Sr, 0.37 kBq (0.01 µCi) of ^{226}Ra and 0.185 kBq (0.005 µCi) of ^{241}Am. These amounts, related to the total activity applied, are so small that even a very slight leakage may lead to serious health hazards. The same is valid if *ALI* values are considered [see Table 8.3 (b)].

8.4.3.2. PREPARATORY MEASURES

Prior to using sealed sources the following radiation protection measures should usually be taken:

— *application for licence* (for details see below);

— *selection or construction of working container* (duty of the manufacturer or institution installing the source) (see Section 8.4.2);

— *obtaining or preparation of the operational instructions*, safety regulations (see Section 8.2.2);

— *training of operators* and instruction of personnel (see Section 8.2.2);

— *transport* of radiation source (see Section 8.2.2).

Requests submitted for registration or licensing based on national regulations generally differ in their form according to the field of application but in all cases the requests must contain data of the radiation sources and instruments, purpose, description and place of application, data on all responsible personnel, and the radiation protection measures to be applied.

8.4.3.3. INSTALLATION AND CONTROL

Radiation hazards arising during the installation of radioactive sources are determined by the type and activity of sources and by the form of application.

At permanently installed sources, e.g., with sources incorporated in instruments, care should be taken regarding permissible dose rate levels around the source where people are working continuously (see Section 8.1.3). In many cases the radioactive source has to be transported in a separate container and transferred from it on the spot. Such operations are hazardous in themselves (the source may become unprotected for a short time; it may roll away and then be handled incorrectly, thereby leading to overexposure) therefore loading operations should by done only by qualified staff.

Sources in portable instruments need a lockable store with proper shielding and/or a fence at safe distances, if necessary. Warning signs (notices, audiovisual warnings) specified by regulations should also be used.

Monitoring with an appropriate radiation measuring instrument (see Section 8.3.2.1) should be done immediately, e.g., dose rates around the equipment ought to be measured in order to show that their values do not exceed the prescribed levels (see Section 8.1.3). This checking may reveal the necessity for additional shielding. Safety measures (signs, locks, etc.) should be inspected, too. The inspection of source integrity, signs, barriers, locks should also be carried out regularly after installation.

8.4.3.4. SERVICE, MAINTENANCE

Service or maintenance of equipment containing the source may be required during its operation. If the source is built in a separate unit then if the beam shield is closed or the source holder dismounted, servicing could be carried out, but the source holder unit should be dealt with *only by qualified personnel.*

Special care should be taken when the source does not have a separate holder but is directly built into the industrial equipment. Those operating such equipment should be instructed which parts may be serviced by them and which need the supervision of an authorized person. It is advisable that the factory *health physicist* is notified about all service, maintenance and modification of equipment because such work is dangerous, frequently leading to overexposure, damage or even to loss of the source.

8.4.3.5. INCIDENTS, ACCIDENTS

The breaking down of industrial equipment involving radioisotopes or any accident (fire, explosion, etc.) in the surroundings, may result in unexpected situations in which radiation hazards (if sources become unprotected or damaged) arise too. Even if the source has remained unchanged there is still the possibility that it may become damaged during emergency operations. In view of this, and provided that the source is in a separate shielded unit, it is advisable to take the unit immediately to a safe place.

If an accident occurs *the operating engineer in charge* should act with all possible speed—as it is unlikely that there would be time to seek a *health physicist's* advice. However, the factory health physicist should be notified immediately since it is his duty to deal with emergency actions. Interests both of production and safety (including *radiation protection*) should be balanced, though obviously the matter is rather complex and difficult. Therefore, it is recommended that operational instructions together with factory regulations include an *emergency plan* taking into account radiation hazard, too. A *radiation protection inspection* is recommended after all factory incidents.

Accidents in which a source becomes unshielded or the environment contaminated may lead to injury to individuals even if such injuries cannot be observed directly. Therefore in such cases *the competent authority must immediately be notified.* Actions to try to prevent the likely consequences of an accident should be taken by experts, the area *must be closed* until their arrival and *persons thought to be contaminated, should not leave the territory.* It is emphasized that decontamination procedures require considerable experience (see Section 8.2.4).

8.4.4. SPECIAL DUTIES

8.4.4.1. INDUSTRIAL NUCLEAR INSTRUMENTS

Activity of sources used in industrial nuclear instruments is generally in the range of MBq–GBq (cca. mCi–Ci); built-in calibration sources have an activity around kBq (nCi–μCi). Activities are determined by conditions of measurements but maximum permissible dose rates, depending on types of application, should also be considered.

The International Organization for Standardization (ISO) *classifies equipments according to its shielding properties*:

— Category I—equipment which itself ensures a high degree of protection;
— Category II—equipment which itself ensures acceptable protection;
— Category III—equipment which needs additional shielding.

The maximum permissible dose rate equivalents are set according to these categories. *The shielding design* of working containers (see Section 8.4.2) may be done according to Section 8.1.4.2. *Barriers or fences* should be placed so that the dose rate behind them will not exceed the appropriate *MPD*, e.g., for Category **B** (factory employees), 2 µSv/h (0.2 mrem/h) (see Section 8.1.3). If there is a permanent working place close to the barriers then a safety factor of two is recommended. Containers and barriers should bear the standard radiation symbol (see Fig. 8.16) and a warning label. If construction of equipment is such that one could place one's hand in the radiation beam, barriers, screens or guards should be used to prevent this.

General aspects relating to industrial nuclear instruments are discussed in Section 8.4.3. *Operational instructions* of instruments are especially important and they should contain in a clear, comprehensible form, the safety precaution rules, too.

8.4.4.2. INDUSTRIAL RADIOGRAPHY

Laboratory work may be carried out with relatively high activity sources because *permanent installations* can be individually designed enabling effec ive shielding to be provided and a high level of safety to be achieved. If the radio-isotope *storage well* (e.g., a steel-lined pipe) is deep enough and the sources are handled by manipulators, operations will be safe and radiation burdens low. The *entrance door* of the irradiation room should be supplied with an automatic locking system which ensures that if the entrance door is open the sources are retained in their storage position and they can be lifted up only after the room has been vacated (see Fig. 7.6).

With the use of medium or lower activity sources there is no need in every case for individually designed installation, the protection of the environment may be achieved by using *suitable storage and working containers* and, if it is necessary, the shielding of the laboratory walls may be strengthened. For design instruction see Section 8.1.4.2.

The maximum permissible dose rate on storage container surfaces is also limited. For example, in some countries it is equal to 20 µSv/h (2 mrem/h). This value for a working container, if it is movable only by hand and by one person, is (in many countries) equal to 0.1 mSv/h (10 mrem/h). However, if it is possible to keep a distance of at least 1 m, e.g., by hanging the container on a bar and having two people to carry it, the value will be 1 mSv/h (100 mrem/h).

With investigations made on site the same or similar portable devices are used as with laboratory work so regulations of working and storage containers are the same. Recently, containers with *remote control operation* (Figs 7.9–7.13) have become more widely used; this is also justifiable from the point of radiation protection. If there is no laboratory assigned for radiographic work then a *lockable store or storage container* should be used. Sources are allowed to be transported in their working containers within the factory because the dose rate requirements of transport containers are not so strict. Not all working containers are suitable for transport outside the factory because most of them do not fulfil all the requirements set for regular transport. In such cases, it is more reasonable to use the company's car for transport but road transport regulations should be considered.

Areas where operation is being carried out should be *fenced off and warning signs placed*. Supervision is needed throughout the whole time of the experiment to keep unauthorized persons from the sources. Barriers should be set up in such a way that the dose rate behind them will be lower than the level permitted for Category **B**, e.g., 2 μSv/h (0.2 mrem/h).

EXAMPLE 8.7

An ^{192}Ir source with activity of 40 GBq (1.1 Ci) is used for on-site work. How far should the barriers be placed from the unshielded source?

If Eq. (8.6) is applied, taking it into account that the maximum permissible dose rate is equal to 2 μGy/h in air (about 0.2 mR/h) and $K_\gamma = 104$ μGy × m^2/(GBq × h) or $K'_\gamma = 4.44$ Rcm^2/(h mCi); see Section 1.3):

$$R^2 = K_\gamma \frac{A}{I_p} = 104 \frac{40}{2} = 2080 \text{ m}^2$$

$$R = \sqrt{2080} = 45.6 \approx 46 \text{ m}$$

or in old units

$$R^2 = 4.44 \times 10^3 \frac{1.1 \times 10^3}{0.2} = 2.442 \times 10^7 \text{ cm}^2$$

$$R = \sqrt{2.442 \times 10^7} = 4942 \text{ cm} \approx 49 \text{ m}$$

8.4.4.3. RADIATION TECHNOLOGIES

Equipment and devices utilizing the *physical effects* of radiation are very different in form and construction so radiation protection problems also vary; in most cases they are specific.

Static charge eliminators use α- and β-sources. The precautions applied are similar to those discussed in Sections 8.4.1–8.4.3. It should be noted, however, that static charge eliminators are used in fields where fire and explosion hazards are considerable. For this reason, the use of very toxic and high activity sources should be avoided and preventive measures, besides other hazards (e.g., fire), should be extended to take into account possible damage of sources and contamination of area.

Besides α- and β-emitters, γ-sources are also used for the *sparking of electric discharges*. Usually, these devices are adequately sealed and their construction elements reduce external radiation to a negligible level, but the intensity of external radiation may become significant if a piece of equipment contains several devices. Similar problems may arise in the storage and transport of the product and especially during production processes when the whole amount of radioactive material is present. If the production of such equipment is on a large scale, detailed safety instructions should be applied regulating the handling, storage, transport, etc. of radioactive material.

Radioactive light sources (see Section 6.1.2.1) can be divided into two groups from the viewpoint of radiation protection. Components of watches or instruments marked with fluorescent paints cannot be regarded as sealed sources but the whole device may be considered sealed. The mounting processes of such devices and especially the washing off of the paints may be very dangerous. Unfortunately many luminescent paints contain

^{226}Ra which besides α- and β- also emits γ-radiation which is in any case inefficient in terms of its light yield. It is more safe to use non-penetrating α- and β-emitters embedded into glass or some other non-dispersible solid material. The use of ^{3}H or ^{85}Kr is especially advantageous because neither of these presents a significant incorporation hazard even with their slow leakage from the luminescent sources.

Radioactive electric power generators also utilize α- and β-emitters. Such batteries are produced for various purposes (e.g., pace-makers) and for devices used in space research (see Section 6.1.2.2).

The chemical and biological effects of radiation are utilized primarily in irradiation facilities containing high activity sources (see Section 6.4).

Small samples may be irradiated in a specially constructed container with appropriately designed shielding to limit external radiation to an acceptable level. However, sample handling should be done carefully; hazards due to contamination, intensive narrow beams and, in first place, the source becoming unshielded should be avoided.

Essential precautions are that there be continuous monitoring in the maze, and that operators are supplied with "*accidental*" *personal dosemeters* (having wide measuring ranges) and with *portable radiation meters*. Checking or replacement of sources may be done only by qualified staff.

Sources of an *irradiation facility* are transferred (usually lifted up) to the irradiation chamber or room from their shielded (usually underground) store only for the time of exposure, therefore materials, products to be irradiated should be positioned in advance.

The operator's room is protected by *suitably designed walls* while the irradiation cell may be reached usually through a *maze* (Figs 6.15 and 7.4). Irradiation facilities generally utilize sources with TBq–PBq (cca. kCi–MCi) activities, thus a few seconds' exposure may be lethal if the sources did not reach their storage position in due time.

Because of the extremely high radiation hazard, besides shielding, an automatic locking and warning system must also be used to eliminate, as far as possible, the consequences of false operation. This *technical safety system* should provide:

— distance handling of sources between the storage and irradiation positions;
— indication of source positions in the operational zone;
— locking of the entrance door: it should be openable only if the sources are in their storage position;
— immadiate *transfer* of the sources to the storage position if the door is opened, if there is a breakdown, if the main's supply is interrupted, etc.;
— the possibility for visual inspection by TV monitor or periscope;
— continuous dose-rate monitoring in the operator's room and at the entrance;
— display of radiation level in the operator's room and the connection of the signal to the locking system.

All information on the source positions and about any false operation should be displayed clearly in the operator's room and, finally, the whole system must be designed and constructed in such a way that accidental exposure of persons becomes impossible.

The integrity of sources should periodically be *inspected* by wiping the parts of instruments in direct contact with sources; moreover, continuous or discontinuous but regular checking of air and cooling water contamination should be carried out.

The loading and unloading of the sources is a very dangerous operation and to make it more safe and convenient, the use of a pool filled with a few metres of water is suggested.

8.5. APPLICATION OF UNSEALED RADIOACTIVE SUBSTANCES

8.5.1. GENERAL ASPECTS

8.5.1.1. METHODS OF PROTECTION

The methods of protection against external radiation are the same as applied for sealed sources (see Section 8.4.3.1), i.e., distance keeping, time shortening and shielding. An additional important danger associated with even the most simple operations regarding unsealed radioactive substances other than those prescribed in Section 8.4.1, is the *incorporation of radioactive material*; this is even more likely in industrial applications than in laboratory use. Protection against undesired contamination and incorporation needs well planned preventive measures (see Section 8.2) and monitoring (see Section 8.3). These special duties will be discussed further for cases when the radioactive material is introduced into the technological process.

8.5.1.2. PLANNING AND DESIGN

When a radioisotope is *selected* for use, isotopes having lower radiotoxicity (Table 8.8), shorter half-life and radiation of lower penetrating ability are preferable with regard to radiation safety. The radiation protection aspects of the experiment should be examined beforehand and *a suitable plan* should be prepared in which the expected qualitative and quantitative transformations of the tracer and products during the technological processes can be calculated, taking into account events, even if their occurrence is rare, which may lead to irregular processes or products.

Table 8.8. Classification of selected radionuclides according to their relative radiotoxicity [8.13*]

Very high ^{210}Pb, ^{226}Ra + decay products, ^{233}U, ^{239}Pu, ^{241}Am, other transuranic nuclides

High ^{45}Ca, ^{60}Co, ^{90}Sr, ^{110m}Ag, ^{125}I, ^{131}I, ^{140}Ba, ^{170}Tm, ^{192}Ir, ^{204}Tl, natural Th

Moderate ^{14}C, ^{24}Na, ^{32}P, ^{35}S, ^{42}K, ^{51}Cr, ^{54}Mn, ^{55}Fe, ^{64}Cu, ^{65}Zn, ^{82}Br, ^{140}La, ^{137}Cs, ^{147}Pm, ^{198}Au

Low ^{3}H, natural U**

* The classification may be somewhat different in other international and national regulations. E.g. ^{90}Sr is in the very high radiotoxicity group in several cases.
** 1 Bq U_{nat} corresponds to 1α disintegration per second (0.489 dps of ^{238}U, 0.489 dps of ^{239}U and 0.022 dps of ^{235}U).

As part of the precautions, precise *material balance* and analysis of physical and chemical processes should be performed. Such analysis gives information on the *expected contamination* of products and materials, about the *incorporation hazard* and on the *necessity for separate handling of products*. Furthermore, all preventive actions and monitoring programs of the tracer experiment can be based on this, too. The operational plan usually involves:

— *preparation of tracer material* and its transport to the place of use;

— preparation of *preventive devices* used against external radiation and contamination on the site;

— *introduction of the tracer* into the technological system;

— *monitoring* during and after the tracer work;

— estimation of *expected* concentration of *contamination* in materials (products);

— *estimation of* external and internal *personal doses*.

All tracer work *should be licensed* by the competent authority.

Trial experiments using non-radioactive substances should be performed in order to get practice and to check the design of the investigation and the effectiveness of the preventive actions. It is suggested that besides laboratory modelling, an experiment *to simulate* the radioactive tracing is carried out in the field.

As a part of the *preparation* procedure, everything necessary for safe handling of radioactive material (manipulators, shielding slabs, plastic foils, containers for wastes, means of decontamination, individual protective devices, barriers, warning signs, etc.) should be located and employees of the factory involved in the tracer work should be given *special instructions*.

If at all possible, *preparation of the radioactive tracer* should not be done on-site but in the laboratory because the introduction of a radioactive isotope into material suitable for tracing is usually a complex operation (involving the possibility of considerable radiation and contamination hazard). The use of unsuitable places for such operations may well lead to serious injury to individuals and a great deal of other damage.

In-plant transport from the laboratory to the field may be done using common transport containers, except for materials with larger volumes for which specially constructed containers should be used. Packaging and transport regulations are discussed in Section 8.2.3.

8.5.1.3. INTRODUCTION OF A TRACER

In industrial investigations, the introduction of a radioactive material into the system has been the *most hazardous operation* of all so far as radiation is concerned. The tracer is at its most concentrated form at the injection site therefore it is here that the hazard of external and internal exposure is the highest.

The procedure for tracer introduction varies from experiment to experiment: it depends on the physical and chemical form of the material, on technological parameters and on the tracer method. *Pulsed injection* and the use of a closed, shielded injector is preferable from the viewpoint of radiation protection. Injection should be carried out after thorough preparatory work together with appropriate safety precautions which can only be specified in the given case.

The objective is to prevent the contamination of the environment and incorporation and injection must be done within the shortest possible time to minimize external exposure of individuals.

For protection of the environment it is reasonable to line surfaces with strippable paints and/or plastic foils, to use trays to collect spillage and to supply personnel with individual protective devices, e.g., breathing masks, gloves, overalls, aprons, shoes.

The selection of the appropriate *breathing mask* is extremely important. The use of *oxygen rescue facilities* or plastic suits supplied with fresh air, providing complete safety, in some cases cannot be avoided. However, exaggerated prudence may hinder work so much that in the last analysis safety will suffer.

In most cases, *dust- and gas-masks or fresh air respirators* are suitable.

8.5.1.4. SURVEILLANCE

Industrial investigations using unsealed radioactive substances should be continuously monitored during the whole experiment. Practical methods of such control are discussed in Section 8.3.

The monitoring of *personal doses* is generally carried out by personal dosimeters and by activity measurement of incorporated materials if internal exposure is likely or may be suspected (see Section 8.3.1.2). Radiological monitoring should be extended to all employees in the factory who are involved in the experiments.

Workplace monitoring (see Section 8.2.2) is based on dose rate and contamination measurements.

Dose rates (see Section 8.3.2.1) should be checked on the accessible surfaces and at places where personnel are usually to be found.

Surface contamination (see Section 8.3.2.2) should be measured after all critical operations (introduction of the tracer, sampling, etc.). Because of the increased background due to the radioactive material, smear tests are usually needed. On completing the experiment direct instrumental control is absolutely necessary.

Air contamination must be measured whenever the tracer material is in the gaseous phase or its dust or vapour may be released to the atmosphere. In general, portable samplers are suitable for the monitoring program.

Checking to see whether *industrial products* (see Section 8.3.2.4) are contaminated may also be necessary during the experiment but the final overall check after completing the investigation is especially important. Decision on the fate of products involved in the experiment is based on such measurements, furthermore the contamination of apparatuses and workplaces can also be revealed.

A careful weighing material balance must be done to check the possible retention of radioactive material in the equipment or elsewhere. In addition, dose rate and contamination measurements should be carried out at all places (traps) where radioactivity might have been concentrated or released unexpectedly. Serious difficulties may arise if the tracer does not dilute completely and radioactivity is concentrated at some places.

It cannot be too highly stressed that the *final surveillance, after completion of the experiment and the removal of the contaminated products and/or wastes, should be extremely rigorous.* There should be a careful check of all equipment and the surroundings, and it is suggested that the results be recorded.

8.5.1.5. RELEASING MATERIALS, PRODUCTS

Materials in which radioactive isotopes may be incorporated during the experiment are:

— *consumer products*, e.g., glass;
— *raw materials, by-products* and *semi-manufactured products* to be reprocessed;
— *materials remaining in the technological process*, e.g., mercury in electrolytic cells;
— *wastes* which are partly collectable, partly released to the environment through liquid wastes and discharged air.

General permissible levels of radioactive concentration of materials (also semi-manufactured products) cannot be established: such levels had to be determined preliminarily during the planning of the experiment assuming that individuals or groups will not receive doses higher than the permissible limit.

If the radioactive concentration of the product is higher than the permissible level, the product or materials must be kept isolated in controlled storage until, due to physical decay, the permissible level is reached or they should be handled as radioactive waste.

Radioactive wastes should be collected and transported to a central *waste processing plant or burial site* (see also Section 8.2.4). Wastes are not regarded as radioactive if their radioisotope concentration does not exceed the limit given by the related standards and in which case they may be handled as common wastes (released into drains, surface water, air).

Radioactive concentration can frequently be lowered by simple on-site methods after which the waste can be discharged. One such method is dilution but in this way the total amount of released radioactivity cannot be decreased. It is more reasonable if the liquid or solid wastes are stored or chemically concentrated, and if gases and vaporous products are filtered.

8.5.1.6. INCIDENTS, ACCIDENTS

During the planning and design of an investigation a detailed emergency procedure taking into account all possible incidents: breakdowns, factory accidents, should be prepared and agreed with the technologists of the factory.

Protective devices (masks, respirators, manipulators, detergents) should be kept on-site and at hand ready to be used even if the likelihood of their being needed is very low.

8.5.2. SPECIAL DUTIES

8.5.2.1. TRACER EXPERIMENTS

The whole of Section 8.5 is relevant to industrial tracer experiments but may also be applied to most *geological and mining investigations*. With these latter types of investigations, however, even greater care should be taken in selecting and introducing radioisotopes and the same is true with regard to monitoring because the contamination of food, drinking water and, in general, man's environment is more likely than with laboratory or with industrial work.

8.5.2.2. RADIOANALYTICAL WORK

Radioactive samples are usually handled in a properly equipped isotope laboratory, but sources used for activation and transfer devices of radioactive samples (e.g., rabbit systems) are often located at separate places in the factory.

When installing neutron sources used for activation analysis not only the external radiation but additional radiation due to the *activation of samples* and structural material should be shielded and contamination avoided, too. Tritiated targets used with neutron generators may lead to significant tritium incorporation therefore their handling should be carried out in ventilated boxes. Environmental contamination can be avoided by using a proper, hard-wearing packaging for activated samples.

The source containers and shielding walls should be designed according to Section 8.1.4.2.

REFERENCES

[8.1] ICRP Publ. 26. *Recommendations of the International Commission on Radiological Protection.* Annals ICRP, Vol. 1, No. 3. Pergamon, Oxford, 1977.

[8.2] ICRU Report 33. *Radiation Quantities and Units.* ICRU, Washington, 1980.

[8.3] ICRU Publ. 30. *Limits for Intakes of Radionuclides by Workers.* Parts 1, 2 and 3 and Supplements, Annals ICRP, Vol. 2, No. 3/4, 1979; Vol. 3, No. 1–4, 1979; Vol. 4, No. 3/4, 1980; Vol. 5, No. 1–6, 1981; Vol. 6, No. 2/3, 1981; Vol. 7, No. 1–3, 1982; Vol. 8, No. 1–3, 1982, Pergamon, Oxford.

[8.4] *Basic Safety Standards for Radiation Protection.* 1982 ed., Safety Series No. 9 IAEA, Vienna, 1982.

[8.5] ICRP Publ. 23. *Report of the Task Group on Reference Man.* ICRP, Pergamon, Oxford, 1975.

[8.6a] Gusev, N. G., Kovalev, E. E., Osanov, D. P. and Popov, V. I., *(Protection against Radiation from Extended Sources)* (in Russian). Gosatomizdat, Moscow, 1961.

[8.6b] Gussew, N. G., *Leitfaden für Radioaktivität und Strahlenschutz.* V. Technik, Berlin, 1957.

[8.7] Hine, G. J. and Brownell, G. L., *Radiation Dosimetry.* Academic Press, New York, 1956.

[8.8] Goldstein, M., *Fundamental Aspects of Reactor Shielding.* Pergamon, London, 1959.

[8.9] *Health Physics Addendum.* Safety Series No. 2, IAEA, Vienna, 1960.

[8.10] ICRP Publ. 21. *Data for Protection against Ionizing Radiation from External Sources.* Suppl. to ICRP Publ. 15, Pergamon, Oxford, 1973.

[8.11] *Regulations for the Safe Transport of Radioactive Materials.* Safety Series No. 6, 1973 Rev. Ed. (As Amended), IAEA, Vienna, 1979.

[8.12] *Neutron Moisture Gauges.* Techn. Rep. Ser. No. 112, IAEA, Vienna, 1970.

[8.13] *Council Directive of 15 July 1980 (80/836/Euratom).* Offical Journal of the European Communities, No. L 246, Vol. 23, Sept. 17, 1980.

[8.14] ICRP Publ. 42. *A Compilation of the Major Concepts and Quantities in Use by ICRP.* Annals ICRP Vol. 14. No 4, Pergamon, Oxford, 1984.

BIBLIOGRAPHY

1. *Manual of Industrial Radiation Protection.* Parts I–V. International Labour Office, Geneva, 1963–1964.

2. *Safe Use of Radioactive Tracers in Industrial Processes.* Safety Ser. No. IAEA, Vienna, 1974.

3. *Radiological Safety Aspects of the Operation of Neutron Generators.* Safety Ser. No. 42, IAEA, Vienna, 1976.

4. Cameron, J. F., Radiation Safety as Related to Radioisotope Instruments. In Proc., *Radiation Engineering in the Academic Curriculum,* Haifa, 1973, IAEA, Vienna, 1975, pp. 259–268.

5. *Radiation Safety and Protection in Industrial Applications.* Proc. Symp. Washington, 1972, Ed. H. F. Klein, U. S. Dep. Health and Welfare, Rockville, 1972 [DHEW Publ. No. (FDA) 73–8012].

6. *Proc. Nat. Symp. Isotope Applications in Industry.* Session VII: Radiation Safety and Training, Bombay, 1977, Bhabha Atomic Res. Centre, Dep. of Atomic Energy, 1977.

7. *Safe Design and Use of Industrial Beta-Ray Sources.* NBS Handbook 66, NBS, Wash., 1958.

APPENDIX

In recent years, the countries accepting the competence of the International Bureau of Weights and Measure (Bureau International des Poids et Mesures, BIPM) have decided in favour of the inauguration of the international system of units (SI = Système International d'Unités) instead of the CGS system earlier commonly used. The exact date of the official inauguration and the days of grace granted for the use of the conventional units vary from country to country, however, not very far in the future the new system of units will certainly be applied all over the world.

The switch-over is proving to be no easy matter; in order to facilitate readjustment, in the following tables a summary will be given on the basic SI units and those derived units which are generally encountered in the application of radioisotopes.

The conversion factors between conventional and SI units are given with an accuracy generally sufficient in engineering practice. Correspondingly, the numerical values were rounded-off thereby ignoring precision for its own sake. The mode of application of Table A/4 is illustrated by two examples.

1. What is the activity of an 80 kCi radiation source in SI units? From Table A/4 it is clear that 1 Ci corresponds to 3.7×10^{10} Bq. Thus, 80 kCi represents an activity of $8 \times 10^4 \times 3.7 \times 10^{10} = 2.96 \times 10^{15}$ Bq $= 2.96$ PBq.

2. What is the K_γ dose constant of a nuclide, in conventional units, for which this value is equal to 7.0 aA m^2/(kg Bq)? As seen from the table, the value of the conversion factor "SI/conventional units" equals 5.163×10^{-1}, that is, converting SI units into conventional ones, the dose constant is equal to $K_\gamma = 7.0 \times 0.5163 = 3.61$ R m^2/(h Ci).

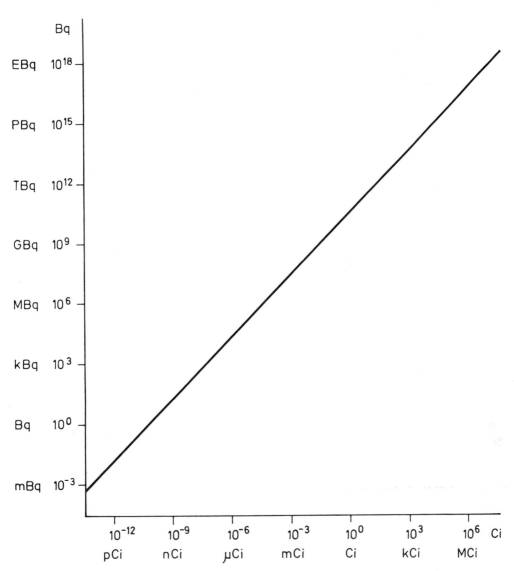

Fig. A/1. Nomogram for the conversion of units of activity, from curie into becquerel and vice versa

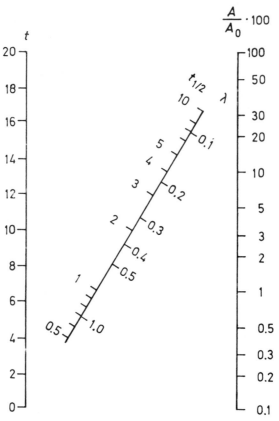

Fig. A/2. Nomogram for the determination of the extent of radioactive decay from the given half-life or decay
constant

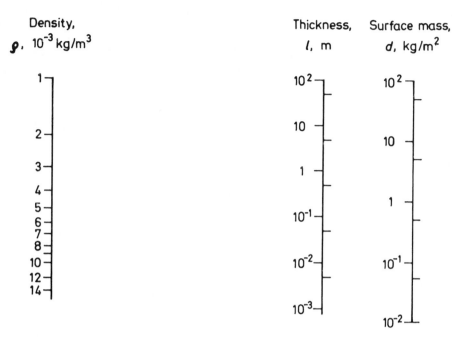

Fig. A/3. Nomogram for the evaluation of the thickness of absorbents in units of m and kg/m², with the density given

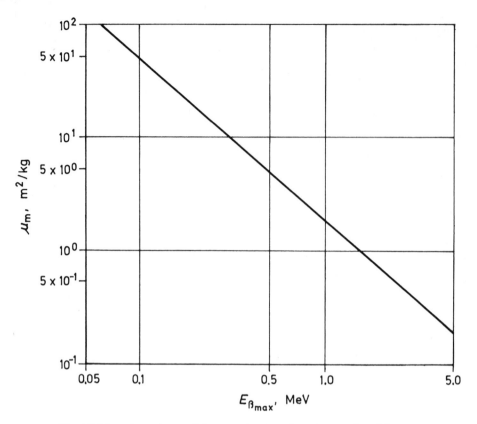

Fig. A/4. Mass absorption coefficient as a function of maximum value of β-energy

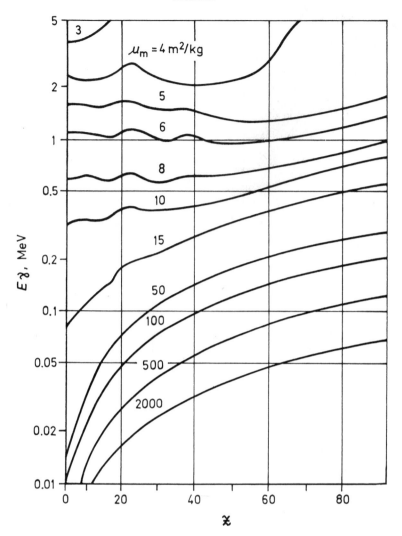

Fig. A/5. Value of mass absorption coefficient as a function of the energy of the γ-radiation and atomic number of the absorbent

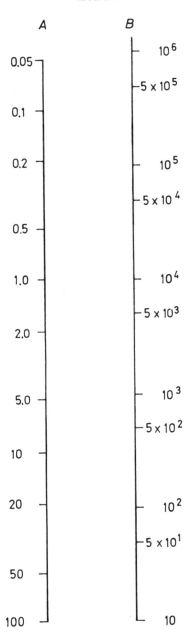

Fig. A/6. Nomogram for the determination of the relative scattering in percent corresponding to the standard deviation (statistical confidence of 0.683). *A*—relative scattering in percent; *B*—counts registered

Table A/1. Basic SI units and their definitions

Physical quantity	Unit	
	Name	Symbol
Length	metre	m
Mass	kilogram	kg
Time	second	s
Electric current	ampere	A
Thermodynamic temperature	kelvin	K
Light intensity	candela	cd
Amount of material	mole	mol

Table A/2. Complementary SI units and their symbols

Physical quantity	Unit	
	Name	Symbol
Planar angle	radian	rad
Solid angle	steradian	sr

Table A/3. Standard multiples and fractions of units*

Prefix		Multiplication factor
Name	Symbol	
exa	E	10^{18}
peta	P	10^{15}
tera	T	10^{12}
giga	G	10^9
mega	M	10^6
kilo	k	10^3
milli	m	10^{-3}
micro	μ	10^{-6}
nano	n	10^{-9}
pico	p	10^{-12}
femto	f	10^{-15}
atto	a	10^{-18}

* In special cases, some other prefixes can also be used, such as hecto (h) $=10^2$, deca (da) $=10^1$, deci (d) $=10^{-1}$ and centi (c) $=10^{-2}$, primarily to respect certain traditions

Table A/4. Units and conversion factors used in isotope techniques

Physical quantity	Traditional unit		SI unit			Conversion factor	
	name	symbol	name	dimension	symbol	traditional/SI	SI/traditional
Force	kilopond (kilogram force)	kp	newton	kg m/s²	N	9.807	1.020×10^{-1}
Pressure	atmosphere	atm	pascal	kg/(m s²)	Pa	1.013×10^{5}	9.869×10^{-6}
	mm of mercury	mmHg	pascal	kg/(m s²)	Pa	1.333×10^{2}	7.501×10^{-3}
Energy (work, heat)	kilowatt hour	kW h	joule	kg m²/s²	J	3.600×10^{6}	2.778×10^{-7}
	calory	cal	joule	kg m²/s²	J	4.187	2.388×10^{-1}
Power	watt	W	watt	J/s	W	1.000	1.000
Electric potential difference	electrostatic unit	e.s.u.	volt	kg m²/(s³ A)	V	2.998×10^{2}	3.336×10^{-3}
Electric charge	electrostatic unit	e.s.u.	coulomb	A s	C	3.336×10^{-10}	2.998×10^{9}
Density	gram/cubic centimetre		kilogram/cu.metre	kg/m³		10^{3}	10^{-3}
Radioactivity	curie	Ci	becquerel	1/s	Bq	3.700×10^{10}	2.703×10^{-11}
Absorbed dose	rad	rd	gray	J/kg	Gy	10^{-2}	10^{2}
Exposure	roentgen	R	coulomb/kilogram	C/kg		2.580×10^{-4}	3.876×10^{3}
Exposure	kerma		joule/kilogram	J/kg		10^{-2}	10^{2}
Equivalent dose	rem	rem	sievert	J/kg	Sv	10^{-2}	10^{2}
Absorbed dose rate	rad/hour	rd/h	gray/second	J/(kg s)	Gy s⁻¹	2.778×10^{-6}	3.600×10^{5}

Quantity	Unit	Symbol	SI unit	SI symbol		
Exposure rate	$\dfrac{\text{roentgen}}{\text{hour}}$	R/h	$\dfrac{\text{ampere}}{\text{kilogram}}$	A/kg	7.166×10^{-8}	1.395×10^{7}
Radiation energy	electronvolt	eV	joule	J	1.602×10^{-19}	6.242×10^{18}
Cross-section	barn	b	square metre	m^2	10^{-28}	10^{28}
Flux density	$\dfrac{1}{\text{square centimetre second}}$	$1/(cm^2\ s)$	$\dfrac{1}{\text{square metre second}}$	$1/(m^2\ s^1)$	10^{4}	10^{-4}
Surface density	$\dfrac{\text{milligram}}{\text{square centimetre}}$	mg/cm^2	$\dfrac{\text{kilogram}}{\text{square metre}}$	kg/m^2	10^{-2}	10^{2}
Mass absorption coefficient	$\dfrac{\text{square centimetre}}{\text{milligram}}$	cm^2/mg	$\dfrac{\text{square metre}}{\text{kilogram}}$	m^2/kg	10^{2}	10^{-2}
	$\dfrac{\text{square centimetre}}{\text{gram}}$	cm^2/g	$\dfrac{\text{square metre}}{\text{kilogram}}$	m^2/kg	10^{-1}	10
Linear absorption coefficient	$\dfrac{1}{\text{centimetre}}$	$1/cm$	$\dfrac{1}{\text{metre}}$	$1/m$	10^{2}	10^{-2}
Atomic mass unit	a.m.u.	u	kilogram	kg	1.661×10^{-27}	6.023×10^{26}
γ-Dose constant	Rhm	$R\ m^2/(Ci\ h)$		$aA\ m^2/(kg\ Bq)$*	1.937	5.163×10^{-1}
β-Dose constant		$rd\ cm^2/(mCi\ h)$		$aJ\ m^2/kg$	7.508	1.332×10^{-1}
Specific ionization	$\dfrac{1}{\text{micrometre}}$	$1/\mu m^{-1}$		$1/m$	10^{6}	10^{-6}
Linear energy transfer (*LET*)	$\dfrac{\text{kiloelectronvolt}}{\text{micrometre}}$	$keV/\mu m$	$\dfrac{\text{joule}}{\text{metre}}$	J/m	1.602×10^{-10}	6.242×10^{9}
G-value	$\dfrac{1}{100\ \text{electronvolt}}$	$1/10^2\ eV$	$\dfrac{1}{\text{joule}}$	$1/J$	6.242×10^{16}	1.602×10^{-17}

* $1\ aA\ m^2/(kg\ Bq) = 121.27\ \mu\ Gy_{air}\ m^2/(GBq\,h)$

Table A/5. Density of some important substances

Substance	Density, kg/m^3
Aluminium	2699
Antimony	6691
Beryllium	1848
Bismuth	9747
Boron	2340
Cadmium	8642
Carbon (diamond)	3513
Chromium	7190
Copper	8960
Iridium	22420
Iron	7874
Lead	11350
Mercury	13546
Molybdenum	10220
Platinum	21450
Silicon	2330
Silver	10500
Tantalum	16654
Thallium	11850
Tin	7310
Tungsten	19300
Uranium	18950
Vanadium	6110
Zinc	7133
Zirconium	6506
Air	1.293

Table A/6. Decimal logarithms of numbers

Number	Logarithm									
	0	1	2	3	4	5	6	7	8	9
10	0000	0043	0086	0128	0170	0212	0253	0294	0334	0374
11	0414	0453	0492	0531	0569	0607	0645	0682	0719	0755
12	0792	0828	0864	0899	0934	0969	1004	1038	1072	1106
13	1139	1173	1206	1239	1271	1303	1335	1367	1399	1430
14	1461	1492	1523	1553	1584	1614	1644	1673	1703	1732
15	1761	1790	1818	1847	1875	1903	1931	1959	1987	2014
16	2041	2068	2095	2122	2148	2175	2201	2227	2253	2279
17	2304	2330	2355	2380	2405	2430	2455	2480	2504	2529
18	2553	2577	2601	2625	2648	2672	2695	2718	2742	2765
19	2788	2810	2833	2856	2878	2900	2923	2945	2967	2989
20	3010	3032	3054	3075	3096	3118	3139	3160	3181	3201
21	3222	3243	3263	3284	3304	3324	3345	3365	3385	3404
22	3424	3444	3464	3483	3502	3522	3541	3560	3579	3598
23	3617	3636	3655	3674	3692	3711	3729	3747	3766	3784
24	3802	3820	3838	3856	3874	3892	3909	3927	3945	3962
25	3979	3997	4014	4031	4048	4065	4082	4099	4116	4133
26	4150	4166	4183	4200	4216	4232	4249	4265	4281	4298
27	4314	4330	4346	4362	4378	4393	4409	4425	4440	4456
28	4472	4487	4502	4518	4533	4548	4564	4579	4594	4609
29	4624	4639	4654	4669	4683	4698	4713	4728	4742	4757
30	4771	4786	4800	4814	4829	4843	4857	4871	4886	4900
31	4914	4928	4942	4955	4969	4983	4997	5011	5024	5038
32	5051	5065	5079	5092	5105	5119	5132	5145	5159	5172
33	5185	5198	5211	5224	5237	5250	5263	5276	5289	5302
34	5315	5328	5340	5353	5366	5378	5391	5403	5416	5428
35	5441	5453	5465	5478	5490	5502	5514	5527	5539	5551
36	5563	5575	5587	5599	5611	5623	5635	5647	5658	5670
37	5682	5694	5705	5717	5729	5740	5752	5763	5775	5786
38	5798	5809	5821	5832	5843	5855	5866	5877	5888	5899
39	5911	5922	5933	5944	5955	5966	5977	5988	5999	6010
40	6021	6031	6042	6053	6064	6075	6085	6096	6107	6117
41	6128	6138	6149	6160	6170	6180	6191	6201	6212	6222
42	6232	6243	6253	6263	6274	6284	6294	6304	6314	6325
43	6335	6345	6355	6365	6375	6385	6395	6405	6415	6425
44	6435	6444	6454	6464	6474	6484	6493	6503	6513	6522
45	6532	6542	6551	6561	6571	6580	6590	6599	6609	6618
46	6628	6637	6646	6656	6665	6675	6684	6693	6702	6712
47	6721	6730	6739	6749	6758	6767	6776	6785	6794	6803
48	6812	6821	6830	6839	6848	6857	6866	6875	6884	6893
49	6902	6911	6920	6928	6937	6946	6955	6964	6972	6981
50	6990	6998	7007	7016	7024	7033	7042	7050	7059	7069
51	7076	7084	7093	7101	7110	7118	7126	7135	7143	7152
52	7160	7168	7177	7185	7193	7202	7210	7218	7226	7235
53	7243	7251	7259	7267	7275	7284	7292	7300	7308	7316
54	7324	7332	7340	7348	7356	7364	7372	7380	7388	7396

APPENDIX

Table A/6. (Cont.)

Number	Logarithm									
	0	1	2	3	4	5	6	7	8	9
55	7404	7412	7419	7427	7435	7443	7451	7459	7466	7474
56	7482	7490	7497	7505	7513	7520	7528	7536	7543	7551
57	7559	7566	7574	7582	7589	7597	7604	7612	7619	7627
58	7634	7642	7649	7657	7664	7672	7679	7686	7694	7701
59	7709	7716	7723	7731	7738	7745	7752	7760	7767	7774
60	7782	7789	7796	7803	7810	7818	7825	7832	7839	7846
61	7853	7860	7868	7875	7882	7889	7896	7903	7910	7917
62	7924	7931	7938	7945	7952	7959	7966	7973	7980	7987
63	7993	8000	8007	8014	8021	8028	8035	8041	8048	8055
64	8062	8069	8075	8082	8089	8096	8102	8109	8116	8122
65	8129	8136	8142	8149	8156	8162	8169	8176	8182	8189
66	8195	8202	8209	8215	8222	8228	8235	8241	8248	8254
67	8261	8267	8274	8280	8287	8293	8299	8306	8312	8319
68	8325	8331	8338	8344	8351	8357	8363	8370	8376	8382
69	8388	8395	8401	8407	8414	8420	8426	8432	8439	8445
70	8451	8457	8463	8470	8476	8482	8488	8494	8500	8506
71	8513	8519	8525	8531	8537	8543	8549	8555	8561	8567
72	8573	8579	8585	8591	8597	8603	8609	8615	8621	8627
73	8633	8639	8645	8651	8657	8663	8669	8675	8681	8686
74	8692	8698	8704	8710	8716	8722	8727	8733	8739	8745
75	8751	8756	8762	8768	8774	8779	8785	8791	8797	8802
76	8808	8814	8820	8825	8831	8837	8842	8848	8854	8859
77	8865	8871	8876	8882	8887	8893	8899	8904	8910	8915
78	8921	8927	8932	8938	8943	8949	8954	8960	8965	8971
79	8976	8982	8987	8993	8998	9004	9009	9015	9020	9025
80	9031	9036	9042	9047	9053	9058	9063	9069	9074	9079
81	9085	9090	9096	9101	9106	9112	9117	9122	9128	9133
82	9138	9143	9149	9154	9159	9165	9170	9175	9180	9186
83	9191	9196	9201	9206	9212	9217	9222	9227	9232	9238
84	9243	9248	9253	9258	9263	9269	9274	9279	9284	9289
85	9294	9299	9304	9309	9315	9320	9325	9330	9335	9340
86	9345	9350	9355	9360	9365	9370	9375	9380	9385	9390
87	9395	9400	9405	9410	9415	9420	9425	9430	9435	9440
88	9445	9450	9455	9460	9465	9469	9474	9479	9485	9489
89	9494	9499	9504	9509	9513	9518	9523	9528	9533	9538
90	9542	9547	9552	9557	9562	9566	9571	9576	9581	9586
91	9590	9595	9600	9605	9609	9614	9619	9624	9628	9633
92	9638	9643	9647	9652	9657	9661	9666	9671	9675	9680
93	9685	9689	9694	9699	9703	9708	9713	9717	9722	9727
94	9731	9736	9741	9745	9750	9754	9759	9763	9768	9773
95	9777	9782	9786	9791	9795	9800	9805	9809	9814	9818
96	9823	9827	9832	9836	9841	9845	9850	9854	9859	9863
97	9868	9872	9877	9881	9886	9890	9894	9899	9903	9908
98	9912	9917	9921	9926	9930	9934	9939	9943	9948	9952
99	9956	9961	9965	9969	9974	9978	9983	9987	9991	9996

Table A/7. Natural logarithms

1.0	0.0000	3.3	1.1939	5.5	1.7047	7.8	2.0541
1.1	0.0953	3.4	1.2238	5.6	1.7228	7.9	2.0669
1.2	0.1823			5.7	1.7405		
1.3	0.2624	3.5	1.2528	5.8	1.7579	8.0	2.0794
1.4	0.3365	3.6	1.2809	5.9	1.7750	8.1	2.0919
		3.7	1.3083			8.2	2.1041
1.5	0.4055	3.8	1.3350	6.0	1.7918	8.3	2.1163
1.6	0.4700	3.9	1.3610	6.1	1.8083	8.4	2.1282
1.7	0.5306			6.2	1.8245		
1.8	0.5878	4.0	1.3863	6.3	1.8405	8.5	2.1401
1.9	0.6419	4.1	1.4110	6.4	1.8563	8.6	2.1518
2.0	0.6931	4.2	1.4351			8.7	2.1633
2.1	0.7419	4.3	1.4586	6.5	1.8718	8.8	2.1748
		4.4		6.6	1.8871	8.9	
2.2	0.7885		1.4816	6.7	1.9021		2.1861
2.3	0.8329	4.5	1.5041	6.8	1.9169	9.0	2.1972
2.4	0.8755	4.6	1.5261	6.9	1.9315	9.1	2.2083
		4.7				9.2	
2.5	0.9163	4.8	1.5476	7.0	1.9459	9.3	2.2192
2.6	0.9555	4.9	1.5686	7.1	1.9601	9.4	2.2300
2.7	0.9933		1.5892	7.2	1.9741		2.2407
2.8	1.0296	5.0	1.6094	7.3	1.9879	9.5	2.2513
2.9	1.0647	5.1	1.6292	7.4	2.0015	9.6	2.2618
3.0	1.0986	5.2	1.6487	7.5	2.0149	9.7	2.2721
3.1	1.1314	5.3	1.6677	7.6	2.0281	9.8	2.2824
3.2	1.1632	5.4	1.6864	7.7	2.0412	9.9	2.2925

Table A/8. Time dependence of the decay factor*

$t/t_{1/2}$	$\exp(-0.693\,t/t_{1/2})$	$t/t_{1/2}$	$\exp(-0.693\,t/t_{1/2})$	$t/t_{1/2}$	$\exp(-0.693\,t/t_{1/2})$
0.00	1.000	0.25	0.841	0.50	0.707
0.01	0.993	0.26	0.835	0.51	0.702
0.02	0.986	0.27	0.829	0.52	0.697
0.03	0.979	0.28	0.824	0.53	0.693
0.04	0.973	0.29	0.818	0.54	0.688
0.05	0.966	0.30	0.812	0.55	0.683
0.06	0.959	0.31	0.807	0.56	0.678
0.07	0.953	0.32	0.801	0.57	0.674
0.08	0.946	0.33	0.796	0.58	0.669
0.09	0.940	0.34	0.790	0.59	0.664
0.10	0.933	0.35	0.785	0.60	0.660
0.11	0.927	0.36	0.779	0.61	0.665
0.12	0.920	0.37	0.774	0.62	0.651
0.13	0.914	0.38	0.768	0.63	0.646
0.14	0.908	0.39	0.763	0.64	0.642
0.15	0.901	0.40	0.758	0.65	0.637
0.16	0.896	0.41	0.753	0.66	0.633
0.17	0.889	0.42	0.747	0.67	0.629
0.18	0.883	0.43	0.742	0.68	0.624
0.19	0.877	0.44	0.737	0.69	0.620
0.20	0.871	0.45	0.732	0.70	0.616
0.21	0.865	0.46	0.727	0.71	0.611
0.22	0.859	0.47	0.722	0.72	0.607
0.23	0.853	0.48	0.717	0.73	0.603
0.24	0.847	0.49	0.712	0.74	0.599

$$* \; \frac{A}{A_0} = \exp(-\lambda t) = \exp\left(-0.693\,\frac{t}{t_{1/2}}\right)$$

Table A/8. (Cont.)

$t/t_{1/2}$	$\exp(-0.693\ t/t_{1/2})$	$t/t_{1/2}$	$\exp(-0.693\ t/t_{1/2})$	$t/t_{1/2}$	$\exp(-0.693\ t/t_{1/2})$
0.75	0.595	1.00	0.500	2.5	0.177
0.76	0.591	1.05	0.483	2.6	0.165
0.77	0.586	1.10	0.467	2.7	0.154
0.78	0.582	1.15	0.451	2.8	0.144
0.79	0.578	1.20	0.435	2.9	0.134
0.80	0.574	1.25	0.421	3.0	0.125
0.81	0.570	1.30	0.406	3.1	0.117
0.82	0.566	1.35	0.392	3.2˙	0.109
0.83	0.563	1.40	0.379	3.3	0.102
0.84	0.559	1.45	0.366	3.4	0.095
0.85	0.555	1.50	0.354	3.5	0.088
0.86	0.551	1.55	0.342	3.6	0.083
0.87	0.547	1.60	0.330	3.7	0.077
0.88	0.543	1.65	0.319	3.8	0.072
0.89	0.540	1.70	0.308	3.9	0.067
0.90	0.536	1.75	0.297	4.0	0.063
0.91	0.532	1.80	0.287	4.2	0.054
0.92	0.529	1.85	0.277	4.4	0.048
0.93	0.525	1.90	0.268	4.6	0.041
0.94	0.521	1.95	0.259	4.8	0.036
0.95	0.518	2.00	0.250	5.0	0.031
0.96	0.514	2.10	0.233	6.0	0.016
0.97	0.510	2.20	0.218	7.0	0.008
0.98	0.507	2.30	0.203	8.0	0.004
0.99	0.504	2.40	0.190	9.0	0.002

SUBJECT INDEX